LONG-RANGE FUTURES RESEARCH

RESEARCH

An Application of Complexity Science

Robert H. Samet

BookSurge Publishing, 7290 B Investment Drive, North Charleston, South Carolina, 29418, in the United States, 2008.

Library of Congress Cataloguing- in-Publication Data

Samet, Robert Hugh, 1943-
        Long-Range Futures Research: An Application of Complexity Science

        Includes bibliographical references and index.

ISBN: 1-4392-1434-4
ISBN-13: 9781439214343
Library of Congress Control Number: 2008910611

Visit www.booksurge.com to order additional copies. Also copies can be ordered from www.amazon.com or www.amazon.co.uk.

ABOUT THE AUTHOR

Robert H. Samet is a civil engineer with forty years experience as a specialist consultant in planning and futures research, development and regional studies, investment appraisal and corporate location strategy. For a decade, he directed a multi-client study to develop the London Megapolis Regional Information System covering some 300 urban centres and 11,000 establishments, to explore the transactional dynamics behind urban investment flows and to explain the changing gradients between locations in investment returns. He was also a consultant of US-based Planning Research Corporation that applied systems science approaches to international development programmes in the Middle East, North Africa, and South Asia. His clients included the Ministries of Planning in both Algeria and Saudi Arabia and also the Saudi Royal Commission for Industrial Cities. He and his wife live in a village in Lincolnshire, England, and have four married sons in the USA, Scotland, France, and Switzerland.

The author can be contacted by readers with feedback at r.h.b.samet@btinternet.com

C O N T E N T S

Page

PREFACE   i

PART 1 - FUTURES RESEARCH AND SYSTEMS SCIENCE

1.   SOCIETAL EVOLUTION AND FUTURES RESEARCH   3
     1.1 Societal evolution   3
     1.2 The systems science origins of planning and futures research   8
     1.3 Projective futures and forecasting   13
     1.4 Prospective futures and scenarios   17
     1.5 Evolutionary futures and landscapes   22
     1.6 United Nations reference scenarios   27
     References   36

2.   THE METASYSTEMS AND DEEP FUTURES   39
     2.1 Coevolution of the planetary metasystems   39
     2.2 Astrophysical metasystem   42
     2.3 Geophysical metasystem   46
     2.4 Physical metasystem   52
     2.5 Biological metasystem   56
     2.6 Civil metasystem   60
     References   71

3.   EMERGENCE   73
     3.1 Metaphors   73
     3.2 Models   75
     3.3 Paradigms   79
     3.4 The evolution of spatial structure   85
     3.5 Complexity science   89
     3.6 Civil emergence   96
     References   102

4.	THE PRINCIPLES OF ECODYNAMICS	105
	4.1 Agents and the classification of households	105
	4.2 Establishments	112
	4.3 Diffusive ecostructures	115
	4.4 Complex adaptive systems	120
	4.5 Territorial colonisation	123
	4.6 Evolutionary potential	127
	References	133

PART 2 - TRANSACTIONAL MICROSTRUCTURE

5.	CIVIL ECOSTRUCTURES	137
	5.1 Metropolis and megalopolis	137
	5.2 Sectoral classification of business space users	144
	5.3 Urban economic structure	147
	5.4 Planning profiles for synthetic ecostructures	154
	5.5 Gradients between locations in property investment returns	156
	5.6 Sustainable civil development	160
	References	168
6.	TRANSACTIONAL COMPLEXITY	169
	6.1 Institutional coordination of urban microsystems	169
	6.2 Spatio-temporal issues	176
	6.3 Transactional complexity	180
	6.4 Interurban linkages and networks	185
	6.5 Perspectives on the informational economy	188
	6.6 Timescale for the informational transition	194
	References	197
7.	CIVIL PHASE TRANSITIONS	199
	7.1 Stages of development	199
	7.2 The workforce and working time	205
	7.3 Societal transitions	209
	7.4 Dynamic investment flows	217
	7.5 Macrolaws of Ecodynamics	223
	References	228

8.   TECHNOLOGICAL EVOLUTION OF INFRASTRUCTURE SYSTEMS   229
      8.1 Technological evolution   229
      8.2 Infrastructure investment waves   233
      8.3 Energy   241
      8.4 Transport   245
      8.5 Telecommunications and information technology   251
      8.6 The exploitation of space   254
      References   261

9.   TRAJECTORY OF THE WORLD SYSTEM OF CITIES   263
      9.1 World urbanisation   263
      9.2 Cities in the continental regions of the world   268
      9.3 Projected number of cities   273
      9.4 The city hierarchy and urban densities   279
      9.5 World cities of the future   283
      9.6 Megacity hazards and civil disasters   286
      References   300

PART 3 - CONTEXTUAL MACROSTRUCTURE

10.   SHAPING FUTURE WORLD DEVELOPMENT   303
      10.1 The capacity of the planet   303
      10.2 Population growth   308
      10.3 Measures of economic and social development   311
      10.4 Alternative routes to a sustainable future   317
      10.5 Future gross world product   321
      10.6 Global investment in the planetary infrastructure 325
      References   333

11.   THE GEOPOLITICAL MACROSTRUCTURE   335
      11.1 Geopolitical perspectives   335
      11.2 Ecopolitan and agropolitan states   341
      11.3 Continental unions   347
      11.4 Continental futures   351
      11.5 Globalisation and transnational corporations   358
      References   364

12. LONG-RANGE GLOBAL SCENARIOS   365
     12.1 Emergence of evolutionary futures and landscapes   365
     12.2 Designed intervention in the global macrosystems   372
     12.3 Coevolution of the global macrosystems   376
           A Planetary governance   377
           B Geopolitical context and military deterrence   378
           C Continental integration and cultural diversity   379
           D Population policies and demographic trends   379
           E Life sciences and medicine   381
           F Land use and food supplies   382
           G Resource conservation and the environment   383
           H High technology and space research   384
           J World system of cities and the planetary infrastructure   385
           K Urban technology and telecommunications   386
           L Community systems and human development   387
           M Societal transitions and employment   389
           N The world production system and transnational corporations 390
           P The world economy and globalisation   392
     12.4 Scenario overviews   393
     References   399

13. EVOLUTIONARY LANDSCAPES FOR A COMPLEX WORLD   401
     13.1 Planetary metatrends and deep futures   401
     13.2 Civil megatrends shaping the long-range future   410
     13.3 Global macrotrends and the intermediate future   413
     13.4 Sectoral microtrends and the near future   424
     13.5 Civilisation in a dynamic balance   427
     References   432

LIST OF TABLES

Page

CHAPTER 5.
    Table 1: The urban hierarchy ... 138
    Table 2: Building area ranges for establishments ... 155
    Table 3: Decentralisation patterns ... 165

CHAPTER 7.
    Table 4: Stages of development for the UK ... 201
    Table 5: Civil phase transitions ... 203
    Table 6: Projected European weekly working hours ... 208
    Table 7: Workforce divisional allocation in an ecopolitan society ... 215

CHAPTER 9
    Table 8: Historical growth of city populations ... 265
    Table 9: Urban population projections ... 274
    Table 10: Projected number of cosmopolises and megapolises ... 279
    Table 11: Typical urban densities by city size class ... 281

CHAPTER 10
    Table 12: Environmental impact in 2050 (Brundtland) ... 319
    Table 13: Environmental impact in 2025 ... 322
    Table 14: Environmental impact in 2150 ... 323
    Table 15: Environmental impact in 2250 ... 324

CHAPTER 13
    Table 16: Environmental impact in 3000 ... 429

PART 4 - APPENDICES

Page

1.	A century of futures research and systems science	434
2.	Verhulst logistic curve (S-Curve)	436
3.	Prognostication of a long-term future state	437
4	Geological and archaeological timescales	438
5.	Civilisations of humankind	440
6.	Meta-timescale for deep futures	444
7.	RICI/SIC sectoral classification coding guide	447
8.	Location index analysis by sector and area	452
9.	Property profiles by urban classification	454
10.	Planning standards for urban establishments	457
11.	Urban area location indices for property investment	458
12.	Transport energy consumption from urban decentralisation	461
13.	Employment structure by stage of development	462
14.	Default planning standards and costs for housing	463
15.	Default planning standards for life expectancy, education and health	464
16.	Default planning standards for energy and transport	465
17.	Default planning standards for communications and information technology	467
18.	Metropolises with populations over 2 million	468
19.	Future urban population by continental region	476
20.	Future number of cities by continental region and urban size class	478
21.	Future number of cities in MDCs and LDCs	479
22.	Past and future number of cities by urban size class (1975, 2000 & 2025)	480
23.	Future number of cities by urban size class (2050, 2100 & 2150)	481
24.	Number of prospective world cities of the future	483
25.	Prospective world cities of the future	484
26.	Future workforce by division and continental region	487
27.	Future land use by continental region	488
28.	Default values for material resource flows	491
29.	Future Gross World Product by continental region	492

30. Future investment stock requirements by continental region   494

31. Future material flows by continental region   497

32. Future world car fleet and global energy use by continental region   498

33. Potential number of ecopolitan states   500

34. Regional trade blocs   501

35. Future workforce and number of large establishments
by continental region   502

36. Long-range scenario profiles   504

37. Glossary of terms   516

38. Conversion of energy units   545

39. Bibliography   546

40. Index   566

PREFACE

SYNOPSIS

Long-Range Futures Research: An Application of Complexity Science explains how the science of evolution and complexity provides an approach for exploring the complex and unprecedented patterns of global change expected in the future. An artificial landscape is created by the evolution of the civil system, which is a complex adaptive system with emergent properties. Unlike biological evolution, there is foresight in civil-societal evolution, and so the system evolves as a combination of causality and purposeful human intervention (teleology). The methodology involves selecting households and establishments as transacting entities, which gives spatial structure to the trajectory of the civil system so that macrolaws may be derived for the evolutionary landscape. Civil ecostructures (towns and cities) form in a far-from-equilibrium system, with emergent behaviour in the macrostructure at the regional, state, continental, and planetary levels. An evolutionary timescale for cities is measured in millennia, in relation to a century for the lifetime of buildings, and the system as a whole survives generations of the transacting entities. There are a number of potential hazards that may impact upon our civilization in the future, and the scale of hazards and disasters can be measured in terms of their spatial scale, ranging from local to regional to global impacts. Long-Range Futures Research takes a planning time horizon of 100-150 years to 2150 to include five future generations and traces the evolutionary trajectory to 2250. However, in the course of a millennium, there will inevitably be discontinuities, and so a meta-timescale for deep futures has been developed.

Long-Range Futures Research: An Application of Complexity Science sets out the scientific basis for futures studies by a unique application of the science of evolution and complexity, for the development of an evolutionary vision of the world. In *Nature's Metropolis: Chicago and the Great West* (1992), historian William Cronon distinguishes between the original natural landscape and the artificial landscape created by urban evolution. The emergence of the metropolis required that a new human order be superimposed on nature until the two became completely entangled. The result is a hybrid-system that conceals its long-standing debts to the natural systems that made it possible. The entire process, which is at least as artificial as it is natural, can be termed second nature in so far as it is as organic and evolutionary as Darwin's model of biological change. The evolutionary vision of society starts from the premise that civil artefacts are a second nature, and, in the development of the planet, the evolution of the car or the aeroplane is just as natural as the creation of a horse or a bird. Kenneth Boulding described these concepts in *Ecodynamics - A New Theory of Societal Evolution* (1978), and artefacts can be classified as material, organisational (i.e., establishments, civil institutions, and states), and cultural. There are two publications, *Hidden Order: How Adaption Builds Complexity* (1995) and *Emergence: From Chaos to Order* (1998) by John Holland, which demonstrate that, in both natural and artificial systems, a small number of rules or laws can generate systems that display real-life complexity. Civilisation is guided by visions of the future, and humans contribute the genetic structure of societal systems in the form of ideas, images, knowledge, blueprints, etc. The phenotypes are the households, establishments or organisations, artefacts, buildings, cities, and states resulting from these plans and information structures.

Futures research is a multi-disciplinary science that addresses human problems; there are four main schools of futures studies, namely the environmental science school, the systems science school, the social and political science school, and the management science school as follows:

- The environmental science school embraces the natural and geoscience disciplines, including land-use, resource use, fresh water, pollutants, emissions, and climate models. The environmental science school is concerned with sustainable development and ecosystem stress, and it includes physicists, chemists, biologists, ecologists, agronomists, meteorologists, geographers, and geophysicists.

- The systems science school deals with hardware, such as the built environment, engineering systems, such as energy or transport and weapons systems, and it includes urban planners, engineers, and systems or defence analysts. The term systems analysis came from RAND in 1948, where it involved strategic decisions on the most cost-effective combinations of military forces and weapons systems. In the civil or public sector, futures researchers do not label themselves as either systems analysts or futurists but generally describe their work as project evaluation, planning research, or policy analysis. Systems science and futures research is a secondary discipline for professionals in the built environment, and it is not yet prevalent in academia. *Long-Range Futures Research* provides the means by which we can change the taxonomy of social science to define a complex adaptive system, where the emergent phenomena are hardware in the form of artefacts and cities that can be analysed using both systems science and urban studies.

- The social and political science school deals with soft systems such as sociology and economics and includes social forecasters and the majority of academic futurists. General systems theory emerged in 1947 based upon laws of natural science, and when Kenneth Boulding published *Ecodynamics: A New Theory of Societal Evolution* (1978), it had the potential for providing a model of social change for futures studies, but it had little applicability for problem solving. This position is transformed as soon as spatial structure is put into his theory, because it becomes an applied science, and that is the intellectual contribution of *Long-Range Futures Research: An Application of Complexity Science.*

- The management science school is concerned with corporate strategy and includes management consultancies, business schools, and major corporations. Management science and sophisticated business forecasting emerged in the 1960s. Management consultancies diversified from work-study and industrial engineering into areas such as factory planning, warehousing and distribution studies, marketing, and computer technology. The principles of industrial dynamics emerged from Massachusetts Institute of Technology (MIT) in the late 1960s. By the 1970s, MBA graduates were recruited into newly formed corporate strategy divisions, and techniques for corporate scenario planning were applied. The interface between the built environment and management science lies in corporate location strategies and site selection, and regional development agencies and city authorities compete to attract businesses and corporations. *Long-Range Futures Research* (L-RFR) is useful to the management science school for environmental scanning.

There are over 125 futures orientated organisations with web-sites and some twenty-five research institutes or think tanks involved in long-range futures research. These include The Frederick S. Pardee Centre for the Study of the Longer Range Future at Boston University, The ACUNU Millennium Project, CSIRO Australia, The Brookings Institution, University of Denver (IFs), Global Scenario Group, Hawaii Research Center for Futures Studies, Institute for Alternative Futures, International Energy Agency, International Institute for Applied Systems Analysis, OECD International Futures Programme, The RAND Corporation, Stanford Research Institute (SRI), Swinburne University (Australia), United Nations Environment Programme, University of Houston Program in Studies of the Future, World Bank Group, World Future Society, World Futures Studies Federation, World Resources Institute, and the Worldwatch Institute.

The demand for *Long-Range Futures Research* is driven by sustainable development studies and climate change models, which involve consideration of at least five future generations with a planning-time horizon of 100-150 years. In the public sector, futures studies tend to have a twenty-five-year exploratory horizon with a fifty-year normative horizon. However, the General Equilibrium Model of economics underpins futures studies undertaken by the social and political science school, and this is only useful for predicting near-term social change. *Long-Range Futures Research* develops an evolutionary model by putting spatial structure into the prevailing model using complexity science as a new tool, so that the notion of economic equilibrium is replaced by the concept of 'far-from-equilibrium stability'. This enables the engineering and scientific communities to deal with the compelling issues of settlement history, path dependency, and endogenous growth, even though this may threaten the intellectual framework of economics. The evolutionary approach provides alternative scenarios for the year 2150 with lower rates of growth in Gross World Product, which impacts on global energy use and carbon dioxide emissions levels. The book is truly unique in addressing this fundamental issue for futures research. The evidence in support of the *L-RFR* outlook is set out in each chapter of the book, and this alternative paradigm has to expose and resolve anomalies in the existing concepts and be tested for its 'goodness' of fit and the ability to predict.

Complexity science may be described as the science of 'emergence', where large ensembles of interacting entities exhibit collective behaviour that is remarkably different from what might be expected by scaling up the behaviour of the individual entities. Complexity science aims to provide an insightful description of the collective properties of a system in terms of attractors of the model dynamics, or macrolaws that are often hard to predict, independently of a mere description of its microscopic entities. There is not yet a general and rigorous theory of complexity science, but complexity research attempts to integrate a diverse set of methodologies that have been developed within the various scientific disciplines. *Long-Range Futures Research* shows how complexity science unifies some thirty diverse features that arise from the evolution of the artificial civil system, and it underlies long-range futures research. The concepts of complexity science are introduced and explained in chapters 2 to 9 and 11 as follows:

2 - Open systems, the system growth parameter, and coevolution
3 - Endogenous growth, path dependence, civil emergence,
     and bifurcation
4 - Transacting entities, complex adaptive system, colonisation,
     and resilience
5 - Microstructure, diffusive ecostructures, and gradient reduction

6 - Information growth, transactional complexity,
    and network dynamics

7 - Civil phase transitions, attractors, chaotic dynamics,
    and macrolaws

8 - Technological evolution, diversity, and the logistic curve

9 - Hierarchical trajectory, equipollence, and the power law

11 - Macrostructure, spatial integration, and decomposition

There are over sixty complex systems institutions with web sites of which 20% relate to life sciences and medicine, 20% to the physical sciences, 15% to computational sciences, 5% to engineering systems science, 10% to complex adaptive systems, 20% to social science, and 10% to management science. The complex systems institutions include the ARC Center for Complex Systems, California Institute for the Study of Complex Systems, Center for Complex Systems Research, Complex Systems Lab, The CNA Corporation (Center for Naval Analyses), Cranfield Complex Systems Management Centre, Centre for Complex Systems (Australia), Florida Center for Complex Systems and Brain Sciences, Illinois Center for Complex Systems Research, Institut des Systemes Complexes Rhone Alpes, International Institute for Applied Systems Analysis, Iowa Complex Adaptive Systems Group, Los Alamos Center for Nonlinear Studies, Michigan Center for the Study of Complex Systems, New England Complex Systems Institute (NECSI), New Hampshire Complex Systems Research Center, Northeastern University CIRCS, Plexus Institute, Portland State University Department of Systems Science, Potsdam Center for Dynamics of Complex Systems, Santa Fe Institute, The Cambridge Nonlinear Centre, UCLA Human Complex Systems, and the UCL Centre for Nonlinear Dynamics.

The study is based upon twenty years of research that focused on the growth of the world's system of cities, the changing urban economic structure in the transition to an informational economy, and the emergence of sustainable city regions or ecopolitan states. One characteristic of the study is that it provides a fresh and positive outlook for the resolution of the predicament of humankind by recognising that, from the year 2000, the global economy is able to generate sufficient capital investment for a demographic and an ecologic transition. From

1976–1986, the author was managing director of a specialist consultancy involved in planning research and project evaluation. For all of that period, his company was engaged by Planning Research Corporation (USA) to apply systems science approaches to international development programmes. Both planning research and futures research have a systems science orientation involving modelling, and the term *planning research* is generally applied to development studies. The term *futures studies* tends to have a social science connotation, and it is more descriptive than the rigorous quantification of futures research that depends upon hard data.

After reading *The Coming of the Transactional City* (1983) by Jean Gottman, whose landmark book *Megalopolis* (1961) identified quaternary information activities, the author decided in 1987 to embark upon on a research programme to explain the transactional dynamics behind urban investment flows. Between 1987–1996, the author directed a multi-client study to develop the Megapolis Regional Information System covering some 300 urban centres and 11,000 establishments with over fifty employees in the London region, which extends to thirty kilometres (twenty miles) beyond the M25 motorway. The research programme revealed the change in investment returns with the evolving economic structure of towns in the transition to an informational economy. Complexity science explains the transactional dynamics behind urban investment flows by a natural reduction in gradients between locations in investment returns.

It would be ungracious to list shortcomings in the official statistical database or institutional obstructions to the classification of towns. However, it is the restrictions of the UK Statistics of Trade Act, 1947, that have made it necessary to build a database of large establishments from extensive business research, to avoid using restricted census of employment data for the classification of urban areas. The methodology that has been developed in this study can be applied to overseas countries in which the statistical database may be deficient, with a view to analysing the economic structure of countries, regions, and cities in terms of the four economic divisions (primary resources, secondary industrial, tertiary commercial, and quaternary information). There is a statistical basis for using the number of large establishments as a proxy for total employment, to give an accurate positioning of a city region with respect to the informational economy and its transactional complexity.

Each continental region has its own political, economic, and social contexts, and the future can be shaped by designed interventions in the global macrosystems. Technological development leads to engineered transformations that can alter the evolutionary trajectory of a system, and investment capital is the system growth or control parameter for the civil system. The aim is to identify critical intervention points at which investment capital, or the release of resource constraints, can achieve a disproportionate and beneficial change in the system. This book provides snapshot pictures by continental region to provide the reader with a clear understanding of the intermediate positions in the years 2000, 2025, and 2150 in the evolution towards a future state. The year 2000 may be taken as a starting point to represent the present condition, and the year 2025 has been selected as a vision of the near future, in which change is likely to be rapid. Within the next fifty years, world development will be running at a peak rate, and it will require a much longer planning horizon than in the last century. The year 2150 is sufficiently far away for the vision to be blurred, and an early planetary civilisation could be reached by the end of the Third Millennium.

With a project of this scale, the intellectual debts to colleagues from Planning Research Corporation, academics, and writers are too numerous to identify. The formulation of a new model involves assembling a number of existing blocks of knowledge and combining them with some new blocks in a novel arrangement. Chronological references that span from 1940-2007 have been listed at the end of each chapter, in order to demonstrate the history and the evolution of futures thinking and to attribute the conceptual components assembled for completion of this research. The sources of pivotal concepts have been clearly identified in the text, and those that have not been highlighted have generally provided useful ideas that have had to be adapted for their application to futures research. The authors of these publications will probably welcome a mention for suggested reading of their work, but they may not necessarily wish to be attributed with the twist or adaption of their ideas. The bibliography in Appendix 39 provides an alphabetical list of authors and their publishers to acknowledge the contributions made in books, academic papers, or in the journal *Futures*. The aim is to integrate the pieces of the mosaic that provide a vision of the future in a way that is consistent with the dynamics of development and observed trends in the real world. Where practicable, the pieces that are attributable to others have been altered as little as possible so that the origin of ideas can be traced. However, these independent pieces do not make up a coherent picture until they are sorted and fitted into position around the gaps that have to be filled by other research. By producing new evidence and establishing a different logic for the arrangement of concepts, the altered taxonomy creates a paradigm shift from an obsolescent world view.

The year 2000 has been taken as a base date for prices and cost information, including per capita incomes and assets, which are expressed in 2000 US$. The information in the Megapolis Regional Information System was updated in 1996, which also coincided with the date of the *United Nations Global Report on Human Settlements* that gave population projections to 2050 and urban trends to 2025. The *L-RFR* prognostication is based upon measurable emergent phenomena, and with a time horizon to 2150 and an evolutionary trajectory to 2250, the approach can out-range global scenario models to 2100 or 2200. These include the *IPCC Special Report on Emissions Scenarios* (2000), *The Stern Review on the Economics of Climate Change* (2006), and the *International Futures (IF) Simulation Model* (2006). The year 2000 is taken as the present day, and MYA and BYA are abbreviations in the meta-timescale for a million years ago and a billion years ago, respectively, and MYF and BYF stand for a million years into the future (or forward) and a billion years into the future. A billion is 1,000 million and in tables or in repetitive use it is abbreviated to bn, whereas a million is written as 1m. To prevent any confusion with metric linear measurements, a metre has not been abbreviated although km is used and areas are given in m^2 and km^2.

The purpose of the book is to set out the scientific basis for futures research by developing an evolutionary landscape for the world, which is based upon a coherent set of hard data that can be applied to other studies by the reader. *Long-Range Futures Research* shows how the science of evolution and complexity (or nonlinear science) provides an approach for exploring the complex and unprecedented patterns of global change expected in the future. To help the reader follow the thought processes, forty tables of data have been provided in *Long-Range Futures Research*, many of which are derived from accepted contemporary sources of information, such as the United Nations, the World Bank, or the OECD. Whilst every care has been taken to derive sound values for the analysis, there is no claim that the values are anything more than realistic enough to demonstrate a methodology. The provision of key empirical or statistical information eliminates the need for any complex mathematics in the text. Also, the reader's attention is drawn to the glossary of terms in Appendix 37, which contains 135 definitions.

Part 1 'Futures Research and Systems Science' (Chapters 2-4) provides a guide to evolutionary futures research and explains how the concept of far-from-equilibrium stability from the science of evolution and complexity challenges the notion of economic equilibrium. Part 2 'Transactional Microstructure' (Chapters 5-9) applies complexity science to a world city region and derives the macrolaws of Ecodynamics. In Part 3 'Contextual Macrostructure' (Chapters 10-13), prescience is applied with a global perspective to describe the geopolitical macrostructure and scenarios for the global macrosystems. Part 4 contains forty appendices with tables, graphs, a glossary of terms, and a bibliography.

At the outset, the author gratefully acknowledges his family background for ensuring the long period of schooling and training, which provided the grounding for a lifetime's journey in search of answers. His grandfather was an Austrian, who emigrated to the USA before the First World War, and his father was born in Lodi, California, in 1918. However, the family left for England at the onset of the Great Depression, and had become naturalised British Citizens by the outbreak of the Second World War. His father and mother married in 1940, and after wartime service as a British Army Officer, his father sadly died in 1947. His mother ensured that he had a continuous education at Hurstpierpoint College, Sussex, from 1952-61, with grandfather providing financial support during the preparatory school years. Her brother H. H. Beaumont, who served as a wartime Officer in the Royal Engineers, strongly encouraged him to become a Chartered Civil Engineer, and he received his professional experience with the international building and civil engineering contractor Wimpey. At the age of twenty-five, the author started his consultancy career in the Project Planning Division of P.A. Management Consultants, which provided a thorough training in project research and evaluation. The author wishes to dedicate the book to his wife Valerie to whom he is eternally grateful, and she has been an inspiration with her enthusiastic support during the decade that the research for *Long-Range Futures Research* has demanded. She has engaged in long discussions at each stage of the writing process that have been both creative and analytical, and in this way she has helped to steer the venture through emerging new evidence en route. He also appreciates the support from H.H.B.'s widow Bea, who has taken a real interest in the literary quest to write and publish this book.

Also, the author thankfully acknowledges the contribution by Geoffrey Brown, the former General Manager of the UK subsidiary of Planning Research Corporation's engineering group, for commenting on a draft outline of *Long-Range Futures Research*. Also he would like to thank former colleagues of PRC, Denis Delaney and David Meggitt, for their own contributions to the know-how involved in developing planning standards and parametric cost models for urban development programmes in the Middle East and North Africa. A debt of gratitude is also due to Aldona Lindstedt for her contribution to the RICI classification of establishments and the database development for the Megapolis Regional Information System. The Investment Property Databank (IPD) is given recognition for the high level of consistency in the calculation of long-term property investment returns for office, retail, and industrial properties within the London Megapolis. Also, Professor John Ratcliffe, Director of the Faculty of the Built Environment at Dublin Institute of Technology (DIT), is acknowledged for providing the venue for a master-class to the Futures Academy, which marked the launch of *Long-Range Futures Research* in December 2003. Sharon Barnes re-created and enhanced the presentation of tables in the text and appendices, in readiness for the formatting and design of the book. The design team at BookSurge has produced a spacious layout for the chapters that reinforces the structure of the book, with tables and charts that have been creatively adapted to give both impact and variety within the overall design concept. Finally, Ronald Danaghe has made a valuable contribution in editing the manuscript in line with British English conventions and spelling, although it is becoming increasingly difficult in technical subjects to distinguish it from American English. To make the text slightly more user friendly for American readers, extra commas have been inserted, particularly before the 'and' at the end of a list. The editorial and design team at BookSurge Publishing have been the catalyst that has enabled *Long-Range Futures Research* to be taken from consultancy in the field to the academic world, and for introducing complexity science to a wide professional group outside academia. However, it will be the readers of this book who will hopefully use complexity science and the emergent properties of the civil system, to explore an evolutionary future for our planet for the long-term benefit of civilisation. The author would appreciate feedback from readers to enhance the publication for future editions, and contributions will be gratefully acknowledged.

PART 1

*FUTURES RESEARCH
AND SYSTEMS SCIENCE*

1.
SOCIETAL EVOLUTION AND FUTURES RESEARCH

1.1   Societal evolution

The essence of the scientific method lies in the description and explanation of phenomena in fields such as astronomy, geology, physics, ecology, and economics. The biologist Ludwig von Bertalanffy was the father of general systems theory (1947), and with economist Kenneth Boulding, physiologist Ralph Gerard, and mathematician Anatol Rapoport, founded the Society of General Systems Research (1954). The society was renamed the International Society for Systems Science in 1988. The original purpose of the society was to undertake interdisciplinary investigations into the nature of complex systems and to tackle problems in the non-physical sciences, such as the biological, sociological, and behavioural fields. In *General System Theory* (1969), Bertalanffy explained the concept of an open system, which cleared the way for a theory of systems that evolve towards complexity and an increase in organisation or order. Open systems enjoy energy and material inflows and outflows across their boundaries. They are able to adapt to their environment and yet maintain stability far from equilibrium. *Long-Range Futures Research: An Application of Complexity Science* sets out the scientific basis for futures studies by a unique application of the science of evolution and complexity, for the development of an evolutionary vision of the world.

The natural sciences describe natural phenomena, and *The Sciences of the Artificial* (1968, 1981, and 1999) by Herbert A. Simon describes artefacts that result from human intervention in the natural world. Understanding the teleological process involves modelling the behaviour of complex adaptive systems in which the organisms, households, and establishments change their strategies in accordance with the feedback they receive from their environment. Information is needed to give spatial structure to a system, and the aim of applied science is to develop a model of the system, with a view to understanding the underlying dynamics or mechanisms for a variety of aspects.

According to Kenneth Boulding, artefacts can be classified as material, organisational (i.e., establishments, civil institutions, and states), and cultural. Cultural artefacts include language, communications, mathematics, music, art, customs, and religion. The production of artefacts involves the factors of production, which are information, energy, materials, space and time, and it is know-how that guides energy to select, transport, and rearrange or transform materials into the complex artefacts produced by humankind. In *Ecodynamics – A New Theory of Societal Evolution* (1978), Kenneth Boulding explained that the evolutionary vision of society starts from the premise that civil artefacts are a second nature, and, in the development of the planet, the evolution of the car or the aeroplane is just as natural as the creation of a horse or a bird. Corporations or organisations retain the know-how or genetic structure to reproduce vehicles or other artefacts in the same way that a fertilised egg contains the know-how to reproduce an animal. In nature, all species are endangered, and selection is an ecological interaction that constantly creates new niches and destroys old ones, and mutation takes the form of invention, innovations, and discovery.

Investment by civilisations for over 10,000 years is the basis for an evolutionary view of societal progress, since it has created an irreversible process of world development through the accumulation of capital stocks that cause constant change and disequilibrium. At the start of this period, the cumulative number of adult hunters belonging to modern man is estimated to have reached ten billion over some 2,000 generations, and, subsequently, a further 500 generations has brought the cumulative total of humankind to have lived on the planet to some fifty to sixty billion by the year 2000. Civilisations develop through an increasing number of transactions involving the transportation and transformation of materials into artefacts, which require both energy to support the transactions and information to provide the organisation and knowledge to direct the energy and resource flows. According to Arnold Toynbee, civilisation is a movement and not a condition, a voyage and not a harbour. Civilisation is associated with urbanisation and technological development, the interaction of humans, language, communication, and culture. The evolution of civilisation involves irreversible and continuing changes over time, which creates long-term development and disequilibrium with increasing diversity, specialisation, connectivity, and complexity in the world system of cities.

In *The Ascent of Man* (1973), Jacob Bronowski states that 'Man has a set of gifts which make him unique among the animals: so that unlike them, he is not a figure in the landscape - he is a shaper of the landscape'. It is true that nature can also provide examples of insects and animals that create artefacts, such as ants or termites, and birds with their nests, as well as beavers with their dams, by which they modify their habitat. However, societal evolution is guided by images of the future, which is different from biological evolution. Humans coevolve with their artefacts to increase the carrying capacity of a territory, to survive in hostile conditions, such as the arctic or the desert, to explore remote regions including the floor of the oceans and to journey to the moon and survey other planets.

Civilisation is an evolutionary process in which the accumulated stock of wisdom is reflected in the institutions, culture, knowledge, information structures, and socio-technological systems. Culture embraces traditions and beliefs, value systems, ways of life, and forms of art, and the ideal of civilisation includes a sense of ethics and a humane approach to life. The cultural heritage of humankind and the culture of *Homo Sapiens Sapiens* is passed on from generation to generation in the form of memory, language, customs, history, religion, philosophy, moral and aesthetic values, learning, literature, music, dance, and art. However, cultural evolution is driven by the exploratory and innovative members of a social group, and a new art in the capture of resources is then learned by the other members of the group, so that its habitual use becomes ingrained in the advancing culture. A closed culture may be ignorant of a better way of life or be a late adopter of new science and technology, and periodically there may be a shift in ideology or scientific theory.

Human behaviour is caused by a combination of anthropogenic factors that arise within the biological metasystem, together with learned cultural responses that are formed in the context of the civil system. Ethnic or genetic differences tend to be preserved through the existence of nations and also through the conservation of urban quarters or ghettos with pockets of racial segregation. In addition, differing levels of intelligence within the metapopulation become reinforced through education and the creation of class barriers between socio-economic groups. Humans specialise in occupations in accordance with their aptitudes, and the conventional wisdom is that cultural variations are phenotypic rather than genetic in origin. It follows that genetic inheritance, cultural inheritance, and the inheritance of wealth are part of the process of societal evolution and the technological development of artefacts. Progression of the human community system is reflected in the stages of economic development, the changing structure of the workforce, and the technical progress of organisations and artefacts. On a daily basis, there are births and deaths in the populations of humans and their artefacts, new technologies replace obsolete ones, new firms are started and others go bankrupt, new workers are employed and others are made redundant, and, as people retire, others are selected or promoted to take their place. The survivors in society find a niche or social need and, with the use of artefacts, they utilise support to create positive transactions and increase their well-being.

According to Boulding, all societies have their own set of human values, and these are coordinated by the political system, the market system, and the ethical values of subcultures. The political system provides legislation, justice, and defence of the state, and a wide range of systems have been observed from dictatorships to democracies. The market system produces a set of relative prices and outputs as a result of the coordination of all the evaluations of the people in it, and its great advantage is that it can coordinate values without agreement. The ethical values of subcultures provide pressure on individuals to conform to the ethos of the group or to be denounced or outcast from the group. Aesthetic, moral, or religious ideas are mental images, which may or may not survive over time as they become tested by reality or as other ideologies emerge. In general, single visions such as nationalism, militarism, communism, socialism, or capitalism are inadequate to describe the enormous complexity of society in a global context.

In Boulding's theory of societal evolution, humankind has developed a set of social organisers identified as threat systems, relationship systems, and exchange systems. Threat systems, with evidence of the capability to deliver the threat, may be followed by submission, defiance, flight, or counter-threat. Relationship systems involve love or hate, benevolence or malevolence, and the concept of identity, family, community, or a group with which one identifies. Exchange systems based upon invitation are powerful in organising the division of labour and for expanding the number and range of human artefacts. Each of the systems may be applied in various proportions, and although the threat system may be the most oppressive, exchange systems can lead to a selfish 'winner-take-all' philosophy. Cooperation in both societal and biological evolution arises from the principle of reciprocity, which involves a Tit-for-Tat pattern of exchange such that the other party is discouraged from defecting from an arrangement for the collective good.

Ecodynamics (1978) points out that social equity relates to a notion of fairness in the distribution of things such as wealth, income, power, position, educational attainment, or access to health facilities. Social equality has come to mean that the rules of the game are equal for all people, such as equality of opportunity or people's rights under the law. In this way, people are not equal, but the granting of votes to classes without property is a device for legitimising economic inequality, since the division of society takes place under fair rules. The growth of democratic societies, with one vote per person, provides an opportunity for wider distribution of power with the inclusion of groups of people of lower status. In an equitable society, it is necessary to recognise and reward real differences in skill, effort, location, danger, and other conditions affecting work. Distributions are the result of equilibrium processes, and, for example, the share of value added between corporate investors and the workforce has been established for many years as a result of negotiations, strikes, and conflicts. In market economies, it is one dollar one vote so that the affluent have significantly more consumer votes than the less well off. Since the basic needs of the less well off should take precedence over the trivial wants of the more affluent, in modern states progressive taxation systematises the transfer of income from the rich to the poor through welfare payments and negative income taxes, etc. The dynamics of the distribution system have to be understood before changes are possible and a new equilibrium position achieved.

1.2 The Systems Science Origins of Planning and Futures Research

The purpose of both planning research and futures research, which are included in the terms *policy analysis* and *systems analysis*, is to inform long-range or strategic planning. By definition, planning is undertaken to design a programme that will create a future state of affairs that is both better than doing nothing and better than the alternatives available within a resource budget. Inevitably, investment planning involves expectations about the future outcomes of alternative courses of action, which will depend upon forecasts of changes in the behaviour of civil systems or of market demand under a variety of circumstances. Forecasts of demand for a service or product are notoriously unreliable in the long term, and complex projects such as airports, power stations, or teaching hospitals often have facility lead times of more than a decade and operational planning horizons of fifty years.

An objective of planning or futures research is to identify the conditions, such as critical resource shortages or resource abundance, which will lead to discontinuities with an irreversible societal transition to a new epoch at a different level of stability. For example, increasing per capita investment or technological evolution will eventually cause a societal jump to a different stage of development, and, conversely, longer life assets and resource conservation in an ecopolitan society (see chapter 7) will also change the dynamic balance of the system. The greatest value from futures research and policy analysis arises from the identification of critical intervention or leverage points in the macrosystems, at which investment capital or the release of resource constraints can achieve a disproportionate and beneficial change to the system. Policy-makers aim to design an adaptive system that minimises the risk of destabilising catastrophes or disasters, which create chaotic conditions for brief periods of evolutionary time. The purpose is to identify desirable changes to the trajectory of the civil system that will be more cost-effective than others. International and corporate policy making is a design process with multistage decisions, and with the constant search for levers that can affect the unfolding future.

A scientific basis is needed for planning research so that strategic thinking can evolve as the future unfolds and changes in the contextual environment are perceived. Sufficient knowledge has to be integrated to understand the detail of the transactional microstructure in which an organisation participates, and to anticipate crucial changes in the contextual environment or macrostructure over which the organisation has no influence. The term 'contextual macrostructure' is used in preference to 'contextual environment', because it is in fact an encompassing system that interacts with, and is linked to, the microstructure. The transactional microstructure for non-primary businesses generally takes place at the urban level, whereas macrostructure arises from the encompassing hierarchy at the regional, national, continental, and global levels.

H. G. Wells founded futures studies in 1902, when he published a comprehensive survey of future developments that surpassed any forerunners in *Anticipations of the Reaction of Mechanical and Scientific Progress upon Human Life and Thought*. In the same year as the publication of *Anticipations,* he gave a lecture to the Royal Institution in which he called for a whole new science for a 'systematic exploration of the future' that could yield a firm inductive knowledge of the laws of social and political development. Wells' work spanned a period of fifty years until his death at the age of seventy-nine in 1946. It was the Second World War that gave futures studies its military orientation in the USA, whereas a cultural perspective emerged in Europe in the 1950s and 1960s with the leading names of Gaston Berger and Bertrand de Jouvenal. This led to a golden era of futures studies, during which the *Limits to Growth* was published, but this had peaked by 1980 when the membership of the World Future Society (WFS) reached 60,000. Edward Cornish had founded the WFS in 1966, and by 2000, the membership had stabilised at 25,000.

In 2004, Edward Cornish, the President of the WFS and editor of *The Futurist* magazine, published *Futuring: The Exploration of the Future*, which reflects popular thinking on the future at the turn of the Third Millennium. Dr Michael Marien, who compiles a monthly and annual *Future Survey* of articles and books for the World Future Society, has prepared more than 20,000 abstracts of futures-relevant literature over the past thirty years and has made a major contribution to the advancement of futures studies. As the Millennium approached, academic futurists focused on defining and establishing a multidisciplinary curriculum for teaching the principles of futures studies. Two notable academics behind this endeavour were Richard Slaughter, who edited the three volumes of *The Knowledge Base of Futures Studies* (1996), and Wendell Bell the author of *Foundations of Futures Studies: Human Science for a New Era* (Volumes 1 & 2 - 1997). These publications provided a thorough overview of the state of the futures field at that time.

There are four main schools of futures studies, namely the environmental science school, the systems science school, the social science school, and the management science school. The environmental science school embraces the natural and geoscience disciplines, including atmospheric and climate models. The systems science school deals with hardware such as the built environment, engineering systems such as energy or transport and weapons systems, and it includes urban planners, engineers, and systems or defence analysts. The social science school deals with soft systems such as sociology and economics, and it includes social forecasters and the majority of academic futurists. The management science school is concerned with corporate strategy and includes management consultancies, business schools, and major corporations. *Long-Range Futures Research: An Application of Complexity Science* creates artificial landscapes from the evolutionary changes in the spatial structure of the civil system, which is an open, complex adaptive system. By combining a built-environment substructure with a social science superstructure, *Long-Range Futures Research* provides a scientific basis for futures research, which enables it to be integrated into undergraduate and postgraduate courses on planning, development, and environmental studies. The chronological development of futures studies is set out in 'A century of futures research and systems science' in Appendix 1.

Prescience is the foreknowledge that arises from futures research, and its purpose is both to explain past evolution and explore potential futures. Foresight is different in so far as it means taking action to avoid harmful situations and to protect us from suffering the consequences of inadequate preparation or errors of judgement. Foresight also implies the ability to plan, in order to seize opportunities when they present themselves. Evolutionary prescience relates to the behaviour of complex adaptive systems, and this differs from both forecasts and the use of prediction, which are more relevant for applied sciences such as thermodynamics, hydrodynamics, or aerodynamics in which fluid flows or structural stresses can be calculated with a degree of confidence. However, forecasts are widely used by government, business, and industry, where there is a reasonable expectation that the past performance of a system will set a trend that may continue for the foreseeable (short-term) future.

There are two terms, *extrapolation* and *prognostication*, that need to be distinguished from one another for futures research. Extrapolation of trends is a useful technique for research into the near future, and the reliability of forecasts by extrapolation decreases with the length of term of the projection. However, with complex adaptive systems, long-term growth of the system will neither be linear nor exponential as it will generally follow the Verhulst logistic S-shaped curve as shown in Appendix 2. Prognostication involves an attempt to make an independent assessment of future conditions, and it is used for long-range futures research where there is a scientific basis or phenomenon that limits the number of possible future states of a system. With many phenomena, including economic development and population growth, the change over from increasing to decreasing slope of a logistic S-curve is driven by both internal dynamics and resource constraints. The critical issue is to understand the dynamics that will result in an inflexion point, when the logistic curve turns over. The intermediate- or medium-term future may be explored by interpolation between the extrapolated near future and a prognosticated long-term future state, as shown in the diagram in Appendix 3.

Planning and futures research have their origins in the systems sciences, which include policy analysis, decision analysis, operations research, systems analysis, systems engineering, and simulation. Systems thinking progressed from the mechanistic models of physical science to operations research, which originated from the wartime secondment of professional scientists to military operations, and then to systems analysis at Rand Corporation for strategic decisions on the most cost-effective combinations of armed forces and military technology. Rand Corporation was a catalyst for the formation of a number of think tanks and advisory corporations for the application of systems analysis. The

System Development Corporation (SDC) evolved from the System Development Division of Rand and became one of the largest advisory organisations concerned with military technology. Planning Research Corporation (PRC) was founded in 1954 by ex-Rand staff with a policy of undertaking fifty percent of its work in the civil sector, and with a systems science corporate culture it emerged as a large, diversified organisation with 6,500 personnel by the 1980s. Since then PRC divested the civil group and was acquired by Northrop Grumman, a global defence company in electronic systems, information technology, advanced aircraft, submarines, shipbuilding, and space technology with 125,000 staff and annual revenues of $25 billion. The historical background to these corporations and other non-profit think tanks involved in long-range planning and forecasting, was documented by Paul Slee Smith in *Think Tanks and Problem Solving* (1971).

Policy analysis emerged from the application of these approaches to public policy issues in healthcare, education, social services, and transport systems. Robert McNamara, the former President of the World Bank from 1968-1981, was formerly US Secretary of State for Defence, and he formalised cost-benefit analysis and world development indicators for the evaluation of programmes. In industry, the principles of industrial dynamics were explored at Massachusetts Institute of Technology in the 1960s and formed a new discipline of management science, and these communication and control engineering approaches contributed to the development of models with feedback loops for simulating the system dynamics of global policy problems in the 1970s. Computer simulations of nonlinear problems, the theory of chaos and complexity research emerged from the Los Alamos National Laboratory (USA), which is the centre for nuclear weapons research and has been a leader in advanced computing since the 1950s. In the 1980s, the *Sciences of Complexity* became a specialised field of study at the Santa Fe Institute in New Mexico and models of complex adaptive systems were developed at the University of Michigan.

The development of evolutionary landscapes for futures research needs to incorporate potential territorial diversity at the outset, and it has to reflect the extensive knowledge base required through a variety of interlinked systems at different levels. As far as a classification of systems is concerned, it is useful to distinguish between natural resource systems (astrophysical and geophysical), non-living physical systems (energy, matter, fluids, and gases), engineering systems (hard technology), information systems (software), and living systems (humans, animals, plants, and bacteria), or vivisystems (establishments, technological 'animats', and the 'animates' of artificial life). Artificial vivisystems are discussed in section 1.5, and animats are described in Chapter 8. There is a continuum for the degree of complexity of systems, ranging from non-living physical systems to biological systems; at the higher end of complexity are human systems. There is a corresponding continuum for complexity in futures research from 'Projective Futures' to 'Prospective Futures', and finally to 'Evolutionary Futures'.

There is also a relationship between the futures research continuum and the three planning levels, i.e., the project planning level, strategic planning, and policy planning. The project level involves the systematic methods of operations research and engineering design for studying system variables for the optimal allocation of scarce resources in accordance with efficiency or economic criteria. The strategic planning level involves examining the viability of successfully undertaking transactions in a complex adaptive system, in which the systemic structure evolves over time, and the criteria for success are effective outcome measures in relation to the capture of resources and adaption to change. The policy level of planning relates to anticipating the total system dynamics in which institutions are created for the normative enhancement or design of the overall behaviour of the system, and the criteria for success are the achievement of social norms or values that are established by consensus. Policies for human systems have evolved from the Greek concept of 'polis', which is used in this book to reflect its wider meaning of a city-state, and the construct of an ecopolitan state enables *Long-Range Futures Research* to encompass geopolitical, economic, social, technological, and ecological issues that affect the contextual macrostructure.

1.3   Projective Futures and Forecasting

Projective futures aim to produce quantitative *forecasts* derived from trend analysis and econometric models, with a relatively short time horizon of up to ten years, which frequently corresponds to the period of office of the authority commissioning the research. They are an attempt to apply a scientific method to explain historical evolution through general laws that may be projected forward to predict a future state. Mechanistic models are used with simplified assumptions and primitive causal relations between variables. It is generally taken that, without external interference to the system, stability will be the dominant condition with fluctuations and cyclical regularities. It is assumed that, within the short timescale under consideration, there will not be significant changes in the causal relations of the model, and that, in the absence of unforeseen events or a catastrophe, the extrapolation into the future is deterministic. Uncertainty is considered by undertaking a sensitivity analysis to the external variables.

Forecasting is generally undertaken to predict the future environment for which strategies should be developed; for example, weather forecasting is important for preparing for tornados, floods, snow-storms, inclement weather, and heat waves. Meteorology not only saves lives and helps to protect property, it also saves business resources as entire industries (such as agriculture, construction, transport, and tourism) are affected by adverse weather conditions. Edward Lorenz, a professor of meteorology at Massachusetts Institute of Technology (MIT), discovered the concept of chaos in 1963, when he found that minute variations in the initial conditions of his fluid dynamic model of weather flow could have a dramatic effect on the actual weather forecasts. He coined the metaphor 'the butterfly effect' to imply that the flapping of a butterfly's wings in Brazil could be amplified to cause a tornado in Texas. Whilst long-range weather forecasting is unlikely to achieve consistently reliable results, climate is rather more predictable since it is in fact average weather.

Econometric models used for forecasting tend to be unreliable beyond relatively short periods of time, because their parameters are generally not evolutionary. Dynamic econometric models are therefore unable to reflect the disequilibrium of evolutionary development, and so their validity is restricted to the near term in which there is limited change. Chaotic behaviour results from the randomness of transactions in the short term, which causes turbulence in the system at the level of the transactional microstructure. However, the dynamic conditions of the real world display an irreducible complexity in the emerging spatial and economic structure, which generates information that is distinctly anti-chaotic, since it creates order from adaption. The evolutionary approach, which is outlined in section 1.5, focuses on the slower dynamics of long-term adaption and evolution in the macrostructure.

Demographic forecasting is dependent upon a law of atrophy, which states that populations of both humans and establishments in a bounded territory or region reach saturation as resources for growth become limiting, in accordance with the logistic curve first identified by Verhulst (1845). The territorial carrying capacity is not constant and evolves over time through investment and technological progress. Once a plateau of capacity has been reached at a particular stage of development, the potential may evolve to successively higher plateaux with further development stages. Demographic changes will be greatest the further the population is from equilibrium, when population growth is unconstrained by resource limits, competition, or selection. The law of atrophy differs from the Malthusian theory of population growth, which predicted that populations would increase geometrically, whereas food would increase arithmetically, with a consequential collapse in population.

The principle tool used for technological forecasting is the logistic S-shaped curve, which gives the 'envelope curve' for technologies as they go through the five phases of inception, technological development, expansion, maturity, and obsolescence. The mature phase with gradually decreasing growth rates is generally due to either limited possibilities for further technological advances or limited scope for market penetration above a certain level. The Delphi method of forecasting uses a panel of experts in related technologies to predict the most likely variant of the life-cycle growth curve, such as the standard logistic S-curve, incremental S-curves, or multiple S-curves, depending upon levels of substitution by competing technologies. These experts apply other techniques, such as cross-impact analysis, correlation studies with past curves from similar technologies, and relevance trees. Readers are referred to *Technological Forecasting for Industry and Government* (1968), edited by James Bright, and

Looking Forward: *A Guide to Futures Research* (1983), by Olaf Helmer, for the origins and applications of these techniques. Generally speaking, it takes the convergence of several different technologies to set off an investment wave in a new industry.

Social forecasting draws heavily upon historical and cultural trends to overcome the difficulty of developing hard concepts for an essentially soft system, which is dependent upon images, ideas, and concepts of equity, equality, gender, threats, and power. Since the publication of *The Coming of Post-Industrial Society – A venture in social forecasting* by Daniel Bell (1974), social predictions have been regularly made by sociologists, politicians, journalists, and writers. Social trends and themes emerging in our society were identified by Alvin Toffler with *Future Shock* (1970) and *The Third Wave* (1980), John Naisbitt with *Megatrends* (1982) and *Global Paradox* (1994). More scholarly publications on social trends include the United Nations Research Institute for Social Development (UNRISD) publication *Social Futures, Global Visions* (1996), edited by Cynthia Hewitt De Alcantara, the trilogy on *The Information Age* (1996-1998), by Manuel Castells, and *Tomorrows People* (2003), by Susan Greenfield. Trends for the near future have been described in *Megatrends 2000* (1990), by John Naisbitt and Patricia Aburdene, *Megatrends 2010* (2005), by Patricia Aburdene, *Next – Trends for the near future* (1999), by Marian Saltzman, and *The Future of Men* (2005), by Marian Saltzman, Ira Matathia, and Ann O'Reilly. William Sherdon describes the limitations of current approaches to econometric, technological, and social forecasting in *The Fortune Sellers* (1998).

An analysis of environmental trends is published annually in the Worldwatch Institute publication *State of the World* and also *Vital Signs – The environmental trends that are shaping our future*. These publications draw on information from national governments, industry, scientific research institutes, and international environment and development organisations to bring together the latest data, statistics, and analysis on the state of the global environment. There are other guides to the future, such as *The Atlas of the Future,* edited by Ian Pearson, which covers population, economic development, health and disease, communications, technology, military capability, urbanisation, and the environment.

1.4   Prospective Futures and Scenarios

Prospective futures research involves building alternative *scenarios* with geographical coordinates and an intermediate time horizon of up to twenty-five to fifty years, which may be within the expected lifetime of the participants. A reduced reality is simulated in computerised models with a given set of initial conditions, performance characteristics, and exogenous variables that reflect the system's environment. The system dynamics incorporate feedback loops that bring results from the past action of the system to control future action. This method accepts that there are a multiplicity of possible futures, and the purpose of building scenarios is to explore the prospects for a viable future state and to anticipate the implications of potential threats and disastrous scenarios. It is recognised that the future is a product of both causality and teleology through intervention, and the discovery of alternative paths to a desirable future makes the approach normative. Prospective futures research acknowledges that the future is unpredictable over time due to changes in the causal relations in the model, and periodically the taxonomy needs to be reformulated. Uncertainty is dealt with through the construction of different scenarios.

Prospective futures research and the art of scenario building have developed considerably since its military origins and later civilian applications by the Rand Corporation. Herman Kahn, a physicist from RAND, founded the Hudson Institute to establish a framework for speculation and to undertake studies on alternative world futures. In 1967, he published *The Year 2000,* and from this base, scenario planning was adopted by a number of corporations. Two characteristics of Herman Kahn's work are his innovative thinking in making extrapolations of unconstrained technological progress and his outlook that 'if something is worth doing well it is worth doing poorly'. In other words, if a question is important, it should be dealt with as well as possible in the absence of theory or lack of data, and this distinguishes policy research from academic research, which tends to avoid questions that cannot be dealt with well. The Hudson Institute encountered hostility from the limits-to-growth movement, and in *World Economic Development* (1979), by Herman Kahn, it was recognised that growth cannot be exponential and that the finite size of the planet induces S-shaped growth curves.

An alternative approach for speculating on future world development was applied in a study commissioned by the Club of Rome, *The Limits to Growth* (1972), and undertaken by an international research team at Massachusetts Institute of Technology (MIT). The study was based upon the system dynamics simulations developed by Jay Forrester with feedback-loops that incorporated the main relationships between the components of a global model. The world model represented an ambitious attempt to bring together forecasts of population growth, resource depletion, food supply, capital investment, and pollution. However, the *Limits to Growth* model and its sequel, *Beyond the Limits* (1992), not only ignore spatial structure and the differences between continental regions, but it also clearly states in *Beyond the Limits* that the model cannot begin to represent the evolutionary dynamics of a world system that is structuring itself in a new way. In effect, the '*Limits*' model cannot transform itself to a new stage of development, and it is, therefore, unable to reflect the structural transitions that take place in societal evolution, such as the agricultural, industrial, or informational stages of development. So, in spite of the fact that the authors of '*Limits*' were able to explain the causal dynamics with unusual clarity, the model was being used outside its validity range. The teleological process by which the civil system responds as a complex adaptive system to extend life expectancy and the lifetime of a civilisation is missing from the model, so that it will have a tendency to predict collapse unless the modellers intervene to make changes to save the outcome.

Constructive criticism of the *Limits to Growth* model indicated that it was unduly pessimistic, and a second report to the Club of Rome entitled *Mankind at the Turning Point* (1974), by M. Mesarovic and E. Pestel, was based upon a multilevel regionalised model of the world system to analyse its future evolution. The regional subsystems were essential to account for the variety of political, economic, and cultural patterns prevailing within the world system. This multilevel model of the world involved storing 100,000 equations or relationships in comparison with several hundred in the '*Limits*' model. However, for the behaviour to be evolutionary, the model dynamics would have needed to be generated as a complex adaptive system, for which the mathematics would have been intractable from the bottom up. So the evolution of the system was analysed with reference to a set of scenarios, where a scenario was defined as a sequence of possible events and socio-political choices. Scenario analysis attempts to consider the future behaviour of a system in the context of different scenarios. Since the Club of Rome reports it is considered intellectually and morally indefensible for futurists to neglect issues of sustainable development, and if these concerns are overstated, it is necessary to put forward the alternative arguments.

The UN provides Long-range World Population Projections to 2300 and a series of Global Reports on Human Settlements to 2030. It is also involved in global scenarios for the UN Intergovernmental Panel on Climate Change (IPCC), which require a 100-year horizon. A fifth generation of the *International Futures (IF) Global Simulation Model* has been developed at the University of Denver by Barry Hughes as a predictive model with a 100 year horizon, for application to the United Nations Environment Programme's *Global Environment Outlook 4* (UNEP GEO-4). The World Bank, the International Institute for Applied Systems Analysis (IIASA), and the World Energy Council are all involved in global and regional energy perspectives, which are driven by urbanisation and the concentration of future growth in the developing countries. The Global Scenario Group of the Stockholm Environment Institute has produced a series of reports with a fifty-year perspective that have received considerable attention, including *Bending the Curve: Toward Global Sustainability* (1998) and *Great Transition* (2002). Also, the Rand Pardee Center and the Pardee Centre for the Study of the Longer-Range Future at Boston University are engaged in studies of the forces shaping global society from thirty-five to two-hundred years into the future. Rand have published *Shaping the Next One Hundred Years* (2003) by Robert Lempert, Steven Popper, and Steven Bankes, which sets out methods for testing the robustness of long-term scenario models. The American Council for the United Nations University has developed global scenarios for the Millennium Project, and Jerome Glenn and Theodore Gordon produce an annual *State of the Future* report, which takes a twenty-five-year exploratory horizon and a fifty-year normative horizon.

There are also popular publications on the astrophysical and geophysical sciences that describe the long-range future of our planet with visions of a spacefaring civilisation, such as *The Next 10,000 Years* (1975) by Adrian Berry or *Entering Space* (1999) by Robert Zubrin. A positive view of long-range futures is given in *Deep Futures* by Doug Cocks, and a 100-year *Long-Term Plan for Greater Vancouver* has been produced by The Sheltair Group. A number of publications identify potential hazards in the future, such as *Crucibles of Hazard: Mega-cities and disasters in transition* (1999) from the United Nations University Press, which highlights the impact upon the built environment of natural hazards, such as fires, floods, earthquakes, severe storms, or technological hazards, such as transport accidents, toxic spills and nuclear disasters, and social hazards, such as crime, terrorism, and wars. The scale of hazards and disasters can be measured in terms of the spatial scale ranging from the local impact of earthquakes, the regional impact of tornados or tropical cyclones (hurricanes and typhoons), and the global impact of climate change. Other publications give scenarios for eco-catastrophes, biohazards, pandemics, and asteroid impacts. Examples of these visions include *The Coming Plague* (1995) by Laurie Garrett, *Our Final Century* (2003) by astronomer Sir Martin Rees, and *A Guide to the End of the World* (2002) by geophysicist Bill McGuire. A simplistic way of avoiding serious consideration of the long-term future is to assume that some catastrophe will inevitably destroy a civilisation so that no further serious thought need be given to the matter. For that reason, a meta-timescale for distant and deep futures is set out in Chapter 2. However, for the purpose of the research for *Long-Range Futures Research*, the planning time horizon has been taken as 150 years, and it is not anticipated that there will be a devastating catastrophe on a global scale within this period. In the course of a millennium, there will inevitably be discontinuities that could be utopian or dystopian.

The prospective futures methodology 'La Prospective' sets out to identify visions of possible futures for a system such as air transport, energy supply, or an industry like information technology and to proactively influence or shape its future development. La Prospective was initiated in France in the 1950s by Gaston Berger and described by Bertrand de Jouvenel in *The Art of Conjecture* (1967). The technique was further developed by Michel Godet in *The Crisis in Forecasting and the Emergence of the 'Prospective' Approach* (1979). Michel Godet has updated his approach in *Creating Futures* (2001). The methodology currently represents the management science school in which scenario planning is used as a strategic management tool. However, the approach embraces elements of systems analysis, as it uses technological forecasting with the Delphi survey and a cross-impact matrix to incorporate human judgement in the identification of the explanatory variables that characterise a system, the relationships between them, and their possible future values. A blend of qualitative and quantitative techniques links planning to forecasting by identifying the best option for a system under multiple scenarios. A morphological analysis identifies for each subsystem the various possible states or configurations, and potential combinations of possible states or configurations give rise to scenarios for the future state of a system or an industrial sector.

The corporate scenario artists are strategic planners whose objective is to create a more adaptive organisation, which can respond to change and uncertainty. The methodology was first explained in *The Art of the Long View* (1991) by Peter Schwartz. The planners are involved with the process of scenario-based planning that has been developed over thirty-five years within transnational corporations, such as Royal Dutch/ Shell. *The Shell Global Scenarios to 2025 – The future business environment: trends, trade-offs and choices* (2005) by Jeroen Van der Veer, portray three plausible futures with contrasting economic, political, and regulatory features with distinct implications for the energy systems. *The Long Boom* (2000), by Schwartz, Leydon, and Hyatt, reflects current thinking in 'Prospective Futures'. The latter book depends upon investment-wave theory for its plot. *Inevitable Surprises* (2003) by Peter Schwartz focuses on the turbulent and volatile short- and medium-term over the next twenty-five years, although in the longer-term evolutionary adaption puts these surprises into perspective. Another approach for reducing avoidable surprises is set out in *Assumption Based Planning* (2002) by James Dewar and published by RAND.

The publication *Scenarios: The Art of Strategic Conversation* (1996), by Kees van der Heijden, provides an exposition of the contribution that scenario planning can make within a unified theory of how real-world organisations practice strategic management. In essence, the organisation is considered to be a learning organism with its own internal model of the world, which is concerned with both survival in a hostile world and self-development in a benevolent environment. The term 'scenario' is used in the external sense to describe the contextual environment over which the organisation has limited influence, and this is distinguished from the transactional environment in which the organisation is a significant player. These definitions are used in *Long-Range Futures Research* to distinguish the transactional microstructure from the contextual macrostructure.

1.5 Evolutionary Futures and Landscapes

Evolutionary futures research develops *landscapes* with spatial structure for a long-range horizon of up to 125 years that involves the consideration of future generations. It has its roots in the study of complex adaptive systems, in which the fast dynamics of the flows at the transaction and interaction level are linked to the slower dynamics of long-term adaption and evolution in the macrostructure. Evolutionary futures research involves the selection of a description of reality and spatial structure, and the application of nonlinear models that contain endogenous variables, together with the variable relationships of the system's macrostructure. Complex adaptive systems show perpetual novelty, although there are persistent features in far-from-equilibrium systems that indicate the form of their evolutionary trajectory. The patterns that form are diffusive civil ecostructures, in which civil engineering infrastructures shape the urban microstructure and give a degree of permanence to the transformations in the macrostructure. A coherent pattern emerges at the scale of the world urban system within a long-range or macro-timescale, compared to the meta-timescales of geological or biological evolution.

Evolutionary landscapes are developed to show the unfolding trajectory of the system, which are prescribed by the macrolaws of Ecodynamics (see Chapter 7) that govern the behaviour of the system. The system is irreversible, and there are a variety of possibilities for different landscapes in the future, which can be influenced by human vision. Investment and technological progress lead to civil phase transitions that create structural change and give rise to stages of development in the macrostructure or epochs in the metastructure. The macro-stages of development towards a planetary civilisation are explained in Chapter 7, Section 7.1. These can be anticipated from the ecodynamic succession of establishments within ecostructures, so that there is some degree of predictability during the stable periods. But there are discontinuities as the system shifts between stable states at macrosystem transitions, with relatively short-term chaotic dynamics, and so the system is indeterminate in detail.

The evolutionary approach focuses on the slower dynamics of long-term adaption and evolution in the macrostructure. For example, World War Two and the aftermath created a decade of chaotic conditions on a continental scale, so that the chaotic dynamics lasted for approximately one percent of an evolutionary timescale of a millennium. All the bombed cities have now been rebuilt, and the evolutionary path is continuing. This is an important point in dealing with chaos theorists who claim that nothing is predictable, which is only true in the turbulent short term. Local examples of chaotic conditions arise in a mining community with a mine closure, or when a new superstore opens and creates short-term chaotic conditions for local traders and customers, but this settles down and life continues. Similarly, civil and natural disasters create chaotic dynamics at the local and regional levels in the short-term. Uncertainty is dealt with by using a backcasting approach to a morphological analysis of the evolution of the major systems that will impact upon scenarios for planetary civilisation. Evolutionary approaches have been adopted in a variety of disciplines and these are outlined below.

Models of societal evolution are inspired by ecological analogies to explain the evolutionary dynamics of complex interactions between individuals from which the social system emerges. Societal transitions have been studied by Ervin Laszlo of the General Evolution Research Group (USA), who draws on ideas from non-equilibrium thermodynamics, evolutionary biology, cybernetics, and systems science theory. He identifies crucial bifurcation epochs in human history, and he points out that, whereas scientists refer to bifurcations as essentially random, civilisations can exercise foresight and take purposeful action. Ervin Laszlo has been the author of several of the books on the evolutionary paradigm, such as *Evolution: The Grand Synthesis* (1987), *The New Evolutionary Paradigm* (1991), and *The Choice: Evolution or Extinction?* (1994). Richard Coren is the author of *The Evolutionary Trajectory - The Growth of Information in the History and Future of Earth* (1995).

Cities provide an historical and future continuity in the evolution of civilisations, and from *4000 years of Urban Growth* (1987), by Tertius Chandler, it is possible to trace from 2250 BC the successive stages of growth for early cities of over 20,000 population. The city population estimates are derived by taking families, households, and homes as the institutional units of urban settlements. The general structure of settlements tends to remain stable for long periods of time, and they evolve in S-shaped cycles of growth. In *Ecumenopolis: The Inevitable City of the Future* (1974), by C.A. Doxiadis and J.G. Papaioannou, the future evolution of the urban system is explored during the next two centuries, in which a period of accelerating growth is followed by a deceleration phase that is still far from stable, before reaching a dynamic balance in a sustainable form for 20-50 billion population. It is important to take a time horizon that extends beyond the ascending phase of the population curve, since the use of resources is linked to the ultimate size of the urban system. *Ecumenopolis* referred to sustainable densities for human settlements, and the authors referred to Ecumenopolis as a final and sustainable form, which was a forerunner of the concept of a sustainable society. It was envisaged that there would be a massive increase in functional linkages between separate urbanised regions, together with a commensurate increase in the physical continuity of urban settlement.

Territorial colonisation is a dynamic process involving the interaction of populations and invasions that is not evoked by the static perception of spatial structure. Ecologists think in terms of ecosystems and bioregions, whereas planners and economic geographers relate to urbanisation and economic regions. The spatial mosaic of the future will increasingly encompass areal descriptions, geographic regions, and numbers of states that will far exceed the number of nations. Remote sensing techniques will feed geographic information systems with overlapping data sets covering demography, land-use, economic indicators, natural resources, energy, and materials. This emerging texture cannot be captured in top-down global or regional models; complex adaptive systems have to be modelled from the bottom up. The irony is that, traditionally, economists have neglected the spatial dimension and, conversely, economics has tended to be neglected in the micro-simulation of city structures and the bottom up explanations for the self-organisation of human societies. Physicists such as Peter Allen have observed that, in living systems, social and biological evolution is associated with flows of matter and energy, which maintains them out of thermodynamic equilibrium as dissipative structures. Allen's paper, *Why the Future is Not What it Was - New models of evolution*, was published in *Futures,* Volume 22, Number 6, July/August 1990.

Historical path dependence is well understood by development economists and economic historians who undertake empirical studies on national economic development, the origin of industries, and the history of technologies. Their works include *The Conditions of Economic Progress* (1940), by Colin Clark, *The Stages of Economic Growth* (1960) and *The World Economy: History and Prospect* (1978), by Walt Rostow, and *World Economic Primacy 1500-1990* (1996), by Charles Kindleberger. Economic development takes place through investment waves of successive layers of economic infrastructure rather like geological strata in the natural world, and this generates structural changes in the economy with increasing per capita assets and transitions in the composition of the workforce. Historical events are clearly not irrelevant, and the path-dependent advantages of the MDCs (more developed countries) over the LDCs (less developed countries) arise from their high quality infrastructure. Contemporary thinking by geographers and urban and regional planners concerning the spatial impact of new technologies on urban form are contained in a series of publications sponsored by the International Council for Building Research (Australia), including *The Future of Urban Form* (1985), *The Spatial Impact of Technological Change* (1987), *Cities of the 21st Century* (1991), and *Cities in Competition* (1995), edited by J. Brotchie, M. Batty, E. Blakely, P. Hall and P. Newton.

Technological evolution is a process of creative destruction according to Joseph Schumpeter (1942), and 'the essential point to grasp is that in dealing with capitalism we are dealing with an evolutionary process'. Evolutionary economists such as Richard Nelson and Sidney Winter, the authors of *An Evolutionary Theory of Economic Change* (1982), take Schumpeter's insight as a useful analytic starting point. If through historical or path-dependent circumstances, a particular technology gains advantage over its competitors, there is additional investment available to make further new technological advances, which deprives the rivals of the necessary resources to catch up and seriously compete.

The coevolution of technologies follows the Ecodynamics of interacting species in which the entry of new species changes the dynamics, and old systems go into decline until they become extinct. The Santa Fe Institute publication *The Economy as an Evolving Complex System* (1997) explains the perspective of the economy operating far from any optimum or global equilibrium.

Artificial vivisystems are computer animations of flocks of birds, shoals of fish, and plants that grow and develop as 'animates' on the screen, together with other simulations of synthetic biology. This is very much a field of study at the Santa Fe Institute that was set up to study the sciences of complexity. In particular, the aim was to establish a common analytic framework for 'adaptive computation' and to develop a set of mathematical and computational tools that could be applied to the sciences of complexity. John Holland of the University of Michigan has invented a mathematical method known as genetic algorithms for describing evolutionary adaption. He developed a model called ECHO (ECHO is short for ecosystem) of a highly simplified ecological community in which digital organisms roam the digital environment to capture a meal in order to survive and reproduce. John Holland is the author of two important books, entitled *Hidden Order: How Adaption Builds Complexity* (1995) and *Emergence: From Chaos to Order* (1998). Lifelike behaviour is the result of simple rules unfolding from the bottom up. Tom Ray, an ecologist, developed an artificial world named Tierra after studying self-replicating computer viruses to start a new science of experimental evolution and ecology. Artificial life was developed by Christopher Langton as an outgrowth of artificial intelligence to study a broad range of biological phenomena. Stuart Kauffman has been involved in developing genetic networks to understand the origin of life.

The approach to evolutionary futures research in *Long-Range Futures Research: An Application of Complexity Science* starts with the perception that the economy is a complex adaptive system of transacting households and establishments. Communities of establishments are villages, towns, and cities that may be bounded at a subnational level by a territorial construct to be termed an ecopolitan state. The urban vivisystems can be designed synthetically from a knowledge of the planning factors for the number and size range of establishments, depending upon the level of economic development (defined in terms of assets per capita) of the ecopolitan state and also the urban category and size class. The transactional microstructure gives rise to a contextual macrostructure, and macrolaws provide a macroscopic description for the evolution of the world system of cities. The organic economy is maintained away from equilibrium by dynamic investment flows, which involve the transformation of resources in a complex trophic web. Complexity is defined by the increasing informational content within the urban environment. The major transformations or bifurcations that arise within a complex adaptive system are described by stages of development that arise from technological evolution and sectoral waves of investment in converging energy, transport, and communication infrastructures.

In the absence of a model, a normative scenario can be produced using the backcasting approach. It is a technique for policy-orientated analysis in which future goals and objectives are defined to generate desirable scenarios, which can then be evaluated in terms of the political, economic, social, technological, and ecological implications. Backcasting is exploratory and systems design orientated, rather like La Prospective, where the aim is to identify critical intervention points that will improve the future state of affairs in a cost-effective way. Both methods are explicitly normative and may be applied to complex societal problems, but the essential difference is that backcasting is non-predictive. An important philosophical point about evolutionary futures research is that the most appropriate system technology can be evaluated and selected in advance, without the trial and error process of Darwinian natural selection. However, competition between alternative technologies will lead to mutations and innovations that will produce the most fitting designs in the long term.

1.6 United Nations reference scenarios

Scenarios are alternative images of the future for policy evaluation, and they are neither predictions nor forecasts but a means of designing more robust strategies. Scenarios generally link qualitative narratives or storylines about the future to quantitative data in the form of tables and figures, which are often generated by computer models. The majority of model-based global studies start with models, and then scenarios are derived on the basis of the models' output. When models are used in this way, the parameters that drive the model are defined, and assumptions are made about the values of the parametric variables that are used by the model but not computed by it, such as population growth, economic growth, or resource use. Choosing default values for these parametric variables involves judgement on the part of the modeller, and default values should be based upon accepted contemporary sources of information, such as the United Nations, the World Bank, or the OECD. Although scenarios are rarely value free, it is useful to distinguish between exploratory and normative scenarios. Exploratory scenarios are open-ended paths into the future that could turn out to be utopias or dystopias, whereas normative scenarios are explicitly value based and teleological routes to preferable end states.

It is generally recognised with global scenarios that, although future events are inherently unpredictable, it is possible through futures research to generate long-range default values for future conditions when better information is unavailable. Scenarios should be updated and policies re-evaluated every ten years to provide a rolling, long-term perspective. It is helpful to select a base case scenario but not necessarily a business-as-usual (BAU) scenario, because the drivers of change, such as economic development and population growth, follow a logistic S-curve with an inflexion point from increasing to decreasing slope, which changes the trajectory from the BAU path. Futures researchers often select four scenarios so as not to complicate the evaluation with too many alternatives, which may include a utopian or a dystopian scenario, to set the boundaries for the study. The selection of an even number of scenarios avoids the impression that there is a central or most probable case, and disaster scenarios are saved for testing the robustness of the selected sets. The driving parameters are grouped into domains, and parametric variables may be selected on the basis of performance indicators for a number of the following parameters:

- Governance - Freedom, Democracy, International Equity
- Economy - GWP and GRP, Income per capita, Investment
- Population - Life expectancy, Education, Health

- Society - Family Life, Culture, Employment, Welfare
- Resources - Land-use, minerals, fossil fuels
- Environment - Climate, Ecosystems, Water Stress, Pollution
- Agriculture - Food demand, Diet, Crops, Livestock
- Technology - Settlements, ICT, Transport, Renewable Energy
- Security - Crime, terrorism, war
- Hazards - Natural, Astro, Geo, Bio, Techno, Socio.

Scenarios for the UN Intergovernmental Panel on Climate Change (IPCC) require a 100-year horizon. These are described in the IPCC *Special Report on Emissions Scenarios* (SRES), 2000. The four families or sets of SRES scenarios are descriptions of possible, rather than preferred, developments, and they represent pertinent and plausible alternative futures. The storylines represent the playing out of certain social, economic, technological, and environmental paradigms, which will be viewed positively by some people and negatively by others. Six models were used to quantify forty SRES emissions scenarios represented by six illustrative or marker scenarios, since the economic/global scenario (A1) is further divided into three scenarios. One is fossil-fuel intensive (A1F1), one is balanced between fossil and non-fossil fuels (A1B), and one will eventually make a transition to non-fossil fuels (A1T). A large uncertainty surrounds future emissions and the possible evolution of the underlying drivers, such as population growth, civil development, and technological progress, together with their impacts upon energy and land-use. The uncertainty is further compounded in going from emissions paths to climate change, from climate change to possible impacts and, finally, to adaption policy analysis, which tends to be focused on a shorter time horizon of say thirty years. Since most of the people who will have children during that period have already been born, realistic population projections can be made within that timescale. The scenario sets selected are as follows, where A1 and B1 involve global integration, and both A2 and B2 have a regional orientation:

A: Economic rather than environmental
  A1: Rapid Economic Growth with Convergence
  A2: Slower Growth in a Heterogeneous World

B: Environmental rather than economic
> B1: Convergence with Global Environmental Emphasis
> B2: Intermediate Development with Local Sustainability

The A1 and B1 scenario sets are based on a low, world-population projection, which combines low fertility with low mortality, and after peaking at 8.7 bn in 2050 it declines to 7.1 bn in 2100. The B2 scenario family is based on the UN median population projection (1998), with a world population of 9.4 bn by 2050 and 10.4 bn by 2100. The A2 scenario set is based on the UN high-population projection (1998) of 15 bn by 2100, which assumes a significant decline in fertility for most regions and stabilisation at above replacement levels. The range of Gross World Product (GWP) for 2100 across some 150 scenarios reviewed by SRES was concentrated between an equivalent 2000 US$180-360 trillion, with a peak in values of around $280 trillion. This represented a consensus amongst modellers of an economic growth rate from 1990 in the range 1.1%-3.2%, with a median value of 2.3% per year. By comparison, the Review by Sir Nicholas Stern, Head of the British Government's Economic Service (2006) takes a 2.4% annual growth in GWP to give an increase of 9.5 times by 2100. The Stern Report also uses a growth rate in global consumption per capita of 1.3% per year from 2001 to 2200 to give an increase in per capita GWP of around 13 times, although this is unlikely to be sustainable.

For 2050, the B2 scenario gives a per capita GDP equivalent to 2000 US$36,000 for the MDCs and US$9,000 for the LDCs. The B2 scenario projection for 2100 gives a GWP of US$265 trillion, with an average annual income equivalent to 2000 US$60,000 per capita for the MDCs and US$20,000 per capita for the LDCs. The B2 scenario assumes a trebling in primary energy consumption with some 50% from renewable sources by 2100. B2 is predicted to give a 2.5°C temperature rise by the end of the century. The GDP per capita values for A2 are a little lower than those for B2 in view of the higher world population, with a predicted temperature rise of 3.5°C.

The American Council for the United Nations University Millennium Project takes a twenty-five-year exploratory horizon and fifty-year normative horizon. Exploratory scenarios depict self-consistent future worlds that could emerge from the present through credible cause, effect, and feedback developments and reach an end state that seems plausible. They are generally developed in sets with an orientation that contains a likely future, but each scenario may in fact be unlikely. Normative scenarios represent desirable future worlds. They are also self-consistent and employ feedback relationships to get from the present to the future state, which represents a goal rather than expected conditions. The Millennium Project uses the scenario to drive the model rather than the other way round. The drivers of change include globalisation, government involvement, security, and technology. The selected scenario sets are as follows:

Normative scenarios to 2050:
      Win-Win: Revamping the world financial system
      Seamless Nations: Global social contract

Exploratory scenario to 2025:
      Cybertopia: The age of communications
      Rich get richer: Competition gets out of hand
      Trading Places: Developing countries flourish
      Passive Mean World: Growth slows and sustainability suffers

The *International Futures (IF) Global Simulation Model* was selected to add quantification to the Millennium Project scenarios. The IF model was developed at the University of Denver by Barry Hughes as a predictive model with a 100-year horizon rather than as a scenario generator, in response to the shortcomings of the World3 model used in the *Limits to Growth* study. The parameters driving each scenario were identified, and scenario-dependent values were assigned to the parametric variables. Judgments were made that reflected the intended spirit of a scenario, where the variables were not the direct drivers of the scenarios. Inevitably, there were a number of possible future situations that the IF model could not handle because of the model structure, or a lack of parameters, and an inability to predict the rates-of-change multipliers in the longer-term.

The Trading Places scenario assumes a UN medium population of 9.5 bn in 2050, with a slightly higher population in a Passive Mean World, and slightly less in Cybertopia. Across the scenarios, the life expectancy at

2050 is taken as 80 for the USA, with Asia 75-85 years, and Africa 60-65 years. In the Rich-get-richer scenario for 2050, the GDP per capita turns out to be USA - US$55,000, China - $22,000, and Sub-Sahara Africa - $1,000. In the Cybertopia scenario, the per capita GDP is USA - $80,000 (4 times South America), China - $23,000, and Sub-Sahara Africa - $2,000. In the Trading-Places scenario, the GDP per capita works out as USA - $34,000, China - $37,000, and Sub-Sahara Africa - $2,500. Finally, in the Trading-Places scenario, world energy demand is expected to rise in the period 2000-2050 by a factor of 1.85, whereas it is expected to decline in the USA by a factor of 0.5 and increase in China by a factor of 6 during the same fifty-year period.

The thirty-year *Global Environment Outlook 3* (2002-32) of the United Nations Environment Programme (UNEP GEO-3) provided four scenario sets, namely *Markets First*, *Policy First*, *Security First*, and *Sustainability First*, with both regional and global perspectives. However, UNEP GEO-3 drew on other scenario initiatives, including those of the Global Scenario Group (GSG) and the PoleStar modelling work of the Stockholm Environment Institute. This group had produced a series of reports with a fifty-year perspective that have received considerable attention, including *Bending the Curve: Toward Global Sustainability* and *Great Transition* (2002). The GSG scenarios addressed critical trends, such as environmental degradation, resource depletion, increasing income disparity, poverty, and marginalization with three scenario sets, namely:

Conventional Worlds: Incremental change without discontinuities
  Market Forces: Global markets drive world development
  Policy Reform: Governance for social equity and sustainability

Barbarization: Civilisation collapses into anarchy or tyranny
  Breakdown: Institutions collapse with conflict and crises
  Fortress World: Protected enclaves with impoverished outside

Great transition: Profound shift in values to equity and sustainability
  Eco - Communalism: Bio-regionalism and local self-sufficiency
  New Sustainability Paradigm: Global ecological transition

For the Conventional Worlds scenario, GSG define a reference scenario based upon business-as-usual assumptions for a market-driven world in the 21st Century, in which current demographic, economic, technological, and environmental trends unfold without major surprise. These assumptions are drawn from international assessments wherever possible, including United Nations projections and the IPCC scenarios. In general, the reference scenario is one of wide prosperity by mid-century in which developing regions approach standards of living enjoyed in Western Europe in 1980 while OECD incomes soar, reaching an equivalent of 2000 US$80,000 per capita in North America. GWP increases from PPP US$33.4 trillion in 1995, at an average growth rate of 2.7% per year to the equivalent of 2000 PPP US$160 trillion in 2050. This wealth would be unevenly distributed with a rich (OECD) to poor (non-OECD) income ratio of 5 and extensive damage to the environment with carbon emissions of 500 ppmv.

In contrast to the reference scenario, the Policy Reform variant of the Conventional Worlds set assumes that coordinated government action is initiated for poverty reduction and environmental sustainability. The pace and scale of technological and social change required for a normative policy reform scenario would be daunting, and there is little political will to follow this route. The selected indicators and targets for policy reform, within the global population and economic totals of the 2050 Conventional Worlds reference scenario, involve a redistribution of income with a reduced expectation of per capita income in North America from 2000 US$80,000 to US$48,000, and an increase in the developing countries from 2000 PPP US$11,500 to $15,000 per capita, to give a rich-to-poor income ratio of 3.2. There would be targets for life expectancy of at least 70 years in all countries, with illiteracy and unsafe water limited to no more than four percent of the population (360 million people), and hunger restricted to no more than two percent of the population (180 million). The final fuel demand would be reduced from 650 EJ to 475 EJ in 2050, with renewable energy resources accounting for twenty-five percent of the total. The objective for greenhouse gases would be to limit the concentrations of carbon dioxide in 2100 to 450 parts per million by volume (ppmv) in relation to 370 ppmv in 2000, so that global temperature change would be kept below 2°C, which is gradual enough for most species and ecosystems to adapt.

The Great Transition scenario expresses the view that increasing consumption beyond a certain point fails to produce an increase in the quality of life. Past a certain point of sufficient comfort ("enough"), increased consumption moves beyond comfort to luxury and then to extravagance and fails to increase fulfilment. Acquisition as an end in itself turns out to be a substitute for contentment. The additional pressure of work to pay for additional material possessions, their maintenance and repair, and their protection and disposal create additional costs that exceeds any incremental satisfaction. In addition to the dematerialisation of the Policy Reform scenarios with clean, efficient, and renewable technologies, Great Transitions couple this with changes in lifestyle that reduce consumption in affluent areas with a change in values to a more just and equitable civil society. The GWP of 2000 PPP US$160 trillion in 2050 would be shared to give a rich-to-poor income ratio of 1.65, compared with ten for the Barbarization scenario. This explores the possibility that the reaction to market forces with business-as-usual will result in a complete collapse to a Breakdown scenario. This may lead to the Fortress World scenario in which authoritarian rulers protect small enclaves of wealth amidst an impoverished majority, with a world population increase by 2050 to 11 bn of which 2 bn are hungry. The environmental consequences result in carbon dioxide levels of 525 ppmv.

The *International Futures (IF) Global Simulation Model* has evolved for more than twenty-five years in support of investigation into the emerging global transition to sustainability. The IFs model uses a dynamic, equilibrium-seeking structure, which allows exploration of near-term futures and also facilitates thinking about the remainder of the 21st Century. During 1998-2002, the European Commission supported IFs for the TERRA 2000 project under the administrative leadership of RAND Europe, and the base case simulation indicated that GWP per capita will increase eleven-fold at market prices and six-fold at PPP by 2100. Subsequently, the US National Intelligence Council (NIC) supported the IFs model for *Mapping the Global Future: The Report of the NIC 2020 Project*. The fifth generation of the IFs model has been supported by Frederick S. Pardee for application to the United Nations Environment Programme's *Global Environment Outlook 4* (UNEP GEO-4), 2007.

UNEP's GEO-4 process builds on the same set of scenarios as GEO-3, but they have been revised, updated, and extended for the fifty-year period 2000-2050, for the following:

Markets first: Market driven development with globalisation
Policy first: Development with social and environmental solutions
Security first: Focus on self-protection and exclusion of have-nots
Sustainability first: Radical shift towards sustainability

Since no overriding "super model" was available for computing future environmental change and the impacts on human well-being, a suite of models were soft-linked with output files from one model being used as inputs to other models. For GEO-4, the IFs model provided population trends and the development in GDP and GDP per capita as well as additional information on value added, household consumption, health, and education. The other models were IMAGE, IMPACT, WaterGAP, EwE, GLOBIO, LandSHIFT, CLUE-S, and AIM as described in the Technical Annex to Section E of GEO-4 (2007), "The Outlook - Towards 2015 and Beyond".

Global population growth reaches 9.7 bn by 2050 for *Security First*, 9.2 bn for *Markets First*, 8.6 bn for *Policy First*, and just under 8 bn for *Sustainability First*. Over the same period, GWP increases five times for *Markets First* and *Policy First*, and in *Security First* and *Sustainability First,* GWP increases by nearly three times. Primary energy use increases from about 400 EJ in 2000 to around 800-900 EJ in 2050 for the *Policy First* and *Security First* scenarios; whereas, under *Sustainability First,* energy use in 2050 reaches 550 EJ. The computed sea level rise for all of the scenarios is about 30 cm in 2050.

Societal evolution and futures research: Chronological references

1. *The Year 2000: A Framework for Speculation* - Herman Kahn & Anthony J Wiener, 1967.
2. *Technological Forecasting for Industry and Government* - James R Bright, 1968.
3. *General System Theory* – Ludwig von Bertalanffy, 1969.
4. *Urban Dynamics* - Jay W Forrester, 1969.
5. *World Dynamics* - Jay W Forrester, 1971.
6. *Think Tanks and Problem Solving* - Paul Slee Smith, 1971.
7. *The Limits to Growth* - Donella & Dennis Meadows, Jorgen Randers, William Behrens, 1972.
8. *Future Shock* - Alvin Toffler, 1972.
9. *The Ascent of Man* - Jacob Bronowski, 1973.
10. *The Coming of Postindustrial Society* - Daniel Bell, 1973.
11. *Mankind at the Turning Point* - Mihajlo Mesarovic & Eduard Pestel, 1974.
12. *Mankind and Mother Earth* - Arnold J Toynbee, 1976.
13. *The Next 200 Years* - Herman Kahn, William Brown, & Leon Martel, 1976.
14. *The World Economy: History and Prospect* - Walt W Rostow, 1978.
15. *Ecodynamics: A New Theory of Societal Evolution* - Kenneth E Boulding, 1978.
16. *The Crisis in Forecasting and the Emergence of the 'Prospective' Approach* - Michel Godet, 1979.
17. *World Economic Development* - Herman Kahn, 1979.
18. *The Third Wave* - Alvin Toffler, 1980.
19. *Megatrends* - John Naisbitt, 1982.
20. *Looking Forward: A Guide to Futures Research* - Olaf Helmer, 1984.
21. *Evolution: The Grand Synthesis* - Ervin Laszlo, 1987.
22. *Megatrends 2000* - John Naisbitt & Patricia Aburdene, 1990.
23. *The Art of the Long View* - Peter Schwartz, 1991.
24. *The New Evolutionary Paradigm* - Ervin Laszlo, 1991.
25. *Beyond the Limits* - Donella and Dennis Meadows, & Jorgen Randers, 1992.
26. *The Choice: Evolution or Extinction?* - Ervin Laszlo, 1994.
27. *The Coming Plague* - Laurie Garrett, 1995.

28. *The Sciences of the Artificial* (3rd edition) - Herbert A Simon 1996.

29. *World Economic Primacy 1500-1990* - Charles P Kindleberger, 1996.

30. *Scenarios* - Kees Van Der Heijden, 1996.

31. *The Knowledge Base of Futures Studies* (3 Vols) - Edited by Richard A Slaughter, 1996.

32. *Social Futures, Global Visions* – Edited by Cynthia Hewitt de Alcantara, UNRISD, 1996.

33. *New Thinking for a New Millennium* - Edited by Richard A Slaughter, 1996.

34. *Foundations of Futures Studies* (2 Volumes) - Wendell Bell, 1997.

35. *The Eleventh Plague* - Leonard Cole, 1997.

36. *The Information Age* (3 Volumes) - Manuel Castells, 1996-1998.

37. *The Evolutionary Trajectory* - Richard L Coren, 1998.

38. *Culture: Beacon of the Future* - Paul D Schafer, 1998.

39. *The Fortune Sellers* - William A Sherdon, 1998.

40. *Bending the Curve* - P Raskin, P Gallopin, P Gutman, A Hammond & R Swart, 1998.

41. *Next: Trends for the Near Future* - Marian Saltzman & Ira Mathatia, 1999.

42. *Future Makers, Future Takers* - Doug Cocks, 1999.

43. *Special Report on Emissions Scenarios* - Intergovernmental Panel on Climate Change, 2000.

44. *The Long Boom* - Peter Schwartz, Peter Leydon & Joel Hyatt, 2000.

45. *Creating Futures* - Michel Godet, 2001.

46. *Great Transition* - P Raskin, P Gallopin, T Banuri, P Gutman, A Hammond, R Kates & R Swart, 2002.

47. *Global Environment Outlook 3* - United Nations Environment Programme (UNEP), 2002.

48. *Advancing Futures* - Edited by James Dator, 2002.

49. *Assumption-Based Planning* - James A Dewar, 2002.

50. *Millennium Project: 2003 State of the Future* - Jerome Glenn & Theodore Gordon, 2003.

51. *Inevitable Surprises* - Peter Schwartz, 2003.

52. *Tomorrow's People* – Susan Greenfield, 2003.

53. *Shaping the Next One Hundred Years* - Robert Lempert, Steven Popper & Steven Bankes, 2003.

54. *The 2030 Spike* – Colin Mason, 2003.

55. *Our Final Century* - Martin Rees, 2003.

56. *Mapping the Global Future 2020 Project* – National Intelligence Council, 2004.

57. *Futures Beyond Dystopia* – Richard A Slaughter, 2004.

58. *Futuring* - Edward Cornish, 2004.

59. *Limits to Growth: 30 Year Update* - Donella & Dennis Meadows, Jorgen Randers, 2005.

60. *Shell Global Scenarios to 2025* – Jeroen van der Veer, 2005.

61. *Megatrends 2010* – Patricia Aburdene, 2005.

62. *Future of Men* – Marian Salzman, Ira Matathia and Ann O'Reilly, 2005.

63. *Sustainable Futures* – Barry B Hughes & Peter D Johnston, 2005.

64. *Exploring and Shaping International Futures* – Barry Hughes & Evan Hillebrand, 2006.

65. *The Economics of Climate Change* – Nicholas Stern, 2007.

66. *Global Environment Outlook 4* - United Nations Environment Programme (UNEP), 2007.

2.
THE METASYSTEMS AND DEEP FUTURES

2.1 Coevolution of the planetary metasystems

The planetary metasystems include the physical, astrophysical, geophysical, biological, and civil systems. The planet's physical system is described in terms of physical energy and chemical dynamics with matter being comprised of atoms and molecules, and electrons circling around an atomic nucleus which, itself, is made up of protons and neutrons. The atom subdivides further into quarks, which has another substructure level of gluons, photons, and neutrinos (vibrating strings). A reversal of reductionist thinking leads to the evolutionary science of astronomy, which studies the solar system of planets, galaxies, and the universe. The planet's geophysical system involves the evolution of the earth from its molten core to the strata of the earth's crust and terrestrial stocks, to the outer reaches of the atmosphere. The geological evolution of tectonic plates, mountain building, earthquakes, volcanoes, and the geophysical phenomena such as glaciers, oceans, and climate intimately relate to each other. The lowest part of the Earth's atmosphere, the troposphere, extends upwards for approximately 10 km and, between the troposphere and the stratosphere, comes the ozone layer, which filters out the harmful short-wave radiation from space and makes possible advanced life forms on Earth. The biosphere is the region of the Earth's crust and atmosphere in which living matter is found. The biological metasystem involves living organisms, with humans and species of animal and plant populations and their ecosystems.

Long-Range Futures Research refers uniquely to the civil metasystem of our planet as the civil system, although there is the possibility that it may form part of an astro-civil metastructure (Section 2.6). The study of these metasystems can be described as the evolutionary sciences, such as astronomy, geology, ecology, anthropology, and archaeology. Each of these sciences has an historical evolutionary path, in which a series of events have been interpreted retrospectively into a pattern to provide an explanation of the phenomena. The same evolutionary processes will also create the future. For example, humans may evolve or genetically engineer themselves into some form of genus intelligens within 100,000 years. In view of the enormous timescale of anthropogenic evolution over many generations in relation to the shorter time span for societal evolution, it is useful for modelling to partially uncouple the biological metasystem containing individual humans from the civil metasystem, in which the household becomes the unit of society. By making the distinction between the ecosystem of which humans are a part and the economy, which should relate specifically to the households, establishments or organisations, artefacts, buildings, cities, and states created by humankind, *Long-Range Futures Research: An Application of Complexity Science* sets out to provide a conceptual framework that may be applied to evolutionary futures research. *Long-Range Futures Research* provides a methodology for understanding the underlying structures and to see the interrelationships and the patterns of change for research into likely future conditions, rather than for the prediction of specific events.

Evolutionary generating phenomena are constrained by the transactional dynamics that arise between three distinct tiers, namely the transactional microstructure of the urban system, the global contextual macrostructure, and the planetary metastructure level. The tiers have some degree of independent dynamic pattern, but all the parts interact, and the so-called environment is always a part of the universal system. It follows that the civil system is an open system, since it cannot be isolated from the interacting dynamics of the other systems. The fast civil dynamics or Ecodynamics of the transactional microstructure not only shapes the transformations in the contextual macrostructure, but the microstructure is itself bounded by the behaviour of the macrostructure as the total system evolves. At the same time, the global macrosystems impact upon the planetary metasystems in an organic or evolutionary timescale, and it is, in turn, bounded by the constraints of the metalaws that drive the metasystems in glacially slow geological time. The evolutionary generating processes are constrained at all levels by the laws of physics, such as the laws of thermodynamics, the law of gravity, the logistical law of atrophy, the instructions encoded in the genome, civil laws, and the laws of economics such as eventual diminishing returns.

The metacycles of astrophysical and geophysical time are measured in billions of years, biological evolutionary timescales are measured in millions of years, with humankind emerging five million years ago. Deep futures involve scientific speculation on the future of life on the planet Earth a million years or more into the future. The civil metacycle for the evolution of a planetary civilisation and the eventual attainment of a dynamic balance is measured in units of millennia, and a possible starting point for civilisation was fifty millennia ago, (one percent of the age of our species) with the emergence of modern man (*Homo Sapiens Sapiens*), when physiological evolution had created a capability for speech, language, and sophisticated communication. The megacycles for the evolution of cities are also measured in millennia, in relation to macrocycles of a century or less for the life of buildings, establishments, technological, and human life cycles. A decade is the unit of measure for the boom and bust of business and property microcyles, and this is generally the horizon for long-range economic forecasts. The macrocycles of 100 years appear to be more determinate than the fast dynamics of transactional microcycles, which are probably indeterminate.

The long-term goal for civilisations is to achieve a dynamic balance between a growing human population, the astrophysical systems of the Earth and the Sun, and the terrestrial stocks of the geophysical system. The terrestrial stocks are geological stocks, such as fossil fuels and mineral deposits to which there are limits and ultimately absolute shortages, and also the agrarian stocks of the ecosystem, which are limited by seasonal units of time and solar flow. A unifying evolutionary paradigm would be to live within the capacity of the planet, which would have the effect of maximising the cumulative number of lives to be lived over time at a reasonable per capita standard and prolonging the existence of civilisation. The consequence of too large a population alive simultaneously is the overshooting situation, which would reduce the carrying capacity of the planet, with fewer people in subsequent time periods and a lower cumulative total over time. The economic system and the ecosystem have to be combined to produce results that enable the needs of the population to be satisfied, taking into account the values and aspirations of society and the possibilities opened up by new knowledge and technological progress.

2.2 Astrophysical metasystem

The Universe was created by a cosmic explosion referred to as Big Bang some thirteen billion years ago, and for convenience, many cosmologists take the age of the Universe to be fifteen billion years. The thermonuclear reaction set off the cosmic evolution of a universe of galaxies in which chaotic hot gases expanded and, as they cooled, the system went through a phase transition in which atoms separated from the cosmic radiation and formed clumps of matter or stars that emerged under gravitational forces in far-from-equilibrium stability to form hierarchies of galaxies, stars, and planets. In the observable universe, there is a cosmological horizon of fifteen billion light years, which contains 100 billion galaxies, each with 100 billion stars. Some galaxies are ten billion years old, which is five billion years older than the Earth, and there are galaxies in which ten percent of the stars could have terrestrial type planets. However, the extinction of extraterrestrial life on numerous early planets will have occurred when the energy producing stars of their planets burned out. But it is quite possible that there are civilisations in the older regions of galaxies, which are more advanced than our own. The local group of twelve galaxies, which includes our Milky Way and Andromeda, lies within three million light years. Gravity is pulling Andromeda towards our galaxy at 100 km per second, and the two disc galaxies will collide in some five billion years. There are fifteen similar groups within thirty million light years, and the Virgo cluster of 2,300 galaxies is fifty million light years away. The Coma cluster with 1,000 galaxies, and the Hercules Supercluster with 10,000 galaxies are both 300 million light years away. It follows that there are 13,500 galaxies within 300 million light years, which is 3,000 times the diameter of our own galaxy.

There are two principal models for galaxies of which sixty percent are ellipticals (three-dimensional ellipses) and thirty percent became flattened discs from the angular momentum in stronger gravitational fields. Our own galaxy, the Milky Way, is a flattened, disc-shaped system that measures 100,000 light years across and 2,000 light years thick (a light year is the distance that light can travel in one year). There are galactic collisions as they pass through one another, from which their configuration becomes distorted, and there can also be galactic explosions that emit clouds of hydrogen gas sufficient to form millions of stars. There are of the order of 10^{11} galaxies, each with 10^{11} stars that are similar to the Sun, and they shine brightly from the heat generated in their interiors by nuclear fusion, in which the Sun's fuel, hydrogen, is changed into helium. There are so many stars of different masses that can be observed at all stages of their life cycles, and their composition can be studied using spectroscopy. Discs of cosmic material spin around young stars, and planets are formed by gravitational accretion of this material. There are perhaps as many planets as stars in all the galaxies, to give the total number of planets as 10^{22}.

The Milky Way is a spiral galaxy of 400 billion stars (suns) in which four spiral arms rotate within a disc of a radius of 50,000 light years. Within a radius of 1,000 light years of the centre, the stellar density is estimated to be ten stars per cubic light year. At the centre of the Milky Way is an elliptical nucleus of a radius of 10,000 light years and a depth of 13,000 light years, in which some 350 billion low mass stars (eighty-five percent of the stars in our galaxy) from the earliest epoch 13 BYA are located at a high density of one star per 10 cubic light years. There is possibly a giant black hole within the central nucleus, which swallows ancient stars and discharges radio, infrared, and high-energy radiation. The nucleus is surrounded by a disc, in which the thickness tapers from say 3,000 light years adjacent to the central nucleus to 1,000 light years at the perimeter, which is 50,000 light years from the centre. The later stars in the surrounding galactic disc contain heavy elements like our sun, so that they could support terrestrial type planets. The spatial density of the stars in the disc are significantly lower than in the nucleus, and there is a decline in spatial density from the centre to the perimeter.

The sun of our solar system circuits around the galactic centre at a radius of 30,000 light years and between the orbital path of the sun of our solar system and the nucleus is an inner blue ring 20,000 light years wide that contains stars older than our own planet. Outside the sun's orbital path lies the outer blue ring that is also 20,000 light years wide to the edge of the disc, with newer stars and the newest stars that are forming today. In our region of the galactic disc, the depth of the disc is 2,000 light years, and there is a density of one star per 320 cubic light years. A spherical halo of a radius of 130,000 light years surrounds the disc, although the density of matter in the halo is much less than that in the disc by a factor of 10,000. The stars in the halo are from the earliest epoch 13 BYA, and as they do not contain elements heavier than helium they could not have had terrestrial planets. The stars in the inner blue ring are younger than those in the central nucleus, and in the outer blue ring are the newer and more massive stars, which are very young compared to the 4.6 billion years of the solar system.

The Solar system orbits around the centre of the disc in 225 million years (an interval known as a cosmic year). The Sun has completed twenty orbits since it was formed five billion years ago and it contains 99.8% of the mass of the Solar system, so that the planets, asteroids, meteorites, comets, gas clouds, and other cosmic debris are held by the Sun's gravitational pull. However, after a further twenty orbits, in five billion years' time, the core of the Sun will shrink as the hydrogen is used up, and it will become hotter and expand to become a star known as a red giant. After a further billion years, the Sun will cease to generate energy and it will cool to form a white dwarf no larger than the planet Earth. As the mass of the Sun diminishes, the orbits of the planets will become extended as the gravitational pull weakens. The moon, which is of similar age to the Earth, revolves around the Earth as the Earth revolves around the Sun. The Earth is the only planet in the solar system that is suited to life of our kind, since it had to be the right distance from the Sun to create the range of life-sustaining temperatures, and it had to be the right size to retain an atmosphere.

Throughout the Earth's history, there have been periodical ice ages, during which climate change follows a cyclical pattern (Milankovitch cycles) as the Earth's orbit around the sun varies from a nearly circular path to one that is more elliptical in a cycle of approximately 100,000 years. A second variable is that the tilt of the Earth's axis varies in its angle of tilt to the sun over a period of 42,000 years. Also, there is a third variable in so far as the Earth gyrates and wobbles with the pole describing a circle so that there is a precession of the equinoxes that causes a shift in the date when the Earth is closest to the sun, in a cycle of 21,000 years. As the Earth's circular orbit degenerates towards an ellipse during glacial periods, the sea levels decline by 1 metre per 1,000 years for 70,000 years, with fluctuations of up to 15-25 metres at variable intervals of around 20,000 years that are caused by slight changes in the inclination of the

Earth's axis. Sea levels then drop sharply at the rate of 5 metres per 1,000 years for 10,000 years. At the commencement of the Earth's circular orbit, sea levels rise rapidly at the rate of 10 metres per 1,000 years for 10,000 years and then a rise of 2 metres per 1,000 years for a further 10,000 years. At present, the Earth is in an interglacial or warmer period between two colder spells, but astronomical calculations show that the Earth's orbit is nearly circular and becoming more so, and that we are close to the end of the Milankovitch cycle. The earth will probably start the descent into another glacial period during the Fourth Millennium, by which time sea levels may rise to perhaps 6 metres higher than current levels.

Astrophysical catastrophes resulting from a collision by an asteroid or a meteorite with the Earth create geophysical changes, and the rule of thumb for estimating the size of an asteroid from the impact crater is to divide the diameter of the crater by ten. The largest crater found so far is 200 km wide, which was caused by an asteroid some 20 km in diameter that collided with the Earth in Ontario, Canada 1.6 billion years ago. Between 300-280 MYA there were eight meteor impacts in North America, including two in Quebec, which formed craters ranging in size from 5 km to 30 km from meteors 0.5 km to 4 km. The Manicouagan crater 70 km wide, also in Quebec and north of the Bay of Fundy, was formed around 210 MYA by an asteroid 7 km wide. There is, however, no evidence of an irradium anomaly in the rocks that would have provided proof that this event was connected to a Triassic extinction of mammal-like reptiles and amphibians. An asteroid 10 km wide is generally considered to be the cause of the Cretaceous mass extinction 65 million years ago, in which the dinosaurs became extinct. The probable site of the impact is under the sea in the Gulf of Mexico at Chicxulub on the Yucatan peninsula, where there is a large meteor crater of 120 km diameter in the subsurface and it is the right age at the Cretaceous-Tertiary boundary. The presence of shocked quartz at the site, of the type produced by high-energy explosions, indicates without doubt that an asteroid was responsible for the event.

An asteroid of 10 km in diameter would have the energy of at least 200 million megatons, compared with a blast of 200,000 megatons from a 1 km metre wide asteroid or 200 megatons from a 100 metre meteorite. Currently, around 200 Earth-orbit-crossing asteroids larger than 1 km in diameter have been identified in the asteroid belt, together with a number of smaller meteorites. It has been estimated that every 100 million years, an extraterrestrial impact by an asteroid over 10 km wide may arise that could cause a global catastrophe, and that every 250,000 years one over 1 km wide will hit Earth. Every 10,000 years there may be an impact with an object 200 metres in diameter that would have global climatic effects. A meteor crater 1.2 km wide and 180 metres deep was created 30,000 years ago in the Arizona desert, by an iron meteorite 100 metres wide that would have caused a blast of 200 megatons. However, several times a century, meteorites of less than 10 metres wide impact with the planet to form smaller meteor craters.

2.3   Geophysical metasystem

The structure of the Earth is deduced by seismology from the time taken by earthquakes to travel through the interior of the earth. The radius of the Earth is 6,400 km and at the centre is a solid metal core, which is surrounded by an outer core of molten iron and nickel whose complex flow generates the planet's magnetic field. The overlying hot solid mantle is in constant motion towards the surface of the planet from convection, where it cools in the upper mantle by the escape of heat through conduction in the rocks and oceans or from volcanic activity. The upper mantle supports the continental and oceanic lithospheres, which, together with the crust, form the tectonic plates. When segments of the Earth's crust slide past each other along inclined faults, the crust is distorted from compression or stretched until the tension is released suddenly by an earthquake. At the mid-ocean ridges, the mantle rises close to the surface and the cooled upper mantle then returns to depth to complete the convective cycle. At the boundaries of the plates, subduction zones form along deep ocean trenches that reach 6-7 km below sea level and lie about 100 km offshore and parallel to the great arcs of volcanoes. At these subduction zones, the plates of the oceanic lithosphere descend several hundred kilometres into the mantle, which balances the creation of new ocean crust at the mid-ocean ridges.

The study of ancient rocks on the continents indicates that, since primordial times, the evolution of the Earth's surface has arisen from a cycle of tectonic plate motion, in which the continents agglomerate into one or several large masses and then rift and break up perhaps twice every billion years. The formation of supercontinents and their splitting up into smaller continents have had a major impact on both global climate and life on the planet. When the continents assemble, there is evidence of a period of continental collision and mountain building, and on fragmentation and dispersal over millions of years, megacontinents eventually drift towards the poles so that the albedo (reflection of solar radiation) of the Earth is increased. Drifting continents change the pattern of ocean circulation and, in combination with glaciation at the poles, an ice age is triggered in which global temperatures drop to 8-12 °C and sea water becomes locked up in the ice caps so that sea levels fall to 90-120 metres below current levels.

A geological timescale is given in Appendix 4, and there is evidence that there were ancestral supercontinental clusterings of which the earliest was Vaalbara in the Archean eon from 3.6-2.8 BYA (billion years ago). It is postulated that in the Proterozoic eon, a supercontinent Kenorland existed from 2.5-2.1 BYA and this was followed by Columbia from 1.8-1.5 BYA. Although it is controversial, a megacontinent known as Laurussia was formed 1.2 BYA from the clustering of the continents of Baltica, Siberia, and Laurentia, for which the evidence of the continental collisions may be found in the eroded roots of ancient mountain belts in North America and Scandinavia. It is also hypothesised that the Laurussian cluster rifted into two megacontinents that were in the region of the poles 900 MYA, together with a smaller equatorial continent. This triggered an ice age, which may have been the coldest period in the history of the Earth. Around 800-850 MYA, there was a further period of continental collision, which resulted in the amalgamation of continents to form a supercontinent Rodinia. Subsequently, 750 MYA, Rodinia fragmented into the separate megacontinents of Laurentia (North America, Mexico, Greenland, Scotland, and Northern Ireland) and Gondwana (South America, Southern Europe, Africa, India, Antarctica, and Australia), and several lesser continents Baltica (Scandanavia, Northern Europe, and Russia), Siberia, and Barentsia. Ron Redfern describes the evolution of continents, oceans, and life in *Origins* (2000), where the maps are based on the palaeography of Dr Christopher Scotese of the University of Texas.

The Precambrian ice age 650-580 MYA was the result of a global cooling when the megacontinent of Gondwana was near the South Pole and the icy land masses reflected more of the solar radiation back into space. In glacial times, when temperatures are low, there is an increased rate of upwelling of the oceans, which effectively draws carbon dioxide from the atmosphere at a faster rate and reduces the level of greenhouse gases. However, the chemistry and biology of the oceans is such that any increase in their acidity, such as from the sulphur discharged by volcanoes, causes a reaction with calcium carbonate deposits at the bottom of the ocean, which eventually releases carbon dioxide from the surface to the atmosphere. The Earth as a whole was subjected to a series of glacial and interglacial periods with fluctuations in sea level until 600 MYA, when carbon dioxide levels reached four times current levels, which caused global warming with the disappearance of glaciers to create the environment for multicellular animals.

In the Cambrian period 500 MYA, there was a semi-permanent rise in sea levels of 240 metres, which is explained by an increasing rate of volcanic activity at sea floor spreading centres, which created ridges of hot rock on either side of these centres that displaced the sea water. In addition, a further rise in sea level arose from a radical warming in the climate from the greenhouse effect as carbon dioxide levels reached over eight times current levels and continental platforms were submerged to a total depth of 360 metres. In the late Ordovician period around 450-440 MYA, there was a change in the pattern of ocean circulation when Gondwana edged over the South Pole and the Earth entered another ice age, and the sea retreated from the continental shelves. The Ordovician event was the first of five environmental mass extinctions. The ice age ended by 430 MYA, and global temperatures increased again until the period 375-360 MYA, when amphibians emerged in the intertidal zone between land and sea before inhabiting the swamps and river banks on land. In the late Devonian period 365 MYA, there was a second mass extinction event on a global scale involving marine invertebrates and vertebrates, which may have been triggered by volcanic activity or earthquakes and a series of tsunamis that caused prolonged turbidity.

At the time of the Permo-Carboniferous Ice Age 330-260 MYA, when the megacontinent Gondwana was again over the South Pole, vast areas of coal forest were absorbing carbon dioxide from the atmosphere so that its concentration declined to present day levels. Sea levels fluctuated by 100 metres in a series of glacial cycles with interglacial periods similar to the climate of the present day. This produced coal-forming cyclotherms from the repeated marine invasions and withdrawals in which shorelines retreated and advanced by 160 km. During the Permian period 250 MYA, the continental land masses began to collide with continental compression in which the coal deposits that had accumulated over the preceding seventy million years were pushed upwards with episodes of mountain formation, and a supercontinent was assembled. Alfred Wegener (see Chapter 3) called this great land mass Pangea.

The unification of Pangea drastically changed the pattern of ocean circulation, and the total length of coastline available for shallow sea marine animals was substantially reduced, and the Permian third mass extinction took place. The volcanoes scattered around the Pangean shores produced one of the largest outpourings of basaltic lava that would have contributed to ocean acidity and atmospheric toxicity, which would have reacted with the exposed coal deposits. As a consequence, a new greenhouse regime was created with carbon dioxide concentrations four times present levels that caused a fundamental change in the Earth's climate. The climate changed so dramatically with temperatures 5°-11°C above present levels that glaciers disappeared, parts of the continental land masses were submerged by rising seas, and a desert age commenced. This brought to an end the Paleozoic era and, in the Triassic period of the Mesozoic era 210 MYA, there was a fourth mass extinction event. Then 200 MYA, the initial rifting and breaking up of Central Pangea began, and this was followed by continental drifting at a rate of 40 km per million years. Alfred Wegener developed the hypothesis for continental drift, which explained the intercontinental dispersion of animals and plants.

By the mid-Cretaceous period 100 MYA, carbon dioxide levels reached five times present levels with a peak in global temperatures that were 11°-15°C higher than today, so that forests existed at the poles. By the late Cretaceous period, sea levels rose to 250-300 metres above the present day, and the land area diminished to half its present extent, as a consequence of the inundated continental shelves and interiors. It is in this context that an asteroid collided with the Earth and precipitated a fifth mass extinction at the end of the Cretaceous period 65 MYA. The Cenozoic era followed, and in the Oligocene era 35 MYA, ice sheets started to form in Antarctica, which developed into permanent ice sheets at the South Pole by 15 MYA. The latter may have been triggered by tectonic plate motions that caused the separation of Antarctica from South America, which set up a circum-Antarctic ocean current that blocked the southward flow of warm water in the Pacific and Indian Oceans. Around three million years ago, tectonic plate movements in the Caribbean put into place the Isthmus of Panama, so that the North Atlantic and Pacific Oceans became separated, and the warm South Atlantic Equatorial Current was cut off. This event contributed substantially to the start of the Northern Hemisphere glaciation 2.5 MYA, which contrasts with 100 MYA when the land area in the latitudes (600-700 N) around the Arctic Circle was only one third of the current area.

The Pleistocene epoch is known as the Ice Age in which a succession of cold 'glacial' periods and warmer 'interglacial' periods with an average overall cycle of 100,000 years, so that there are ten cycles per million years. Within each of the major glacials, minor warmer and colder periods occurred. In glacial periods, the advancing ice sheets locked away enormous quantities of water so that average sea levels fell and humans could walk from mainlands to former islands. Then the ice retreated during the interglacial periods, with the most recent peaking approximately 125,000 YA, when global temperatures were 1º-3ºC warmer than present times. The coastlines were transformed as sea levels rose to five metres higher than today, with subsequent fluctuations of up to fifteen to twenty-five metres below current levels at variable intervals of around 20,000 years during the interglacial period until 80,000 YA. As the continental ice sheets started advancing again 75,000 YA, the sea levels dropped to ninety metres below current levels 50-60,000 YA, and then fell to 120 metres below the present level at the peak of glaciation 18,000 YA. From the peak of the last glaciation 18,000 YA to 6,000 YA, sea levels rose at the rate of one metre per century. In the past 5,000 years, sea levels have risen at the rate of 0.2 metres per century (2 cm per decade), although over the past 1,000 years there has been virtually no rise in sea levels.

Newer theories of plate tectonics differ from the continental drift theory in so far as the operational unit is now a plate with a thickness of some 150 km (the lithosphere) rather than the continental crust that is only 40 km thick. The Earth's surface is divided into seven major plates and several smaller plates. As a result of mantle convection, the plates are always in motion relative to one another, and when they diverge, new oceanic crust is formed and, where they converge, crust is destroyed by transfer back into the mantle. The convergence results in major deep-focus earthquakes and intense volcanic activity. Where the edge of a continental plate converges with the edge of another continental plate, large mountains and plateaus are formed. For example, India collided with Asia around 50 MYA to form the Himalayas and the Tibetan Plateau, and, subsequently, this collision closed the Eastern Tethys to form the Black, Caspian, and Aral seas. Around 18 MYA, the collision of Africa-Arabia with Laurasia closed the Western Tethys, now the Mediterranean, and these major geological upheavals resulted in geographical transformations and environmental changes. This land-bridge allowed African animals, including apes and monkeys, to cross into Europe and Asia, and the drier conditions in East Africa caused the spread of grassland that may have triggered the emergence of the first hominoids.

Oceanic tides are caused by the gravitational forces exerted by the moon and sun on the Earth, and the largest amplitude of the semi-diurnal tidal cycle arises at a full moon and a new moon, when the moon, the sun, and the Earth are aligned. There are episodic tidal waves known as tsunamis, which are caused by displacements of large volumes of water by earthquakes, massive volcanic eruptions, and enormous landslides into the ocean along the coastline. However, the surface ocean currents and waves are driven directly by the wind, together with the effect of the Earth's rotation, and these surface currents transport warm water from the tropics to moderate the climate in the cooler temperate regions. In the northern hemisphere, the movement of the surface water is deflected slightly to the right of the direction of the wind by the rotational forces, and the next deeper layer of ocean is dragged along by the surface layer with a rotational deflection to the right. This process is transferred to deeper and deeper layers, until the forces are not strong enough to drag an even deeper layer. At high latitudes, the dense saline surface-water cools and sinks to the deepest layer, so that there is a stratification of ocean layers according to differences in temperature and salinity. Deep circulation of the oceans arises from the displacement of the deeper water towards the surface, which resupplies the surface with the nutrient elements of phosphorous and nitrogen that are regenerated in the deeper parts of the ocean basin. The rate of upward motion of the water is about four metres per year, and as the mean depth of the oceans is 4 km, the average time to transport a parcel of water from the bottom of the ocean to the surface is 1,000 years.

2.4   Physical metasystem

In the early high-temperature conditions at the formation of the Earth, the hydrogen-rich atmosphere was composed of water vapour that condensed into the oceans with radiative cooling. The Earth's surface was kept above freezing by the rays from the sun, whose radiant output was some thirty percent lower 4.5 BYA. Methane gas and carbon dioxide from volcanic activity then created a greenhouse effect so that temperatures started to rise again until 4 BYA, when the initial high carbon dioxide content of the atmosphere was diminished by reaction with powdered rock in an aqueous medium. By 3.5 BYA the carbon dioxide content had fallen sufficiently for the temperature at the surface of the Earth to cool so that bacteria-like organisms developed from chemical syntheses with available organic matter. In the process of photosynthesis, in which carbon dioxide and water are converted into organic matter and oxygen gas, deposits of calcium carbonate known as stromatolites were formed from biologically mediated rock destruction. The stromatolites contain records of primitive unicellular organisms known as cyanobacteria (algae). For the next billion years, most of the oxygen derived from photosynthesis was used up in the oxidation of iron and sulphide bearing minerals in the lavas erupting from the interior of the Earth, as well as reacting with gases, such as hydrogen sulphide and methane associated with the transfer of material from the mantle to the crust. Also, the sulphate concentration of the oceans increased, which provided the necessary ingredient for sulphate-reducing bacteria to proliferate with the production of hydrogen sulphide gas.

By 2.1 BYA, the atmosphere began to change as a result of the photosynthesis by bacteria-like prokaryotes and green algae. The concentration of oxygen altered the atmosphere irreversibly from its high carbon dioxide composition. This transition to an oxidising atmosphere created a crisis for many organisms that were poisoned and became extinct. However, the oxygen became sufficiently rich for it to form an ozone shield by the dissociation of oxygen gas in sunlight. The ozone layer protects living organisms by absorbing much of the harmful radiation from space, such as certain bands of ultra-violet light. The presence of oxygen would be important for the evolution of life, because cells that developed the capability to handle the gas could exploit the energy released by the oxidation of organic matter. Also, the ozone shield enabled complex organic structures to make use of the solar radiation in shallow water or on land without being burned.

Nitrogenous compounds from living organisms are returned to the atmosphere as nitrogen, as a result of metabolism (denitrification), and around 1.5 BYA, the nitrogen and oxygen concentrations of the atmosphere were seventy-eight percent and twenty-one percent respectively by volume. These concentrations have remained relatively constant to the present day. The internal heat in the Earth's mantle releases Argon from the radioactive decay of an isotope of potassium in rocks, and Argon provides nearly one percent by volume of the atmosphere. In 2000, levels of carbon dioxide in the atmosphere were 0.037% by volume or 370 ppmv (parts per million by volume), and the concerns are that any increase in carbon dioxide will cause a rise in temperature from the greenhouse effect. This would cause polar ice to melt with a reduction in the reflectance of the planet, and this would result in a further temperature rise. As the Earth receives the major part of its solar energy in the tropics and reflects back into space a proportion of the smaller amount received in the polar regions, a meridional temperature gradient exists between the poles and the equator. It follows that the climatic system is maintained in far-from-thermodynamic equilibrium.

Deep sea and ice core records have an environmental memory of several hundred thousand years, and the BBC publication *Earth Story* (1998) explains that the Milankovitch cycles account for at least sixty percent of climate change within an ice age, and the cores also show that the proportion of carbon dioxide in the atmosphere moved in step with the glacial advances and retreats before the burning of fossil fuels became a factor. In the century from 1900 to 2000, the rise in sea levels was less than 20 cm, in relation to a rise in carbon dioxide concentrations from 300ppmv to 370ppmv and a temperature increase of 0.5°C. The Intergovernmental Panel on Climate Change (IPCC), which is sponsored by several UN agencies, estimates that by 2100 global temperatures could rise by some 3°C from a current temperature of 15°C, for a doubling of carbon dioxide emissions to 0.07% by volume (this compares with a 0.5°C rise during the past 100 years). Higher temperatures could cause the ocean waters to expand and some polar ice to melt, raising sea levels by a projected 6 centimetres per decade (one metre by the year 2150), which could cause devastation in low-lying coastal areas of the world. In Chapter 12, the evolutionary approach provides alternative scenarios for 2150 with lower rates of growth in Gross World Product, which impacts on global energy use and carbon dioxide emissions levels, and it is more probable that global warming will lie in the range 1°C -2°C per century. Although the biosphere may successfully adapt to global warming for a considerable period, a sharp discontinuous change may become inevitable such as the disintegration of the West Antarctic ice sheet that would release two million km^3 of ice into the sea. It is important not to exacerbate the rate of rise in sea levels, so that the civil system has time to adapt by relocating cities to higher ground and the installation of flood control structures.

Deep ocean mixing of carbon dioxide from the atmosphere causes an ocean thermal time lag of several centuries and perhaps 500 years, so that observed increases in temperature tend to be less than would be anticipated from estimates of anthropogenic emissions. The *Long-Range Futures Research* assumptions are based upon the estimates by William Cline of the Institute of International Economics that the Earth's current temperature is 15°C, and the expected timescale and extent of long term global warming would give a rise in temperature of 3°C to reach 18°C within the next 250 years. It is anticipated that temperatures will rise by 1°C per century until 2500 when temperatures will peak at 20°C for a fourfold increase in carbon dioxide levels to 0.15% by volume. Temperatures will then decline by 0.5°C per century to 17.5°C by the year 3000, as fossil fuels (coal, oil, and gas resources) will become exhausted during the Third Millennium. The rise in sea levels by 3000 is expected to reach four metres. In the absence of policy changes, an eightfold rise in carbon dioxide levels to 0.30% would result in a peak temperature increase of 7.5^{0}C, with an even greater rise in sea levels of six metres by the end of the Third Millennium.

As the warm waters of the Gulf Stream head northwards they become denser and sink to form a deep, cold, ocean current as part of the overturning Atlantic Conveyor, which heads south again to join the global Ocean Conveyer. However, global warming will cause melting of arctic sea ice and the Greenland ice cap so that there will be large influxes of cold water into the North Atlantic, which will change the salinity of the Gulf Stream. This will weaken the circulation and the warm Gulf Stream conveyor mechanism may cease, with a sharp decline in north European temperatures to near arctic conditions comparable to similar latitudes in Canada and Siberia. According to an article in a publication of the Geological Society of America, *GSA Today* (Volume 9, Number 1, January 1999) by oceanographer Wallace Broecker, who discovered the global 1,000-year ocean conveyor system, models suggest that the earth would have to undergo a 5°C greenhouse warming in order to force an Atlantic conveyor shut-down. This may be approached by 2500, and it could be the trigger for ending the interglacial period around 3000, with the descent to another glacial period taking 70,000 years. This would result in mass migrations towards the south from North America, Europe, Russia, Central and Eastern Asia, into a land area that may be insufficient to sustain a world population as high as 8 billion. However, if temperatures decline in the second half of the Third Millennium, the Atlantic conveyor may slow rather than cease and abrupt climate change is unlikely.

Ozone is constantly created and destroyed in the stratosphere when oxygen molecules absorb ultraviolet radiation (UV-B), which causes them to split into two separate oxygen atoms. The separate oxygen atoms can then bond with another oxygen molecule to form ozone, which consists of three oxygen atoms. This process absorbs most of the incoming short-wave radiation (wavelength of around 300 nanometers) that is most harmful to living matter. The ozone shield permits organisms to occupy habitats that expose them to the direct rays of the sun and also allows greater mobility for territorial colonisation. However, by absorbing radiation of a slightly longer wavelength, the new ozone molecule splits up again and combines with a free oxygen atom that then recombines into two oxygen molecules. In this way, there is a balance between the creation and destruction of ozone in normal circumstances with the absorption of radiation.

Human activity has resulted in the release of CFCs (chlorofluorcarbons) that are very stable and that do not break up until they reach high altitudes. The weather conditions in Antarctica form ice crystals on columns of ozone that provide a surface on which chemical reactions and incoming radiation can break up the CFCs. This then liberates chlorine in the stratosphere, where it acts as a catalyst, and every free atom of chlorine destroys 100,000 molecules of ozone before being deactivated. In addition to stratospheric ozone production, small amounts of ozone are also produced in the troposphere, where it acts as an oxidant and its effects are generally considered to be harmful. Anthropogenic and natural biotic sources at ground level are augmenting the ozone in the troposphere from sunlight-activated reactions with volatile organic compounds (VOCs) associated with petroleum and petrochemicals, and nitrogen oxides (NO_2). City smog and photochemical air pollution are created by increasing ozone in the atmosphere, which causes asthma and other respiratory problems for humans.

2.5   Biological metasystem

Around 3.5 BYA, microscopic organisms emerged in the hot springs that were caused by intense volcanic activity, and they may have formed around the vents of superheated water at the mid-ocean ridges. These organisms have the identical shape of bacteria living today, and bacteria are the most prolific example of a single-celled organism called a prokaryote. The ability to detect hot water led to an evolutionary adaption to harness sunlight through photosynthesis. Then around 3 BYA, green cyanobacteria (algae) began to release significant quantities of oxygen, which is a waste product of photosynthesis, into the aquatic environment. Around 1 BYA, a huge number of large, single-celled oxygen breathing eukaryotes, known as acritarchs, had evolved that were 100 times larger than the prokaryotes from which they had descended. However, between 900-800 MYA, the coldest period in the history of the Earth caused an extinction of acritarchs. The break-up of Rodinia 750 MYA will have affected the ocean currents and the distribution of oxygen and other nutrients, and after the Precambrian Ice Age 550 MYA, the warmer climate precipitated rising sea levels that flooded many coastal regions to provide large, shallow-water environments along the extended coastlines. These factors contributed to the Cambrian explosion of multi-cellular animals, during which the number of orders of animals doubled roughly every twelve million years in an exponential increase. Modern representatives of these phyla include worms, crabs, shellfish, lampshells, sea urchins, sponges, tunicates, and centipedes.

The Permian extinction in the Paleozoic era on the formation of Pangea 250 MYA, resulted in some ninety-five percent of marine species becoming extinct as well as over fifty percent of animal species during a timescale of perhaps five million years. This era was followed by the Mesozoic era (Age of Reptiles) in which dinosaurs came to dominate the planet until the mass extinction at the end of the Cretaceous period. Following the extinction of the dinosaurs, the evolution of large mammals became possible in the Cenozoic era (Age of Mammals), which led to the emergence of bipedal hominids in the Pliocene epoch less than five MYA. Evolution is driven by two agents of natural selection, of which competition leads to increasing specialisation, and episodic environmental destruction favours adaptability with increasing complexity in an organism's nervous system.

Homo Erectus evolved with a brain capacity of 1,000 cm³ around 1.5 MYA, in the Pleistocene epoch (2 MYA-10,000 BC). *Homo sapiens* made his appearance circa 250,000 BC as a process of natural selection from his forbears *Homo erectus* and developed a brain capacity of 1,500 cm³. It is generally thought that whilst the archaic *Homo sapiens* had acquired signalling and communication techniques, it is unlikely that the mouth and throat structure had evolved with sufficient neural development for the coordinated muscle control that speech requires. By 50,000 BC, modern man *(Homo sapiens sapiens)* had emerged from Africa into Asia, then by sea to Australasia, before Europe and the Americas in the late Palaeolithic period. These extensive migrations made humans a widely dispersed species, although the total population was only of the order of one million. Modern man has a brain capacity of 1,500-2,000 cm³, and is known to have the physiological capability for the systematic articulation and comprehension of language. Humans differ from other animals by the extent to which they look ahead to the future, making plans, and undertaking long sequences of related operations and projects.

As the glacial period came to an end in the period 10,000-8000 BC, the world underwent a series of ecological changes that had far-reaching and beneficial effects on human communities. Prehistoric hunters adapted well to the environmental changes, with alterations in diet that reflected the climatic effects on other animal and plant species, and they migrated from the tropics into northern regions whenever suitable conditions arose. Environmental changes were important to the evolution of man, since it favoured higher intelligence and larger brains, adaptability, and organisational skills to capture resources, and good communication with the use of speech. Humans survived by learning and passing on information, knowledge, and skills that resulted in societal and technological evolution.

It has been estimated that of the 500 million to one billion species that have existed in evolutionary time, ninety-seven percent are now extinct. Species extinctions are an essential part of the evolutionary process and, in the 600 million years of multicellular life on Earth, it has been estimated that episodic environmental transitions account for sixty-percent of all extinctions. By the middle of the Second Millennium, there was probably greater biodiversity than at any other time and, according to UNEP (United Nations Environment Programme), of the estimated twenty to thirty million species living today, some 1.75 million species have been classified. The average lifespan of a species in the fossil record is in the range of one-million to four-million years, and the normal background extinction rate has been one species per million species per year. It follows that for twenty million species, a background extinction rate (i.e., forty percent of all extinctions) would not be less than perhaps 2,000 species per century. There is a tremendous diversity of life forms with the number of classified species reaching 4,500 mammals, 10,000 birds, 25,000 fish, 8,000 reptiles, 5,000 amphibians, one million insects (perhaps only ten percent are classified) and myriapods (i.e., centipedes or millipedes), and 250,000 vascular plants. Although species extinction is a natural process, there is concern that natural extinction rates have increased during the past 400 years (1600-2000) as a result of human activity. The recent overall rate of extinction for the 1.75 million species identified above has been 250 per century; so, for the total number of species, the overall background extinction rate is estimated to have risen from say 2,000 to 4,000 per century.

Large-scale ecological communities are known as biomes, which include tropical forest, temperate forest, grassland, desert, tundra, mountains, and marine biomes, such as coral reef, coastal shelf, and open ocean. Ice ages have driven many species from glaciated temperate regions to tropical habitats, where the climate is not as harsh. Consequently, there is a complex vegetational structure with more plant species in the tropics, and these support a greater variety of herbivorous animal species, which in turn supply the energy of carnivorous predators. There is also a vertical structuring in ecological communities, which arises from the competition to intercept light and capture the available solar energy. In areas of abundant water and nutrients, light is a limiting resource, and the richest diversity of organisms may be found in the canopies of tropical rain forests. Habitat destruction accounts for much of the loss of biodiversity, such as the clearance and burning of tropical rain forests that occupy six percent of the Earth's land surface.

There is a nonlinear relationship between habitat loss and species loss, and the rule of thumb is that fifty percent of the species will survive even if ninety percent of the habitat disappears. This is because species can resettle and survive in alternative habitats within continental regions, where habitat loss or modification may arise from climate change, agriculture, forestry, urban development and pollution. The rain forests account for half the species of plants and animals on the planet, say ten to fifteen million species of which the majority are insects, weeds, and fungi, according to Edward Wilson in *The Diversity of Life* (1992), and 0.1-0.2% of these species become extinct annually. This gives a current extinction rate of one or two species per hour, which corresponds to one to two million species per century. This is 500-1,000 times the historical background extinction rate of 2,000 species per century, and it reflects a mass extinction event caused by human economic development. According to the *World Atlas of Biodiversity* (2002), published in association with UNEP World Conservation Monitoring Centre (WCMC), the current rates of extinction for mammals and birds are some 100 to 200 times higher than their background rates. So it is probably best to refer to current extinction rates in general as 100-1,000 times the background rate. Conservation of biodiversity keeps options open for the future and, since some twenty-five percent of prescription drugs are derived from plants, new cures may depend upon the prevention of genetic loss. Also, fresh genetic material may be needed to introduce new strains of plant and animal species if domesticated strains become vulnerable to insects, pests, or viruses.

In *Sociobiology* (1975 and 2000), Edward Wilson explains that an ecosystem functions as a communications network in which the signals are motional, visual, tactile, acoustic, chemical, odorous, and gustatory (taste). The genetic information system passes on information from generation to generation, and environmental changes may result in natural selection that prevents the signal from being repeated unless there is a mutation. Finer tuning to signals results in increasing specialisation and diversification of the parts. As differences develop, each species will achieve a different level of integration for energy or resource acquisition, with the formation of new sub-groups in which the structure and function of the parts are subordinated to the goals of the group as a whole. At a more complex level, the methods and models of nonequilibrium thermodynamics have been applied to ecosystems in which the system growth parameter 'exergy' (the available work content in energy) drives the system away from equilibrium, and is destroyed in the irreversible process of evolution. For example, see *Into the Cool – Energy Flow, Thermodynamics, and Life* (2005), by Eric Schneider and Dorian Sagan. Nature abhors gradients such as measurable differences in pressure, temperature, or chemical concentration, and ecosystems develop that utilise the exergy and thereby reduce the ambient gradients. Ecosystems display a tendency to grow to a limit at maturity, when the system has exhausted all possible dissipative routes and it settles down with an optimal amount of energy captured and degraded within its system.

2.6 Civil metasystem

The Pleistocene epoch commenced two MYA (million years ago), and early man developed technology that included stone tools, the discovery of fire, the building of shelters, the use of marine craft or rafts made from logs lashed together, the wearing of clothes, the making of hand axes, and archery. The Pleistocene epoch marks the start of the archaeological record, which can be subdivided into Lower Palaeolithic 1.5 MYA - 250,000 YA, Middle Palaeolithic 250,000-50,000 YA, and Upper Palaeolithic 50,000-10,000 YA. The geological and archaeological timescales are combined in Appendix 4. *Homo Erectus* lived during the Lower Palaeolithic stage, *Homo Sapiens* lived during the Middle stage, and *Homo Sapiens Sapiens* had emerged by the Upper Palaeolithic stage.

From 50,000 YA (years ago), semi-nomadic hunters and gatherers at dispersed locations constructed compounds with clusters of huts to provide base camps. These settlements were temporary, although they remained for several seasons and possibly several years, before being abandoned when essential materials were salvaged so that they could be rebuilt elsewhere. There were also scattered locations of smaller camps and hunting stations, rock shelters, and quarry workshops for making stone tools and eventually blade tools before the recent Holocene epoch began in 10,000 BC. The nineteenth century prehistorians devised an 'Age' system of Stone and Metal Ages that overlap both the Palaeolithic and Holocene epochs, in which the Middle East was slightly ahead of Western Europe. The Stone Age is divided into three phases, the hunting Palaeolithic, the fishing and agricultural Mesolithic (10-8,000 YA), and the forestry Neolithic (8-6,000 YA) with pottery and timber artefacts, including mastless long boats. The subsequent metal ages also recognise three phases that are referred to as the Copper or Chalcolithic (6,000 YA), the Bronze (4,000 YA), and the Iron (3,000 YA) Ages.

As the last glacial period retreated, bands of palaeolithic hunters expanded the human niche from a population of perhaps one million to a population of some ten million at the start of the agricultural revolution in 10,000 BC. The primitive instinct for human survival led to the formation of families and tribes, and the human community system has evolved from the bands of hunters and gatherers, whose way of life was based on division of labour, on cunning, and on cooperation, and intelligence. Hunting and gathering was the principal economy in the course of human evolution to that date. When the glacial period ended in years 10,000-8000 BC, the world underwent a series of ecological changes that had far-reaching effects on human communities. Temperatures rose and the ice sheets, which had covered almost a quarter of the Earth's land surface, melted, allowing plants and animals to spread northwards into latitudes that for several thousand years had been too cold to support them. At the same time, deserts that had occupied half the land between the tropics receded, as much of the water that had been locked up in the ice sheets was released to fall as rain. In many areas, milder temperatures meant that food resources were more abundant and more varied than before, so that populations grew rapidly in these favourable circumstances, and settlements gradually became larger and more permanent as the agricultural revolution gained momentum.

Between 10,000 BC and 5000 BC, there was a rise in the sea level of some fifty metres, which amounts to one metre per century. By 6500 BC, the rise in sea level cut Britain off from the continent of Europe, much of the eastern coast of North America was flooded, the islands of Indonesia were separated, and Australia was disconnected from New Guinea and Tasmania. The Mediterranean coastline descends steeply, so not much land was lost there, but in the Middle East and North Africa, the coastline of the Arabian Gulf and the lower river valley of the Nile delta were submerged, which created pressures leading to the domestication of plants and animals and the formation of the first traditional cities. In the fertile crescent of the Middle East (Mesopotamia), which stretches from southern Jordan in the west to southern Iraq and Iran in the east, farming villages were becoming widespread by 7000 BC. By 5000 BC, the foundations had been laid for the development of cities and civilisations, and the earliest cities were located in Mesopotamia (around 4000 BC), the Nile Valley (3000 BC), the Indus Valley of the present Pakistan (2500 BC), in the Yellow River Valley of China (by 2000 BC), in South America, Mexico, and Peru (in AD 500), and in the Niger Valley in West Africa (AD 1500). The largest cities such as Ur in southern Mesopotamia and Thebes, the capital of Egypt, may have had populations exceeding 200,000, but in general, the populations of the earliest towns remained in the range 2,000 to 20,000.

From the earliest farming villages (8000 BC), to the building of cities and civilisations (around 4000 BC), and the development of new transport technologies, such as the sailing ship (4000 BC) and the wheel (3000 BC), humankind commenced the development of the world. Civilisation is a social and technical achievement, and it is associated with cities and states, the interaction of humans, language, communication, and culture. Because humans can speak and record their thoughts and emotions in writing and pictures, they can pass on their experience and knowledge from generation to generation. The early civilisations reorganised their societies and economies to cope with the increasing specialisation of occupations, such as craftsmen for tool making, building artisans, merchants for trade and commerce, scribes for recording transactions, priests, and civil and military administrators. Important cities had to be provided with stronger defences, and large armies struggled for land and power with new weaponry. Cities are the largest artefacts created by man, and early societies made an enormous investment of human effort and physical capital in the construction of a city for economic, defence, and spiritual reasons.

The early city-states were based upon intensive agriculture that could support high levels of population, and it is the ecological context that helps to explain the successful development of the early civilisations. Technological developments, such as irrigation, terracing, and plough agriculture increased the efficiency with which food could be produced and allowed agriculture to become established in previously marginal zones. However, where populations were especially high, or the environment particularly vulnerable, the long-term ecological effects of the more intensive agricultural practices soon became apparent. In southern Mesopotamia, large-scale irrigation quickly led to increasing salinity with a substantial decline in crop yields, and successful expansion eventually reached the point of diminishing returns in the productivity of both land and labour. Demographic expansion inevitably encountered territorial resource limits, and the obvious response was enlargement of the resource base through territorial expansion. The rationale for civil-society is to cooperate in increasing the carrying capacity of a territory through investment in settlements that can be protected, which extended the process of urbanisation.

Inevitably, territorial acquisition led to the creation of larger and larger states, with smaller kingdoms being conquered by their powerful neighbours. Where a civilisation exercises political domination over other states, the resulting consolidation of occupied territories is termed a colony or an empire. Some civilisations were empires with a unified political organisation that encompassed diverse ethnic groups over a wide area, which culminated in the appearance of empires in the First Millennium BC of unparalleled size and power. These include the Assyrian empire, the Persian empire, and the Greek empire. A little later, the greater part of the Indian subcontinent was united for the first time under the Mauryan empire, while the Mediterranean world and most of western Europe came under the control of the Roman empire. China also witnessed a period of empire building in the last few centuries BC, when a number of warring kingdoms were combined into a single unified state under the Han dynasty.

Based upon the work of Arnold Toynbee, *A Study of History*, some twenty-five civilisations have been identified, each of which had a duration of over 500 years. For the purpose of this research, civilisations have been identified by continental region, chronological period, and duration of the civilisation, and these are listed in Appendix 5. If civilisations or empires lasting over 200 years are included, there are perhaps fifty civilisations or empires that are significant in the history of humankind. The earliest civilisations between 3500 BC and AD 500 lasted for an average of over 2,000 years and covered an approximate area of some 1m km². In the period AD 500-1500 the average duration of an empire was 500-1,000 years, and the area covered was some 5m km². Since AD 1500, empires have covered areas of 10m km², but the average duration has been 200-500 years. The duration of empires has reduced over time as a consequence of accelerating growth and rate of change, with increasing urbanisation, enhanced communications, a rising amount of travel by citizens, and the advancing knowledge of populations. It would appear that the size of empire has increased with progress in both the transport technology (i.e., tracks, roads, ships, railways, and air transport), and the military technology (i.e., infantry, cavalry, artillery, bombers, and intercontinental missiles). Continental unions of 5m-10m km² seem to be appropriate to the current range and speed of transport and military technology.

Studies of past civilisations by Joseph Tainter, in *The Collapse of Complex Societies* (1988), show that economic explanations are structurally and logically superior to other explanations for collapse. The collapse of civilisations, such as the Western Roman Empire in the 5th century AD, ancient Egypt, the Indian Mauryan Empire, Chou China, and the Mayan civilisation in the 9th century AD suffered from economic failure, which led to exhaustion of natural resources and the eventual decline of the civilisations. Diminishing returns from complex societies manifest themselves in excessively large administrations, increasing taxation, a depreciating currency, disrepair of public works, depletion of resources, declining military strength, internal conflict, and population decline. It is apparent from the recent collapse of the USSR that even primary world powers do not have sufficient financial strength to finance diminishing returns indefinitely. It may be concluded that, as civilisations or empires exceed a certain size, they become too complex and politically unstable and parts of the empire break away until a smaller and more stable continental grouping forms.

Our civilisation on Earth has developed artefacts and settlements, but it may not be unique and it may form part of an astro-civil metastructure, in which an astro-civilisation uses artefacts and lives in a socio-technological society to extend the lifetime of its planetary civilisation. Astro-technological civilisations are probably dependent upon the existence of metals for the development of technology, and certainly life as we know it has an atom of iron in every haemoglobin molecule in the bloodstream. The discovery of electricity and magnetism was dependent upon copper and iron, and it would be difficult to produce radio telescopes or television without electricity. It follows that the search for astro-civilisations would be best concentrated in the blue rings of our solar system, away from the core region where red giants and low mass stars predominate.

There are also habitable zones for planets around stars, in which there is a range of distances from the sun that permit life-sustaining temperatures and enable liquid water to exist for four to five billion years. An important factor is that the planet has to be of the right size to retain an atmosphere. In 1961, radio astronomer Frank Drake of Cornell University proposed a formula containing cosmic, ecological, and technological parameters for determining the number of astro-technological civilisations in our own galaxy, for which the descriptions have been modified as follows:

$N = F \times S \times E \times L \times C \times A \times T$, where

N is the number of astro-civilisations
F is the annual number of stars forming in our galaxy
S is the fraction of stars (suns) with a solar system
E is the fraction of planets at a suitable distance from their star to permit the formation of an ecological system
L is the fraction of planets with the appearance of life
C is the fraction that develop into mature civilisations with settlements
A is the fraction that become astro-technological civilisations (ATCs) with at least a radio telescope to communicate with other civilisations
T is the lifetime of an astro-technological civilisation

The Cambridge Atlas of Astronomy, edited by Jean Adouze and Guy Israel (1989), offers a scenario in which ten new stars are born each year with planetary systems, and all have one planet with an ecological system where life evolves on one in five planets that always develops to a mature civilisation with settlements. If half of the civilisations become astro-technological, then F x S=10, E=1, L=0.2, C=1 and A=0.5 and the multiplication of these values for the parameters is equal to unity. Under these assumptions, N=T, so that the number of astro-technological civilisations is equal to the average lifetime of a civilisation measured in years. It follows that if astro-technological civilisations had a lifetime of fifty millennia, then there may be 50,000 civilisations within our galaxy. In a steady state, N/T would be the rate at which astro-technological civilisations disappear from the galaxy.

On the assumption that life on Earth is the result of a natural evolutionary process that would be applicable elsewhere in the universe, alternative values for the parameters could be F=10, S=0.8, E=0.25, L=0.2, C=0.2, A=0.5, and T=250,000 years. In effect, these values would indicate that ten stars similar to the sun are born each year in the galaxy, of which eight have planetary systems with only two suitably placed for the appearance of life. However, life only appears on twenty percent of these suitable planets, and only one in five of these develops a mature civilisation. Since only one half of these become astro-technological civilisations, there would be 10,000 ATCs for a civilisation lifetime of 250 millennia.

The Drake equation is a compact model that incorporates a wide range of uncertainty, and it does not allow for the fact that civilisations can collapse and then rise again. The probable number of galactic civilisations is dominated by the value for the lifespan of an astro-technological civilisation, and this critical issue will be considered in *Long-Range Futures Research: An Application of Complexity Science*. In *Entering Space* (1999), Robert Zubrin points out that if the duration is short (say ten millennia) then ATCs are so rare that contact would be very unlikely. If the duration of advanced civilisations reaches fifty millennia, then occasional contacts are likely to occur. If the duration of astro-technological civilisations exceeds 250 millennia then frequent contact is likely to be made.

According to Einstein's Theory of Relativity, nothing can traverse space faster than the speed of light, although radio waves do in fact travel at the speed of light and galactic communications with another astro-technological civilisation could be achieved within a timescale of millennia. It is assumed that any ATCs would be distributed equally between the four concentric bands of the galactic disc, each of which are 10,000 light years wide, as the spatial density of the ATCs is expected to decline from the inner band to the outer band in relation to the number of stars. Since no radio contact has been made with another astro-civilisation to-date, the most probable explanation from the Drake equation is that any astro-civilisations at a similar stage to our own are 1,000 to 5,000 light years away, which would indicate that the number of astro-technological civilisations in our galaxy may not exceed the range 100-10,000. If a spacecraft could travel at 0.5 percent the speed of light, it would take 200 millennia to travel to a planet 1,000 light years away. This would require extraordinary advances in aerospace engineering, as the fastest spacecraft launched so far, Voyager 1, used a gravity-assist manoeuvre at Jupiter to leave the solar system at 0.005 percent lightspeed, which is in excess of 30,000 miles per hour.

However, there may only be a single habitable band in the galaxy that is centred on the orbital path of our Sun and 10,000 light years wide. This is because there are four spiral arms containing interstellar matter that float as waves in the galactic disc in accordance with the laws of hydrodynamics, rather like the wake behind a ship, and these take 200 million years to complete one revolution. The sun and any ATCs within its orbital band revolve around the centre in 225 million years, at a speed similar to the spiral arms, and so they remain in between the spiral arms and would only cross them once every seven billion years. The Cambridge University Press publication, *Extraterrestrial Intelligence,* by Jean Heidmann (1997), points out that the risk of crossing the luminous spiral arms, in which ancient stars are recycled and enriched with heavy elements, is that they contain supernovae that could increase the cosmic ray flux by a factor of 100, and any living population could face extinction from exposure to the radiation. This astrophysical phenomenon could apply to all spiral galaxies, and this is an additional factor that could reduce the number of ATCs by a further factor of four, so that even for a civilisation lifetime of 250 millennia, the upper limit for the number of astro-technological civilisations in our galaxy may be in the range twenty-five to 2,500. If the life of a civilisation is only fifty millennia, then it is unlikely that the number of astro-technological civilisations exceeds the range five to 500.

For the purposes of deep futures and astrofuturism, an interplanetary timescale for an astro-species to explore other planets within its solar system can be measured on a scale one to five million years, which reflects the period that it has taken for humans to become an astro-technological civilisation and to achieve a landing on the moon. However, travelling across the galaxy, or to Andromeda within our cluster of galaxies, could not be accomplished in a galactic timescale of less than five to twenty-five million years. Intergalactic travel would require a universal timescale of five to twenty-five billion years, and it is in this context that a meta-timescale for an evolutionary future needs to reflect the galactic spatial structure as shown in Appendix 6.

In relation to this meta-timescale, it may be expected that during the near Interplanetary Era, an asteroid over 1 km wide will hit Earth. The current Ice Age on Earth is likely to come to an end in the near Galactic Era in 20 MYF (million years into the future). In the far Galactic Era, it is probable that an asteroid over 10 km wide will impact with the Earth and in 250 MYF there will be a clustering of continents. The new science of astrobiology charts the ultimate fate of our world in *The Life and Death of Planet Earth* (2002), by Peter Ward and Donald Brownlee. In the Intergalactic Era, 1 BYF (billion years into the future), the biosphere on earth will expire in line with increasing solar energy, which is expected to increase by one percent per 100 million years, since the levels of carbon dioxide will fall below the concentration for photosynthesis by plants. By 2.5 BYF the Sun's brightness will have increased by forty percent and the oceans will recede and the water level will drop by several thousand meters as they evaporate into space. The loss of the oceans is likely to bring about the cessation of plate tectonics with the ending of the subduction process. Our solar system has moved from the Sagittarius arm where it was formed, to more than half way towards the Perseus arm, which it will cross during the Intergalactic Era in 3.5 BYF. There will also be a collision and distortion of the Milky Way and Andromeda galaxies in 5 BYF, before our Sun burns out in 6 BYF.

It has been explained above that there is possibly a maximum evolutionary period of seven billion years in our galaxy for advanced species that are not radiation resistant. The more advanced astro-civilisations would be in the inner blue band over 10,000 light years away, so that the first interplanetary radio signals may be expected to arrive within 10,000 years time by the end of the Planetary Era. If a Supercluster such as Hercules, with 10,000 galaxies at a distance of 300 million light years, is a probable source of signals from an advanced astro-technological civilisation, then contact is likely to be made in 300 million years time at the end of the Galactic Era. Finally, if radio signals were transmitted five billion years ago from the oldest and most advanced galaxies near the cosmological horizon ten to fifteen billion light years away, the species on our planet are likely to be destroyed before for any messages could be received in the Intergalactic Era in 5-10 BYF.

Astro-species are likely to have evolved from bacteria, which are highly radiation resistant and can survive in a vacuum at extremes of temperature. They can therefore be transported across interstellar space within cosmic debris or asteroids as the early colonisers of planets. However, their subsequent evolution is likely to be vastly different from those on Earth, as a result of alien environments and mass extinctions in the past, as have occurred on Earth. Surviving astro-species may possess a different range of cognitive facilities than the humans, animals, birds, and marine species on Earth. Communications with intelligent astro-populations, or extraterrestrial intelligence (E.T.I.), may be achieved in the late Planetary Era to create a galactic cyberspace with a population of twice that of the Earth at twenty-five billion. This may then be extended by interstellar space probes and robotics to other civilisations within our own galaxy to achieve an ATC population of one trillion by the time of the Galactic Era in 10 MYF. However, if it is possible to transmit and receive signals to 100 of the 100 billion galaxies (i.e., one per billion) by the Intergalactic Era, the ATC population would reach 100 trillion. Ultimately, the ATC population may be measured in quadrillions (1,000 trillion).

The metasystems and deep futures: Chronological references

1. *A Study of History* - Arnold J Toynbee, 1962.
2. *Sociobiology* - Edward O Wilson, 1975.
3. *The Next Ten Thousand Years* - Adrian Berry, 1975.
4. *Ice Ages* – John & Katherine Imbrie, 1979.
5. *Atlas of World History* - Times Books, 1984.
6. *The World as a Total System* - Kenneth E Boulding, 1985.
7. *Past Worlds Atlas of Archaeology* - Times Books, 1988.
8. *The Collapse of Complex Societies* - Joseph A Tainter, 1988.
9. *The Cambridge Atlas of Astronomy* - Edited by Jean Adouze & Guy Israel, 1988.
10. *The New Atlas of the Universe* - Patrick Moore, 1988.
11. *A Wonderful Life* - Stephen J Gould, 1989.
12. *The Economics of Global Warming* - William R Cline, 1992.
13. *The Diversity of Life* - Edward O Wilson, 1992.
14. *Ecology* - Peter D Stiling, 1996.
15. *Global Environmental Change* - Karl K Turekian, 1996.
16. *Extraterrestrial Intelligence* - Jean Heidmann, 1997.
17. *Earth Story* - Simon Lamb & David Sington, 1998.
18. *Entering Space* - Robert Zubrin, 1999.
19. *Are We Alone in the Cosmos?* - Edited by Ben Bova & Byron Preiss, 1999.
20. *What if the Conveyor were to Shut Down?* – Wallace Broecker, 1999.
21. *Origins* - Ron Redfern, 2000.
22. *The Two-Mile Time Machine* - Richard Alley, 2000.
23. *The Skeptical Environmentalist* – Bjorn Lomborg, 2001.
24. *Cassell's Atlas of Evolution* – Dougal Dixon, Ian Jenkins, Richard Moody, & Andrey Zhuravlev, 2001.
25. *World Atlas of Biodiversity* - Brian Groombridge & Martin Jenkins, 2002.
26. *A Guide to the End of the World* - Bill McGuire, 2002.
27. *Deep Futures* – Doug Cocks, 2003.
28. *Life and Death of Planet Earth* – Peter Ward & Donald Brownlee, 2004.

29. *Maps of Time* – David Christian, 2004.

30. *Surviving Armageddon* – Bill McGuire, 2005.

31. *The Role of Ocean in Climate* – Wallace Broecker, 2005.

32. *Into the Cool* – Eric Schneider & Dorion Sagan, 2005.

33. *Climate Change 2007* - Intergovernmental Panel on Climate Change, 2007.

3.
EMERGENCE

3.1　Metaphors

A metaphor is the imaginative application of a descriptive term or phrase to a phenomenon, for which it is not literally applicable. Fruitful metaphors and analogies are invested with meanings or explanations so that they assist our conceptual system to transmit ideas from one mental arena to another. An example is provided by a conversation between an economist and physicists at the Santa Fe Institute (SFI), from the popular book *Complexity* (1992), by M. Mitchell Waldrop, where Brian Arthur is explaining why economics is not simpler than physics. He pointed out that, in economics, they have 'agents' instead of particles, and these agents react to other agents just as particles react to other particles. In so far as the spatial dimension is not generally considered in economics, it would make it simpler. However, the big difference is that the particles in economics are smart, whereas in physics they are dumb. In physics, particles respond to forces in accordance with universal laws, but in economics, agents act on the basis of expectations and strategies. This is an interesting metaphor in so far as individual members of society could be considered to be smart particles; however, it reveals the source of an immediate weakness in economic concepts, because the urban spatial structure is ignored.

Scientific communities tend to develop their own metaphors, which become systematically encoded within the particular branch, such as mechanics, thermodynamics, hydrodynamics, biology, or ecology. The abstract concepts of systems capture some of the essential features of real-world processes, and scientists such as Maxwell, Volta, and Ampere selected a source system such as fluid mechanics to establish a collection of imaginary properties for their target system involving magnetic and electrical flows. In this way, well-developed theories and mathematical properties from a field that has been well studied and tested in the real world can provide an early description or conceptual framework for developing theories or models for an entirely new system. At the same time, it is important not to overlook essential differences between complex systems, and, for that reason, ideas from physics may not easily be translated to societal systems.

Hypotheses about the civil system may be formed from empirical and statistical data or computer simulations, so that an initial model may be derived from a preliminary set of rules for the phenomena being studied. These early rules eventually evolve into a more complex set of laws that take account of any shortcomings in the initial formulation. Generally, induction is used to infer a set of rules from the specific instances that generate the fragmentary empirical data or the results of artificial simulation, and, in this way, useful internal models are developed for ill-defined problems or situations. Induction is based upon rules of thumb, analogies, the adaption of ideas, the application of experience, evidence on what is working elsewhere, and it does not depend upon logical deductions from the underlying dynamics. As the hypothetical models become validated through the successful prediction of emergent phenomena, the more primitive explanations are discarded.

Inspired research or experimentation is the traditional approach for acquiring an organised set of data from which new phenomena are discovered. The researcher decides what should be included or excluded from study at the outset, although by definition what is likely to be found is generally unknown. However, when the excitement of that 'Eureka' experience from a new insight has subsided, an applied scientist may then have to devote several years of work to providing a full explanation for the phenomena that he has discovered. The aim is to establish the elementary mechanisms that will generate the empirical data or describe the dynamics of the system. This phase may involve substantial desk research to obtain further statistical information on related changes to the contextual environment and may involve exchanges of concepts and metaphors with scientists working on complex systems in other fields.

Every so often, an innovative reformulation of concepts is made that is so useful in describing real-world phenomena that it becomes an accepted part of the culture. The most significant metaphorical or analogical discoveries are made by applied scientists who are extremely familiar with the building blocks of ideas from their own and adjacent disciplines, so that they can, with effort and imagination, combine and recombine the blocks until they can produce a new explanation or conceptual formulation. Metaphors and analogies enable us to change our perception or point of view, and conceptualisation involves the development of models with assemblies of building blocks to represent different structural elements of a complicated system.

3.2 Models

A model is a coherent set of descriptions about the relevant relationships of some aspect of the world, which is intended to clarify our understanding of a problem or the behaviour of a system. It is a simplified conceptual or physical image that may be used to investigate the implications of making changes to a system without actually implementing those changes or altering the system in any way. A model may take a variety of forms, including a mental image of the situation, a set of statistics or tables, a series of mathematical equations, a computer programme, or a physical simulation. In most systems analyses, models are the methodology used for predicting a future state in which alternative interventions are proposed, together with an assessment of the corresponding consequences. Intuition and mental models are generally inadequate to handle the large number of factors and the causal relationships.

Michael Batty and Bruce Hutchinson (editors) demonstrated in *Systems Analysis and Urban Planning* (1983) how systems models have been used to explore the dynamics of urban systems. However, a partial understanding of the way that societies evolve leads to instabilities in the system, and attempts to relieve one set of problems may create a new mode of system behaviour with unexpected or counterintuitive side effects. It follows that if critical influence points are identified at which the behaviour of the system can be changed, there is still a relatively high risk that the system may be altered with unforeseen or undesirable consequences. It is generally acknowledged that the release of resource constraints and changing the rate of capital investment are sensitive influence points, but it is also necessary to look for potential instabilities and to build resilience into the system. Another characteristic of the dynamics of societal systems is that measures to cure symptoms of a problem appear to work in the short term, but the same action can produce completely different consequences in the long run. In effect, political decisions taken in the context of an election within five years are more likely to favour short-run improvements than policies and programmes aimed at long-term results, which could depress the behaviour of the system initially. Finally, changes that appear good from a micro or local point of view may produce a solution that is inferior from a macro or global perspective.

In *Hidden Order - How Adaption Builds Complexity* (1995), John Holland explains that the modelling of complex adaptive systems such as ecosystems, the urban system, or the economy has common characteristics. Transacting entities such as organisms, households, or establishments survive or face extinction on the strength of their ability to acquire resources within a territory, and evolution involves remembering combinations of useful tactics that increase their fitness or propensity to survive. The entities receive, process, and store information so that they are able to learn and remember, adapt and evolve to a state of maturity before senescence and their eventual demise. The transacting entities use their own internal models of the world to explore future possibilities so that strategies can be selected by the entities to prescribe their transactions and interactions as they progress through their life cycles. The entities are unable to reliably foresee what contingencies will arise or to predict their opponent's strategies, and their inability to anticipate situations will result in surprises. Evolution or the discovery of major new configurations or interactions may create instability or irreversible disturbances such that there is a discontinuity in the system.

The behaviour of a basic transacting entity or unit of analysis, such as an establishment, can be observed at either a lower or a higher level of aggregation. For example, interviews or surveys may be conducted with the divisional directors or managers of organisations, who would become the observational units, and from their responses it may be possible to understand better the behaviour of the establishment. An analysis of an entity from an examination of units at lower levels is termed a structural analysis, and it is associated with philosophies of reductionism or constructionism. When the analysis is undertaken by observation of

Long-Range Futures Research

the interactions or transactions with other establishments at a similar level, the transacting environment within the microstructure is termed a network or trophic web. Where an analysis is undertaken by observation at higher spatial levels of aggregation, such as by civil ecostructure, state, or continent, it is termed a contextual analysis. Where establishments are aggregated by industry, the industrial observations are known as a sectoral analysis. The level at which an analysis is undertaken is sometimes known as the granularity of the analysis, and there is a trade-off between the coarseness of the grain at higher levels for manageability of the information, and grain fineness at lower levels for a more detailed picture.

The complex behaviour of a population of interacting or transacting entities in vivisystems emerges from the bottom up. The macrostructure phenomena that emerge in complex adaptive systems are dependent upon the microstructure transactions. The diversity of individual establishments at the microstructure level can be modelled at the macro level by grouping them in terms of an economic classification, such as business sectors and economic divisions (i.e., primary, secondary, tertiary, and quaternary) or by spatial aggregation into towns, states, and continents. In moving from the micro to the macro level, there is a simplification of reality with some loss of the detailed information. To build a dynamic model of a complex adaptive system, it is necessary to select a level of detail for which it is possible to derive the macrolaws for the trajectory of the system from that level of detail. In effect, the four economic divisions need to be able to reflect the evolutionary properties of economic structure (see Chapter 5). Also, the size of ecopolitan states needs to represent an appropriate level of territorial diversity to reflect both the growth in the number and spatial distribution of metropolises, together with the evolution of geopolitical structure at the continental and global levels (see Chapter 11).

The purpose of modelling an evolutionary landscape is that it should be easier to investigate than the real system. If new macrolaws can be derived to describe the behaviour of the emergent phenomena, without requiring the detail generated by microstructure models, then the macroscopic description of the system behaviour becomes much simpler to understand and to explain. This leads to the search for macrolaws to explain the emergence of ecostructures and to prescribe the trajectory of the urban system. In the real world, situations have a history, which introduces the concept of path dependence, but there is also the opportunity for designed interventions to improve the trajectory of the system. However, the evolutionary features of the model must more than compensate for any impairment to its functioning as a result of any reduction to its complexity or data requirements, when exploring the future state of the global macrosystems.

Conventional economic models tend to simulate precisely the behaviour of systems based upon unrepresentative assumptions, whereas computer models for complex systems may accurately represent initial conditions but are unlikely to anticipate changes in the causal relations over time. Evolving systems are subject to variation by external variables, and, typically, the change is gradual, but occasionally, dramatic change such as the appearance of a new species or sector, or a catastrophe, punctuates dynamic stability. Only in the final phases of maturity will the system achieve a dynamic balance, and even then, some parts will develop and others will decline. The range of possibilities for the emerging system is too vast for transactional entities to recognise an optimum solution. This new scientific paradigm treats the world as a complex system, dynamic and unpredictable in detail, in which the system is always unfolding and always in transition.

However, one of the interesting features of models of nonlinear systems is that the emergent behaviour can sometimes be captured in a relatively small number of equations. It is unnecessary to model the millions of transactions that take place within a region, since it is towns and cities that evolve, and predictions can be made of the future number and classification of establishments. State governments and local authorities collect establishment data for both employment statistics and property taxation, and either official sources or empirical research can provide both historical and initial conditions for the number, size, and sectoral classification of establishments for a city region. Modelling the evolution of an artificial landscape, such as the world's system of cities, is more like making long-range predictions for a relatively stable environmental characteristic such as climate, rather than an erratic and unpredictable feature such as weather.

In the allocation of funds for capital programmes such as municipal infrastructure, transportation, housing, education, or health, etc., many projects are in an early conceptual or planning phase. Planning factors are used to determine the number and sizing of facilities, and parametric cost models are used to provide order-of-magnitude cost estimates in the early stages, with successive improvements in the accuracy as more detailed project engineering information becomes available. In system-based cost models, each of the engineering systems is described in terms of parameters such as the area, length, or size of the system, its capacity, or its output. The costs are generated in a parametric equation by quantification of the resource requirements in terms of the selected parameters and the application of resource unit costs. The allocation of default values for the parameters enables a preliminary cost estimate to be made prior to the specification of the technology for each of the engineering sub-systems.

3.3   Paradigms

A scientific paradigm is the accepted wisdom of a specific discipline, which includes its concepts, theories, laws, models, and applications that are used for the explanation and prediction of phenomena that reinforce the theory. In due course, anomalies are discovered, which the existing body of knowledge cannot readily explain, and so, initially, these inconvenient facts are set aside or ignored and treated as bad data, trivia, noise, or interesting exceptions to the general rule. However, with time, a new conceptual framework or model is proposed that may highlight further anomalies that had remained unrecognised within the reigning paradigm, which may then be dethroned and a new paradigm comes into being.

Paradigms emerge from the application of 'The Scientific Method', which involves observation, hypothesis, the discovery of a mechanism, and the development of a model that will assist in predicting the evolution of a system. However, the explanation of a set of phenomena leads to posing new problems that emerge with the cumulative advance of science. In solving these new problems, the extension of the science may lead to a reformulation of the underlying principles, which is termed a paradigm shift. Scientific research is undertaken to enhance the understanding of evolving new fields; however, there may come a point when the order of magnitude of the change in intellectual outlook or explanation is so great that it may require a new science. These events in the history of science fall within the classification of scientific revolutions by Thomas Kuhn in *The Structure of Scientific Revolutions* (1962).

A geophysical example of a paradigm change concerns the German scientist and meteorologist Alfred Wegener (1880-1930), who published a book in 1915 with a title that translates as *The Origin of Continents and Oceans*. He proposed that, in the past, there was one enormous supercontinent known as Pangea, that subsequently broke up, and the fragments drifted apart to their current positions. This permitted the oceans to flow into the gaps between the separate continents. His evidence was the close fit of the shapes of the continents, together with their identical geological formations along the adjacent edges. The metaphor he used for describing these phenomena was a comparison with the refitting of torn pieces of a newspaper by matching their edges and then checking that the lines of print run smoothly across. Alfred Wegener was thoroughly discredited by the eminent geophysicist Sir Harold Jeffreys who thought that his ideas were nonsense. Wegener lost his life at age fifty on his fourth major expedition to Greenland, whilst attempting to cross from the central ice cap to the base camp on the west coast. He was unable to demonstrate from surveying measurements that Greenland was moving westward relative to Denmark, and he died with his credibility destroyed. Mainstream geologists ignored the evolutionary evidence, and the adoption of his theories had to wait until the 1960s, when oceanographers had mapped the sea floor, to identify the driving mechanism as a spreading mid-ocean ridge of volcanic activity.

The Serbian mathematician and meteorologist Milutin Milankovitch (1879-1958) provided another example when he published a paper in 1914, entitled the *Astronomical Theory of the Ice Age*. This idea was first advanced in 1867 by James Croll (1821-1890). Milankovitch based his theory on the fact that there are slight changes in the geometry of the Earth's orbit around the Sun, which varies from a nearly circular path to one that is more elliptical in a cycle of approximately 100,000 years, with equally slight changes in the inclination of the Earth's axis and its angle of tilt to the sun. Since the intensity of sunlight received each year varies with the Earth's distance from the sun, ignoring fluctuations in the output of the sun from year to year, the Milankovitch cycles force climate change and can accurately predict the glacial and interglacial periods within an ice age. Until the 1970s, there were serious doubts about the validity of the Milankovitch theory, but then scientists were able to compare the climatic oscillations revealed in deep-sea cores with the amount of solar radiation calculated by Milankovitch. The present Cenozoic Ice Age is expected to last another 20 MY (million years), during which Milankovitch Cycles will continue at a frequency of about ten complete cycles per million years. However, it is the movement of land masses to the polar regions during continental drift and changes in the flow of ocean currents, that change the Earth's climate regime into and out of an ice age and its susceptibility to astronomical cycles.

In the course of scientific progress, there have been numerous examples of paradigm shifts in the beliefs of the specialist practitioners in a discipline. This creates the need for a change in representation and taxonomy, or new mathematical tools such as trigonometry, coordinate geometry, calculus, mechanics and, now, complexity science, to improve the prevailing model. The dissatisfaction with an existing model often arises from practitioners, engineers, technologists, and innovators, which motivates academic research into new theories or techniques, and occasionally defines a new science. Scientists acknowledge that, in applied sciences, idealised abstractions such as 'perfect gases' and 'homogeneous' media are used to simplify the complexities of the real world. Similarly, in economics, the concept of 'perfect information' was adopted prior to the idea of bounded rationality. By means of these simplifications, different disciplines endeavoured to develop a simplified model or Laplacean description in which a set of equations can be formulated to describe the behaviour of the system over time.

The unity offered by classical science and neoclassical economics was based upon the predictable model of celestial mechanics, in which the system had reached equilibrium in space-time at a mature stage of evolution to achieve imperfectly determinate solutions for a local condition close to equilibrium. This provides a false sense of predictability about the system, and there is a reluctance to adopt an evolutionary perspective for fear that economics will somehow become unscientific if it loses the facility for prediction. However, economic theory is an assemblage of a large number of building blocks that incorporate hard-won concepts and distillations of ideas since *The Wealth of Nations* was published in 1776. The real anomalies in economic theory can only be fully perceived after compelling explanations are provided within a new conceptual framework.

Four Thousand Years of Urban Growth (1987), by Tertius Chandler, documents the historical evolution of cities, which has resulted in the irreversible development of spatial structure through the accumulation of capital stocks that has caused constant change and disequilibrium. Economists Alfred Marshall, Colin Clark, Joseph Schumpeter, Walt Rostow, Kenneth Boulding, Richard Nelson, Sidney Winter, and Brian Arthur have each contributed to an evolutionary economics paradigm, which departs from the metaphors of mechanics used in neoclassical economics. Inevitably, in the further development of these pioneering ideas, which have spanned a period of some fifty years, other writers and insights from fresh empirical research have contributed to conceptual advances in dealing with complex adaptive systems.

Alfred Marshall observed the organic nature of economic development in his *Principles of Economics* (1890), and he acknowledged the value of biological concepts, although they are more complex than economic dynamics. From his investigations of increasing returns, he observed that movement along the long-period supply curve is irreversible, although he frequently made use of the static concept of equilibrium. In 1940, Colin Clarke published *The Conditions of Economic Progress* and, in his classic analysis of the relationship between economic growth and changes in the structure of the labour force, he was one of the first people to observe that the economics of the services or tertiary division had yet to be written. Many of his contemporaries were reluctant to admit that the service industries ever existed. Why is it that, today, exactly the same situation exists with the quaternary information division of the economy? Traditionally, business establishments have been classified by their end products, and it is necessary to redefine the tertiary division as a commercial division before the quaternary division can be defined.

In the *History of Economic Analysis* (1954), Joseph Schumpeter points out that every scientific venture starts with a "pre-analytic" vision, in which a distinct set of coherent phenomena is selected as a worthwhile object for our analytic effort. Research commences with a study of the work of our predecessors, with a view to identifying the original work that needs to be done to acquire the missing pieces in the mosaic of our vision. Concepts are then formulated with their logical relationships to provide a complete system of logic that generates a new model for thinking about the future of world development. Joseph Schumpeter provided the insight in *Capitalism, Socialism and Democracy* (1942) that development was a process of creative destruction and that clusters of technological innovation are responsible for the longer-term economic fluctuations. An extension to the analysis reveals the importance of infrastructure investment in canals, railways, electricity distribution, motorways, air transport, satellite communications, and technology-intensive office developments as a causal factor in long waves. In *The Stages of Economic Growth* (1960), Walt Rostow provided the concept of stages of development as a simplification of some extremely complex processes, with leading sectors being major innovation cycles in their growth phase. However, if the age of high, mass-consumption corresponds to a distribution stage, will the future stages be informational, ecopolitan, and in due course a planetary civilisation?

Kenneth Boulding, former President of the American Economics Association, laid the foundations for evolutionary economics in *Ecodynamics – A New Theory of Societal Evolution* (1978). He realised that one of the most important aspects of any form of human knowledge is taxonomy or classification. In particular, he was concerned that, throughout a large part of the history of economic thought, there has been a tendency to confuse stocks with flows and also a tendency to pay too much attention to flows and not stocks. For example, in *The Wealth of Nations* (1776), Adam Smith identified wealth, which is a stock, with the value of the annual produce of a nation or the national revenue and, ever since, economists and governments have used Gross National Product (GNP) or Gross National Income (GNI) per capita as a measure of wealth. Income is to capital what births and deaths are to population. Real capital includes household capital, such as houses, furniture, appliances, cars, etc.; social capital such as schools, hospitals, and public amenities;

industrial capital that includes factories, machines, equipment, and trucks; and civil capital that includes utilities and transport systems. There are also finance and financial instruments, which are rights to purchase goods or money, such as coins, bank notes, loans, bonds, stocks, shares, and futures contracts. These are very important for determining the distribution of the total value of real capital, which is the total net worth, within society.

Nelson and Winter have developed evolutionary models of technological change, which is a dynamic process, and the disequilibrium dynamics are interesting and capable of analysis. The system may only be close to equilibrium for a small portion of time. Their work is published in *An Evolutionary Theory of Economic Change* (1982), which led to a surge of interest by economists in evolutionary models. The geographical and spatial distribution of successive technologies displays patterns similar to those found in the succession of biological species in ecosystems. Brian Arthur published his research into positive feedback mechanisms in *Increasing Returns and Path Dependence in the Economy* (1997), and further ideas were developed from his work at the Santa Fe Institute in New Mexico on applications of the complexity sciences. The Santa Fe Institute publication, *The Economy as an Evolving Complex System* (1997), edited by Arthur, Durlauf, and Lane, explains the perspective of the economy operating far from any optimum or global equilibrium, and techniques are being developed to deal with out-of-equilibrium dynamics. A recent publication, *The Origin of Wealth – Evolution, Complexity and the Radical Remaking of Economics* (2006), by Eric Beinhocker, provides a survey of the ideas that are reshaping economics.

Evolutionary economics does not answer the question 'What evolves?' and even Kenneth Boulding felt that what evolves is something very much like knowledge. Instead of defining an analogue for the species, contemporary evolutionary economists have concentrated on the process of natural selection by exploring the algorithms behind artificially limited simulation models that explain the dynamics of competing technologies. Evolution of the civil system involves increasing complexity, which arises from enhanced information and hierarchical structures, and increased connectivity between both urban centers and establishments. The process of evolution realises potential, and development of one system or subsystem may generate new potential in another system, which leads to further evolution. The concept of equilibrium as the full realisation of potential in a vivisystem is only a temporary phenomenon like a climax in an ecosystem, and a natural disaster or creative destruction will open a new path for evolutionary development.

Controversial ideas that challenge conventional and widely held notions have to be convincing and fully substantiated, as the logic and any analogies will be thoroughly scrutinised. Analogies enable connections to be made between different scientific disciplines, and they guide investigators in where to search for relevant phenomena and their explanations with respect to the system under study. Facts alone do not overthrow theoretical concepts, and so an alternative paradigm must be put forward that both exposes and resolves any anomalies in the existing theory. The model therefore has to explain an ample number of facts, and the pragmatic tests for appraising a scientific model are both the 'goodness' of its fit and the ability to predict. A new paradigm is a major step in the advance of a science, and it was Sir Karl Popper who stated that, although science may not be provable, it is improvable. The initial evaluation of a new paradigm will include examination of the relevant evidence and verification by physical observation, and a model that stands up to these tests can be accepted on the basis of a comparative assessment against competing models, where it represents an improvement in the current state of the art. However, if a model repeatedly fails to predict, it may be falsified and require modification, so that in due course a superior model may be developed to overcome any shortcomings in the proposed new paradigm.

3.4   The Evolution of Spatial Structure

Economics Nobel Laureate Paul Krugman has shown his concern for the neglect of spatial structure in economic theory by pointing out that the world is explained in terms of forces that they know how to model. They have generally ignored spatial structure because it involves increasing returns and imperfect competition, which are harder to model than constant or diminishing returns. In his book, *Development, Geography, and Economic Theory* (1995), Paul Krugman points out that a principal economics textbook such as the well respected Joseph Stiglitz's comprehensive 1100 page *Economics*, contains no reference in the index to location or spatial economics and the single reference to cities relates to rural-urban migration in less developed countries. Also, the popular 900-page textbook on economic principles by William Beaumol and Alan Binder contains no reference to cities, location, or space in its index.

Mainstream economists ignore spatial structure, because it does not fit the model of Competitive General Equilibrium theory, although it is acknowledged by leading economists, such as Nobel Laureate Kenneth Arrow, that a number of the assumptions in the General Equilibrium model are false. Economists working within the orthodox theoretical framework have long recognised that there might be multiple equilibria, and that a particular equilibrium or equilibria may not be stable in the face of perturbations. But these issues threaten the underlying intellectual framework, and they have been set aside until there are compelling reasons to deal with them. For the engineering and scientific communities, settlement history, path dependency, and endogenous growth are convincing enough reasons that an evolutionary model with non-linear dynamics needs to replace the equilibrium model. However, until a new conceptual framework is presented that both exposes and resolves the anomalies in the existing theory, economics cannot undertake the transformation into a science.

Classical economics takes place in an unreal world without spatial structure, because this is inconsistent with the achievement of equilibrium. Spatial structure introduces locational interdependence with multiple equilibria solutions, and the cumulative and progressive change from the transformation of investment capital prevents the economy from equilibrating. Also, until recently, mainstream economists have resisted informational explanations of agglomeration economies and the concept of a Quaternary Information Division (see Section 5.2), since this is upsetting to the idea of perfect information on which the equilibrium theory is based. Kenneth Boulding (1963) remarked that "the very concept of a knowledge industry contains enough dynamite to blast traditional economics into orbit." Equilibrium seeking models, which include the General Equilibrium model of economics, incorporate feedback mechanisms for modelling system dynamics, but the range of validity is restricted to the near-term to limit the impact of longer-term nonlinear dynamic change.

Since 1990, there has been a resurgence of theoretical and modelling work on the spatial aspects of the economy, which is described in *The Spatial Economy* (1999), by Masahisa Fujita, Paul Krugman, and Anthony Venables. In this publication, the authors state that the defining issue of economic geography is the need to explain the concentrations of population and of economic activity, such as the distinction between the manufacturing belt and farm belt, the existence of cities, and the role of industry clusters. The reader of *The Spatial Economy* only has to read to page 6 of the book to discover that economists' historical unwillingness to address issues of economic geography was mainly due to a sense that these issues were technically intractable. Therefore, the authors are only mildly apologetic that their analysis has an air of unreality as a result of modelling tricks, which reveals the difficulty in developing a mechanism for the evolution of a hierarchical urban system in a full, general equilibrium model. Krugman has pointed out that any theory on the size distribution of cities will need to explain the empirical law, referred to as Zipf's rank-size rule, which has been valid in countries such as the USA, Brazil, and India despite changes in the number and average size of cities (see Chapter 9 – Section 9.3). This has not yet been explained by any plausible economic model.

Ecodynamics is the term used in *Long-Range Futures Research* to describe the evolution of spatial structure in the civil system, together with the changes in the population of artefacts, which arise from the interactions between households and the transactions with and between establishments. The adoption of Ecodynamics as an economic model would represent the type of paradigm shift described by Thomas Kuhn in *The Structure of Scientific Revolutions* (1962), but although it may be perceived that there are increasing limitations to existing theory, the alternative has to be demonstrated as superior. One of the lessons from the Alfred Wegener case is that, even when convincing evidence is presented, the new perspective is generally ignored until the scientist can explain the mechanism by which the phenomena arise. In Chapter 4, the complete mechanism for the organic nature of economic development is explained. Empirical evidence for the mechanism involved the alignment of the spatial structure from the London Megapolis Regional Information System that included 11,000 establishments of over fifty employees (see Chapter 5) with an independent databank of institutional property investment returns (see Chapter 5 - Section 5.5).

The origin of the term 'economy' is the Greek word 'oikonomia', which means literally the management of a household or an establishment, such as a small business or shop. From the 7th century BC coins were brought into circulation in Greek cities, which enabled the archaic oikos or household of merchants to grow into establishments with as many as 100 or more mercenaries or employees. It is quite consistent to simplify reality in economics by the aggregation of individual humans into households, so that households and establishments become the basic transacting entities, and firms or corporations are merely an aggregation of establishments. These aggregations do not result in any significant loss of detailed information, compared to the selection of the disparate levels of the individual and the firm in the neoclassical model. However, the benefit in selecting households and establishments as transacting entities is that they can be aggregated vertically to form business sectors and economic divisions, or horizontally to form residential and business districts with further aggregations into towns or cities and states. This change in representation makes the spatial structure obvious in Ecodynamics, although it has always been true but ignored in the neoclassical model. The urban structure of cities can therefore be defined as a collection of transacting entities, and a fundamental property of these entities is that they are located in space and time. Smart human agents provide the transacting entities with their mental capabilities for goal setting, prediction, and adaption to other transacting entities and hazards as explained in Chapter 4.

An evolutionary economics paradigm is now emerging with nonlinear dynamics, which is more representative of the real world than an equilibrium model. The evolutionary and cumulative disequilibrium caused by endogenous changes does not fit the neoclassical theory that the economic system is ahistorical and only adapts to exogenous change. In fact, neoclassical economics has difficulty in dealing with a number of issues raised in this research, such as the constraints of the biosphere, the concept of settlement history and path dependency, spatial structure, and the disequilibrium produced by regional investment, the distribution of high technology industries in clusters, increasing returns in the growth phase of the life cycle for a new technology, and the demand for an increase in informational activities, resulting from the uncertainty of dynamic conditions.

3.5 Complexity Science

When Kenneth Boulding published *Ecodynamics: A New Theory of Societal Evolution* (1978), in which he provided his vision that human artefacts such as vehicles, cities, or nation-states follow the same laws as nature, it had the potential for providing a model of social change for futures studies but, in reality, it had little applicability for problem solving. In order to model long-run change, it is necessary to put spatial structure into the model to transform it into an applied science, and that is the contribution of *Long-Range Futures Research: An Application of Complexity Science.* The evidence in support of the *Long-Range Futures Research* (L-RFR) outlook is set out in each chapter of the book and with a time horizon to 2150 and an evolutionary trajectory to 2250, the approach can out-range global scenario models to 2100 or 2200. L-RFR can provide alternative futures to the models with a 100-year time horizon such as the *Limits to Growth* (1972), the *IPCC Special Report on Emissions Scenarios* (2000), *The Stern Review on the Economics of Climate Change* (2006), and the *International Futures (IF) Simulation Model* (2006). The pivotal point to this book is that these models are dynamic but not evolutionary. The rival *Long-Range Futures Research* prognostication is based upon measurable emergent phenomena that can be verified by physical observation, and this has been an essential ingredient in the classic examples of major turning points in scientific development.

The development of Ecodynamics is following a path that is not dissimilar to the advent of Thermodynamics in the Nineteenth Century. In 1824, Sadi Carnot, a French military engineer, published a popular book on steam engines called *Reflexions sur la puissance motrice du feu,* in which he made the first statement of what was to become known later as the second law of thermodynamics. Although Carnot's work was based upon the then current misconception that heat was some kind of massless fluid or caloric, he pioneered an intellectual revolution. In the 1840s, James Joule laid the experimental foundations for the first law of thermodynamics, and it took until the 1850s for the two laws to be stated in explicit mathematical form by Rudolf Clausius. The first law of thermodynamics states that energy can neither be created nor destroyed. The second law of thermodynamics states that in any processes involving energy transformation, free or high-grade energy ('exergy'), which is a measure of the available work content, is degraded to the bound or low-grade form that is equivalent to heat.

Clausius adopted the Greek word 'entropy' meaning transformation, as a measure of the energetic cost. Entropy is only strictly defined for equilibrium conditions in closed systems, and in all natural and technological processes, the availability of the energy decreases and there is a corresponding increase in entropy. Loosely defined, entropy is a measure of the disorder or chaos in a thermodynamic system. Chaos and gas have the same Greek root 'khaos' as the term is appropriate for a cloud of randomly moving particles that collide and bounce off whatever they hit. A system gaining entropy is also losing information and, when the system reaches maximum entropy and minimum free energy or exergy, it will be in an equilibrium state.

The closed system thermodynamics of Clausius was concerned with the linear behaviour of the system close to equilibrium, and the increase in entropy is a characteristic of the system's capacity to evolve irreversibly in time. However, in open systems, there is a flow of energy or material in the system that creates nonlinear dynamic behaviour far from equilibrium, and dissipative structures form that oppose the disorder implied by the second law of thermodynamics. The Belgian physicist Ilya Prigogine was awarded a Nobel Prize in 1977 for his work on nonequilibrium thermodynamics. Open systems are able to self-organise into complex structures while exergy is being dissipated to create far-from-equilibrium stability; however, if the flow parameter is increased sufficiently, the system moves from organised complexity to chaos. The emergence of complexity corresponds to an increase in information in the system, and the dissipation of energy in an open system replaces the closed system concept of entropy. Information has anti-chaotic properties, which creates order out of disorder in a far-from-equilibrium system.

Nonlinear science focuses on a specific class of behaviours encountered in many different contexts. In a nonlinear system, the combination of elementary actions can introduce dramatic new effects through positive feedback mechanisms. This can give rise to unexpected structures and events whose properties can be quite different from those of the elementary laws, in the form of abrupt transitions or an irregular and markedly unpredictable evolution in time and space, referred to as deterministic chaos. For example, by applying energy in the form of heat to a highly ordered and solid structure such as ice, a phase transition occurs and water molecules that were rigidly locked into a crystal lattice become vibrant and form a fluid (water). If the energy parameter is raised by increasing the heat, then a further phase transition arises when the water boils and forms dissipative structures on the surface before the water vaporises into steam, where the molecules are in chaos. The self-organising dissipative structures of nonequilibrium systems studied by G. Nicolis and I. Prigogine in *Exploring Complexity* (1989) are at the lower end of complexity.

Nobel Laureate Philip Anderson describes complexity science as the science of 'emergence', where large ensembles of interacting entities exhibit collective behaviour that is remarkably different from what might be expected by scaling up the behaviour of the individual entities. A physical example that he put forward is the emergent property of the liquidity of water, which is in no sense an extrapolation of the liquidity of individual water molecules. When hydrogen burns in oxygen, the populations of hydrogen and oxygen molecules diminish and that of water molecules increases. The evolutionary sciences have one thing in common, and that is the emergence of spatial structure or hierarchy. The spatial structure of cosmic evolution manifests itself through the planets, stars, galaxies, galactic groups, clusters, and superclusters, right up to the encompassing universe. In the case of geology, spatial structure arises from the principle of superposition, which states that deeper undisturbed rocks are older than the strata that lie above. Geological hierarchies are established by dividing time into intervals such as ages (10^5 years), epochs (10^6 years), periods (10^7 years), eras (10^8 years) and eons (10^9 years). In the biological metasystem, the elementary particles, atoms, molecules, macromolecules, organelles, cells, tissues, organs, organisms, and species are localised spatially within the larger structure that extends to ecosystems and biomes. Spatial structure in the civil system emerges from the combination of households and establishments into districts, towns and cities, regions, states, nations, and continents on the planet.

Ecology is the branch of science that deals with the interactions of living organisms with each other and with the physical and chemical components of their enveloping environment. In living systems, evolution arises from the ecological or 'quasi-ecological' interactions and transactions between entities of a complex adaptive system or vivisystem. Information gives an advantage to an agent or organism and sets in motion a symmetry-breaking process with the emergence of species. Asymmetry in information or energy creates gradients that are essential for information or energy flows, and it is the flows that initiate change. Evolutionary change results from the interplay of genetic variation and natural selection, with the diminution of some species to extinction and the expansion of others under constantly changing parameters in the environment. Genetic information flows unidirectionally from the genotype to the phenotype, so that in biology, random genetic variations are blind to their phenotypic consequences. There will also be mutations and an accumulation of genetic changes that lead to diversity.

The ecological energetics of living systems with their diffusive ecosystems differ from the non-living dissipative structures, and increasing diversity and emergent complexity arise in an ecosystem through the life cycle phases of inception, growth, maturity, and senescence in the organisms, species, and the ecosystems themselves. It should be recognised that for any living organism, the state of equilibrium corresponds to death. The literature on ontogeny and phylogeny indicates that development stages are the most satisfactory basis for describing the relationships between the increase in complexity and organisation of societies, lineages, and descendants. Evolutionary lineages go through the same sequence of youth, adolescence, maturity, senescence, and death as the individuals. In *The Cycle and Meaning of the Existence of Humankind* (1994), V. N. Kompanichenko explains that ontogenesis can provide the biogenetic relationship between the life cycle of a human and the evolutionary cycle of humankind, so that a correlation can be made with respect to rates of physical growth, levels of intellectual development, and the stage of maturity for civilisation.

Bounded territories provide closed environments, and a law of atrophy governs the population dynamics of humans, establishments, artefacts, animals, or plants, so that growth follows the logistic curve derived by Verhulst. Population dynamics cause the instability that drives evolution, and this generally arises in conditions that are not in equilibrium. Evolution is a mixture of continuous evolution (gradualism) and discontinuous events (bifurcations) that are generated by irregular instabilities with short-term chaotic dynamics. The many populations in one ecosystem may begin in a state of relatively stable balance until one species gains an advantage and enters a stage of accelerating growth until the limits of the advantage are reached and the growth rate declines. Eventually, a new phase of relative stability and balance is reached, in which the population tends toward a steady-state ceiling or a condition of very slow growth. The cyclical pattern of stability-acceleration-deceleration-stability is characteristic of the growth of populations within a bounded state. The traditional view in ecology was that communities were on a trajectory towards a climax in which they reached equilibrium and available niches determined species diversity. A major shift in ecological thought away from equilibrium stability arose with the advent of powerful personal computers, which permitted the exploration of disequilibria with nonlinear equations.

It is important to avoid the confusion that arises from taking the concept of 'self-organisation' from the physical science of nonlinear thermodynamics and applying it superficially to vivisystems or civil systems. In a closed physical system, entropy is defined as the lack of available energy to do work, and entropy increases as energy becomes exhausted in an equilibrium system. In 'real world' problems, the planet Earth is an open system in which biological evolution is powered by the sun and vivisystems are governed by the law of atrophy. Economists describe these atrophic forces by means of the concept of diminishing returns, whereas environmentalists imply that human societies transform food and energy from the environment and misuse the entropy concept by using this as the explanation for the emergence, growth, and eventual decline of civilisations.

In addition to the examples outlined above, evolution arises in chemical reactions, technology, adaptive computation, artificial life, and economic development, etc. Complexity science is not limited to biology, and preconceived ideas based on Darwinian biological evolution are not necessarily applicable to societal evolution. A crucial difference between *The Origin of Species* (1859) and pre-Darwinian evolutionary theories was the new perspective that there is no plan to guide and direct the biological evolutionary process and there is no foresight or goal in biological evolution. The Darwinian view was that random mutations and non-random natural selection are responsible for the continuous and gradual emergence from primitive origins of more elaborate, better-adapted, and more specialised organisms. In the newer theories of evolution by Stephen Jay Gould and others, evolution occurs when a dominant species is destabilised in its habitat by contingency events or by a new species that may have developed at the periphery, and there is a discontinuity or bifurcation point with an evolutionary leap. The dynamics of evolution apply to ecosystems in which the living species interact with the natural milieu. In societal evolution, individuals and organisations develop complex visions of the future, and this encourages foresight and teleological or goal-directed behaviour in which non-random evaluations are made of competing images. A decision is a choice amongst competing images of the future, which are considered to be realistic at the time.

In ecology, the methods and models of nonequilibrium thermodynamics have been applied to ecosystems in which 'exergy' is the system growth parameter that drives the system away from equilibrium and is destroyed in the irreversible process of evolution. Ecosystems emerge from the dissipation of solar energy gradients and, in this way, nature creates order from disorder in an open system that may be far from equilibrium. For example, see *Complexity and Thermodynamics - Towards a new ecology,* by Eric Schneider and James Kay, an article in a special issue on complexity in the journal *Futures,* Volume 26, Number 6, July/August 1994. Also, see *Into the Cool* (2005), by Schneider and Sagan. Ecosystems develop to enhance their ability to utilise the exergy, and the larger biomass of mature ecosystems degrades the solar energy gradients more effectively than less mature ecosystems. However, new diffusive pathways emerge, as less mature ecosystems evolve with an increase in biomass from biological growth, greater diversity of organisms, and an increasing trophic web.

Critics of the thermodynamic approach suggest that biological evolution is a more complex, multifaceted, "survival enterprise", and, as explained previously, population dynamics are also governed by a law of atrophy. These concepts provide a useful analogy for Ecodynamics, in which investment capital is the system-growth parameter. Civil ecostructures diffuse investment capital to create endogenous growth with an increase in the asset stock (equivalent to biomass), so that the concept of far-from-equilibrium stability challenges the notion of economic equilibrium.

Long-Range Futures Research: An Application of Complexity Science provides a proper explanation for the evolution of civilisation from the analogical realisation that investment capital has the equivalent properties to 'exergy' in an ecosystem and that investment flows drive the economic system away from equilibrium. The civil system is an open system, since it cannot be isolated from the interacting dynamics of the other systems, and this leads to an economic explanation for diffusive civil ecostructures (towns and cities) that is based upon the concept of far-from-equilibrium stability. Complexity science explains the Ecodynamics of complex adaptive systems, such as ecosystems or ecostructures, when more energy or investment capital is available than is being used. Evolution involves a process of spatial organisation that results in an increase in complexity to enhance the diffusion of energy or investment in the system. Investment capital is the system growth or control parameter and, by controlling the rate of flow of investment into a region or an ecostructure, economic decline can be reversed, or a stagnant economy can be made increasingly vibrant. In effect, the investment parameter determines how stagnant or chaotic the system becomes. In the far-from-equilibrium hypothesis of Ecodynamics, the composition of ecostructures is viewed as constantly changing and never in equilibrium, and this change in taxonomy actually changes the science so that Ecodynamics adopts evolutionary principles and opens up a new path for economics. By modelling the emergent phenomena, planners and engineers can get on with their work by the application of Ecodynamics, which is a significant improvement on the existing model. However, since evolution requires long periods of time, the science of complex adaptive systems and dissipative structures may be less suitable for modelling near-term social change than models involving equilibrating or homeostatic mechanisms.

It is the growth in the populations of artefacts that creates concerns for sustainable civil development. The evolutionary model follows a logistic S-curve with declining rates of economic growth. The civil system dynamics expressed in the laws of Ecodynamics (see chapter 7) provide a scientific explanation that is different from Adam Smith's concept of "the invisible hand" in relation to the unintended benefit achieved by society from the freedom of individuals to pursue their own ends. A new conceptual framework is developed in the following chapters of *Long-Range Futures Research* to enable readers to understand the dynamics of development and the linkages between the major systems that will impact upon a planetary civilisation. The evolution of civilisation creates order out of chaos, with adaptions to the design through invention and discovery, and humankind is motivated to realise a state of higher potential to counter the atrophic forces of decline.

3.6 Civil emergence

The evolution of the civil system is driven by the transactional dynamics of urban investment flows, and the patterns that form are diffusive civil ecostructures (towns and cities), which have long-term stability within a continental urban hierarchy. In *Nature's Metropolis: Chicago and the Great West* (1992), historian William Cronon distinguishes between the original natural landscape and the artificial landscape created by urban evolution. The emergence of the metropolis required that a new human order be superimposed on nature until the two became completely entangled. The result is a hybrid system that conceals its long-standing debts to the natural systems that made it possible. The entire process, which is at least as artificial as it is natural, can be termed second nature in so far as it is as organic and evolutionary as Darwin's model of biological change.

Long-Range Futures Research is concerned with the future evolution of civilisations, and cities are used to provide an historical and future continuity. Since over two-thirds of the world's metropolises with over 1m population in 2000 were already important cities 200 years ago, and a quarter have been important for at least 500 years, this approach extends our visibility 100-150 years into the future. What actually evolve are cities and civilisations of increasing complexity, which are treated as civil ecostructures for the diffusion of investment capital, knowledge, and culture. Cities evolve over a thousand years and more, with houses that last a century, in relation to infrastructure technologies that go from ten percent to ninety percent penetration within half a century. With long planning horizons and accelerating change in the world around us, it is increasingly important to develop insights as to how the fast dynamics of the transactional microstructure is linked to the longer term adaption and evolution of the global macrostructure. The enormous range of transactions in the microstructure generates hierarchical complexity in the macrostructure, as long as there is a payback in the form of reduced transaction costs from the extra investment in the structural complexity. A civil society has the potential for transforming the planet into an integrated civilisation, and as the potential for it is realised, the remaining potential diminishes unless it is regenerated. The full realisation of potential is a dynamic balance, and whilst some traditional societies remained stable for generations, disequilibrium is the current global condition for civil society.

Ecodynamics describes the emergence of complex civil ecostructures through the diffusion of investment capital, in which the urban system is transformed endogenously and undergoes continuous and irreversible change. Ecodynamics is a more expressive description than the term nonlinear science for its application to the civil system, and it forms the basis of an evolutionary model for economic development. The macroscopic properties of a complex adaptive system emerge from the totality of the co-effects of the interactions or transactions within a collective regime. In evolutionary or complex systems, there is an elaboration of structure as the system adapts to its environment as a condition of future viability. This leads to persistent features in far-from-equilibrium systems that may indicate the form or envelope of the evolutionary trajectory of the system, although it is inherently difficult to predict the path in detail. Civil Engineers design infrastructures and structures with lifetimes that may exceed 125 years, and civil engineering infrastructures shape the urban microstructure and give a degree of permanence to the major transformations in the macrostructure. Unlike other applications of complexity science, civil engineering infrastructures leave a concrete trail that creates a pattern that can be described by macroscopic laws, without having to deal with the intractable mathematics of the underlying dynamics.

The metaconcept of systems science is the existence of hierarchy, such that the different levels in a system are a complex of successively more encompassing sets, with transacting entities, such as households and establishments, then residential, business and industrial districts, with towns and cities, regions, states, nations, continents, and the planet. Each level is more complex than the level below, and it is characterised by emergent properties that are not apparent at the lower level. If the major subsystems within a system are relatively independent of the lower level internal structure of the subsystems, a theory of the system and its behaviour can be formulated at the particular level that is being observed, ignoring both the substructure at the next level down and the longer time scale and more gradual change in the interactions occurring at a higher level.

The development of an evolutionary perspective for the civil system requires recognition at the outset that we are dealing with a complex adaptive system that has emergent properties. The methodology entails selecting a level of detail from which it is possible to formulate macrolaws that describe the behaviour of the emergent phenomena, as described in *Emergence – From Chaos to Order* (1998), by John Holland. The macrolaws for the emergent phenomena are complementary to the microlaws that determine the underlying dynamics, and they are not a substitute for research that provides a coherent explanation for the causal relationships of the microsystems. Since the emergent phenomena arise from the combined effects of both causality and teleology, the macroscopic description encapsulates the resulting evolution of the system without having to apportion the effects to either of these constituent processes. The developmental perspective ('developmentalism') is used by designers, engineers, computer scientists, and entrepreneurs to examine how new phenomena can be created in complex systems that will persist in an evolutionary environment. This is the reverse of scientific reductionism, in which research is concentrated on the micro-level relationships, although it should be noted that the macroscopic description is additional to the laws that determine the underlying dynamics. In this way, realistic evolutionary results can be derived for exploring the future, which has the benefit of being much simpler to understand and explain, with a significant reduction in data requirements.

In the early 19th century, the Institution of Civil Engineers (UK) defined its body of knowledge as the art of directing the great sources of power in nature for the use and convenience of man. This included irrigation schemes, bridges, water supply, sanitation, power supplies, industrial projects, cities, military defences, transport, and communication systems. Engineers and applied scientists strive to develop the necessary theory and technology to provide solutions to human problems as determined by their social importance. The civil system evolves as a combination of causality and purposeful human intervention (teleology), and so at the foundation of futures studies lies both a knowledge of the causal relations between the variables of the system and also an understanding of the teleological process. *Long-Range Futures Research* adopts the view that the prime purpose of the civil system is human survival and reproduction by the creation of artificial communities of households and establishments, together with the use of artefacts to extend both our life expectancy and also the lifetime of a planetary civilisation. Competition for resources results in spatial organisation, which is accompanied by an increase in information that manifests itself in the adaption, coevolution, initiative, imitation, and migration that takes place. As territory is colonised by human populations and their artefacts over time, an escalation of human needs arise that relates to the physical quality of life, such as rising standards of living with increasing per capita assets, social relationships, equitable treatment, comfort, and self-realisation. Since futures research is concerned with exploring and shaping the future, it includes selecting a preferable course for societal evolution in the context of the cultural values that create a vision of desirable futures.

At the macro level, continental unions, civil institutions, political ideologies, and religious beliefs provide continuity factors in cultural change and are slow variables. At the micro level, the household is institutionalised as the primordial unit of society, in which the kinship system requires a stable and intimate relationship between young children and adults over nearly two decades of learning. However, within the institutional and cultural contexts there are rapid changes in lifestyles, consumer tastes, and fashion. Archaeology and anthropology explain the historical evolution of civilisations in terms of the hardware of technology and the software of culture and language.

From a study of history and the rise and fall of civilisations, it is inevitable that some continue to grow with technological progress, whilst others collapse or decline from the diminishing returns of socio-political complexity or exhaustion of their territorial natural resources. Susantha Goonatilake makes the observation in her article *Reconceptualizing the cultural dynamics of the future* (1992) that there are periods of stability when the ruling classes of the existing social structure are able to apply sanctions to groups that question the legitimacy of the existing order, and thereby prevent changes to the system. When it is recognisable to a civilisation that the current state of affairs or the socio-technological system no longer matches the reality of the situation, the sanctioning signals lose their credibility and society achieves a new consciousness. If the societal integrating factors such as culture, beliefs, and values are strong enough to enable the society to adapt to new circumstances, there will be a major shift and reordering of the system, which is termed a societal transition. If the response by the existing elite to the counter-signals from the society is excessively retarded, then a revolution takes place rather than a transition, and valuable parts of the old order of things may well disappear with the bad. Finally, if the warning signals relate to an impending disaster, such as a potential invasion or a famine, then the catastrophe creates a bifurcation point, and the society makes a non-linear jump to a different level of stability. Some may revert back to an earlier state of stability when life seemed simpler, yet others transcend the limitations of the present and transform to a new type of society. The different stages of history corresponded to the socio-technological systems of the time and changes in the political, economic, and environmental conditions.

Strategic thinking by multinational, multi-site, and multiproduct corporations is driven by the objective of gaining access to and controlling both markets and the required resources, with the propagation of establishments to serve different territories. The term *resources* include finance, human skills, facilities and equipment, materials and supplies, and information and communications systems. Continuous growth of the corporation is achieved through the realisation of full potential by strategic business units (SBUs), which are able to generate investment capital at a faster rate than could viably be reinvested in them. Consequently, the new investment is put into the less mature SBUs of the corporation, which can benefit from a much higher investment level than they could possibly self-generate. Diverse corporations allocate resources for the achievement of strategies for sustainable growth and profitable performance in a variety of geopolitical contexts, sectoral markets, and customer segments. Product managers develop differential strategies for products at different stages in their lifecycles from inception to growth, maturity, and obsolescence, since products follow the logistic curve applicable to territorial saturation by populations of artefacts.

A corporation's ability to control its future development by influencing the shapes of its product life-cycle curves is affected by the consumer's reactions, the marketing effectiveness, corporate efficiency, the competition, and technological substitutes for the products. Strategic investment planning involves an evaluation of options and their tradeoffs to determine the most appropriate scale, location, and time phasing for the reallocation of available resources. Tactical reallocations may be made for increments in capacity or for re-equipment to improve productivity. Strategic allocations are made for market research, R&D for new products, and technological innovation, the exploitation of new territories, and entry to new markets. Investments that can be deferred without impairing current market positions or long term growth may remain uncommitted as a potential fighting fund for opportunistic investment or for the defence of market segments, where an adversary's capability and intentions will be unknown beforehand. Strategic intelligence and an understanding of complex adaptive systems should enable a corporation to identify critical intervention points, at which a predetermined resource allocation will produce a designed change to the system dynamics.

The need for a change in the prevailing development model has arisen from the demand for better explanations of observed phenomena, with an improvement in representation and taxonomy by using complexity science as a new tool. Pathfinding in an evolutionary landscape requires a map and *Long-Range Futures Research: An Application of Complexity Science* is a guide that makes accessible an evolutionary systems approach, so that readers can understand the linkages between the complex systems that affect planetary civilisation. *Long-Range Futures Research* provides an answer in the final chapter to the question 'where are we now?' in relation to the life cycle of planetary development.

Emergence: Chronological references

1. *The Conditions of Economic Progress* - Colin Clark, 1940.
2. *Capitalism, Socialism and Democracy* - Joseph A Schumpeter, 1942.
3. *A History of Economic Analysis* - Joseph A Schumpeter, 1954.
4. *The Structure of Scientific Revolutions* - Thomas S Kuhn, 1962.
5. *Ontogeny and Phylogeny* – Stephen J Gould, 1977.
6. *An Evolutionary Theory of Economic Change* - Richard R Nelson & Sidney Winter, 1982.
7. *Systems Analysis in Urban Policy-Making and Planning* - Edited by Michael Batty & Bruce Hutchinson, 1983.
8. *Handbook of Systems Analysis* - Hugh Miser & Edward Quade, 1985.
9. *Exploring Complexity* - Gregoire Nicolis & Ilya Prigogine, 1989.
10. *Nature's Metropolis* - William Cronon, 1992.
11. *Complexity* - M Mitchell Waldrop, 1992.
12. *Reconceptualizing the cultural dynamics of the future* – Susantha Goonatilake, 1992.
13. *Economics and Evolution* - Geoffrey M Hodgson, 1993.
14. *The Cycle and Meaning of the Existence of Humankind* – V N Kompanichenko, 1994.
15. *Complexity: Metaphors, Models and Reality* - Edited by George A Cowan, David Pines & David Meltzer, 1994.
16. *Increasing Returns and Path Dependance in the Economy* - W Brian Arthur, 1994.
17. *Evolutionary Economics* - Esben S Anderson, 1994.
18. *Evolutionary Concepts in Contemporary Economics* - Edited by Richard England, 1994.
19. *Evolutionary Economics and Chaos Theory* - Edited by Loet Leydesdorf & Peter Van den Besselaar, 1994.
20. *Complexity and Thermodynamics* – Eric Schneider & James Kay, 1994.
21. *Hidden Order: How Adaption Builds Complexity* - John H Holland, 1995.

22. *Development, Geography, and Economic Theory* - Paul Krugman, 1995.

23. *Introduction to Nonlinear Science* - Gregoire Nicolis, 1995.

24. *At Home in the Universe* - Stuart Kauffman, 1995.

25. *The Sources of Economic Growth* - Richard R Nelson, 1996.

26. *The Economy as an Evolving Complex System* - Edited by W Brian Arthur, Steven N Durlauf, & David Lane, 1997.

27. *Emergence: From Chaos to Order* - John H Holland, 1998.

28. *The Spatial Economy* - Masahisa Fujita, Paul Krugman & Anthony J Venables, 1999.

29. *The Origin of Wealth* – Eric Beinhocker, 2006.

4.
THE PRINCIPLES OF ECODYNAMICS

4.1   Agents and the classification of households

Interacting human agents are the units of genetic information in a household or an establishment, such that the combination of agents contributes the genetic diversity of the organisation. In effect, the human agents act as genes and they provide the blueprints or genetic know-how that shapes the organisational form and controls its future development. Agents constrain the self-organisation processes to those that have a high probability of success, and, in effect, the establishment develops a corporate culture, which is the accumulated wisdom within a technological regime. An establishment learns what works and what does not within the realities and constraints of an ecostructure connected to the global system, and it develops the capabilities for environmental scanning, modelling and prediction, and adaption to other establishments and sectors (or species). The strongest competition for resources will generally come from other establishments within the region, which drives evolving sectors to occupy distinct niches in the environment and leads to specialisation and sectoral change.

Households are consumer decision-making units that are constrained within an institutional and cultural context. Empirical data on consumer demand generally takes the household as the unit for data collection, although in conventional economic theory, the unit of analysis is the individual, who is assumed to maximise utility from the consumption of goods, services, and leisure time subject to a budget constraint. In principle, the members of a multi-person household achieve greater utility than they could in a single-person household through sharing accommodation costs, household artefacts, and services, and an adult member may choose to leave and form another household if the benefits are greater. This is explained in *The Economics of Household Behaviour*

(1997), by Peter Kooreman and Sophia Wunderink. In Ecodynamics, the household is taken as the unit of analysis, where altruistic behaviour within the family not only provides love, companionship, and care but also the sharing of work, income, and leisure to maximise the household utility.

Households evolve continuously through different life stages as the number and ages of the household members change over time. Household size is dependent upon a number of factors, including the number of births, life expectancy, and household stability as measured by divorce and separation rates, the number of generations within the same household, other relatives, and domestic help. In households with children, a proportion of household income will be used for meeting their needs, such as food, clothing, healthcare, education, and entertainment so that the amount available for adult goods and services will be less than in households without children. In Western societies, there is a tendency for the elderly to live apart from their adult children in apartments, bungalows, retirement homes, and residential care, where their economic independence has been assured through savings, pension schemes, and social security systems until finally they die.

Household lifestyles are affected by the total household income, where levels of earnings are a function of education, training, and work experience. However, the hours of work depend upon prevailing employment conditions, and they may not be the preferred choice of the family. The conditions may result in working away from home, shift work, part-time employment, working in the informal or 'black' economy, involuntary unemployment, or early retirement. Also, married women and females with children inevitably have lower participation rates in the workforce than either married men or unmarried women. Different members of a household create different effects on household demand, and therefore, household income is not the only explanatory variable. In fact, high earnings generally involve longer working hours and possibly commuting from suburban areas to a metropolitan centre, which results in less leisure time for the income earners. This introduces technological choices for households in terms of public or private transport, day-care centres for children, convenience foods, time-saving household equipment, and contractors' services. Household behaviour depends upon both the social group of the household and psychological aspects of consumer behaviour that are discussed below.

In the distributional stage of development in the MDCs (more developed countries), the mass marketing segmentation of households was by socio-economic groups A, B, C1, C2, D, and E with respect to the chief income earner. This classification is based upon occupational groupings with Group A for top-level civil servants, army ranks of colonel and above, top management in industry and commerce, and partners in professional practices. Group B includes civil service executive officers, army ranks of captain and major, middle management, and chartered or qualified professionals. Group C1 includes clerical officers, junior managers, staff nurses, teachers in primary schools, and non-manual technically qualified staff. Manual grade occupations in Group C2 include supervisors and skilled workers that have served apprenticeships, and Group D includes semi-skilled and unskilled workers. Group E is reserved for senior citizens living on state benefits, unemployed workers, and disabled and sick people dependent upon state or welfare payments. Self-employed, non-manual, and manual occupations are classified in a similar way to employed occupations; however, if they are responsible for one to four employees, they are graded one group higher, and for five to twenty-four employees, they are graded two groups higher, and responsibility for over twenty-five employees grades them three groups higher.

Geodemographic classifications give spatial structure to the socio-economic grouping of households within urban centres, in the context of the urban classification by size, class, and category, with postcodes defining the residential areas. The principle behind geodemographic systems is the clustering of residential areas with similar characteristics to provide a first-order indication of household consumer behaviour. This is described in *The New Marketing Research Systems* by David Curry (1993). The demographic information on residential areas is obtained from census data, which is released at district level in terms of the percentage of households with various attributes to preserve anonymity for individual households. Household information is then compiled with the allocation of names and addresses from electoral role data and the classification of house types. This household information is then enhanced with credit information, transaction data from credit cards and bar code scanners from retail stores, and other electronic data capture from utilities, the internet, cable and satellite television viewing, as well as magazine subscription lists, travel bookings, customer loyalty schemes and promotion schemes such as 'Air Miles'.

Advances in information and communications technologies have accelerated the shift from mass-marketing to micromarketing, and from broadcasting to narrowcasting. Proprietary geodemographic systems classify some ten household types into supergroups that represent respectively a notional ten percent of the population. The household information can then be enriched substantially by responses to market and social research surveys, to identify twenty household clusters or groups that can be further segmented into life-stage and lifestyle profiles that correspond to, say, one to two percent of the population. The seven household life-stages are young, single people, single parents, young couples with no children, young couples with the youngest child under five years, couples with dependent children, older couples with no children at home, and older, single people. The table below gives a composite classification of twenty household clusters, in line with the main geodemographic systems:

Classification of household clusters

H.1. Conspicuously wealthy
H.1.1. Exclusive 4/5 bed detached homes
H.1.2. Country dwellings

H.2. High income homeowners
H.2.1. Individual 3/4 bed detached homes
H.2.2. Estate executive detached houses

H.3. Prosperous urban professionals
H.3.1. Prosperous urban enclaves
H.3.2. Town houses and apartments

H.4. Owners in mature neighbourhoods
H.4.1. Spacious semi-detached houses
H.4.2. Bungalows or dwellings in comfortable retirement areas

H.5. Institutional communities
H.5.1. Nursing homes, boarding schools, or monasteries
H.5.2. Military bases and detention centres

H.6. Smaller estate houses
H.6.1. Smaller detached houses
H.6.2. Smaller semi-detached houses

H.7. Private lower income accommodation
H.7.1. Cosmopolitan or multi-ethnic neighbourhoods
H.7.2. Terraced properties or rural dwellings

H.8. Public housing tenants
H.8.1. Low-rise council or Housing Association dwellings
H.8.2. Council flats and sheltered housing

H.9. Overcrowded slum areas
H.9.1. Overcrowded tenements or flats
H.9.2. Sub-standard or deteriorated dwellings

H.10. Serviced sites and low-cost shelter
H.10.1. Mobile homes and houseboats
H.10.2. Low cost or prefabricated housing units

Life-stage and lifestyle profiles are compiled from household questionnaires that obtain responses on household size (number of persons), marital status (including the length of a couple's relationship); age groups in ten-year bands from 15-24, 25-34, etc., to 65+; children's ages (under 2 years, 2-4, 5-9, and 10-14 years); genders, educational levels, employment status and occupation; the number of earners and household income (nine bands); years of residence (mobility factor), and newspaper readership and television viewing. This information is collated with data from surveys, such as the National Family Opinion poll in the US or the Target Group Index (TGI) in the UK, which is conducted from a national sample of 24,000 adults. This is an annual survey relating to several thousand household products, services, media, leisure, and opinions on a variety of issues. Additional financial status information (mortgage, loans, credit cards, savings, investments, and pension plans), sports and leisure pursuits, and motoring and commuting patterns, enable socio-graphic descriptions to be given to household lifestyles and patterns of consumer transactions.

Demographic projections involve assumptions on the total fertility rates for women, which need to exceed an average of two births per woman to provide the replacement rate required to stabilise populations in the long run. In the more developed countries, fertility rates have declined to two or less, with the USA at 1.9, France at 1.8, the UK at 1.6, Germany and Japan at 1.4, and steep falls to 1.2 in Italy and 1.1 in Spain. The United Nations and the OECD assume that total fertility rates will return towards replacement levels by 2025; however, an explanation is required for the reason that fertility has declined during an era of rising living standards. The general consensus is that the investment needed to provide more than two children with an extended period of education, to acquire high-level skills, exceed the benefits. For many families, the preference would be for car ownership and a single child if it came to a trade-off. In addition, it is gender issues and the changing role of women in society that lie behind the decline in fertility rates. Firstly, equal pay for women raises the opportunity cost of the loss of earnings if they leave employment or work part-time, or alternatively maternal care may be replaced by the payment of child minders. Also the contemporary rates of divorce provide younger women with the incentive to maintain their financial independence by becoming qualified and establishing themselves in careers before having any children. Consequently, many women have their first child in their thirties rather than their twenties, and there is a correlation between family size and the woman's age at the first birth, which affects fertility rates. Research studies also corroborate the finding that higher educational achievement by women results in lower average fertility.

Projection of the age profile of a population can be done with a reasonable degree of confidence, because mortality rates are more consistent than fertility rates. Patterns of behaviour are associated with the differences in age structure, and for example, regional or national migration is most prevalent for younger people in their twenties. First time house buyers are predominantly in the age range twenty-five to thirty-four, and at this stage, a twenty-five to thirty-year mortgage is taken out, and consumer durables are purchased on credit. In the age range thirty-five to forty-four, growing families trade up to larger houses as they increase their space requirements, and this is often combined with an outward move from the city centre to a suburban location. At this stage, family responsibilities are reflected in the purchase of life assurance policies and long-term investments in pension plans. Consumer spending and the rate of accumulation of assets tend to reach a peak for forty-five to fifty-four-year-olds, and, commonly, funds become available for direct investment in the stock market. In the age range fifty-five to sixty-four, when the adult children have left home, there are often the first signs of health problems and, in declining industries, early retirements and redundancy packages are offered as companies down-age their workforces. There is a trend for this age group to down-size their properties and to relocate to rural towns and villages where property prices are lower. They tend to purchase a new car and to buy new furnishings for their smaller homes, from which they work in part-time employment and self-employment to maintain living standards for as long as possible. Finally, after age sixty-five-plus, full-time retirement is the norm, and during old age, there is a run-down of a large proportion of the savings that were accumulated during the years of employment.

In *Socio-Styles* (1990), Bernard Cathelat explains that the diversity of household behaviours within socio-graphic descriptions, arises from the personalities of the human agents that comprise a household. Psychographics is the term used for labelling behaviours that can be mapped within a matrix, where the horizontal axis ranges from habitual to exploratory, and the vertical axis ranges from rational to instinctive. Groups with habitual and rational lifestyles are referred to as traditionalist, whereas rational and exploratory lifestyle groups are termed venturers. Instinctive and habitual groups are described as humanistic, whereas instinctive and exploratory lifestyles are labelled sensualist. Each of the four groups – humanistic, sensualist, traditionalist, and venturers have some twenty attributes that are reflected in outlooks that are local, cosmopolitan, provincial, and international and, in respective behaviours, such as the search for security, identity, status, and concepts. Humanistic characteristics include romantic, natural, prudent, economical, and simple. Sensualist traits include stylish, high-touch, liberal, extravagant, and novel. Traditionalist attributes include responsible, material, moral, value for money, and utilitarian. The venturers tend to be enterprising, high-tech, culturally aware, interested in customisation at a price, and can cope with complexity.

Successful transactions and the capture of resources in a complex adaptive system involve the application of all these different patterns of behaviour. In the short term, habitual and rational behaviour will succeed in achieving a predictable outcome from the immediate exploitation of current information. Establishments and households that react to the positive feedback of known supplies of resources will initially achieve higher returns from their efforts until eventually the resources become exhausted. The habitual behaviour of traditionalists leaves them unaware of the opportunities available in new territories, and their strategy is likely to fail in the long term. The exploratory and instinctive behaviour of the venturers and sensualists will introduce an element of randomness or fluctuations into the system from new discoveries and innovations. It follows that for long-term survival, an establishment has to strike a balance between ignoring existing information and the generation of new information from the search for resources and ideas in other territories. A vivisystem will adapt to exploit new discoveries and innovations until, eventually, territorial saturation creates the negative feedback that provides the motivation for further exploration and invention.

4.2   Establishments

An establishment is the basic transacting entity of a human settlement, and it may comprise a single- or multi-person household or organisation. Although industrial plants, businesses, professional practices, and households may be differentiated functionally as producers, traders, informers, and consumers, respectively, all establishments are transacting entities. Establishments are located at facilities and form part of a complex adaptive system in which the environment is altered when establishments relocate to new or existing facilities. Central Government and Local Authorities maintain establishment databases for employment censuses, trade and industry statistics, as well as land and property details for rating valuations as explained by Michael Healey in *Economic Activity and Land-Use* (1991). The establishment embraces the stock of assets and artefacts that it uses in the course of its operations. Agencies that are responsible for infrastructure assets, utilities, transport, or mines, for example, are the appropriate establishments for deployment of these assets.

Establishments are organisms that are highly bounded in performance by their genetic structure, which is an encoded informational process involving the agents. Groups of establishments form clusters or districts at level (N+1) and ecostructures at the town or city level (N+2), in the same way that, in ecology, groups of organisms are members of species that interact to form an ecosystem. Organisms and establishments are economisers in the cost of transactions. These establishments transact with other establishments in the system, anticipating situations and taking investment decisions, to create a complex adaptive system or vivisystem. Establishments may represent different levels in the system, such as regional or national, and the control of a complex adaptive system is highly dispersed. Establishments are classified into economic divisions, business sectors (species), and sub-sectors, to give generic names to each species of establishment, and changes in the economic structure evolve over time. The species are located at specialised facilities, and their organic nature is already in the language with the term industrial plants. Subsidiaries of corporations are known as branches or branch plants, with in-migrants being termed transplants and deaths resulting in plant closures. Part of the adaption process is that establishments change their habitats to improve their well-being, and in relocating to a new or existing facility the environment is altered and another establishment will reoccupy the vacated facility. Over a period of time, clusters of similar establishments tend to inhabit the same district, and this adaptive process also leads to the reproduction of additional establishments at diverse locations. Space is colonised by the propagation of establishments in which numerous city and town centres have identical branches of retailers, leisure facilities, restaurants, hotels, banks, building societies, and insurance companies. This phenomenon is known in the U.S.A. as 'Generica'. The pattern for period of occupancy within the London Megapolis is that fifty percent of establishments with over fifty employees have been in their current location for ten years or less (sixty percent for central London), and thirty-seven percent have been in their current location for over fifteen years (twenty-seven percent for central London).

The pay-off or stimulus for successful transactions is the capture of investment capital (business profit or domestic savings) so that the organism or establishment can grow and reproduce. The combinatorial complexity resulting from the incorporation of establishments into multi-site corporations is a morphogenetic process in Ecodynamics. In effect, the evolving form of a corporation will depend upon the investment allocations made to the individual or subsidiary establishments, and the corporation will control transactions and select establishments for replication and also those for closure or extinction. Mutations in the agents produce differences in the form and function of the organisation and, conversely, the form and behaviour of the organisation will determine the agent changes that have survival value. Mutations may often be adverse, and agents that produce losers in the fight for survival and reproduction will themselves suffer a loss of resources. Evolution involves a process of increasing informational complexity, and the capabilities stemming from more advanced sub-routines, programmes, models, or information and communication systems enables an organism to outwit, rather than outrun, a competitor. Adaption is a change in strategy by an establishment based upon system experience.

Species may also be complementary, such as the energy, transport, and telecommunication sectors, and they coevolve without competing in any vital manner. Species fitness arises from the dance of coevolution, as each sector constantly tries to adapt to all the others, and natural selection within species applies to the establishments and not directly to the agents. Each species has its own life cycle, and changes in the system are likely to favour some species and to diminish others, with a consequential change in the regional landscape. The evolving regional landscape provides a silhouette of the survivors of the preceding populations of organisms or establishments, facilities, infrastructures, artefacts, and terrestrial stocks. Feedback from both the environment and transactions undertaken between establishments results in oscillations in the various populations, with migration, births, and deaths of establishments, business cycles, property cycles, and investment waves.

4.3   Diffusive ecostructures

At the foundation of evolutionary theory is a space symmetry-breaking process that results in the formation of spatial structure. In effect, there is an inherent instability in a homogeneous landscape with evenly spaced economic activity, and the potential for change is generated as soon as a disturbance establishes economic or informational gradients between one location and another. This is recognised in the concept of comparative advantage in international trade and is described by Michael Porter in *The Competitive Advantage of Nations* (1990). Similarly, major ports induced the formation of agglomerations of establishments that became primate cities for the redistribution of large shipments. Economies of scale attracted complementary activities such as transport, warehousing, insurance, banking, commerce, and retail markets from which traders endeavoured to capture a share of the gross profit to provide investment capital for their own enterprises. Early theories on economic landscapes were developed in Germany, which included the Central Place theory of Walter Christaller (1933) for the emergence of markets and theories on the economics of industrial and commercial location by August Losch (1954). Regional science became established when Walter Isard published *Location and Space Economy* (1956).

Geographers and regional or urban planners do not neglect spatial issues, and there is a long tradition of urban geography that provides a logic rather than predictability to spatial location, in which communications and transport technology act as constraints. In *A Communications Theory of Urban Growth* (1962), Richard Meier introduced a new element into spatial structure that moved away from explanations that depended upon the minimisation of transport costs. However, neither regional science nor the urban economics studies of the 1960s and 1970s were able to provide an evolutionary model for urban development. In *The Geography of the World Economy* (1989), Paul Knox and John Agnew pointed out that the achievement of equilibrium is inconsistent with spatial structure. The presence of agglomeration economies and any increasing returns to scale creates endogenous change that results in the transformation of civil ecostructures, with evolutionary and cumulative disequilibrium that does not fit the neoclassical theory that the economic system only adapts to exogenous change.

Regional and urban planners recognise the presence of the information economy and *Cities of the 21st Century* (1991), edited by John Brotchie, Michael Batty, Peter Hall, and Peter Newton, contained forty references to different aspects of information. In this publication, Richard Meier was the author of a chapter, entitled "The Transition to Ecumenopolis – Ecological Planning for New Urban Growth", in which he acknowledged that Boulding's Ecodynamic system emphasised the primacy of energy and information as flows. Meier also coined the term 'Ecostructures' to reflect the ecological anatomy of cities in *Cities in Competition* (1995), and in the same publication, Michael Batty produced a chapter on "Cities and Complexity". At this point, it was decided that, for *Long-Range Futures Research,* households and establishments would provide the basis for spatial structure, and that spatial structure should be incorporated into Ecodynamics to describe the evolution of the civil system. In *Cities and Regions as Self-Organising Systems* (1997), Peter Allen developed Prigogine's ideas associated with thermodynamic systems into a theory for the growth of cities, and this concern with dynamics attracted funding from the US Department of Transport. A special issue of the journal *Futures*, Volume 29, Number 4/5, May/June 1997 *Time and Space – Geographic Perspectives on the Future*, edited by Michael Batty and Sam Cole, summarised the state of the field at that time. In *Cities and Complexity* (2005), Michael Batty describes cellular automata modelling of urban development and agent-based models of urban movement, which reflects the application of complexity science. A summary of current approaches for modelling cities as complex systems is given in *The Dynamics of Complex Urban Systems* (2007), edited by Albeverio, Andrey, Giordano, and Vancheri.

Civil ecostructures are information-rich structures that evolve through selective changes and movement by households and establishments, with immigration and emigration as they respond adaptively to spatial and temporal variations in the environment. An ecostructure is defined as a complex adaptive system of transacting establishments that utilise a background flow of investment capital to create order by the formation of urban islands of transactional complexity. Investment capital is diffused in the system of connected ecostructures through an exchange of resources in a series of transactions, and it may be transformed into a variety of artefacts or transparts (see Chapter 8) that are transferred between establishments in a complex trophic web. Investment capital is dissipated in sustaining, maintaining, and reproducing internal order and coherence, which has spatial and temporal dimensions. Within the civil system, ecostructures diffuse investment capital, knowledge, and culture, and human societies are also dissipative structures that stay in a highly organised state by exchanging matter, energy, and information with their environment.

Ecodynamics explains how the transformation of investment capital into establishments and infrastructures under the central direction of agents, creates a complex adaptive system that evolves into the world system of cities. The autonomy of establishments to transact with other establishments throughout their life cycle is not independent of their environment, since they are affected by natural selection as part of the Ecodynamics within ecostructures. By concentrating on features such as the evolution of the urban system, rather than the endless turnover of the constituent establishments, the changes are very much more predictable than endeavouring to forecast the economy itself. The pattern of the urban microstructure and the major transformations in the macrostructure are shaped by the civil engineering infrastructures.

Ecodynamics may be described as the 'quasi-ecological' evolution of the civil system, which encourages the growth of urban areas, with the creation of a social environment in which the population eventually stabilises with an overall improvement in its welfare. The term urban system is used in its widest sense to include cities, towns, villages, and isolated establishments, which are connected through infrastructures even though the coefficient of connectivity may be very low in remote areas. Within a civil ecostructure, a balance needs to be achieved between the interacting populations of humans and the transacting community of establishments. The populations occupying urban areas comprise humans, households, and business establishments, facilities, infrastructures, vehicles, capital assets, such as machinery and equipment, inventories of materials that are transformed into stocks of goods, and reserves of food, energy, and natural resources, which are consumed. The populations are classified so that they may be counted (census) or quantified (stock-take) at a point in time. Each establishment and artefact has a life cycle, and changes in the system are likely to favour some and diminish others.

Societal development results from an increase in the range and number of interactions by humans or households and transactions between establishments. Transactions utilise resources and generate traffic flows involving the transport of people and goods or the transmission of information and energy within the urban system. Natural disasters and unsatisfactory transactions, such as conflict or business failures, create changes to the system with emigration of households and establishments to locations where there are better prospects for survival or growth. However, the emissions, exhausts, and disposals from the populations degrade the environment. The evolution of urban systems lead to metropolitan regions or states with a complex infrastructure, in which the establishments are differentiated into sectors and sub-sectors and the human agents have specialised functions. This permits an increasing number of elaborate transactions that push back the limits to development and facilitate the advance to a sustainable society.

However, societies are not only affected by the diffusion of investment capital, since cities or ecostructures simultaneously facilitate the cultural diffusion of new fashions, consumer habits, and lifestyles through retailers, the advertising media, and television. The appetites and habits that are stimulated serve the corporations by the development of consumer markets for their products, and the motivation of society to seek employment that will provide an income to purchase the desirable standard of living. Households organise themselves within urban space to enhance their well-being through a selected pattern of consumption and the generation of household investment capital, so that humans can reach their full potential. Human ecology may be explained by social science as the interactions that take place within the quasi-ecological evolution of civil ecostructures, which are a consequence of the actions of agents within society. The options available to society become more complex with the accumulation of capital and the evolution of infrastructures that are designed to achieve economies of scale in the provision of water, food, housing, health care, energy, and information.

Households and establishments capture investment capital (savings or profit) in successful transactions so that they can grow and reproduce. This increases the net worth of a household or an establishment, which is the difference between its assets and liabilities. Assets include capital assets, securities, money, and debtors, whereas liabilities include loans and creditors. Business saving is the net increase in net worth after dividends and interest have been paid, and the balance sheet will generally show an increase in both assets and liabilities. If the balance sheets of all households and establishments in the civil system were consolidated, creditors and debtors would match each other as would securities and loans. Therefore, the consolidated savings represent the increase in the stock of capital assets plus the growth in investment capital (money stock).

Economic development arises from the accumulation of capital stock, and Chapter 8 explains that there is a close correlation between growth in the stock of a country's infrastructure and its economic output. However, traditional economic thought is bound by the concept that economic development revolves around raising per capita real income. The redistribution of investment is more likely to be achieved than the redistribution of wealth, and social inequality can only be addressed by the provision of civil and social infrastructure assets in the less developed countries. Income is really a surrogate measure of wealth, and if artefacts had a constant length of life, which would be determined as the time between their production and consumption or obsolescence, then the ratio of capital stock to income, would be reasonably constant.

The regional assets, defined as the stock of capital that has accumulated into and within a city-region, determine the efficiency of the city-region as a setting for production. The use of urban land by the wide range of facilities that make up the regional assets includes housing and community buildings, health, education, recreational, retail, commercial, industrial, and transportation facilities. The capital stock comprises both public capital, which originally provided transportation facilities, water, and sewerage, or public facilities such as parks and libraries, all of which may become privatised with time, and private capital, which funds the investment in resources development, industry, commerce, and information services. Infrastructure imposes spatial structure on a region and creates economies of connectivity at locations, which encourages complementary investments that generate additional advantage (see Chapter 8).

4.4   Complex adaptive systems

A complex adaptive system is a community of interacting or transacting entities, such as organisms, households, or establishments that change their strategies in accordance with the feedback they receive from their environment. Complex adaptive systems generate complexity from the interactions and transactions of the interacting organisms or transacting entities, and emergent behaviour arises from the evolving dynamics. However, observable regularities within a complex system, such as the introduction, reproduction, and extinction of entities, are disturbed in episodes of punctuated equilibrium, which means that the system is indeterminate in detail and there will be a degree of unpredictability about the future state of the system. A defining characteristic of a complex adaptive system is that the evolution of the system as a whole survives generations of the transacting entities, which may have relatively short life-spans. Although evolution is inherently unpredictable, there are persistent features in far-from-equilibrium systems that may indicate the form or envelope of their evolutionary trajectory.

The search in science has two distinct forms of explanation, which may be described firstly as nonevolutionary that are locally or partially determinate and secondly as evolutionary or complex systems. Ecodynamics recognises that evolutionary change is caused by changes in system structure, whereas nonevolutionary system dynamics represents change within a given structure. In systems terms, changing the structure refers to a change in transactional complexity or information links, with additional hierarchical levels that enhance the information flow in the system. In time, a new information structure enables the civil system to transform itself and evolve to a new stage of development. The term 'economic growth' is reserved for quantification of the scale of the physical dimensions of the economy, whilst 'economic development' means structural transformations or qualitative change in the evolving civil system. Economic development consists of the expansion of knowledge and the application of new technologies, which continually push back the limits of the limiting factors. Sustainable growth is a self-contradictory expression, because, eventually, growth becomes unsustainable.

In evolutionary development, households and establishments are the basic transacting entities of a human settlement (level N), and spatial structure emerges since several establishments cannot be in the same place at the same time. Different levels in the system are a complex of successively more encompassing sets. Groups of households and establishments form residential, business, and industrial districts (level N+1), and groups of districts make up towns and cities (level N+2). Groups of towns and cities occupy a territory known as a region (level N+3), and groups of regions create a state (level N+4). States make up a nation (level N+5), groups of nations form a continent (level N+6), and continents make up the planet (level N+7). All levels are linked by resource transfers with some assembly or combination, which augments or transforms resources to create a complex adaptive system with feedback processes. The augmentation or transformation of resources affects both the development of the urban system and the depletion of resources.

By definition, entities cannot be studied as vivisystems in isolation, and the reduction of an organism or an establishment (level N) to its sub-systems or substructures, such as divisions (level N-1), departments (level N-2), sections (level N-3), functions (level N-4), or tasks (level N-5) is a negative direction that will only reveal increasingly microscopic detail rather than emergent behaviour. A morphological analysis of an organism means the study of its organic form, and the term is used in systems engineering for the conception of subsystem parameters that can be combined to produce a system design with high performance characteristics. Morphogenisis is the development of the complex form of an adult organism from a simple beginning, such as an egg or a bud, and is the source of emergent evolutionary properties. To distinguish between the complexity of morphology and the dynamic complexity of emergence within vivisystems, the term transactional complexity is used for the latter.

The transactional microstructure at the urban level generates emergent behaviour in the macrostructure at higher levels, such as the regional or continental levels, that is not deducible from the features of lower-order transactions. At each hierarchical level, the states at levels above and below act as bounding conditions for the behaviour of variables at the level under consideration. It follows that, as the total system evolves, the coevolution of two or more levels creates thresholds of change beyond which entirely different resource conditions and configurations emerge. Within the local, regional, national, and international space economy there will be regional differentiation and specialisation in economic sectors, and there will be a distribution of transnational headquarters, regional offices, or production plants, and local branches or branch plants. While the overall system may remain stable, chaotic dynamics could prevail at the town or city level where, for example, investment by a national supermarket chain could result in the closure of a number of shops, or a switch by an energy utility from coal to gas-fired power stations could destroy the economies of a number of mining communities. When these dynamics propagate between levels, behaviour patterns may be altered significantly at other levels for relatively brief periods of evolutionary time.

The evolution of civilisation follows ecodynamic macrolaws in which the generation of investment capital by the transaction system results in territorial colonisation through adaption and the formation of ecostructures, which diffuse the investment flow in order to decrease gradients in the return on investment. The rise in the level of complexity corresponds to an increase in the information content relating to the flow of resources necessary to sustain the system. Progress is reflected by the centripetal forces of agglomeration with increasing complexity and heterogeneity. As investment capital is transformed, the law of atrophy applies as an increasing number of sectoral establishments chase a finite supply of resources. The declining potential in the more mature city regions causes an evolutionary spiral with natural selection, adaption, coevolution, and relocation, which is punctuated by periodical transitions with sectoral investment waves and establishment extinctions. These waves of creative destruction to the environment integrate or simplify the system to achieve a reduction in transaction costs and more effective resource utilisation. Alternative pathways are provided for the diffusion of investment in the centrifugal direction of dispersal, standardisation, and homogeneity. A hierarchy can only exist and maintain its integrity if there is a state of tension between the tendency toward homogeneity and an asymmetric condition of heterogeneity; otherwise, the system would enter a runaway condition and come to rest at a new position.

In any evolving system, bursts of simplification often cut through increasing complexity and establish a new base from which complication can grow again. Evolution generally wins over an organic timescale with the regeneration of potential through new diversity. The emergence of macrostructure from the integration and recombination of the constituent parts of the system forms a hierarchy of complexity, which subordinates and regulates the parts of the system and creates the conditions for a subsequent increase in complexity and the transformation to a new stage of development. The successive invasion of ecostructures by new establishments and the extinction of old ones continues until the potential for further ecodynamic succession is exhausted and no further complexity can be achieved, at which point a climax will be reached. To maintain a constant configuration, the power of the few at the upper level in a hierarchy is counteracted by the power of the many at the lower level. Complete symmetry can never be achieved, since there will be a residual asymmetry inherent in the conditions essential to maintain the energy or resource flow. When the planetary system of ecostructures reaches the phase of maturity at which no further evolution or technological change will increase the overall diffusion of investment, a dynamic balance will be achieved if the regenerative capacity of the biosphere is sufficient to counter the atrophic forces of decline.

4.5 Territorial colonisation

The Verhulst logistic curve lies behind the waves of territorial colonisation that arise when populations in any given place increase until the limits of environmental resources are approached. From that point in time, the population will stabilise through a reduction in birth rates or an increase in mortality rates, unless the excess population departs to a new region. Colonists tend to settle in the nearest location that offers sufficient resources, and the new regional population will follow the logistic curve until the carrying capacity of the territory is reached and the emigration process is repeated. A colonisation wave is propagated out from the original region, and new colonies form in unoccupied and under-occupied territories. The young profile of early settlers and the pioneering of new establishments is gradually superseded by more mature and larger establishments that absorb the increasing flow of investment. Evolution arises as smaller establishments are forced into more specialised niches, which involves fine-tuning on information channels to capture resources. The growth rate of the establishment population will accelerate towards a stage of maximum diversity, until, eventually, the diversity declines at maturity. In the later stages of maturity, when dominant sectors emerge using the broadest bandwidth for resource capture from mass markets, there is an increase in homogeneity and a decline in the diffusion of investment.

The assembly rules for evolving ecostructures not only depend upon the stage of development and rates of urbanisation but also the city size and urban category. Planning factors for synthetic ecostructures are discussed in Chapter 5, and infrastructure planning factors are dealt with in Chapter 8. Geography and path dependence will determine the special characteristics of an ecostructure, such as a commercial centre, an industrial city, a transport node, a military or naval base, a university town, an historic tourist centre or a resort. The size class of an ecostructure will determine the market size for the number and size of competitive establishments. Chance events, such as the pre-emption of the best locations and sites, may provide an establishment with the initial advantage for entering a market segment, and occupation of a niche gives an incumbent a distinct advantage. A thin market will support few specialists, while the generalist survives by obtaining small amounts of resources from multiple sources. A richer market can be subdivided into competitive segments that will provide sufficient scope to be dominated by specialists.

The concept of succession in ecological communities is predictable and orderly change, where primary succession occurs when plants invade an area in which no plants have grown before. This may arise after a disturbance caused by lightning or fire, grazing or erosion when exposed ground emerges or patches of sky appear in a forest canopy. In Ecodynamics, the introduction of establishments from the primary resources division precede establishments from the secondary, tertiary, and quaternary divisions. The invasion is not sequential by division, and assembly of the local economy takes place by the random arrival of establishments. The ecodynamic succession in ecostructures leads towards a climax community with a reduction of establishments in the secondary industrial division and an increasing proportion of establishments in the quaternary information division. It is this succession that leads to the stages of development in the macrostructure from traditional to agropolitan, infrastructural, industrial, distributional, informational, ecopolitan, and then planetary. For civilisations, climax communities may be disturbed by major catastrophic events such as destruction by war, earthquakes, or floods, and minor disturbances to the economy by plant closures or the appearance of a new supermarket and the consequential closure of local shops. However, after the Second World War, numerous cities were rebuilt by the reproduction of new establishments, and similarly derelict or vacant urban sites become occupied by new species such as the development of retail parks on former industrial locations. In this way, there is evolutionary change in species diversity and the ecostructure recovers from bifurcation points to a different level of stability.

Diversity is essential to ecostructure stability, and ecostructures with a multiplicity of interrelated species of establishments are more resilient to the external changes and shocks that could destroy a vital link in more simplified systems. The Darwinian phrase 'survival of the fit' was erroneously escalated by Herbert Spencer to 'survival of the fittest' and this has led business and industry to engage in fierce competition, whereas collaboration and coevolution is more survival-positive than competition, since surviving species find a niche and utilise support from other species. In urban systems, a variety of business establishments have complex interdependencies that reproduce themselves into newly created organisations and also produce hybrids and mutate into entirely new kinds of employment. Urban systems are vulnerable and, where there is no real loss of technical efficiency, cities and regions need to provide a significant share of the goods locally to avoid the disruption and destruction that arises from being too dependent on the national and global economic systems. Also, communities need diversity in engineering and technological systems, such as energy or transport, to be more resilient to possible failure from a large power plant or railway system, etc. Resilience and the length of time for which a community has existed are the most suitable measures of dynamic stability, and this is achieved where there is a high level of diversity. Where there are low levels of environmental disturbance, and establishment densities are near the carrying capacity of an ecostructure, competitively dominant species will out-compete all other species and reduce the diversity. Conversely, where environmental conditions are unfavourable and only good colonists can survive, there are low population densities, which also result in low diversity.

Ecological metaphors have a use in strategic thinking for providing an evolutionary vocabulary to describe predatory, competitive, and coevolutionary interactions. The classification of species determines the most likely type of interaction between each other and, for example, in Ecodynamics, power generators will coevolve with ecostructures and the energy demands of their residential, commercial, and industrial districts. Retail superstores are predators amongst local shops and have a dominant-competitive relationship with suppliers. Industrial supply chains tend to be more dominant-cooperative and logistics companies thrive on mutual cooperation. Mutual competition is strong in financial services where any differences are frequently distinguished by cost alone. Maintenance contractors work in a symbiotic relationship with a diverse range of establishments, whereas brokers and agents have a parasitic relationship with their host sectors. Second-hand dealers and recyclers have a resemblance to scavengers. Advertisers invade our consciousness, whereas academic establishments pollinate fertile minds. Finally, exotic species may form part of the tourism or entertainment business. However, real business case studies are rather more interesting for strategic planning than ecological analogies.

The four sequential phases for the changing composition of establishments in an ecostructure are as follows: 'exploitation' when tracts of land are initially colonised for urban development at annual growth rates in excess of three percent; 'integration' when an established urban area extends its boundaries and develops into an urban or regional centre at growth rates of over two percent per year; 'maturation' when a city approaches a largely stable climax with annual growth rates at less than one percent; 'senescence' when a city goes into decline that cannot be arrested by urban regeneration. Immature ecostructures tend to have relatively short trophic networks linked to primary industries, with a less diverse range of establishments that tend to be smaller in scale. In mature ecostructures, there is a high diversity of larger establishments, with narrow specialisation in an extensive trophic web that involves recycling. The increased complexity of more mature ecostructures requires a larger quaternary information division that permits feedback control and generally more stability.

Strategies for the exploitation phase include innovation, invasion of markets, colonisation by the propagation of branches of service establishments, business diversification, and proliferation of products. Entry strategies require close attention to factors influencing the flow of marketable products and revenue generation. The integration phase involves centralisation, coalitions, coevolution, market differentiation, and customisation, imitation, mutation, and predation. Integration strategies have an emphasis on gaining market share as rapidly as possible and driving the growth curve as high as possible by investment and supplying cash flow requirements. The maturation phase is accompanied by deregulation, standardisation, rationalisation, decentralisation, and transformation through re-engineering. Although the sales volumes may be high at maturity, strategy objectives include delaying the decline in profit per sale with a series of cost reductions to perpetuate this phase as long as possible.

The final state for a corporation is the decomposition phase, which is associated with diminution, devolution, disposals, defection by personnel, and depreciation of assets. Decomposition means decomposing the business back to the business units and establishments so that they can be reassembled or recombined with other parts of the industry. The evolution of industrial sectors through their life cycle stages results in a spatial distribution of investments as establishments form sub-regional clusters during the technological development stage, regional agglomerations in the expansion or market penetration stage, and dispersal in the mature stage of market saturation, followed by standardisation and continental rationalisation with plant closures and extinctions.

4.6 Evolutionary Potential

Civilisations and cities are complex adaptive systems in which institutions and information structures create order out of chaos. The rise and fall of civilisations is explained by the evolution of diffusive ecostructures within a specified spatial and temporal range, which reach a peak of organisational and transactional complexity as a temporary phenomenon, and then decline through simplification and reduction in the cost of complexity. This reverse process only goes as far as is necessary for the benefit/cost ratio to be restored. The accumulation of capital stocks, knowledge, and irreversible changes in the course of evolution prevent the decline proceeding below a base level that will probably be rather higher than the starting point. The process of change that both destroys and releases potential, is captured by Schumpeter's term 'creative destruction'. The diffusive variables involved in evolutionary development include the following:

4.6.1 Natural advantage and unequal distribution

Geological evolution of the planet, in which the continents, oceans, mountains, valleys, and arable lands were formed, created uneven potential or natural advantage in the distribution of nature's resources. The terrestrial stocks of the geophysical metasystem are the geological stocks that are not renewable within a human time span. The terrestrial stocks include potential energy in the form of coal, oil, and gas fields. In the case of minerals or inorganic materials, the planet is virtually a closed system. Humankind's use of immediately available high-grade resources results in the sequential exploitation of increasingly inaccessible or lower-grade resources, which are more expensive. This encourages the use of sustainable substitutes and recycling of materials, which reduces the rate of depletion. The emissions, exhausts, and disposals from the populations tend to degrade the environment when the rate of pollution exceeds the absorptive capacity of the biosphere. The losses to the biosphere will be unevenly distributed, and environmental destruction will be minimised by dispersal of the burdens of production.

4.6.2 Energy transformation and dissipation

The planet Earth is part of the solar system, and the process of evolution has been powered by the sun, the gravitational forces of the moon, the Earth's rotation, climatic forces of the biosphere, and magmatic and hydrothermal forces within the earth's crust. All types of food trace their origins to solar energy that has been captured by green plants through the process of photosynthesis. The plants are then consumed by herbivorous animals, which may be eaten by carnivorous animals or human beings. The replacement of agrarian stocks is limited by the seasonal units of time and solar flow. The realisation of potential through the transformation of natural energy into mechanical energy results in the dissipation of energy resources, which has to be resupplied through regeneration by renewable resources.

4.6.3 Population equilibrium and decline

The evolution of populations including humans, animals, plants, organisms, and artefacts follows a life cycle of birth, growth, internal development, maturity, senescence, and death or decomposition. A growing population gradually increases the adversity of the condition of a species within a territory through overcrowding, poverty, malnutrition, or disease, which results in atrophy with reduced survival rates, decreasing fertility, or net emigration. The process of selection and competitive advantage arises in species of interacting populations, which involves procreation or reproduction, mutation, and innovation for occupation of a 'niche' with an equilibrium population. However, the stock of an equilibrium population is likely to change in character (i.e., by age, condition, performance, or other characteristics), even though the quantity of the defined population may remain the same. Genetic engineering and technological innovation result in species that give higher performance or lower unit costs, which displace the existing populations and result in a decline of particular populations, such as varieties of plants or categories of artefacts.

4.6.4 Social utility and the diffusion of technology

Civilisation is achieved by the accumulation of knowledge, technology, and stocks of capital, which reduce the vulnerability of the human species to external forces. Societal evolution and information networks counter atrophic tendencies that lead to the creation of islands of complexity at dispersed locations. Knowledge has the ability to direct energy towards the selection, transportation, and transformation of materials into appropriate forms. The development of the world system of cities, towns, and villages and the stock of human artefacts has been realised by increasing specialisation, differentiation, and structural complexity, which proceeds in well-defined stages of development until there is decreasing social utility. Diminishing returns lead to the diffusion of superior technology and the dispersal of populations from the mature centres to the less mature parts of the system, which leads to the further generation of potential.

4.6.5 Potential difference and disorder

Geopolitical potential is achieved through the continental union of states or territorial expansion. Political regimes determine how resources are exploited for the benefit of their own populations, which are protected by boundaries and military deterrence. Closed civilisations or societies tend to limit their human and natural potential since they are not recharged through the free flow of people, ideas, technology, capital, and goods. The geopolitical system increases the size and diversity of the resource base, and the primary markets that can be served. The realm of influence is dependent upon the information networks and the range and speed of transport and military technology, and also the potential difference between adjacent states. The greater the potential difference between parts of the global system, the greater the disorder and chaos that will prevail in the territories of least potential. This is because a regional or local level of complexity is achieved at the expense of the temporal global investment budget.

Geopolitical power is accumulated in the same way as wealth. A nation-state's influence beyond its borders is a combination of its total capital stocks and its military strength. Rich countries have substantial accumulations of knowledge and capital and access to large energy and material sources. They have become rich because of cumulative infrastructure development and productivity increases over a long period of time. It is very hard for poor societies to get richer because all their time and effort has to be spent replacing expired knowledge from deaths and the food and goods consumed or worn out. The poor countries rely on human and animal muscle as energy sources, and are more dependent upon local materials. In time, there will be a redistribution of investment from the low-utility uses of the more mature territories of the world system to the high-utility uses of the less developed regions. World potential increases when there is an increase in the world stock of assets per capita, and social utility will be maximised through the provision of assets to the less developed countries.

The potential difference between two regions creates gradients in the rate of return on investments, so that investment flows to the less mature parts of the system, which strive to rise towards the potential of the more mature parts. At the same time, the more mature parts of the system use their sophistication to mobilise investment capital and information to resist a loss of their own potential. Although there is a direction and an end state for the process when there are equal capital stocks per capita and the system is in dynamic stability, even then some states may rise and others decline. The realisation of potential will use up the available investment capital, unless it can be regenerated in an open system through natural, solar, biotic, genetic, societal, or political processes. When all the planetary potential is realised in the final phases of maturity, the system may achieve a dynamic balance if the planetary metasystems (astrophysical, geophysical, physical, biological, and civil) can regenerate sufficient potential to counter the atrophic forces of decline.

Human societies are initially attracted to regions of higher potential, where the potential of an ecostructure is a measure of its capacity to diffuse investment capital within a specific spatio-temporal range. The evolutionary potential of a state will reflect the scale of available natural resources and the absolute number of large establishments, together with number per 10,000 population. However, a diversity of states provides individual markets with unique microclimates in the form of resources, industries, employment, and culture. States will adapt in order to increase the potential for their ecostructures and subsystems to evolve.

The states will capture available investment capital for the development of infrastructures, which will attract additional establishments and enable new and existing establishments to grow and reproduce, and create new diffusive pathways through human settlement expansion. As a state matures, it develops at all hierarchical levels in the urban system, with more complex ecostructures, greater connectivity between them, and a diverse range of establishments.

If the potential of the world system of cities could be modelled as a dynamic network of nodes for the diffusion of investment and the attraction of human populations using adaptive computation, then the diversity offered in the transactions between some 15m large establishments (see Chapter 10), or 10,000 cities of over 100,000 inhabitants (see Chapter 9), within 1,000 ecopolitan states (see Chapter 11) would provide a robust and resilient system. Complex adaptive systems include redundancy with a multitude of establishments acting in parallel so that minor failures are absorbed in the vivisystem, and major failures may be contained by becoming less significant failures at the next highest level in the hierarchy. This characteristic would create a considerable 'damping' effect in the system dynamics models that ignore the evolutionary behaviour of a complex adaptive system and exaggerate the prospect of global collapse.

As the total world population rises, an increase in investment capital will raise the equilibrium level for the population. Designed interventions to facilitate adjustment for both social and ecological issues may prevent catastrophes in the long run. If a nonlinear graph is drawn of total population against global investment, then points on the graph will give the per capita assets. When the carrying capacity of the planet is approached, a bifurcation point may be reached when any further increase in the investment parameter may turn the system chaotic and the total population level could fluctuate between a wide range of values. At the peak of planetary potential, there is likely to be a greater diversity of artefacts, higher overall efficiency in per capita investment, and energy use, and probably a lower human population than at its maximum.

The principles of Ecodynamics: Chronological references

1. *A Communications Theory of Urban Growth* - Richard L Meier, 1962.
2. *Issues in Urban Economics* – Harvey Perloff and Lowdon Wingo, 1968.
3. *Human and Energy Factors in Urban Planning: A Systems Approach* - Edited by P Laconte, J Gibson and A Rapoport, 1982.
4. *The Coming of the Transactional City* - Jean Gottmann, 1983.
5. *Entropy, Information and Evolution* - Edited by Bruce Weber, David Depew, & James Smith, 1988.
6. *The Geography of the World Economy* - Paul Knox & John Agnew, 1989.
7. *The Competitive Advantage of Nations* - Michael E Porter, 1990.
8. *Socio-Styles* - Bernard Cathelat, 1990.
9. *Cities of the 21st Century* - Edited by John Brotchie, Michael Batty, Peter Hall & Peter Newton, 1991.
10. *Economic Activity and Land-Use* – Michael Healey, 1991.
11. *Contemporary Urban Sociology* - William G Flanagan, 1993.
12. *The New Marketing Research Systems* - David Curry, 1993.
13. *Out of Control: The New Biology of Machines* - Kevin Kelly, 1994.
14. *Cities in Competition* - Edited by John Brotchie, Mike Batty, Ed Blakely, Peter Hall & Peter Newton, 1995.
15. *Cities and Regions as Self-Organising Systems* - Peter M Allen, 1997.
16. *The Economics of Household Behaviour* - Peter Kooreman & Sophia Wunderink, 1997.
17. *Space and Time* – Edited by Michael Batty & Sam Cole, 1997.
18. *Understanding Agent Systems* – Mark D' Inverno & Michael Luck, 2001.
19. *Integrating Geographic Information Systems and Agent-based Modeling Techniques* – Edited by H Randy Gimblett, 2002.
20. *Cities and Complexity* – Michael Batty, 2005.
21. *Dynamics of Complex Urban Systems* – Albeverio, Andrey, Giordano & Vancheri, 2007.

PART 2

TRANSACTIONAL MICROSTRUCTURE

5.
CIVIL ECOSTRUCTURES

5.1   Metropolis and Megalopolis

The definition of municipal boundaries varies widely, and in the United States the general concept of a Metropolitan Statistical Area (MSA) is one of a large population nucleus together with adjacent communities, which have a high degree of economic and social integration with that nucleus. In Europe, the tendency is to view metropolitan areas as encompassing all the continuously built-up area of a metropolis, but to exclude large and small urban centres and rural settlements nearby. In Europe, the term "functional urban region" or metropolitan region corresponds to the US definition of an MSA. A metropolis is defined as an urban area of over 1m inhabitants and a megapolis or megacity region (MCR) is defined as a large metropolis with a megapolitan region in excess of 10m population. In the context of a spatial hierarchy for the world system of cities, there is a need for an intermediate category of large metropolises of 5m population that correspond to a number of cosmopolitan capital cities. The term 'cosmopolis' is used to define this subset of metropolitan cities.

The United Nations Centre for Human Settlements classifies the smallest settlements as urban, where there are 2,000 inhabitants with urban characteristics such as streets and urban utilities. Smaller rural settlements are classified as villages. For the purpose of *Long-Range Futures Research*, the term towns or urban centres will be used to describe urban areas of less than 100,000 inhabitants, and above this threshold, they will be described as a city or a contemporary polis. An urban hierarchy of fifteen levels has been classified in accordance with the city or town population in Table 1.

TABLE 1: THE URBAN HIERARCHY

Population range	Classification
100m+	Metapolis (supercontinental)
50m+	Metapolis (continental)
20m+	Megalopolis (subcontinental)
10m+	Megapolis / Megapolitan region
5m+	Cosmopolis
2m+	Metropolis (national)
1m+	Metropolis (subnational)
500,000+	Provincial city
250,000-499,999	Principal city
100,000-249,999	Polis or regional city
50,000-99,999	Urban centre
20,000-49,999	Micropolitan district centre
10,000-19,999	Micropolitan local centre
5,000-9,999	Service centre
2,000-4,999	Township

In examining the dynamics of the urban system, it is important to understand the emergence of the megalopolis and its novel characteristics. It made its appearance by 1950 when the populations of the three largest cities in the world were New York 12.3m, London 8.9m, and Tokyo 7.5m. The term 'megalopolis', which literally means 'a great city' (Oxford Dictionary), was introduced by Jean Gottman to describe major urbanised areas containing populations of over 25m. He used the term to delineate large regions in which several individual metropolitan areas, with each in excess of 1m inhabitants, had become mutually adjacent. Based upon populations in 1960, five emerging megalopolitan regions were identified in the world at that time: 'Boswash' (Boston, New York, Baltimore, and Washington), 'Chipitts' or Great Lakes (Chicago, Detroit, and Pittsburg), 'Tokaido' (Tokyo, Nagoya, Kyoto, and Osaka), 'Rhine' (Randstad and the Rhine-Ruhr), and the UK megalopolitan axis. However, the average density of these megalopolises, where some 25m inhabitants would be contained within an area of some 50,000 km², amounted to only 500 inhabitants per km².

Subsequently, the megalopolitan concept was modified by Constantinos Doxiadis to describe a megapolitan city region with a minimum population of 10-20m, in which the metropolitan centre is surrounded by outer centres that are linked by orbital and radial routes to form a diffused urban system, where the density declines from 2,500 inhabitants per km² in the LDCs to 1,250 inhabitants per km² in MDCs. This decline arises from a reduction in the number of persons per household at each stage of economic development and also from the extension of the metropolitan region with new perimeter development. In the LDCs, a population of some 15m-18m will occupy an area of some 6,500-8,000 km², whereas in the MDCs, populations of 10m-20m will be contained in areas of 12,500-25,000 km². This concept has more recently been described for the USA by journalist Joel Garreau in *Edge City* (1991), in which new urban centres of less than 1m population emerge spontaneously around metropolitan centres. As cities become larger than 10m population, they tend to evolve from a circular to an elongated form. A circular region 120 km (70 miles) in diameter will give an area of 11,500 km², compared with an elliptical shape 160 km (100 miles) by 120 km, which results in an area of 15,000 km². Hence, the megacities of 10m-25m population have a tendency to become 100-mile cities, and megapolitan city regions will tend to lie in the range 10,000-30,000 km².

A Metropolitan Statistical Area (MSA) is centred on a single, large city, and in 2000, the MSA population for New York was 17.8m (area of 17,405 km²) at a population density of 1,020 inhabitants per km², whereas the Consolidated Metropolitan Statistical Area (CMSA) population for New York is 19.3m (area of 30,671 km²) and includes New York, Newark, and Bridgeport at a CMSA population density of 630 inhabitants per km². By comparison in 2000, the population of the Greater Los Angeles MSA was 12.7m (area of 12,562 km²) at a population density of 1,010 inhabitants per km², while the wider CMSA population of 17.6m (area of 87,941 km²) includes five counties. These include the sparsely populated eastern areas of Riverside and San Bernadino counties, with an overall CMSA population density of 200 inhabitants per km². By 2000, the population of the Tokyo Metropolitan Government was 11.9m (one prefecture with an area of 2,165 km²) to give a population density of 5,500 inhabitants per km², whereas the CMSA population of Greater Tokyo was 27.9m (four prefectures with an area of 13,900 km²) and includes eighty-seven surrounding cities and towns including Yokohama, Kawasaki, and Chiba, with an overall population density of 2,000 inhabitants per km².

In Western Europe with the emergence of the European Union, international development pressures have increased in the area encompassed by the major urban concentrations centred on London, Paris, Brussels, Randstad, the Rhine-Ruhr, and an extension through Switzerland to Northern Italy. These concentrations form the Megalopolitan axis of North-West Europe. Also from the London Megapolis in the South East of England, there is a belt of urban areas that extends across Central England to the North West, along the routes of the M1 and M6 motorways. This belt contains six Metropolitan Areas, and with a population of 30 million, it may be described as the Megalopolitan axis of England. The London Megapolis corresponds physically to the Greater London area, plus the Outer Metropolitan Area, and for thirty years (1961-1991), these combined areas were recognised in the Census of Population. The Greater London population of 8.2m in 2000 is contained within 1,620 km² to give a compact density of 5,100 persons per km². In 2000, the London Megapolitan Region contained a total population of 12m in an area of 11,000 km², giving an overall density of 1,100 inhabitants per km². Eurostat defines a 'Larger Urban Zone' (LUZ) for London composed of Greater London and forty-two surrounding districts with a total population of 11.6m. In 2000, Randstad in the Netherlands had a population of 5.1m located in an area of 5,250 km², at a population density of 970 inhabitants per km². The Rhine-Ruhr megapolitan region had a population of 9.1m within an area of 9,000 km² to give a population density of 1,010 inhabitants per km².

In the less developed countries, the populations of the major cities tend to be exaggerated because the metropolitan boundaries have been established for polynucleated megapolitan regions or megapolises. Since large populations reside outside the city's built up area, predictions made in the 1980s that cities such as Mexico City and Calcutta would have 30-40m inhabitants by the year 2000 were untrue, and Mexico City had 18m and Calcutta had 13m inhabitants. See the *Global Report on Human*

Settlements 1996 by the United Nations Centre for Human Settlements (Habitat). The developing countries are now experiencing rapid urban growth rates, and megapolises such as Mumbai (Bombay), Cairo, Mexico City, Sao Paulo, and Shanghai have recently challenged the megapolises of the developed world. In 2000, the metropolitan region of Mexico City had 18m population and Sao Paulo 17m inhabitants within a regional area of 8,000 km^2, to give population densities of some 2,250 inhabitants per km^2. Similarly, the population of Cairo is 10.4m, of which 6.8m live in the old city (215 km^2) at a density of 31,000 inhabitants per km^2, whereas the greater Cairo metropolitan area (6,353 km^2) includes the Governorates of Cairo, Giza, and Qalubeya in which a population of nearly 14.8m gives a population density of 2,330 inhabitants per km^2. The average household size is 4.4 persons.

The megalopolises, megapolises, and metropolises of the world are being linked together by air travel, telecommunications, television, and information networks that are facilitating the internationalisation of commercial and business services. The principal world cities such as New York, London, and Tokyo are strengthening their positions in the globalisation of world financial markets, and the multinational corporations located there rely extensively on advanced information services (i.e., financial, legal, business services, insurance brokers, management consultants, marketing, and advertising) to enable them to operate globally. It follows that there is an increasing concentration and specialisation of informational services in large cities at the expense of regional and urban centres. Cities are likely to depend increasingly on informational villages or specialised office districts for knowledge generation, information exchanges, and international transactions.

The transformation of investment creates irreversible structural change to the world system of cities, which increases the complexity of civil ecostructures over time and raises the level of connectivity between the nodes. Generally, the highest capacity transmission links connect the largest cities first, and then interconnections follow down the urban hierarchy where the traffic volumes are lower and the unit costs are higher. Within the dynamic system, the connectivity between diffusive ecostructures may range from low, medium, or high, to super connectivity. The degree of resilience of the system will be determined by the evolutionary stage of the connected ecostructures. Ecostructures develop at all hierarchical levels of the urban system, and there is increasing diversity in the number and type of establishments as they mature with more complex structures and greater connectivity.

During the rapid expansion phase in the evolution of the urban system, investment in infrastructure development and industrialisation create the initial concentration in metropolitan areas arising from economies of scale. However, this phase is followed by the distributional stage of development with suburbanisation in the metropolitan regions and decentralisation towards more attractive locations in new towns. In the informational stage of development, the diseconomies of congestion and industrial restructuring lead to counter-urbanisation and a more thorough diffusion of investment throughout the entire urban hierarchy. As the city system becomes more complete, investment capital flows from the vibrant regional cities to the higher-order primate city for investment outside the region in less mature urban centres, as well as to the numerous lower order centres in the regional subsystem. In the mature phase of urban evolution, the ageing population is swollen by early retirements with consequential investment and growth in rural areas, resort towns, and lower-cost locations.

As the major world city in Europe, London is a well-studied and accessible location where the establishments may be considered the equivalent of a Pompeii to an archaeologist or the fossils of the Burgess Shale to a biologist. Londinium was founded nearly 2000 years ago in AD 50 and it became the capital of the Roman province. In AD 200 Londinium was designated as the capital of Britannia Superior and a wall was built around the city with gates at Ludgate, Newgate, Bishopsgate, and Aldgate. At the time of the Norman invasion in 1066, the population was around 15,000; by 1811, the population had reached 1m. The population of Greater London reached a peak of 8.7m in 1939 and declined to a low point of 6.4m in 1991, although by 2000 it had recovered to 8.2m. However, the population of the megapolitan region at 12m is broadly equivalent to CMSA (Consolidated Metropolitan Statistical Area) in the USA.

However, implicit in the designation of London as a megapolis is that it functions within a world urban system, and its global role is fulfilled by the international banks and multinational corporations, which are supported by the full economic base of the region. The aim in selecting London as a real-world city region is to obtain data for proof of the conceptual framework and to undertake research that is replicable, by using a comprehensive establishment database in conjunction with an independent databank of property investments. This is preferable to a computer simulation as it enables discoveries to be made without knowing at the outset what is likely to be found. Also, a simulation has the disadvantage of requiring prior knowledge of the causal relationships, and it may introduce mapping errors between the simulation and the phenomena being studied.

The London Megapolitan Region comprises Central London as its core, the London Metropolis (or Metropolitan Area), which is bounded by the M25 motorway, and a commuter belt extending some 30 kilometres (20 miles) beyond the M25. The orbital motorway has contributed to the integration of the London Megapolis, by linking the radial routes that serve the urban centres and airports outside the M25. The region extends to Reading (M4 corridor), Luton Airport (M1 corridor), Stansted Airport (M11 corridor), the Medway towns and Maidstone (M2 and M20 Channel Tunnel corridor), and Gatwick Airport (M23 corridor).

The London Megapolitan Region contains over eighty regional or urban centres of which twenty-eight centres have London postcodes and serve populations of 200,000 or more. Within the M25 motorway ring, there are a further twenty-four urban centres, each with a population of over 50,000, which serve urban areas in excess of 100,000 inhabitants. The outer boundary of the London Megapolis encompasses a region in which there is an eighty-five percent level of commuting self containment, and the commuter belt surrounding the M25 contains a further thirty centres serving urban populations of a similar size to those within the inner M25 area. Outside the metropolitan centre, there are some seventy district centres, sixty local centres, seventy service centres, and over forty township communities within the London Megapolis.

5.2 Sectoral classification of Business Space Users

Establishments are aggregated into business sectors and sub-sectors to simplify analysis of the urban, regional, and national economic structures. The same classification of establishments into sectors also provides the means of naming the species of establishments that inhabit the urban environment, so that the principles of Ecodynamics can be applied to the populations that colonise the territory. However, a further aggregation into four divisions (primary resources, secondary industrial, tertiary commercial, and quaternary information) is an essential prerequisite for understanding the evolution of ecostructures and regional economic development. The increasing complexity of ecostructures will result in both an increase in the informational content of the transactional microstructure and the emergence of large-scale transformations in the contextual macrostructure with evolving stages of development.

The existence of a fourth economic division, or an information division, has generally been accepted during the past decade; however, it is necessary to redefine the tertiary division as a commercial division before the quaternary division can be defined. The approach recognises that international corporations have introduced substantial economies of scale to the provision of commercial and information products, as happened to industrial products in an earlier era, and it is the business establishment that is classified according to its end product. This differs from the sociological perspective of a postindustrial society with its emphasis on occupational structure and the information society. 4-Scene Development Corporation has developed a sectoral classification system (©RICI code) based upon research since 1986, which provides a contemporary classification of some ninety business sectors that has been cross-referenced at the establishment level to the International Standard Industrial Classification (400 classes) in Appendix 7. A brief description of each division follows:

- The Primary Resources Division includes agribusiness, mining, energy, and water supply industries.
- The Secondary Industrial Division includes construction, property, and manufacturing industries.

- The Tertiary Commercial Division provides tangible economic services including banking, finance and securities, insurance, transport, wholesale and retail trades, travel, hospitality, and consumer services.
- The Quaternary Information Division encompasses the information service industries, such as media, business and legal services, telecommunications and information technology, design, technical services, research and development, education, health, welfare organisations, associations, and government.
- A Quinary Residential Division currently encompasses all households. These are classified in Chapter 4 - Section 4.1.

It may be noted that the inclusion of healthcare in the quaternary information division is based upon the high level of knowledge and research in the life sciences and medicine. However, in Chapter 7 - Section 7.3, it is proposed that for an ecopolitan society, a Quinary Lifespan Division should be defined to include residential establishments and both voluntary and salaried or wage employment in human care and life science.

The approach to the informational economy adopted in this study is founded on a metropolitan perspective and the evolution of the world system of cities. Jean Gottman proposed in *Megalopolis* (1961), that the tertiary division as defined by Colin Clark should be differentiated from quaternary information activities, such as newspapers, publishing, broadcasting, performing arts, advertising, the law, engineering, research, education, and government. He recognised that the commercial organisation of Megalopolis included banking, insurance, and other financial transactions, in addition to transport, distribution, retail, and other services, such as restaurants, beauty salons, and service stations. Unfortunately, the tertiary division as defined by Colin Clark simply included all those service activities that could not be conveniently grouped into the primary or secondary divisions.

In redefining the tertiary division as a commercial division, activities such as utilities, construction, property, maintenance and repairs, cleaning and sanitary services, and healthcare have to be allocated to the most appropriate divisions. For that reason, the primary resources division includes agriculture, fishing and forestry, mining, the energy industries, and water supply. The secondary industrial division includes construction and property, together with building maintenance, cleaning, and sanitary services. Information services include media, communications services, such as postal services and telecommunications, security services, which generally include surveillance (CCTV) and intelligence exchange, and government services, which essentially provide regulatory and control functions. It was considered to be more practical to include brokerage services, such as insurance and shipping brokers, within the commercial insurance and shipping business categories, and generally they receive commission from their own sub-sector. Similarly, estate agents have been allocated to the property sector since, in many cases, they are acting as agents of the property sector in both managing and marketing properties. Certainly, research and development, engineering design and architecture, which involve innovation, blueprints for new artefacts, and creative fertilisation of ideas are appropriately classified in the quaternary division.

The Information Economy (1977), by Marc Porat, provided a four-sector aggregation for the US labour force by occupation, which confirmed the concept prior to the RICI classification by establishment. The sociological perspective of an information society proposed by Marc Porat and termed the 'postindustrial society' by Daniel Bell is based upon a study of information occupations. This was developed to take account of the increasing proportion of the labour force working in information occupations across all divisions, including primary, secondary, tertiary, and quaternary. Clearly, both occupational and industrial analyses contribute to the historical assessment and future evaluation of the role of information in economic development. However, the Standard Industrial Classification (SIC) is based upon the category of activity undertaken by the employer and not the kind of work the employee performs. The growth of white-collar office employment is evidenced by the increasing number of office establishments, which are replacing industrial sites in the metropolitan centres.

In the RICI Sectoral Classification, establishments are classified by both sector and facility type. In other words, a distinction is made between employment in the office headquarters of manufacturing or retail corporations and employment at industrial sites or retail outlets. This may be simply achieved through the coding system so that the total stock of office establishments can be quantified for a city, even though the activities may relate to the primary, secondary, tertiary, or quaternary divisions. It is important for economic base studies to retain the sectoral classification for an establishment, whether an office or an industrial site.

Regional economic-base theory requires data on the sector and facility type of an establishment because of the need to take account of the impact of transnational corporations, the international division of labour, and the emerging global economy. Basic business sectors serve customers or clients from other geographical regions, and non-basic business sectors serve customers from the same region. In the London Megapolitan region, some of the basic business sectors include manufactured exports, international construction and consultancy, international banking, insurance, shipping, air travel, tourism, entertainment, and retailing to visitors. The notion of services as a residual division outside the primary and secondary divisions, which may be differentiated as producer or consumer services, is an obsolete and confusing concept since a number of business sectors such as banking, insurance, airlines, legal and accounting services, telecommunications, and government serve both industry and households. In *The Global City* (1991), Saskia Sassen states that much of the work on producer services is an unwitting, neoindustrial response to a postindustrial economy.

5.3 Urban economic structure

The research in this book is based upon the Megapolis Regional Information System©, which provides a sectoral classification for the hierarchy of urban centres within the London World City region. It examines the spatial distribution of large establishments that occupy nearly two thirds of the business space with a similar proportion of total employment. Unique urban area profiles are produced from the number and size range of each type of property including offices, industrial, distribution, retail, leisure, hotels, hospitals, universities, and passenger terminals. The specialisation necessary for a first-order world city is reflected in the mix of large establishments (over fifty employees) present in the megapolitan region.

The business-space users include UK and foreign owned companies, partnerships, associations, public sector organisations, education and health authorities, research institutions, and national and local government offices. For each urban area, business district and the land-use category, establishments are classified by sector, facility type, and size (both office and production or operational space), with business information that includes the address, postcode, telephone number, contact name, and bands for the period of occupation, number of corporate employees, and annual sales. It follows that an assessment can be made of the office workers engaged in manufacturing or commercial activities, to form a view of the occupational structure of a business.

From an analysis of the 1981 and 1991 Censuses of Employment, employment in the quaternary information division falls from nearly 50% in Central London to 35% in the Outer M25. The tertiary commercial division declines from some 40% in Central London to 35% in the Outer M25. The secondary industrial division rises from less than 10% in Central London to over 25% outside the M25, and the corresponding increase for the primary resources division is from about 2% to 3%. The comparative figures from 1981 to 1991 show a decline for all regions in the primary and secondary divisions and an increase in the share of the information and commercial divisions. Within the Megapolis, 40% of employment is in the quaternary division, compared with 38% in the tertiary division, 20% in the secondary division, and 2% in the primary division.

The London Megapolis has undertaken the transition to an informational economy and a postindustrial society; however, there are spatial disparities between the inner areas and the outer areas, and the eastern and western zones. The overall picture is that there are high levels of quaternary division activities in Central London, with the industrial division concentrated principally in the outer perimeter. There are also nearly twice as many large establishments in the west as there are in the east, with differing numbers of establishments per 10,000 population in the various zones of the Megapolis. Companies that are located in the same business district, or area, contribute to each other's success by clustering together so that they can share in the availability of specialist occupations, advanced business services, transport links and airports, and other locational benefits including retail, leisure, hotel, and hospital facilities. An increasing range of property types and sizes appear in towns with larger populations, as would be expected within the urban hierarchy (see Appendix 9).

The economic structure of regions, areas, zones, towns, and districts is defined in terms of the sectoral characteristics of the businesses located there and may be derived from the proportion of employment in each of the four divisions. As a proxy, a useful way of measuring the relative concentrations of businesses is to relate the proportion of large establishments in a particular sector to the proportion of large megapolitan establishments within the same sector. The ratio between the two gives a location index, since the establishment data and facility-size thresholds are consistent across the database. An index higher than unity means that the sector is concentrated in that locality. Where the analysis is undertaken in terms of employees by sector, the index is called the 'location quotient'. From the Census of Employment, it may be determined that there is a close correspondence between the percentage of total employment by sector (SIC 1980 Divisions) and the percentage of large establishments (over fifty employees) by sector. It follows that the use of the location index provides a non-spatial classification of towns and districts and gives a powerful insight into the stage of development of the local economy.

Once urban areas are classified by both size and economic structure, the two most dominant sectors will indicate whether a town or district is a quaternary office centre, a financial or commercial centre, an industrial centre, or a transport and distribution centre. Quite often, the underlying basis for the local economy may be obvious from the characteristics of an urban area, such as a metropolitan or regional centre, a location with a major international airport, a county town or a new town, a military town, a university town, or an historic tourist centre. The Megapolis regional information system calculates location indices for each town. By plotting the towns on a grid with the secondary industrial location index (SLI) on the Y-axis and the quaternary information location index (QLI) on the X-axis, the relative position of each town can be seen with respect to the central position (unity on the X-axis), at which transition to a post-industrial town will have taken place.

There are well known districts within Central London in which businesses cluster together, such as the City of London, which is the commercial and financial centre with banking, securities dealing, and insurance, and the City of Westminster, which is the centre for government. Corporate headquarters are concentrated within the perimeter of the central area, mid-town is the district for legal and business services, and the West End is the shopping area with fashion boutiques, jewellery, and accessories. The advertising industry is concentrated in the western part of the central area, together with the entertainment industry, restaurants and cultural facilities such as theatres, the opera house, and concert halls, museums, libraries, and art galleries. London districts W1 and W2 contain the major hotels, which serve both business and the tourist industry. Further west are concentrations of the television, film, video, and record industries. The newspaper industry has relocated eastwards to London Docklands and is now being joined at Canary Wharf by major banks requiring trading floor areas that are not currently available within the confines of the City of London. Finally, London also contains a concentration of quaternary activities, such as the major teaching hospitals, universities and colleges, publishing, information services, engineering consultants, architectural practices, and membership organisations.

Appendix 8 shows that manufacturing industries have generally decentralised from the metropolitan centre with numbers increasing from Outer London to the Inner M25 and rather more beyond the M25. Apart from textiles, clothing, footwear, and leather products, of which over half the large firms are within the London postcode area (the metropolitan centre), the outer urban centres with populations of 50,000-99,000 and district centres with populations of 20,000-49,000 account for over half of the megapolitan industrial sites, and fewer than twenty percent remain in the metropolitan centre. The Outer Metropolis (i.e., within the M25) has attracted the transport and wholesale distribution businesses and also motor services, with a clustering of air transport related business in the west towards Heathrow Airport. Construction companies are also predominant in the Outer Metropolis, together with cleaning companies and security services.

The location of large retail sites tends to correspond to the distribution of population, and there is a trend for greater concentration of large retail stores at urban centres of 50,000-99,000 population than at regional centres. However, whilst there may be twenty-five percent more large retail establishments per 10,000 population at urban centres, regional centres offer retailing economies of scale with an average size of store that is some twenty percent larger. In deprived areas, such as the Inner London boroughs, there is less purchasing power and a consequential reduction in retail floorspace. Also, with lower car ownership, there tends to be smaller retail units that are located within walking distance of the residential areas.

Finally, quaternary division establishments in telecommunications, computer systems, and office technology are predominant in regional and urban centres outside the M25 in the western crescent of high-tech establishments, which include technology and research and development. The information division is the most dominant division within the London Megapolis, with knowledge-based establishments supporting the national and international activities of the commercial, industrial, and resources divisions. The growth in the information division may be attributed to the increase in intelligence-gathering and research, and the sub-division of labour and specialisation required to gain a competitive advantage in the European Union and the world economy.

Towns and districts that are quaternary-dominant, tend to have more large establishments per 10,000 population than industrial towns with secondary sector domination. In the western zones of the Megapolis, the corporate headquarters of multinational corporations have been attracted to towns within easy reach of Heathrow Airport and, in addition, high-tech industries in telecommunications and information technology have settled in the area. The employment multiplier effect from the addition of new establishments to an urban area, or the expansion of existing facilities, is derived from backward linkages when new employment is generated in the supplying establishments, or forward linkages where other establishments expand their purchases of goods from the new establishments and create additional employment. Where the aggregate expenditure of income by employees is increased, then additional commercial sites are attracted to the locality where the earnings are spent, with their own requirement for extra staff.

An imbalance between the population of a town and the number of large places of employment may be partly attributed to radial commuting from an outer centre into the metropolitan centre, or conversely by orbital or lateral commuting from one outer centre to another. The choice of residential location is often a trade-off between a combination of house prices and environmental considerations compared with the cost of commuting. It is generally accepted that the high number of large establishments in the western zones surrounding Heathrow Airport result in higher house prices and encourage long-distance commuting and orbital travel. As a consequence, there is heavy congestion on the western segment of the M25 motorway, and in the long run, firms will tend to relocate to urban areas where the locational benefits at least offset the costs of congestion.

Within the Megapolis, the overall pattern for the location of corporate headquarters tends to be within metropolitan, regional, or urban centres, depending upon the need for easy access to specialised information services and the facility for international or intercity travel to the decentralised sites of production or routine operations. However, eighteen percent of headquarters are at district centres, and in the electrical and electronic engineering industries, the proportion rises to twenty-five percent.

5.4 Planning profiles for synthetic ecostructures

For each class of urban area, the pattern of the number and size of major establishments provides a unique urban profile in which the local economic structure is reflected in the property types occupied. With the evolution of cities, the number and size of facilities change to match the demand for commercial and information services. The planning profiles for each urban category are derived from an average profile for each category type, and, for that reason, they are termed synthetic ecostructures. As a town evolves with the changing economic structure, certain types of establishments decline in numbers or even become extinct. It follows that the urban landscape provides a silhouette of the survivors of the preceding populations of establishments, although it does not reveal the sequence by which the local economy was assembled.

Planning standards may be derived to give the number and scale of facilities per 10,000 population, and, generally, it is the category of urban area that determines the number of facilities and the urban class (population) that affects the scale of facilities, taking into account any work travel flows. For each category of centre, it has been found that the number of commercial establishments is a function of the total number of both secondary industrial and quaternary informational establishments. This should probably be expected since the number of banking, insurance, transport, hotel, and restaurant facilities are dependent upon the number of establishments in the other sectors. This dispels the myth that earlier sectors, such as the primary or secondary sectors, are the basic economic motor and that the later tertiary and quaternary sectors are dependent upon earlier ones.

To establish planning standards for urban facilities, it is necessary to investigate the following:

- Minimum space per large establishment
- Number of large establishments per 10,000 population
- Floorspace of large establishments as a percentage of total floorspace for the facility type
- Floorspace per inhabitant
- Floorspace per employee

An approximate threshold of fifty employees (full-time equivalents) has been used to give a consistent measure of economic activity to distinguish between large and small establishments for all types of facility. However, in the case of the distributive trades, retailing, and warehousing, it would be necessary to reduce the threshold for the number of employees to twenty-five to cover fifty percent of the total floorspace. For the purposes of this research study, the higher threshold of fifty employees has been used for all facilities, and the minimum space per large retail establishment is 1,000m², and for large warehouses is 2,000m².

The socio-economic characteristics of a location may be assessed from the land use and the number and size of establishments per 10,000 population, including offices, industrial, distribution, retail, leisure, hotels, hospitals, universities, and passenger terminals. Deprived areas have fewer large places of employment, resulting in lower per-capita incomes and less retail floorspace. Typical land-use categories for the various types of property are industrial and trading estates, central business district or business park, shopping centre or retail park, etc.

Planning standards for the Megapolis have been derived in Appendix 10 in terms of gross floorspace per inhabitant and gross floorspace per employee, for both the metropolitan centre and an urban centre. Gross floorspace is normally defined in terms of the net internal area of buildings, as used by the DoE in the Commercial and Industrial Floorspace Statistics, 1995. The public area of certain facilities is also given, and this relates to the public areas such as retail sales areas, public areas and bedrooms in hotels, patient areas in hospitals, and student areas in universities (excluding halls of residence). It should also be noted that, in the case of warehouses, the standard of 200m² per employee relates to the storage area only, and the addition of ten to twenty percent adjacent office space at 12.5m² per person will tend to give an overall employment density of 50m²/employee. The industrial-space standard of 40m² per employee is a default value that assumes thirty percent of employees are office staff and seventy percent are production operatives. All the planning standards given in Appendix 10 are an indicative guide for planning research, so that a range of development possibilities can be reviewed quickly with a small research team. However, there is no substitute for a full evaluation of a specific development opportunity before making an investment.

The gross building area occupied by establishments is given in Appendix 9 in terms of the parameter L, which includes the space for the combined office, production and ancillary areas in size ranges set out in Table 2.

TABLE 2: BUILDING AREA RANGES FOR ESTABLISHMENTS

Local space code (Gross building area)	Category range (m²)
L	Unidentified
L1	200 - 499
L2	500 - 999
L3	1,000 - 1,999
L4	2,000 - 4,999
L5	5,000 - 9,999
L6	10,000 - 19,999
L7	20,000 - 39,999
L8	40,000 - 99,999
L9	100,000 - 199,999
L10	200,000 +

The facility types identified in Appendix 9 are generally self-explanatory, but a brief description may be useful in the case of depots, passenger transport, trading units, justice, and leisure facilities. Depot facilities provide an assembly point for vehicles, goods, and operatives, which are despatched to make collections and deliveries, or to undertake maintenance or inspection tasks, and generally require areas for storage and for parking heavy goods vehicles and associated workshop areas. Local authorities, utilities, post and courier services, security firms, dairies, plant and vehicle hire, fire stations, military bases, and airports all have depot facilities. Passenger transport facilities include terminal facilities and ground facilities for passengers at airports, railways, bus stations, and motoring facilities for car passengers such as car parks, motorway service areas, and forecourt areas at petrol stations. Trading units, which are generally on industrial or trading estates, are used for storage, showrooms, garage workshops, distribution, etc. Criminal justice facilities include courts, police stations, prisons, and other criminal justice institutions. The public areas for sports, leisure, and cultural facilities such as theatres, museums, and libraries, are defined in terms of the gross area of the buildings. However, in view of the diverse range of facilities, the parameter for public areas, F, is sometimes expressed in terms of the number of employees (exclusive of performing artistes).

From Appendices 9 and 10, it is possible to determine for each urban classification the number of large establishments by facility type and size range, so that the total floor area for each type of facility can be assessed by knowing the percentage of the total floorspace that has to be allowed for the proportion of smaller establishments. The area of school and further education facilities are calculated from the planning standards per child for the various age groups. By applying the appropriate construction cost per m^2 for each facility type, the non-residential building costs for a regional centre of 100,000 population amounts to $1.5 bn, in relation to residential building costs of $3 bn. A further $1.5 bn is required for infrastructure and utilities so that, as a default value, the capital cost of a new European city of 100,000 population would amount to $6 bn or $60,000 (2000 US$) per inhabitant.

5.5 Gradients between locations in property investment returns

Since the majority of empirical data on profitability at the establishment level is not publicly available, it is difficult to explore the locational advantages of urban centres at different points in time. Measures of sectoral profitability are heavily weighted by the performance of multi-site corporations, and corporate financial statements reflect their diversified portfolio of products, which obscures the profitability of specific lines as they evolve in their life-cycle phases. The smaller, single-site firms with a limited range of products only represent a minor share in sectoral statistics. The productive scale for any product is limited by the size of the market, and the difference between larger and smaller companies is the difference between producing hundreds or thousands of varieties compared with tens. Large corporations, in theory, derive stability from their diversity of products, although they lack growth potential. Smaller companies have, in theory, the prospect of episodes of high growth from a success or a 'hit' with one or more of their specialised products. There is every incentive for single establishments to capture investment capital so that they can diversify their product range and evolve to a higher level of complexity with increased survival prospects.

In the absence of establishment profitability data, an alternative approach would be to measure sectoral employment changes over time by location, to identify the growth phase of an economic sector. Similarly, the sectoral characteristics of floor-space changes can be monitored over time. However, the local impact of the rationalisation of mature companies of over 500 employees can be devastating if one of the large local establishments is selected against. A local economy is dependent upon a diversity of companies by sector and size, since the growth of medium-sized companies in their market penetration phase will need to counteract the job losses of the mature firms. Small firms of less than twenty employees have sometimes been credited with the generation of any net increase in employment, but this overlooks the major contribution of the medium-sized firms.

The returns on corporate investment change with variations in growth rate, so that viable units become profitable small establishments, of which a proportion become highly-profitable, medium-sized establishments in the market penetration phase. At maturity, large establishments become less profitable, and establishments decentralise to lower-cost locations within the region, and production costs are reduced through standardisation. As the market becomes saturated, unprofitable industries are relocated to the lowest-cost locations such as the less-developed countries. The oldest and least efficient establishments are closed in favour of the newer establishments that are set up in the decentralisation phase. Corporate acquisitions and mergers may affect the pattern of closures and result in the rationalisation of headquarters to a single site. During a regional transition from an industrial to an informational economy, the profit cycle of economic sub-sectors will lead to a change in land use, with derelict industrial sites becoming retail parks or brown sites for urban housing. Also, commercial back-office operations may relocate to out-of-town business parks, and this creates space for retail expansion in town centres.

The coevolution of the industrial, commercial, and informational divisions of the economy is reflected by changes in the long-term property returns that can be achieved at different urban centres. However, the transformation of towns and cities as a result of cumulative investment over time is interrupted by business cycles, with variations in interest rates.

It follows that a comparison between the long-term returns on property investments at the various categories of urban centre has really to be made over the same time-frame. The contemporary timescale for the transition of a town from industrial to postindustrial is a span of forty to sixty years. By accurately positioning each town of the London Megapolis in its transition to an informational economy, contemporary data for total long-term returns on property investment, rental levels, and rental growth may be superimposed on the graphical representation of the towns at the various stages of the transition. The paragraphs that follow examine the returns on property investment in relation to structural change within the megapolitan urban centres.

As towns undergo the transition from an industrial town to a postindustrial society, the number of large office buildings at an urban centre (population 50,000-99,000) may increase from, say, fifteen to forty-five over a period of thirty to forty years. New generations of school leavers enter office employment as older generations leave industrial employment. During the first phase of the transition, there is a doubling in the number of offices from fifteen to thirty, whereas in the second phase, the increase from thirty to forty-five only represents an increase of 50%.

The level of investment is sensitive to the increase in the number of establishments per 10,000 population, and when the rate of increase slows down, the long-term total returns on office property investment decline. The long-term total returns on property investments are taken to include the rental income received and the capital growth over a fifteen-year period 1980-1995, as measured by the Investment Property Databank. It can be seen in Appendix 11A that the pattern of long-term returns from office investment follows an S-curve with increasing returns as a town takes off from being industrial based (secondary location index, SLI of 1.5-2.0 and quaternary location index, QLI of 0.5) and an acceleration during the transition (QLI 0.6-0.8), and a slowing down as the postindustrial state is reached (QLI 0.9-1.1), with a further decline in long-term returns as the information division becomes dominant (QLI 1.2-1.4).

The number of establishments per 10,000 population gives an indication of per capita incomes at the location, similar to GNI per capita, and it appears that rental levels tend to increase with the number of establishments per

10,000 population. Rental levels in the quaternary-dominant towns to the west of London tend to be higher than the east.

The pattern of long-term returns on retail investment also follows an S-curve, and this is shown in Appendix 11B, where towns are positioned using both the tertiary commercial division location index (TLI) and the QLI. The maximum long-term returns arise as a town becomes postindustrial (QLI 0.9-1.1) and continue at a slightly reduced level through the period QLI 1.2-1.4 and TLI 0.9-1.1, after which there is a further decline in the long-term returns in the informational economy.

There is a relationship between the rental values of industrial properties and the rental values of business space in a locality. It follows that industrial rents tend to be higher in the quaternary-dominant towns and lowest in the industrial towns, where there is often a surplus of industrial space. Appendix 11C shows that the trend for total long-term returns for industrial space tends to reach a maximum at QLI 1.2-1.4 before declining with the top of the S-curve. It should be noted that new towns, such as Basildon, Harlow, Crawley, Hemel Hempstead, and Stevenage have a well-balanced economic structure with a TLI of at least 1.1. Enfield also has a TLI of 1.1, and, with twenty-nine offices, has a strong office base, but with twenty-five industrial sites, it has a low QLI of 0.3-0.5, which belies the fact that it is undergoing a transition in its economic structure, with higher, long-term returns than might be expected from the location indices on their own.

The long wave of investment is extremely valuable for both analytical and predictive purposes and should probably be taken into account in property valuations. This new approach enables property investors to not only take a view on the right place for investment, but also the right time to acquire or dispose of properties at those locations. With new derivative-led investment products, property investors can now hedge risk and take positions on future performance as measured by a fund-performance benchmark such as the IPD Index. However, if the property investments are treated as bonds, then the dynamics of the urban transition should be reflected in the classification of property assets by urban category; otherwise, investing in synthetic properties in virtual cities is likely to be rather more risky than real bricks and mortar.

However, the most important implication of this research is that the proportion of the labour force involved in the quaternary information division in a metropolitan region is a measure of the quantity and variety of information requiring processing, and this is a reflection of the level of complexity of the ecostructure. The long-term returns on property investment for office, retail, and industrial facilities were obtained for each urban centre or district based upon institutional property investments. This provided the empirical evidence for the change in investment returns with the evolving economic structure of towns, so that gradients are set up between locations in the rate-of-return on investment. It was conclusively shown that the total long-term property investments follow a logistic S-curve with lower returns in industrial towns, increasing returns in the transition to an informational economy, and then diminishing returns as a town becomes postindustrial.

In conclusion, ecostructures of households and establishments are complex adaptive systems for the diffusion of investment capital, and they evolve with increasing complexity to economise on the unit cost of transactions. Capital formation drives the urban system away from equilibrium, which sets up gradients between locations in the rate-of-return on investment. More complex cities require an increasing per capita cost to maintain them, and investment in transactional complexity within a region or state will reach a point where there are diminishing returns from further investment in complexity. The high costs of complexity can be reduced through decentralisation or dispersal to locations with lower property and staff costs. So an ecostructure reaches a spatio-temporal peak of diffusive capacity, as less mature ecostructures emerge to provide alternative pathways for the diffusion of investment. The redistribution of investment to less mature parts of the urban system tends to reduce the gradients in the rate of return on investment. This encourages the development of a world city hierarchy that will, in the long term, increase diversity and reduce inequality between major urban centres.

5.6 Sustainable civil development

Sustainable development has been defined by the World Commission on Environment and Development (the Brundtland Commission, 1987) as "meeting the needs of the present without compromising the ability of future generations to meet their own needs". However, a more specific definition for sustainable civil development is that it involves extending the lifetime of a planetary civilisation, by leaving viable human settlements for future generations at appropriate planning standards using longer-life assets, clean technologies, and renewable resources to achieve equality in life expectancy for all citizens. Sustainability involves conservation of the overall stock of assets, which includes artefacts, wisdom, natural resources, and genetic material. If human societies are to evolve to their full potential with a reasonable per capita standard of living, development of the world's cities will involve substitution of artefacts for natural resources and the other constituents of the asset stock. There is, therefore, an upper limit to the capital stock, the population, and the operational resource requirements that can be sustained, if losses to the life-supporting capabilities of the biosphere are to be minimised. For this reason, it is vital that a reasonable per-capita standard is defined in terms of sufficient stock of assets and wealth, rather than a flow of income or consumption.

The overall asset stock (S) comprises human settlements and artefacts (S_H), human wisdom (S_W), natural resources (S_R), and genetic material (S_G). The total asset stock will be given by the equation:

$$S = S_H + S_W + S_R + S_G$$

Sustainable development involves conservation of the asset stock, in which substitution may take place within the constituent parts as explained by David Pearce and Jeremy Warford in *World Without End* (1993). For example, the accumulation of artefacts or the world system of cities may be substituted for natural resources, as long as the operational resource requirements are sustainable. The stock of wisdom is reflected in the indigenous communities, institutions, culture, knowledge, and information structures.

In the transition from an industrial economy to a postindustrial society, the increase in GNI per capita enables investment to be made for the abatement of pollution, which gives rise to an ecologic transition. The aim of sustainable civil development is to respect nature's boundary conditions and to intervene in the process of growth to reach a position of dynamic stability in the long run. However, there is a tendency for a catastrophe or disorder to upset the balance of systems, and these events may occur as a result of unstable systems, beyond the boundaries of a city region. It is envisaged that the new communication technologies will eventually make it possible to develop an efficient and ecologically conscious global community, in which countries cooperate to reach dynamic stability with a planetary civilisation.

Infrastructure services that help the poor can also contribute to environmental sustainability. Clean water and sanitation, non-polluting sources of power, safe disposal of solid waste, and better management of traffic in urban areas provide environmental benefits for all income groups. The urban poor often benefit most directly from good infrastructure services, because they are concentrated in settlements subject to unsanitary conditions, hazardous emissions, and accident risks. However, in many rapidly growing cities, infrastructure expansion lags behind population growth, causing local environments to deteriorate. By raising the productivity of farms and of rural transport, both an increase in the incomes of rural workers and a reduction in food prices for the urban poor can be achieved. The benefits of transport and communications include the access they provide to other goods and services, especially in cities. The construction and maintenance of infrastructure such as roads and waterworks contribute to poverty reduction by providing direct employment.

Cities draw on the natural capital of other regions to support their inhabitants and to absorb their waste, and the aggregated land area is known as the city's "ecological footprint". This concept is explained in *The Earthscan Reader in Sustainable Cities* (1999), edited by David Satterthwaite. The extent of that footprint depends upon the city size, the economies of scale that can be achieved from urban concentration, and the average standard of living of the inhabitants. It follows that the cities may achieve local sustainability by placing unsustainable demands on resources elsewhere, and, consequently, the regional and global impact of the world system of cities may increasingly result in global warming, climatic change, acidification, and stratospheric ozone depletion. Sustainability needs to be addressed within the permeable boundaries of an ecopolitan state and the civil ecostructures needed to support human populations of some 5m-20m.

In the long term, quality of life and living standards are threatened by over-exploiting or over-polluting the natural systems, because of the technological costs of reversing or overcoming the damage. Environmental standards are increasingly being applied in relation to air pollution, water and river quality, waste and toxic substances, and wildlife conservation. Energy supply and consumption involves the combustion of fossil fuels and releases carbon dioxide, sulphur dioxide, and nitrogen oxides into the atmosphere. Carbon dioxide is the main contributor to greenhouse gases and global warming, whereas the other two cause acid rain, and nitrogen oxide also contributes to tropospheric ozone depletion. In addition, solid fuels contain high proportions of wastes and pollutants such as sulphur, heavy metals, and ash. The transport sector accounts for a high proportion of major air pollutants, and in the European Union, transport emits some twenty to thirty percent of the region's carbon dioxide, fifty percent of the nitrogen oxide pollutants, and up to ninety percent of carbon monoxide. Transport methods also emit some thirty-five percent of Europe's volatile organic compounds (VOCs), which magnify the build up of tropospheric photochemical oxidants.

The dynamics of urban development are facilitated by the evolving energy, transport and information technologies, and the population movements arising from the changing employment opportunities in the evolution of towns and cities. The decentralisation of urban populations from monocentric to polycentric city forms, made possible by the motorway network and the growth in car ownership, results in increased use of energy, together with increased emissions of carbon dioxide. This process has led to the debate on the most efficient urban form for environmental and resource sustainability and improvements in the quality of life. Policies for overcoming automobile dependence are described by Peter Newman and Jeffrey Kenworthy in *Sustainability and Cities* (1999). A fundamental issue on the future use of the car is the trend over the past fifty years towards urban decentralisation and counter-urbanisation, where population growth is focused on the least urbanised areas. Production has tended to move away from traditional urban locations to cheaper peripheral locations, where new industrial property is available at lower densities in higher amenity environments. Also, there are the housing market factors, which have caused a popular shift in favour of small town or rural life, away from the problems of congestion and social deprivation of the older metropolitan centres.

The European Commission Green Paper on the Urban Environment advocates high-density compact cities and comes out very strongly against the idea of new towns or new settlements, with the recommendation that all future development should be within the boundaries of existing urban areas. The Commission argues that the compact city provides an intensive milieu that would encourage urban creativity and provide a superior quality of life in European cities and would solve the problem of environmental sustainability. It is questionable that high urban densities correlate with high urban quality of life or a reduction in congestion, and the environmental sustainability aspect needs to be properly evaluated. The early reduction in the use of hydrocarbon fuels through engine efficiency, and a shift away from hydrocarbon-based transport technology to hydrogen, natural gas, various bio-fuels, or electric motors would lead to a different conclusion. The use of carbon taxes to make the new technologies competitive with the price of petrol, and legislation to force the switch to fuels that do not create atmospheric carbon dioxide and global warming, could, within fifty years, enable an alternative clean energy infrastructure to be developed for road vehicles. The freedom for people to use their cars, even at a high price, is highly valued, and it is envisaged that saturation levels for car densities in Europe could be one car per economically active person, i.e., forty-five cars per 100 population. In the London Megapolitan Region there are forty-four cars per 100 population, and in Greater London 39m² of land per inhabitant is devoted to transportation uses.

Between 1961 and 1991, the population of South East England increased by five percent, and, with urban decentralisation, the Greater London population fell from 7.99m to 6.39m, whilst that for the Outer Metropolitan Area increased from 4.39m to 5.45m. However, if the population of the metropolitan centre and each class of urban area had increased by five percent without decentralisation, an alternative pattern of transport energy consumption would have arisen. It follows that the energy efficiency of the megapolitan urban structure may be assessed by comparison of the two scenarios. This methodology is explained by Michael Breheny, the author of a chapter entitled "Counter-urbanisation and Sustainable Urban Forms" in the publication, *Cities in Competition*" (1995).

In Table 3 below, the population figures are given for the areas of the Megapolis, together with a 1991 population projection, assuming no decentralisation. Using rates of energy consumption by car and other modes estimated for a European Commission study by ECOTEC (1993), it is shown in Appendix 12 that weekly transport energy consumption in 1991 was less than two percent higher than it would have been without decentralisation.

TABLE 3: DECENTRALISATION PATTERNS

	Megapolis Population 1961-1991 (000s)					
	Pattern of decentralisation				Without decentral-isation	Population difference by area
Area	1961	1971	1981	1991	1991	1991
Greater London	**7992**	**7452**	**6696**	**6394**	**8392**	**-1998**
Inner London	3493	3032	2497	2343	3668	-1325
Outer London	4499	4420	4199	4051	4724	-673
Outer Metro-politan Area	4390	5153	5379	5447	4610	+837
Megapolitan Region	**12382**	**12605**	**12075**	**11841**	**13002**	**-1161**
Outer South East	3610	4326	4657	4953	3790	+1163
South East Total	**15992**	**16931**	**16732**	**16794**	**16792**	**+2**

In general, the following principles have emerged for a sustainable urban form:

Urban size

- Larger urban settlements are more energy efficient.
- Public transit systems are economic for larger cities.

Urban densities

- Higher urban densities are consistent with lower travel demand.
- Efficiencies in the integration of transit systems arise at densities of over 4,000 inhabitants per km².

Business location

- Commercial property and land costs rise with higher urban densities, which increase the incentive for companies to decentralise to reduce both property and staff costs.
- Manufacturing companies are attracted to the modern factories in the industrial areas at the outer centres, where there is scope for expansion and easier access to the motorway network.

Residential location

- There is a residential preference for lower urban densities away from the congestion, crime, social deprivation, poorer schools, and the physical decay of the older cities.
- House prices are lower at suburban or decentralised locations.

Employment location

- Urban areas with a balance between employment and houses will be more fuel-efficient.
- Urban decentralisation increases radial work trip lengths, but lateral and orbital commuting reduces average journey times.

Energy consumption

- Counter-urbanisation movements down the urban hierarchy over the past thirty years have resulted in a 2% increase in energy consumption.

- The highest rates of energy consumption arise in the rural areas, which have shown the largest population increase.

Information technology

- Telecommuting by 15% of the workforce for one day per week would reduce energy costs by 3%. By the year 2010, some 20-25% of the workforce may telecommute for three to five days per month.
- Telecommunications and information technologies enable corporations to decentralise back-office operations.

The London Megapolis is an efficient urban form with an overall density of 1,100 persons per km², containing a compact Greater London at a density of 5,100 inhabitants per km², which has adapted to modern urban conditions with maximum economies of scale and other advantages of concentration, without losing the capacity to provide liveable environments. The metropolitan centre is served by the railway, London Underground, and other forms of public transport. The outer centres provide lower-density residential areas with easy access to the metropolitan centre for cultural events, entertainment, and shopping. It appears quite feasible for the small increase in vehicle fuel consumption from a decentralised way of life to be offset by telecommuting by a small proportion of the workforce.

In due course, it is likely that road-user charging will increase the occupancy of cars, people will be encouraged to live closer to work, and more fuel efficient vehicles will be developed. By the year 2050, alternative technologies such as the electric car, or hydrogen and methanol fuels, are expected to provide substitutes in over half the vehicle population. The contribution from urban planners will be the concentration of development at urban and regional centres, ensuring that the number of establishments per 10,000 population meets the planning factors or norm for the urban category and restrictions on urban area development in township communities, service centres, and local centres. There is a range of trade-offs between the ecostructure systems, which include energy, transport, urban land, human time, environmental regulations, and the relative prices in the transaction system.

Civil ecostructures: Chronological references

1. *Megalopolis: The Urbanised Northeastern Seaboard of the US* - Jean Gottmann, 1961.
2. *Sleepers Wake! Technology and the Future of Work* - Barry Jones, 1982.
3. *World Labour Report* - International Labour Office, 1984.
4. *The Future of Urban Form* - Edited by John F Brotchie, Peter Newton, Peter Hall & Peter Nijkamp, 1985.
5. *Our Common Future* - WCED Brundtland Report, 1987.
6. *The Metropolis Era (Volumes 1 & 2)* - Edited by Mattei Dogan & John D Kasarda, 1988.
7. *Metropolis* - Emrys Jones, 1990.
8. *Regional and Urban Policy in the United Kingdom* - Roger Prestwich & Peter Taylor, 1990.
9. *London - A New Metropolitan Geography* - Keith Hoggart and David R Green, 1991.
10. *The Global City* - Saskia Sassen, 1991.
11. *Edge City* - Joel Garreau, 1991.
12. *An Urbanising World: Global Report on Human Settlements 1996* - UNCHS (Habitat), 1996.
13. *Sustainability and Cities* - Peter Newman & Jeffrey Kenworthy, 1999.
14. *Earthscan Reader in Sustainable Cities* - David Satterthwaite, 1999.
15. *Spon's Architects' and Builders' Price Book 2002* - Edited by Davis, Langdon & Everest, 2002.

6.
TRANSACTIONAL COMPLEXITY

6.1   Institutional coordination of urban microsystems

Regulatory systems emerged in the ancient city-states of Greece where the term *polis* referred to both the state and the city. The cities became centres of local production and long distance trade and supported merchants and craftsmen with an army for defence. Aristotle wrote of the need to control man's social instinct and of the requirement for a state as a means of establishing law and justice. He defined citizens as those who have the power to take part in the bureaucratic or judicial administration of any state, and the state was defined as a body of citizens sufficing for the purposes of life. Aristotle's concept of citizenship appeared to be democratic, although slavery existed and society became divided into classes of people comprising those with the gifts, ability, and leisure to participate in matters of state and the many mechanics and labourers who were residents of the city-state but who did not have time to be involved.

In republican Rome, citizenship initially referred to the political rights and obligations of patrician Romans, who paid the taxes, fought the wars, and influenced the government. Citizenship was eventually extended to large portions of the population in the provinces of the Roman Empire. The system of Roman law gave citizens the right to vote in the Assembly of the Roman People and also the right of appeal to the same Assembly or to the Emperor. Today, the concept of democracy includes citizenship, human rights, constitutions, voting and elections, economic freedom, and self-determination. However, in matters of international legislation, military security, and the democratic processes of many nations, the regulatory system depends primarily on legitimated threat. The challenge for the future is to achieve a change in democratic systems so that a more equitable society emerges with increasing concern for future generations and humankind.

Institutions are created to provide the structure that societies impose upon the interactions between humans and the transactions between establishments to prevent chaotic conditions. According to Douglass North in *The Economy as an Evolving Complex System 2* (1997), institutions perform an anti-chaotic role as they define the formal and informal rules of the game and regulate the interactions and transactions through legislation, codes of practice, convention, and behavioural norms. Institutions are also responsible for the normative enhancement or design of the urban microsystems and also of the global macrosystems (see Chapter 12). The institutional architecture of the civil system has four main components, and these are governance or regulatory systems, commercial systems, human development systems (social capital), and infrastructure systems (civil capital).

There are institutions of government, such as legislatures, regulatory organisations, and local government, which require a well-run civil service and an effective judiciary with laws to ensure that contracts are enforced, property rights are honoured, bankruptcies are settled, and monopolies are controlled to maintain competition. Commercial institutions include banking, financial and insurance, corporate institutions for professionals, and trade unions. Human development systems include public information systems, an education system with academic institutions, a health system, a community system in which the institution of the family or household is protected by society and reinforced by the state, and a social security system. The infrastructure systems include transport, telecommunications, utilities networks, sewerage and waste disposal with protection of the environment. Institutional reforms specify new rules or alter old ones to change the behaviour of the system. However, the difficulty in altering institutional paths means that institutions also act as constraints with respect to societal choices.

In the 21st Century, the governance systems will face two counterbalancing forces: globalisation with continental integration of nations and localisation with the devolution of power to states at the subnational level. Evolution of the civil system involves increasing complexity, which means additional hierarchical levels at the planetary, continental, and subnational levels. New institutions will evolve at both the supranational and subnational levels and, in a bicameral democratic system (i.e., with two legislative chambers), seats in the upper house are often allocated to give equal representation to states or provinces, which gives preference to norms of territorial representation over norms of proportional representation by population. A highly integrated and socially equitable international economic order will require a change in the balance of power, with multiparty politics, subnational political units, and the involvement of community groups and nongovernmental organisations (NGO's) in governance.

Human development institutions are essential for raising the standard of living in the poorest countries of the world and also for the deprived segments of society in the MDCs. In the western countries, the views of the Washington Consensus (The IMF, the World Bank, and the US Treasury Department) are disseminated by political leaders and economic decision-makers through public information channels to institutionalise the global economy. The term 'Washington Consensus' was originally coined in 1989 to describe a relatively standard set of reform packages for the recovery of Latin America from the financial crises of the 1980s, but the term has since been augmented to include the WTO and neoliberal globalisation policies. Critics of this collusion query who will benefit and whose rights will be protected by rules that fuel unchecked and unequal growth. Economic regionalism is seen as a defence against global capitalism, in which the power of the many at the subnational level can counteract the few at the supranational level.

The concept of a global civil society has evolved for forging new transnational links and the presentation of demands to governments in many countries. Global citizenship has three central propositions as follows:

- Equality in individual and human rights
- Free and universal political participation
- State responsibility to ensure adequate standards of human welfare

The growing network of associations, interest groups, and non-governmental organisations (NGOs) that make up Global Civil Society are effectively using the new communications and information technologies to criticise those governments, international agencies, and transnational corporations that ignore or harm the interests of the poor.

The internet is providing a new communications channel for coordinating local interests groups and NGOs. Institutions for the accumulation of knowledge include Research and Development establishments, licensing agreements, together with educational establishments, such as schools, colleges and universities, and broadcast learning programmes. Urban centres are expected to offer better access than rural areas to educational services and healthcare, with clean water supplies and sewerage. Rapid urbanisation, in which access to shelter is deteriorating with 100m people without permanent homes, requires social safety nets to replace the moral institutions offered by a rural hierarchical society.

Public information systems include newspapers, radio, television, telecommunications, and information technology. The newspaper industry has reached saturation levels in the more developed countries, and the markets are becoming increasingly dominated by larger newspaper groups that produce local newspapers with common feature articles. Television is the preferred source of news for the majority of people on account of its timeliness and reliability, and it is only on some very local issues that local newspapers are the primary news providers. By 1990, some ninety-eight percent of households in the US had colour television sets, and sixty-percent of households had access to at least twenty television channels. Two-thirds of households had a video recorder whose prime use was to 'time-shift' the recording of programmes so that they could be viewed at more convenient times. There is evidence of 'narrowcasting' and media de-massification in the US, in which cable networks cater to large Hispanic minorities or ethnic groups in particular communities. Broadcasting standards, film censorship or classification, press standards authorities, and libel laws form part of the institutional fabric in an increasingly permissive society.

In a less developed country, such as India, over ninety percent of the circulation of daily newspapers is in cities with over 100,000 population, and the papers reach fewer than ten percent of the population. The sheer size of India means that the country has more university graduates than developed countries the size of the UK. Similarly, the country has a large number of people working in urban areas in the quaternary information division. As William Mitchell explains in *E-topia* (2000), the imaginative use of satellite technology has facilitated the rapid advance of broadcast mass communications in India and has overcome the limitations of the rural dispersal of the population and the inadequate educational infrastructure. The Indian government has promoted high-speed satellite earth stations at Bangalore, Hyderabad, Pune, Noida, Thiruvananthapuran, and Chandigarh, with 24-hour international connectivity to software parks and enterprises within a 20-30 km radius. India is now the second largest exporter of software. However, India's 'infostructure' remains well below the standard of the more developed countries, and this creates a handicap for the country's industry and commerce.

New public works for highways, railways, air transport, telecommunications, broadcasting, electric power, and nuclear energy, involve a wide range of legal and regulatory matters before they can proceed and decisions taken, which creates institutional path dependence in the evolution of the technology. Institutional requirements include international regulatory authorities and transport regulations for road vehicles, aircraft and maritime vessels. A system of public roads and an infrastructure of petrol stations had to be organised for a mass market of cars, together with signs, signals, and rules of the road. The railway gauges and the size of railway cars selected in Britain were not the most efficient system and created a penalty for taking the lead in rail technology. Similarly, aviation research and development, the airport infrastructures for air transport, air traffic control systems, and other examples, such as the allocation of the radio spectrum or telecommunication channels, require institutional mechanisms, standards, and regulation.

Supranational, national, and local governments are allowing entry of private enterprises into previously monopolistic infrastructure markets, as a result of the visible physical deterioration of infrastructure assets and services, a lack of spending on new facilities, and a substantial backlog in maintenance and rehabilitation. The obsolescence and physical delay of urban infrastructure threatens to slow and even reverse the economic growth of the older industrial cities. Powerful transnational alliances between network operators in transport, telecommunications, energy, and water are intensifying patterns of ownership. In the telecommunications sector, investment in national and international grids involves alliances of private telecom, internet, media, and entertainment companies. This is resulting in new customised infrastructure networks being linked and overlaid on established 'sunk' infrastructure, which involves the unbundling of complex unitary networks so that private investment is attracted to low-risk infrastructure elements that can be 'splintered' off from the whole. The outcomes from this development are increasingly adaptive nonlinear networks that add to the connectivity and complexity of socio-technological systems.

Network unbundling can take the forms of vertical segmentation, horizontal segmentation, and virtual segmentation. For example, in the electricity sector, vertical segmentation involves the separation of generation, transmission, and power distribution. Horizontal segmentation can take the form of geographical unbundling, such as the breakup of the British Rail network into regional franchises and the UK water industry into regional monopolies, with an industry regulator using a range of performance comparisons to monitor the efficiency of the different regions. In the telecommunications sector, cellular and satellite-based services have been horizontally segmented from the traditional landline services. However, it is the new information infrastructure that permits virtual network segmentation by handling the millions of transactions between customers and multiple suppliers. For example, virtual segmentation of local gas, electricity, and telecommunications distribution networks allows multiple suppliers to compete for the right to supply customer's premises. Also, virtual network segmentation of highway networks offers different lanes, such as privately funded toll express lanes, high-occupancy vehicle lanes, and commercial truck lanes that can coexist with public access motorways.

Large socio-technological systems arise from the evolution of local systems by entrepreneurs, with the support of social, political, and institutional processes, into standardised, national, and widely-accessible infrastructures. There are also interconnected secondary technical systems such as scientific research, electronic finance, credit cards, international banking systems for currency exchange, and government regulation. In *Splintering Urbanism* (2001), Stephen Graham and Simon

Marvin give the example of a petrol filling station where an ATM or credit card can be inserted to pay for the fuel, and the mechanism of the pump seamlessly integrates at least six large socio-technical systems, namely the highway and vehicle systems, computer and telecommunications systems, the banks' financial and ATM or credit card system, and the global oil production and distribution system. This perspective shows how infrastructure networks accrue on an incremental basis, creating denser and more elaborate systems that are linked over wider distances. However, the stability of these networks is threatened by liberalisation and the unbundling of infrastructure systems in a global economy.

Financial institutions absorb the surplus investment from the primary and secondary sectors for reinvestment in property development in the office, retail, industrial, and housing sectors. In the distributional stage of development, government and the institutions collaborate in adapting the urban system through substantial public expenditure in road systems and new towns to facilitate industrial relocation and the suburbanisation of the population. This extends the opportunity for private home ownership through mortgages, encourages car ownership, creates further investment returns from high-interest charges on easy credit, changes lifestyles and consumer habits through advertising, and increases energy consumption through central heating systems, a range of domestic appliances, and personal travel. This leads to a further round of capital accumulation, and businesses decentralise to business parks, superstores are created on retail parks, and leisure facilities flourish. As the metropolis evolves into a megapolis, the institutions supplement their investment funds through private-pension schemes until urban growth slows down. Property investment is then diverted to other urban centres with the intention of generating another virtuous cycle of capital accumulation. This process is described by David Harvey in *The Urbanization of Capital* (1985).

It was stated in Section 4.3 that, within an ecostructure, a balance needs to be achieved between the interactions of different populations, which include humans, establishments, facilities, infrastructures, and vehicles, each with their own characteristic life cycles. The real activity in the ecostructure is found in the transactions within the community, which are initiated between households and other establishments in the area with the aim of enhancing their state of well-being. The greater the number and range of transactions, the greater is the resultant quality of life for humans. The community microsystems include institutions and facilities for health, education, welfare, social services, recreation, religion, and culture.

6.2   Spatio-temporal issues

The microsystems of the civil system relate to the organisation of civil ecostructures, the regulation of the inhabitants (i.e., law enforcement system), a distribution system for exchange or transactions, and a community system for human development. There are also systems for the control of resource flows and urban space-use to reflect the time-use of human activities. Zoning is the allocation of land uses within an urban area to residential, retail, commercial, industrial, and other various mixed uses. Planning zones and development control regulations are essential in rapidly growing urban areas to coordinate the private configurations of land-use with the public sector, such as transport facilities, utilities, and health and education services. Societies also require information systems, infrastructure, and energy systems that are developed in conjunction with those for industrial production and the transport of goods and people. The microsystems within a civil ecostructure include the following:

- Regulatory system
- Community system
- Cyclical time patterns
- Transaction system
- Resource flow system
- Urban space-use system
- Information system
- Energy system
- Industrial system
- Transport system
- Distribution system

When rural populations migrate to the cities, one of the most noticeable changes in their lives is the allocation of human time to different activities.

In rural areas, daily activities are scheduled by the seasons with people walking long distances to the fields and pastures, or carrying water to the household, and long hours are spent working the soil, tending animals, and collecting fuel. Family life, public meetings, trading, religion, education, entertainment, and rest take up the balance. In urban areas, there is also a common cyclical tendency in human activity, such as a typical day, week, season, or year. Peak periods or seasonal fluctuations are reflected by the academic year in education, the holiday period in tourism, and the promotional sales in retail.

The human life cycle provides a revealing framework for the allocation of time by a person. If an average person lives to seventy-eight years, they would spend 680,000 hours alive, of which thirty-five percent may be asleep. A forty-year working life with four weeks annual holiday, and minor periods of sickness or absenteeism, will only account for ten percent of an urban lifestyle. Personal services, which include washing, cleaning, household activities, care of children, and purchased personal services, probably account for some twenty percent of human time. Play amounts to perhaps ten percent, and much of this time is spent in childhood. Eating and drinking probably take up some seven percent of a person's time. A further six percent is spent on audiovisual entertainment including television. Education and reading may account for a further five percent. Travel, including walking, cycling, and transport in vehicles, may account for four percent of the human time budget. The balance of three percent is taken up with shopping, meetings, parties, conversation, and personal reflection. Detailed statistics of the use of human time can be compiled from labour-force statistics, school and higher education enrolment, travel surveys, television viewing surveys, circulation audits of newspapers and magazines, library lending rates and book sales, attendance at sporting events and cinemas, shopping profiles, and restaurant expenditures.

It is clear that, over a period of time, the dynamics of civil ecostructures will change through changes in the populations of households, establishments, and artefacts within the system, and the level and scale of transactions within a specific territory. For a typical European city of 1m inhabitants, the daily flow of inputs will include 11,500 tons of fossil fuels, 320,000 tons of water, and 2,000 tons of food. The daily outward flow includes 300,000 tons of wastewater, 25,000 tons of carbon dioxide, and

1,600 tons of solid waste. Inevitably, institutions are needed for the control of food safety, water quality standards, and permissible pollution levels. If society is successful in raising the limits of growth through economic and technical adaptions, then a new limit will be encountered in due course. It would seem that time is generally the ultimate limit, because if the period for effective action is reduced, systems that can cope with slower rates of change finally begin to fail. The systems determine the limits to the factors of production or reproduction as described below.

The pushing back of space limitations has been vital in the process of expansion and development. Planning zones create efficient land-use patterns that will support well-designed urban transport systems, which are capable of incremental expansion and extension. Zoning may define the intensity of use through minimum or maximum limits for plot sizes, floorspace, or floor area ratios. In agriculture, there has been the increase in product per hectare, and in urban development the technology of high-rise buildings and multi-storey structures has increased the utilisation of metropolitan land. However, higher yields from both rural and urban land involve additional energy and material inputs. Multi-storey buildings require deeper foundations, stiffer frames to resist wind loadings, lifts, heating, and ventilation, so that the cost per m² of net floorspace increases with height, and the net floor area is reduced by the provision of extra lifts and plant rooms. Building regulations and codes of practice for the design of building systems are combined with specifications and standards for the building elements. However, the higher land values in central areas and the increased costs of construction result in higher population densities with less space per person. This leads to increasing suburbanisation and decentralisation in which residents and business seek to obtain a better environment with improved space standards, at a lower total cost.

Although the market for land may be equivalent to trading a commodity in certain times and places, it is the material and locational properties of land that change its value. The quality of the soil, the level of the water table, rock formations, and mineral composition, oil or coal deposits will all have a significant bearing on its value. There is also a trade-off between the limiting factors of space-use and time-use, so that remote areas of fertile land may provide economies of space with diseconomies of time. Space and time can also be complementary where an increased yield of crops diminishes the time needed to harvest them. This would arise when an irrigation system or a greenhouse structure permitted two or three crops per year, to economise both space and time and provide an increased yield per hectare.

The car, which serves to economise time, has led to the requirement of extensive road networks and car parks and an increase in the use of materials and energy. However, underground railways or air transport save both space and time, and consequently increase land values in the vicinity of underground stations or airports. Communications technology that provides a substitute for human transportation is another example in which space-use and time-use are complementary.

During the process of economic development, trade-offs take place between factors of production such as space and time, or industrial technology and employment. For example, transport infrastructures that reduce journey times permit suburban development and reduce population densities with a relief of congestion. The primary, secondary, tertiary, and quaternary factors of production, together with their infrastructures, are summarised below:

- Primary (extraction)
  - Natural resources and energy
  - Pipelines and electricity grid infrastructures

- Secondary (production)
  - Industrial technology
  - Manufacturing establishments and urban infrastructure

- Tertiary (transaction)
  - Commerce, transport, and distribution technology
  - Finance, capital markets, and commercial infrastructure

- Quaternary (interaction)
  - Information technology
  - Media, communications networks, defence, and social infrastructure.

Infrastructure systems relating to industry, energy, and transport are described in Section 8. Distribution systems evolved with the earliest civilisations for trade and commerce between geographically separated buyers and sellers. Since the industrial revolution, the development of integrated physical distribution systems has had the aim of both increasing the size of markets to achieve a fall in the unit cost of production and reaching a trade-off between the level of service offered to customers and a reduction in the unit cost of distribution. The ability to achieve profitable transactions depends upon the logistical network of a company, and the distribution performance has to be complementary to the other corporate functions of production, marketing, and finance, so that a just-in-time delivery capability can be achieved using advanced information and communication technologies.

6.3 Transactional complexity

In the transactional or economic system, the aim is to satisfy human wants through the provision of goods and services. These are supplied by production and exchange and limited by scarcity of resources and technology. Economic efficiency means going as far as possible in the satisfaction of wants within resource and technological constraints. In a market economy that uses money or credit, terms of trade are usually specified by prices, which take into account property rights, search or transaction costs, and the ability to enforce a contract. The strength of the bond between parties to a transaction takes the form of a preference for one opportunity amongst several options, and this selection has to be reciprocated by the other party. Each party acts purposefully to maximise the expected utility or gain from the transaction.

Complexity describes the stage of evolution or level of maturity of an ecostructure in which investment capital or energy is diffused with increasing transactional complexity. This involves high sectoral diversity within an ecostructure, larger establishments, narrow niche specialisation, business networks rather than supply chains, higher levels of energy efficiency, an increase in information, and stability or resilience in the presence of external disturbances. In the context of complexity science, any movement towards complexity involves an enhancement of the information structures, through hierarchical structures with increasing numbers of levels between the microstructure and the macrostructure, increasing connectivity between nodes of a network or web, knowledge structures involving internal models (including computation, rules, metaphors, and paradigms), informational systems such as software, and, finally, technological or morphological complexity with subsystems, elements, components, assemblies, sub-assemblies, and parts.

The term transactional complexity as used in this book describes the level of informational complexity in a complex adaptive system, and it may be measured for an ecostructure through the proportion of establishments in the quaternary information division. Information is pervasive at all levels in a hierarchy, and, according to information theory, there has to be an element of novelty or, by definition, it is not information. Increasing interdependence of establishments in the transactional system involves increasing specialisation as each has to contribute to the vivisystem as a whole. Complexity increases as new arrangements are made for maintaining the internal coherence of a larger system with each incremental step of increasing interdependence. Eventually, the evolution of the system will reach the limits of cost-effectiveness, and increasing transaction costs will lead to the defection of establishments from the system, which may proceed to decompose until a new, stable level is achieved. Evolution is a disequilibrium process in which periods of stability are temporary, since improbable events will in due course happen. Discontinuity and punctuated equilibrium can arise from expanding a vivisystem beyond some indeterminate threshold, and infrequent occurrences of major power system failures, computer system crashes, and stock market collapses are typical examples. Technological evolution is discussed in Chapter 8 in which increasing sophistication and informational complexity evolves to increase the performance of a variety of mechanical or electronic assemblies.

The Structure and Dynamics of Networks (2006), by Mark Newman, Albert-Laszlo Barabasi, and Duncan Watts explains the new science of networks. Networks that are intended to serve a single coordinated purpose, such as transport networks, telecommunications systems, and power grids, are evolving structures built over long periods of time by a variety of independent agents and authorities. Networks have both topological and dynamical properties, which depend upon the traffic and the pattern of connections. Many networks involve the connection and disconnection of linkages and nodes (or vertices), so that the connectivity of the network as a whole depends upon processes operating at the local level, which both constrain and are constrained by the network structure. In a highway network, the nodes are cities and the links are the highways connecting them. In an air-traffic system, a large number of airports are connected via a few major hubs, which are highly connected nodes.

The concept of connectivity and the evolution of transport and utilities networks are important in the application of complexity science to infrastructure economics. This will be explained in relation to the growth of railway networks in Europe and the networks for underdeveloped regions, as outlined by Kevin Cox in *An Introduction to Human Geography* (1972). The phases of geographical growth for railway networks are an initial phase of localised linkage, a subsequent phase of network integration, a phase of intensification and, finally, the mature phase in which the unprofitable or less profitable lines are closed down. In the initial stages of network development, there is a predominance of short links and isolated network segments, a prevalence of links that are complementary to the previous network of navigable waterways, and branching or tree-like networks. This approach results from the uncertainty of demand for the construction of a railway in a new region, and initial links are selected so that an optimal benefit-cost ratio for traffic revenues to investment costs in the context of capital constraints to minimise the debt burden of future expansion.

At the integration stage of network evolution, isolated links were integrated into a network to circumvent the earlier complementary waterways, so that they become more competitive than complementary. The increase in traffic flow leads to both greater connectivity and density, which is defined as the length of route per km^2. In the intensification phase, the branching networks evolve into circuit networks formed by triangular, delta shaped, or rectangular structures. This involves connecting places that are already linked by a more circuitous route. In addition, low-cost, single-track feeder lines are added to increase revenues by drawing traffic from rural areas or small ports. The coevolution of the road network and rural bus services, which are more competitive over short distances, leads to the final phase of maturity. There is selection between the more profitable parts of the network and the less profitable branching routes and feeder lines, with the closure of redundant tracks. The resulting network becomes more completely dominated by circuit elements than before. Also, the development of containerisation and freight terminals tended to concentrate goods traffic onto fewer routes to raise the level of demand, so that economies of scale could be achieved in the handling and transit of freight.

The dynamics of network evolution and route location is a learning process in which the initial levels of demand lead to more knowledge and understanding of urban development and the profitability of projected links. As the initial investment yielded increasing returns, over-optimism caused entrepreneurs to over-expand, and diminishing returns came with maturity of the system. The connectivity of nodes may be calculated as the number of links with other places or nodes in the network divided by the maximum possible for a network linking that number of places. More simply, a hierarchy of connectivity for a network as a whole can be determined for places by comparing the actual number of direct connections. Urban connectivity with respect to transport routes tends to be higher for larger cities, and the diffusion of network infrastructure corresponds to a socio-economic gradient in which the less developed regions lag behind the more developed.

Civilisations develop through an increasing number of transactions involving the selection, transportation, and transformation of materials into artefacts, which require both energy to support the transactions and information to provide the organisation and knowledge to direct the energy and resource flows. Communications comprise a significant share of the interactions within a society's population, and this contributes to the successful completion of transactions. Complex societies have developed interlinked subsystems that evolve with time to achieve economies of scale and a reduction in the unit costs of transactions, principally through the world system of cities and urban growth. These economies of scale are realised through civil administration; efficient extraction and utilisation of resources; investment in agriculture, such as irrigation, mechanisation, and storage; industrial development; commercial organisation for trade and distribution; public protection with police and fire services, defence forces, health and education services; and the provision of public works including maintenance of the capital stock.

Societies depend upon order and regulation for their cohesion, and increasing connectivity within a global economy causes a rise in the number of unpredictable factors that may threaten the well-being of citizens. These include invasion by other ethnic groups, reduced employment opportunities, and a fall in consumption, financial collapse, poverty, and epidemics, crime and violence, or natural disasters. The role of state government is to regulate the class conflict between the corporate producers and the human consumers, which is exacerbated by the mobility of capital and the relative immobility of the less-skilled labour in the MDCs. The coevolution of the transnational corporations and ecopolitan society will be dependent upon reciprocity between both classes, in which investment and employment opportunities are spatially located, so that consumption patterns and corporate revenues reach an equilibrium position at a new level. Too little connectivity at a location can result in too little change and adaption, and beyond a certain threshold, too much connectivity results in standardisation and uniformity with a reduction in complexity and a decrease in adaptability. By focussing on the world system of cities within the protective context of a continental union, a suitable framework may be provided for urban analysis at the level of the state.

In the course of time, nations may rise and decline, only to recover again with waves of investment in new technologies or changing patterns of trade and commerce. Similarly, cities grow on the strength of the profitability of their industries and businesses, which leads to further investment and employment creation. The prospect of a higher standard of living attracts migrants and a positive cycle of residential development, so that urban expansion may continue with the creation of new service and transport networks, an increase in the transactions per time period, a rise in per-capita energy consumption, and, in time, diseconomies of scale. In complex societies, increasing amounts of information are processed to search for innovative methods to reverse a declining economic trend. The informational economy evolves with greater specialisation and new information and communication technologies, which permit cities to alter their spatial structures by outward expansion, polynucleation and, in the largest agglomerations, by the growth of a megapolis.

A megapolis is a feature of a postindustrial society, in which the proportion of employment in the quaternary information division reaches forty percent. This provides a measure of the level of complexity of a megapolitan region, and an objective of this research is to identify the point where increasing transactional complexity yields no increased benefits. It may be expected that there will be declining marginal productivity in a number of sectors in a postindustrial economy, including education, research and development, agriculture, minerals and energy extraction, manufacturing and, in turn, information processing. The marginal cost of evolving to a higher level of complexity is likely to be less attractive than remaining at the current level. However, any further economic stress is likely to result in reductions in local government budgets, a decline in services to the population, and eventually, a decrease in the level of complexity. It follows, therefore, that employment in the informational economy is likely to reach an upper limit in advanced economies, when inferior rates of economic growth will result in stagnation and a levelling off in the demand for new information.

6.4   Interurban linkages and networks

In addition to the constraints imposed upon a country by its physical geography, including climate, mountain ranges, land area, rivers, and coastline, early city-system growth is profoundly affected by a colonial history and its early economic development. For example, since American independence, the study of *American Metropolitan Evolution* (1961), by John Borchert, shows that, prior to 1830, the location and spread of American Cities was primarily influenced by wagon and sail technology, in which the extension of transport networks westward created the initial skeleton of the national city-system and several regional city-systems. From 1830-1870, canals, railways, and steamboats were the dominant transport technologies and, prior to 1850, postal services, newspapers, and business travel were the principal means for the interurban diffusion of information. Ports with large rail territories grew to prominence as a result of their trading activities, and from 1870, it was industrialisation that provided the impetus for urban growth. Telecommunications developments with the telegraph and telephone linked the major industrial centres, so that information diffusion proceeded in a horizontal linkage to other primate metropolitan centres with their own regional subsystem of cities large enough to provide a suitable market or specialised products. From these origins, a rank-size hierarchy of urban centres has arisen from interurban transaction linkages and, in the 20th century, the car, air transport, and advanced telecommunications further shaped the regional networks. In the 21st century, the new high-speed digital telecommunications infrastructure will reshape the urban patterns that emerged from the earlier networks of piped water, sewage disposal, gas mains, the electricity grid, telephone lines, and transport systems.

The Central Place theory of Walter Cristaller (see Section 4.3) related to the distribution of wholesale and retail services between cities of different sizes. The smallest class of rural towns provides the basic high-frequency services needed by rural communities. Larger towns are fewer in number, but they provide higher-order services to satisfy the needs of a larger market from several townships and villages. In turn, metropolises and cities provide even more specialised services to the other towns in the regional urban system. There are two mistaken assumptions associated with this theory, which are, firstly, that it is applicable to manufacturing and, secondly, that economic growth will trickle down the urban hierarchy in a predictable way, without taking into account upward and lateral transactions. In *City Systems in Advanced Economies* (1977), Alan Pred demonstrates how forward and backward linkages through business networks generate economic growth in the urban system, and examples are given in Section 5.3 for the London megapolis.

A metropolitan region operates in the context of the world-city system of which it is part, and the evolution of city regions arise from positive feedback in which a web of interconnected transactions and interactions causes an increase in the number and diversity of resource, industrial, commercial, informational, and lifespan establishments. The benefits of an establishment of belonging to the larger ecostructure are derived from a reduction in unit transaction costs, and a well-developed commercial sector effectively diffuses the investment that accumulates in the economy. Saskia Sassen explains how global city networks are related to both intraurban and interurban information infrastructures in *Global Networks – Linked Cities* (2002). In *World City Network – A Global Urban Analysis* (2004), by Peter Taylor, world cities are viewed as global service centres, connected into a single worldwide network. An alternative approach is outlined in *Global City-Regions* (2001), edited by Allen Scott, where world cities are considered as complex urban regions, encompassing several cities, networked in a polycentric structure. There are currently more than three-hundred city-regions around the world with populations greater than one million.

The revenues received by multinational headquarters from exporting contribute to the local and non-local multiplier effect and the expansion of other sectors of the economy. However, the investment decisions by corporate headquarters and the multi-site allocation of resources to the various divisions, may result in a spatial distribution of new employment that is widely dispersed depending upon the location of the branch plants. Urban areas may be seriously affected by the rationalisation that follows the acquisition or merger activities of multinational corporations, where production is transferred to low-cost developing countries or moved to the country of origin of a foreign company. In particular, the closure of older plants in the metropolitan centre has resulted in high levels of unemployment amongst unskilled workers in the inner-city areas.

There is a trend towards the transformation of corporations into networks in which investments are made to establish sets of relationships in different environments. Global competition requires local information on markets with highly diverse dynamics, and decentralised collection of information at the micro-business unit level has to be integrated into corporate strategic thinking at the macro level. A top-down strategy would introduce decision-making errors at local levels that would be amplified through the network with potentially serious consequences. This approach enables small and medium businesses to link up with major corporations in cross-border networks that can continuously adapt.

The complexity of the web of strategic alliances, flexible production, and subcontracting arrangements increased significantly during the 1990s with digitalisation of the telecommunications network, information networking technology, advances in computer microelectronics, and software developments. Continuous data flows from bar-coded sales at retailers to distribution centres and production plants have dramatically reduced delivery cycles. Also, flexibility in production processes permit smaller batches and large-scale customisation of products. Transaction costs are reduced by achieving lower thresholds for economies of scale, with the ability to disperse the risk of demand fluctuations throughout the trophic web. Smaller premises at lower-cost locations serve customers directly or through franchises that can coevolve with customer use. Virtual organisations are emerging rapidly in which brand names are given to products that are entirely outsourced.

The Internet will change the pattern of transactions in financial services, commerce, retailing, entertainment, and communication. It provides a distribution channel to every household and has the potential for eliminating brokers and intermediaries. However, the sheer volume of information will create opportunities for intermediaries to sort and filter incoming email and to locate protected and ex-directory prospective customers. Reference agencies will be required to vouch for on-line services, and detection systems will be needed to isolate fraudsters and racketeers. Also, the sale of services and goods on line are expected to enjoy a period of rapid growth, with goods being dispatched to the customer rather than the customer collecting them from the vendor's premises. Certainly, this is the route for ordering replacements and spare parts where the consumer is a repeat purchaser. Recruitment services and employment networks may accelerate a shift towards commodification of skills with intense competition for work or consultancy opportunities. The constraint on Internet transactions will be an individual's limited span of attention and a reaction against 'infoglut'.

6.5   Perspectives on the informational economy

Human societies are problem-solving organisations concerned with acquiring the resources necessary to sustain the society, reproduce themselves in the next generation of citizens, and defend the civilisation from invasion. They invest in technological development and systems that extend their capabilities by creating artificial shells for our species with four-wheels, robotic arms, jet wings, telephonic voice, television eyes, a computerised brain, engineered genetics, and hydrogen-energy supplies. When a particular problem is to be solved, the evaluated data for a specific situation generates information. Progress occurs when the supply of information or knowledge is increased in scope, validity, and in opportunities for application. Information therefore has a value as a direct demand from the problems to which it can provide solutions. *Knowledge* is the term used for the overall stock of information that has future value in application.

The producer of an information service does not part with the knowledge in the sense that a producer of physical goods does. Intellectual property can be sold to a number of customers, who may apply it for their own use but may be prevented from diffusing the knowledge through legal protection and licensing. An information producer may in fact add to his knowledge from the applications of his customers at the same time as he is receiving payment for it. However, the information producer and his clients lose the exclusive ability to derive benefits from the information as the circulation is increased. In the market that operates between the demand from problems and the supply of informational solutions, the value of the information in use arises from either an increase in opportunities or revenues for the recipient or a reduction in hazards or costs. The assembly of information for problem solving follows a path of decreasing returns to effort, since the generalised data may be acquired at a relatively low cost but the specialised derivative work is acquired at a substantially higher cost. The cost and benefit curves for the development of information reach a point where the net value of the information passes through a maximum, before diminishing returns set in as the cost of obtaining further information increases faster than its value. If the additional information required would cost more than is economically justifiable to pay for it, then a decision may be taken with an information gap, i.e., with imperfect information. It follows that the value of information in use will establish a breakeven point for the number of customers required to recover the informational costs.

Transactions and interactions between establishments and households are mediated through the relative value of a transaction in relation to the cost. Money is a transaction medium and, as a means of communication, a currency is rather like a language in so far as a common language or currency is required for an exchange to take place. In the economic system, the market mediates exchanges through the structure of relative prices and profits. The market price is in equilibrium when there is neither an excess of demand nor an excess of supply. Capital may be defined as the stocks or populations of all economically significant objects, and there is potential for investment or capital to accumulate within a territory until saturation approaches and there are diminishing returns from any further growth. Ecological limits will constrain the total stocks per capita that can be maintained. Each artefact or commodity in the economy is a progeny of an establishment, and over time there is a change in the stock or population of all commodities. Stocks are increased in a period by production exceeding consumption, or conversely, are decreased by excessive consumption.

Neoclassical economic theory is dependent upon the notion of market equilibrium, with near perfect information. In a model of the economy in which conditions are static, agents would gradually obtain the information that is necessary for them to operate in the economic system, and there would only be a limited demand for new information. The demand for information arises from the uncertainty of dynamic conditions that are changing over time. These concepts are described in *Communication Economics and Development* (1982), edited by Meheroo Jussawalla and D. M. Lamberton. The assumption of perfect information in neoclassical economics is accommodated by means of the concept of transaction costs, which include costs for search and information, screening and signalling, bargaining and decision-making, and performance monitoring and contract enforcement. Organisations develop their capability to undertake exchanges in which the transaction costs are effectively reduced and absorbed by the price mechanism.

The demand for information can fall to indifference or disinterest and could even be reduced to zero. Also, invalid information would contaminate the stock of knowledge, and if the rate of attrition or depreciation of knowledge is greater than the rate of accumulation, then knowledge will be lost. There is, however, a time-span dimension in knowledge in the context of searching for information to solve future problem situations. It follows that there needs to be a continual supply of new knowledge for the evolution of society, but it is inevitable that this information will be acquired at increasing cost. In the neoclassical framework, it is taken that the economy is constrained by labour and materials resources, not by its limited capacity to acquire increasingly advanced information to facilitate the innovation process. The definition of an information division leads to the significant point that tradable information is a scarce resource.

Information also has the characteristic properties of a public good. Whereas a stock of private consumption goods is exhausted through allocation to consumers, a stock of public goods may be consumed equally and totally by every individual subject to capacity limits. Society expands its flow of information through education and culture, and organisations or individuals search for those alternatives that yield more of the temporarily scarce commodities. Information therefore has potential for creating possibilities that are important for the evolution of humankind, which results in continual change, economic development, and disequilibrium. This disequilibrium gives rise to the need for further informational activities, which leads to an evolutionary systems paradigm for analysis of the economy. However, economists have traditionally made adjustments and refinements to the neoclassical model of market competition, rather than developing a systems view of the economy.

Economists have different understandings of an evolutionary approach, for which the key conditions are time irreversibility and endogenous change. A further condition that has to be addressed is that there is foresight in societal evolution, and the quasi-ecological evolution of the urban system results in organic change that is compatible with concepts of path dependence and multiple equilibria. Designed interventions can alter the evolutionary trajectory of the system and, in the context of the civil system, technological development leads to engineered transformations by means of the investment-flow control parameter. In view of the institutional coordination of urban microsystems, the economy is not a process of emerging spontaneous order. As explained in Chapter 3, the evolutionary approach focuses on long-term adaption and evolution in the macrostructure. When establishments propagate or reproduce, the agents provide the genetic information that will shape the organisational form and control its future development. Depending upon an establishment's ability to capture resources or investment capital, it will survive or be selected against by competitors.

With complex patterns of global change, households and establishments develop their own representations or models of future world conditions, which are built with unequal access to information on the underlying dynamics and the evolving global macrosystems. Organisations are information-processing units that evaluate the available knowledge according to different perspectives that depend upon their own stage in the life-cycle phases. Households and establishments explore promising alternatives when it is apparent that the payoff from current activities falls below expectations or aspiration levels. In order to capture investment capital (profit), a new venture organisation has to engage in exploratory behaviour that involves innovation, a search for novelty, risk-taking, and information-evaluation activities that are orientated towards discovering new opportunities. This requires accuracy in selecting the right market segment in which to operate, and in defining, refining, and redefining its positioning until breakeven is achieved and the establishment becomes a viable entity.

In the growth phase of an establishment, the technological capability has to be developed to achieve higher levels of efficiency, with standardisation of the output for larger volumes to exploit the market demand created. Companies develop a hierarchy, together with competencies, internal governance structures, corporate cultures, and operational manuals, which makes organisational change path-dependent. Activities become coordinated with decision rules that will coevolve with market penetration, and, at the same time, the frame of reference for a world-view will require modification so that the organisation adapts to the changing external environment. At the phase of maturity there is continual pressure to reduce transaction costs and to increase market share so that a declining trend in resource capture and profitability will be averted as long as possible. Since mature market size is stable or declining, the competition becomes intense, and profits become reinvested in other less mature business operations.

The behaviour of households and establishments can range from habitual to exploratory. One of the functions of habits is to reduce the amount of complex information required for rational decision-making, and expenditure on repetitive consumable goods is subject to inertia. Habits are a mechanism for dealing with information overload and for relegating specific actions from continuous rational assessment. One of the dangers of habitual patterns of thought is the lack of scrutiny of the assumptions, axioms that are being used, and persuasion by evidence or argument without adequately considering all aspects of the problem. Geoffrey Hodgson points out in *Economics and Institutions* (1988), that habits are regarded as rational activities because the cost of changing them is perceived to be too great. When information processing becomes a routine without exploratory behaviour, a transacting entity will become increasingly maladapted to environmental change and, in due course, it will face a reduction in resources. However, a regular level of household income, together with the cultural norms of a residential area, establishes a lifestyle with a pattern and level of spending for a family at a particular life-stage. The institution of family and household routines provide reliable information concerning the likely actions of others, so that in a highly complex world with information overload, regular and predictable behaviour is possible.

As telecommunication channels achieve speeds in the range of megabits per second, good audio and video are possible with sophisticated graphics. These transfer rates have been provided for some time to households by one-way cable television networks, but now a multitude of sites can transmit data at these rates. One consequence of this is information overload, so that information merely becomes noise, and recipients are likely to switch off and disregard information that may be valid. This will result in a deterioration of the knowledge stock and an inaccurate perceptual model leading to decision-making errors. Also, it becomes increasingly difficult to challenge the existing stock of knowledge and to dispel myths, in an information society that increasingly provides fragmentary analyses in sound-bites. There is an 'infoglut' of competing messages on cable television with perhaps 200 channels, TV advertising, the internet, and e-mails, faxes, correspondence, newspapers, journals, books, CD-ROMs, videos, radio, audiocassettes, posters, and junk-mail. Also multi-media artefacts that display violent or pornographic images result in negative behaviour or emotions. Whilst the best of television provides a window on the world with an unparalleled opportunity for learning, perhaps pay-TV or other forms of user charging will be applied to provide a filter and a rationing of the worst of the excesses.

6.6　Timescale for the informational transition

In industrial societies, the urban population exceeds fifty percent, and over thirty percent of employment is engaged in the production of goods and an increasing percentage in services. In advanced industrial economies, urban population stabilises at around seventy-five percent, and industrial employment declines from a peak of around forty percent to less than thirty percent in the distributional economy, where less than ten percent is in primary resources and over sixty percent in services (of which thirty-five percent are in the tertiary commercial division and twenty-five percent in the quaternary information division). At this stage, substantial economies of scale are introduced in the tertiary commercial division with containerisation, new patterns of distribution to serve a retailing revolution by an expansion of retail floorspace with town centre shopping precincts, superstore developments and retail parks, and banking innovations with the transfer of funds by credit cards, etc. In the final transition to an informational economy, in which forty percent of employment is in the quaternary information division, some thirty-five percent in the tertiary commercial division, and less than twenty-five percent in the secondary industrial division, there are an increasing number of corporate relocations from the metropolis to medium-sized cities, so that the increase in employment in the outer centres broadly matches the decline in the metropolitan centre. It is the long-term trend of the transition of industrial towns to postindustrial that provides the dynamics for commercial property development.

The timescale for the informational transition is dependent upon the growth in Gross National Income (GNI) and limits to the factors of production at varying points in time, which cause a rise in the price of resources that are in short supply. New technologies are stimulated by the increased prices and provide an economic substitute for the earlier technology. Technical progress then permits further economic development, which leads to social development with improved standards of living, education, and health. The transition to an informational economy is particularly dependent upon increases in knowledge and educational capacity, together with the necessary information infrastructure.

Population and demographic changes, such as the number of births, the age structure, retirement, and deaths not only affect the number of households but also the potential labour force. The United Nations publication, *World Population Ageing 1950–2050* (2002), forecasts an increase of nearly 1.7m in the population for Great Britain from 59.4m in 2000 to 61.2m in 2025. Between 2000 and 2025, the numbers in the 0–14 age group are expected to decline by 2m from 11.3m to 9.3m, in the 15-59 age group by 2m from 35.9m to 33.9m, and the number in the 60+ age group will increase by nearly 5.7m from 12.3m to 18.0m. In the period 2025–2050, the total population is expected to decrease by 2.3m to 58.9m. The numbers in the age group 0–14 are expected to decline by 0.5m to 8.8m, in the 15–59 age group a further decline of 3.9m to 30m, and the number in the 60+ age group will increase by 2m to 20m. The increases in the proportion of older age groups between 2000-2050 will be 60-64 from 20.6% to 34.0% of the total population, 65–79 from 15.8% to 27.3%, and 80+ from 4.1% to 10.8%.

However, limits to growth such as finite natural resources or the employment-to-population ratio, provide constraints on the rate of technical progress. Productivity increases in the primary division, and immigration originally provided the manpower surplus for the secondary industrial division. Automation and robotics in manufacturing, economies of scale in the commercial division, and an increasing female contribution have permitted the significant growth in the number of knowledge workers in the quaternary information division. However, it is predicted that the share of employment in the information division may reach an upper limit of some fifty percent of employment, on the basis of the declining trend in the industrial division.

Increases in the level of demand in an economy lead to increased investment, with growth in the GNI per capita. An increase in the size of an industry leads to economies of scale, and productivity improvements are achieved with the renewal and expansion of the capital investment per employee. However, the growth curves for GNI per capita are logistic curves, and in advanced industrial economies, there is a slowdown in the rates of growth, because known technologies have already been applied, and the countries must rely for subsequent growth on the rate at which new knowledge is created. The S-shaped path of technological absorption with respect to GNI per capita is reflected in a wide range of indicators, which are discussed in Chapter 8.

It may be expected that the quaternary division employment will also follow a logistic curve with rising per capita output. The increase in the proportion of quaternary employment from 30.7% to 35.4% in the period 1981–1991 for Great Britain amounted to 1.5% per year, whereas the increase from 36% to 40% for the Megapolis was 1.0% per year. It follows that, for a town to increase its proportion of employment in the quaternary division from 30% to a maximum of 50% (quaternary location index QLI from 0.75 to 1.25), it may take twelve years to move to 40% with an average annual growth rate of 1.25% and a further twenty-eight years to move to 50% with an annual growth rate of 0.8%, giving a forty-year period for the transition. An industrial town with only 20% employment in the quaternary division (QLI 0.50) may expect to take an additional twenty years to become postindustrial, depending on the trajectory of development, so the contemporary timescale for the informational transition may span forty to sixty years.

For Great Britain, these rather crude assumptions indicate that the period 1960–1980 was the start of the process in moving from an industrial to a distributional economy, and the full transition to an informational economy would take from 1980 to the year 2025. The half-century of world development from 1975 to 2025 is likely to be known as the information age.

Transactional complexity: Chronological references

1. *An Introduction to Human Geography* - Kevin R Cox, 1972.
2. *The Information Economy* - Marc U Porat, 1977.
3. *City - Systems in Advanced Economies* - Allan Pred, 1977.
4. *Communications Economics and Development* - Edited by Meheroo Jussawalla & D M Lamberton, 1982.
5. *Communication and Information Economics* - Edited by Meheroo Jussawalla & Helene Ebenfield, 1984.
6. *The Urbanization of Capital* – David Harvey, 1985.
7. *Trends in the Information Economy* - OECD, 1986.
8. *Economics and Institutions* - Geoffrey M Hodgson, 1988.
9. *Collapsing Space and Time* - Edited by Stanley D Brunn & Thomas R Leinbach, 1991.
10. *e-topia* – William J Mitchell, 2000.
11. *Splintering Urbanism* – Stephen Graham and Simon Marvin, 2001.
12. *World Employment Report 2001* - ILO Publications, 2001.
13. *Global City-Regions* - Edited by Allen J Scott, 2001.
14. *Global Networks - Linked Cities* - Edited by Saskia Sassen, 2002.
15. *World Population Ageing 1950–2050*, United Nations, 2002.
16. *Spying with Maps* - Mark Monmonier, 2002.
17. *ME++* - William J Mitchell, 2003.
18. *World City Network* – Peter J Taylor, 2004.
19. *Complexity and the Limits of Knowledge* – Edited by Peter Allen & Paul Torrens, 2005.
20. *The Structure and Dynamics of Networks* – Mark Newman, Albert-Laszlo Barabasi & Duncan Watts, 2006.

7.
CIVIL PHASE TRANSITIONS

7.1 Stages of development

At each stage of capitalist accumulation, technological innovations generate economic development with a new stable position at higher levels of per capita assets. Schumpeter described this as a process of creative destruction in which economic development involves the periodic disruption of the capitalist production system and its creative renewal by innovations. However, definitions of capitalism have arbitrary elements, and capitalism can be classified in a variety of ways that include economic organisation, economic function, physical characteristics, or the production structure. A classification of capitalism by the economic function of capital would relate to resource infrastructure, civil infrastructure, social infrastructure, and information infrastructure. The ideological definitions of capitalism include the concept of private ownership, rather than state ownership, or a free-market economy with minimum regulation, or democracies compared with dictatorships. There are, however, many examples in western capitalism where state corporations were the owners of the capital of utilities and transportation.

Since the stages of economic development are accompanied by the addition of capital stock, it is preferable to identify the stages of capitalist development by the phases in which new technology was introduced to give substantial increases in societal productivity. For example, towns and cities were formed that were dependent upon extractive industries and materials, once agricultural productivity had increased to the point that a surplus could be produced. These traditional activities in construction and crafts led to the expansion of towns as a function of trade or commerce. In due course, merchant capitalism has led to the transition for countries by the acceleration of infrastructure development, and industrial capitalism follows. In the subsequent era of mass consumption, a distributional capitalism evolves with utilities networks, containerisation, and a retailing revolution. Finally, evolution in communications and informational technologies is likely to lead to a dispersal of activities that will produce the preconditions for an ecopolitan stage.

In the agropolitan stage of development, a mainly self-sufficient rural region focuses on supplying the urban population, and it is described as *mercantile* when it engages in national trade and commerce with moves toward international trade. At that level of development, the Ricardian comparative advantage of regions lies in pre-given natural endowments for the production of foodstuffs and raw materials for export. The preconditions for industrialisation are based upon the ability to raise the rate of investment and the per capita stock of capital and to meet the enormous capital expenditure requirements for both economic and social infrastructure development. However, in the later stages of industrial development, it is the competitive advantage of regions based upon collaboration that becomes significant in terms of technological development and innovation. These factors include university-based researchers, concentrations of suppliers for specific industries, and a continental market to give economies of scale. It follows that the basis for achieving sustained economic development in the more advanced countries is to build regional agglomerations of economic activity, in which the engines of growth are multi-national corporations acting in concert with high-technology local companies.

The stages of economic development for a country are accompanied by waves of infrastructure investment that contribute to increasing urbanisation of the population. The life cycle of major infrastructure investments follows an S-shaped curve through the five phases of inception, technological development, expansion, maturity, and senescence. The mature phase with gradually decreasing growth rates of output is generally due to either limited possibilities for further technological advances or limited scope for market penetration above a certain level. Infrastructure investment waves have tended to span a period of fifty years. In Table 4, below, the stage of development for the United Kingdom, the proportion of urban population at the start of the period, and infrastructure investments are aligned by period, to provide an historical perspective on the current economic structure.

TABLE 4: STAGES OF DEVELOPMENT FOR THE UK

Period	Stage of development	% urban population	Investment waves
1500-1749	Mercantile	10%	Trade in slaves, precious metals, artefacts, commodities.
1750-1849	Infrastructural	25%	Canals, turnpikes, coal mining, docks, steam power.
1850-1959	Industrial	65%	Railways, steam-ships, electricity distribution, broadcasting.
1960-1979	Distributional	80%	Motorways, air transport, satellites, telecommunications.
1980 -	Informational	90%	Technology intensive offices, global information technology networks, cable TV.

The traditional stage of development is followed by an agropolitan or mercantile stage, in which the proportion of employment in the tertiary commercial division increases from some sixteen to twenty-five percent. The approach follows Walt Rostow's articulation of economic development in terms of well-defined stages, in which the infrastructural stage, where historically the urban population exceeded thirty-five percent, is proposed in place of his preconditions and 'take-off', and the industrial stage could be achieved when some twenty percent of national income was steadily invested and the urban population exceeded sixty-five percent. This research distances itself from his take-off stage because the predicated rate of rise in net investment has not generally been supported in practice. The important point is that the less developed countries (LDCs) require additional investment transfers from the more developed countries (MDCs) to achieve the infrastructural stage of development with a per-capita GNI of $1,001-$2,500 (2000 US$).

By the infrastructural stage of development, the energy and transport infrastructure per capita increases by a factor of two to three times that at the early agropolitan stage. The age of mass consumption has been renamed the distributional stage (or service economy), and this leads to the informational stage, which has previously been labelled postindustrial by Daniel Bell and Herman Kahn before the informational economy was clearly definable. Whilst the use of stages of development is a simplification of a complex process, in *The World Economy, History and Prospect* (1978), Walt Rostow emphasises the dynamics of growth in a world perspective through a succession of leading sectors, which are major new technology life cycles in their growth phase. He explains long-wave fluctuations by the forces set in motion by changes in the profitability of producing foodstuffs and raw materials and the stimulation of investment in new territories with accelerated urban growth.

For the purposes of futures research, the following eight stages of civil development are defined – traditional, agropolitan, infrastructural, industrial, distributional, informational, ecopolitan, and planetary, with an indicative Gross National Income (GNI) per capita. In the System of National Accounts used by the World Bank for Development Indicators from the year 2000, GNI replaces Gross National Product (GNP). These civil phase transitions form the core of the evolutionary approach, and the system growth or control parameter for moving conditions from one stage to another is the flow of investment capital. The cumulative investment at each stage of development gives rise to an increase in the urban proportion of the population, together with an increase in the proportion of establishments in the quaternary information division as shown in Appendix 13. Table 5 below shows the period required to increase the GNI per capita by a factor of 2.5 at each stage of development on the basis of contemporary growth rates:

TABLE 5: CIVIL PHASE TRANSITIONS

Stages of civil development	GNI per capita 2000 US$	Annual growth of GNI/cap (1990-1999)	Contemporary duration of stage in years
Traditional	Under 400	Under 1.2%	75 to 150+
Agropolitan	401 - 1,000	Under 1.6%	60 to 120+
Infrastructural	1,001 - 2,500	Under 1.8%	50 to 100+
Industrial	2,501 - 6,000	Under 3.0%	30 to 60+
Distributional	6,001 - 15,000	Under 3.7%	25 to 50+
Informational	15,001 - 40,000	2.0%	45
Beyond Year 2025			
Ecopolitan	40,001- 100,000	Under 1.0%	100 to 200
Beyond Year 2200			
Planetary- World Ave	25,001- 60,000	0.11%	800
Beyond Year 3000			

The annual rate of growth in Gross World Product reached a peak of 5% during the decade 1960-69, when the world annual growth in GNI/capita also peaked at 3% per year. Since then, there has been a progressive slowdown in the world average annual growth of GNI/capita for each decade in relation to the preceding decade, with 1970-79 at 1.9%, 1980-89 at 1.4% and 1990-99 at 1.2%. This indicates that world development has passed the phase of accelerating growth, and it is following an S-shaped logistic curve on which the growth rates will level off. The decade 1990-1999 is characterised by an increasing proportion of countries with negative rates of growth in the earlier stages of development. However, within each decade, the annual growth in GNI per capita increases in the agropolitan, infrastructural, industrial, and distributional stages of development and then declines at the informational stage. On the basis of the performance by the majority of countries that are showing growth, it will take 240 years for a successful country to progress from the traditional stage of development to the informational stage. According to Paul Wallace in *Agequake* (1999), the OECD model *The Macroeconomic Implications of Ageing in a Global Context* (1998) shows that the impact of ageing populations in the more developed countries, together with a projected decline in workforce participation rates, will result in a reduction in growth rates to 1.0% in the EU and 1.5% in the USA by the year 2025, which corresponds to the ecopolitan stage of development. During the period 2250-3000 an early planetary civilisation will form with stable continental unions and diminishing rates of growth.

The inverse of intensity of use of a given material (or energy) can be expressed as a ratio in the form of GDP output per kilogram, which increases with per capita GDP for all economies, although the decline in material intensity tends to level off once the informational stage is reached. Additional incentives to achieve eco-efficiency will be required for countries to progress to the ecopolitan stage of development. This decoupling of economic output from materials use has given rise to the concept of "dematerialisation of the economy". For example, at the industrial stage, the GDP in Purchasing Power Parity (PPP US$) per unit of energy reaches an average of PPP $110 per Exajoule (EJ) and then rises to PPP $135 per EJ (excluding the USA at PPP $100 per EJ) by the informational stage (combining stages two and three) as set out in Appendix 28. Also, the GDP per metric ton of materials flow rises from PPP $250 at the industrial stage to PPP $400 at the informational stage. Oliviero Bernardini and Riccardo Galli describe *Dematerialisation: long-term trends in the use of materials and energy* in an article in *Futures* Volume 25, Number 4, May 1993.

The changing nature of demand for material flows arises as economies develop from agricultural communities, through the subsequent stages of development with the building of material-intensive cities and related water supply, sewage facilities, utilities networks, factories, transport, and distribution infrastructures. This accounts for a significant proportion of industrial activity in the LDCs and is accompanied by accelerated industrial growth, followed by industrial maturity and saturation of bulk products demand. Different waves of infrastructure development tend to follow each other, and this also applies to consumer durables such as the diffusion of the car and household appliances, which have prolonged the phase of the intensity of use of many metals. With saturation of demand, materials use becomes more confined to replacement markets and, at the postindustrial stage, recycling acquires a significant role in decreasing the material intensity of economic activities.

Over time, there is a gradual shift in a nation's output towards the production of goods with higher unit value and lower material content, or the production of services that may be expected to lead to a declining intensity of use of materials as a whole. Similar trends are also evident within high energy and material intensive sectors such as manufacturing, where there has been a shift in investment in the different categories of capital goods. The share of investment in information technologies has risen from ten to thirty-five percent of the value of equipment purchased in the more developed countries, whereas investment in industrial equipment, transport, and other goods have all declined to a similar share of twenty percent each. In affluent societies, marginal income is increasingly being spent on a broad range of products characterised by a low ratio of material content to value, including home entertainment, recreation, physical fitness, and medical services. Value added comes increasingly from sophisticated new service sectors centred on non-material factors, such as design, image, quality, flexibility, safety, reliability, and environmental compatibility.

There is clearly a need for further growth in the developing countries to meet at least the basic needs of civilisation, even though it is not likely to be feasible to raise their material living standards to the levels of the rich countries. It follows that there is a geographical time phasing or synchronisation in the process of development, so that the wealthy countries should reduce growth to permit the developing countries to increase their standard of living to the distributional stages of development at which a more equitable society can be reached. Achievement of a dynamic balance in the world economy may arise with a late planetary civilisation.

7.2   The workforce and working time

In Appendix 13, estimates have been made of the world's workforce by stage of development and country income group, and in 2000 the workforce amounted to 2.95 billion or some 49% of the world population.

The distribution of the world's workforce between the primary, secondary, tertiary, and quaternary divisions has been estimated for twelve income groups. In the developing countries, the urban population is nearly forty percent of the total population, compared with seventy-five percent for the developed countries. The size and distribution of a city's workforce are determined by a number of factors that include the natural population growth in the city, the net migration to (or from) the city, the participation rate of the labour force, and the human capital embodied in the labour force (i.e., health, education, and skill levels). Immigration tends to swell the ranks engaged in the service sector (principally commercial) until labour can be absorbed in the industrial division. It should be noted that the proportion engaged in the quaternary information division is generally less than six percent for the low-income economies where the urban population is only thirty percent of the total. A true comparison with the developed countries can be made by noting that the quaternary division in developing countries amounts to twenty percent of the urban workforce. Also, it should be noted that, in the less developed countries, the total workforce in the quaternary division is 235m, which is larger than the quaternary workforce of 211m in the more developed countries.

Workforce estimates are based upon the proportion of the population in the age group fifteen to sixty-four, which varies between fifty-five percent in poor countries and sixty-five percent in the OECD countries, giving an overall average of sixty-three percent of the population. On average, the labour force participation rates for males are around eighty-five percent, compared with sixty percent for females, so that the overall participation rate is around seventy-five percent. It follows that the workforce amounts to some forty-nine percent of the total population. However, these overall figures conceal significant variations in the female participation rates, which are highest in the poorest and richest countries, and lowest in the lower-middle and upper-middle income countries. In the poorest countries, people work because they have to in order to survive, and in the richer countries they work because it is linked to life style issues. In Islamic countries, a very small proportion of women work, with eight percent in Bangladesh, ten percent in Saudi Arabia, sixteen percent in Pakistan, and thirty-eight percent in Indonesia. By contrast, female participation rates are thirty percent for India, thirty-three percent in Latin America, sixty percent for the UK and USA, and eighty percent in China.

The productivity gains of the industrial revolution in the 19th century enabled the working week in Europe to be reduced from eighty hours to sixty hours. By the end of the 19th century the basic working week was fifty-four hours, and this was reduced to forty-eight and forty-four hours following the First and Second World Wars, respectively, to a forty-hour week in the 1960s. At the end of the 1960s, the average holiday entitlement was less than three weeks, and by 1980, this had risen to four weeks and then to five weeks by 1990. Since 1980, the standard working week has been reduced to thirty-nine hours, but union demands for a thirty-five-hour week have been resisted by industry and government as not being affordable. Instead, the trend has been to combine reductions in working time with an increase in the operating time of facilities and plant through shift work and weekend working to achieve a reduction in unit costs. The decrease in the length of working time has been achieved through the reorganisation of work and a systematic review of all elements of working conditions as follows:

- Length of the working day
- Amount of overtime
- Length of the working week
- Spread of shift work
- Increased part-time working
- Flexibility of working time
- Short time working
- Annual hours contracts
- Working life cycle (age of entry, retirement, and continuation training)

In effect, the initiatives have come from management for decentralised negotiations, the reorganisation of working time, and from standardisation to diversity.

In Europe, an average of forty-six weeks are worked per year (allowing for annual and statutory holidays), with an average of thirty-eight hours per week, giving a total of 1,750 hours per year. However, the proportion of overtime and the number of part-time workers varies considerably between countries. The north European countries such as the UK, Netherlands, and Sweden tend to have high levels of female participation, with twenty-five percent of part-time employment. In Southern Europe, such as Italy and Spain, the female participation rates are lower at thirty-seven and twenty-six percent, respectively. In a special issue on Working Time in the Journal *Futures* (1993), edited by Hugues de Jouvenal, the hours generally worked in the UK and the EU in 1990 are given in Table 6 below, together with future projections.

TABLE 6: PROJECTED EUROPEAN WEEKLY WORKING HOURS

	UK	EU	*Year* 2025	*Year* 2050
Full time	44	41	35	30
Part time	17	19	15	15
Overall	38	38	30	25
% of employees working part-time	23%	14%	25%	30%
Annual hours	1,750	1,750	1,400	1,100

The conjectural future projections are based on assumptions from several global studies on the future of work, including *The End of Work* (1995), by Jeremy Rifkin. By 2025, it is envisaged that the average number of hours worked per week will be reduced to thirty, through a reduction in the working week and an increase in the proportion of part-time workers to twenty-five percent, of which twenty percent are likely to be female and five percent male. A twenty percent decrease in the annual hours worked from 1,750 to 1,400 would enable unemployed workers to be absorbed into the workforce to provide a more equitable distribution of employment. In the longer term, the introduction of a four-day week of thirty hours, and thirty percent part-time workers doing two days per week of fifteen hours, would give an average working week of twenty-five hours. On the assumption that forty-five weeks were worked per year, the average annual hours would be 1,100. It is envisaged that, in the future, employers and employees will agree to quarterly-hours contracts of say 400 hours, which will be reduced during the 21st century to 300 hours. This approach will enable establishments the flexibility of agreeing to local working time schedules that optimise the output of the facilities whilst ensuring an improvement in the balance between work and leisure and increasing the overall participation of the age group fifteen to sixty-four years.

7.3   Societal transitions

Both capitalism and communism employed technology to achieve a considerable improvement in humankind's condition than is provided in a state of nature. Communism was historically a response to the excesses of capitalism in the industrial revolution and it advocated that the means of production should be vested in society, for the benefit of all, through state ownership. In communism, political control was achieved through a hierarchy of elites, which placed restrictions on free communications and effectively excluded these regimes from participating in the global economy. Closed civilisations tend to limit their human and natural potential, and the communication technologies of the information age served mainly to increase the awareness of the failure of the communist system.

The trilogy on *The Information Age* (1996-98), by Manuel Castells, provides a comprehensive sociological perspective on the current stage of development in the western countries, which is expected to last from 1980-2025. His first volume, *The Rise of the Network Society* (1996), has set a trend in the social sciences for referring to urban networks rather than the spatial concept of urban hierarchies, which emphasises the connectivity between the nodes. However, the notion of a network society does not provide a sufficient description of urban structure for an evolutionary model, as an increase in the complexity of the urban system involves an increase in the number of hierarchical levels. This is explained in *The Polycentric Metropolis – Learning from mega-city regions in Europe* (2006), written and edited by Peter Hall and Kathy Pain.

In his third volume, entitled *End of Millennium* (1998), Castells stresses the importance of being liberated from political ideology, and complexity science does indeed cut right across ideology. For example, there are three widely held social philosophies in the MDCs that may be summarised as follows:

- A socialist society, with a philosophy of social equity and justice, with people's expectation of a right to wages from work or the provision of welfare benefits from the government.

- A consumer society, with a philosophy of free market economics and competition, with people's expectation of a right to consume and rising standards of living with social safety nets.

- An ecopolitan society, with a philosophy of societal evolution for the development of human potential, with a sustainable future that will prolong the existence of civilisation. A shorter working life with fewer working hours will result in a decline in lifetime income. There will need to be more emphasis on long-life artefacts, interregional cooperation for the investment of savings and income, with compensation for reduced income by means of an increase in leisure time and social transfer payments from ecotaxes.

The two images of the future of 'an informational economy' and 'a sustainable society' are generally considered separately by writers, without the realisation that one leads to the other. The evolutionary potential of information, and the importance of knowledge and information for identifying sustainable options when commodities and natural resources become scarce, indicates that information systems contribute to increasing complexity with the creation of a macrostructure that is anti-chaotic.

The aim of social policies for an ecopolitan society would be to shift expectations from expansion and growth to human development, ecological consciousness, and quality of life. Socialism or a centralised approach to the organisation of society becomes too ordered and stagnant; whereas, in the extreme, an unregulated free-market or laissez-faire system can lead to chaotic conditions. An ecopolitan society with complexity and diversity would be a complex adaptive system that keeps order and chaos in balance, so that a vibrant and equitable economy can reach higher potential than either of the two extremes. A combination of these ideologies at the government, corporate, and civic levels would seem to meet the needs of the future. A regulatory system is likely to evolve, in which a redistribution of income is achieved through socialist policies such as progressive income taxation, consumer taxation via value-added taxes, and ecopolitan policies such as ecotaxes to achieve environmental goals and taxes on common resources such as land and minerals.

An ecopolitan society will also need to adapt to climate change in conjunction with mitigation policies. The damage suffered by New Orleans from hurricane Katrina demonstrated the potential damage that can be caused by a rise in sea level. However, the cost of adaption is relatively high because of the huge sunk-costs in the existing urban and industrial infrastructure. Adaption policies include land-use planning and zoning to avoid low-lying coastal areas and river plains, with the provision of an adequate capacity of storm water systems. Also smart-growth strategies need to be developed that avoid car-dependent urban sprawl, which cannot economically be served by public transport (see *Sprawl Costs* by Robert Burchell et al, 2005). At the same time, a fifty-year renewable energy investment wave from 2025-2075 will be a major source of industrial employment as in previous infrastructure investment waves.

The stimulus for a societal transition in the more developed countries is likely to come during the period 2000-2050 from a persistent set of pressures that include slow economic growth, rising unemployment and underemployment, the increasing social and health costs of ageing populations, debt constraints on government spending, increasing levels of crime, and other social problems, and declining reserves of low-cost fossil fuels. In Europe, there will be a fall in the number of workers and taxpayers per retired person from four in 2000, to three in 2025, and two in 2050, and it is an ageing population that provides a social limit to growth. Carbon taxes could provide a partial subsidy to part-time workers over sixty years of age undertaking fifteen or less hours of work per week, and a transfer payment for pension provisions for an ageing population. By 2100, the birth rate of the world population of ten billion will just exceed the death rate, and life expectancies are expected to increase by one year per decade after the year 2000. This is in line with United Nations projections, and an average lifespan of 100 years is probably two centuries away.

Transnational corporations and major national companies are likely to be located at regional centres where there will be a significant proportion of employment in the quaternary information division. However, outside the metropolitan centres, there is likely to be a movement towards less complexity with less centralisation, standardisation, and synchronisation. Extended retail opening hours, shift working, flexitime, off-peak travel incentives, and cheap rates for night-time communications and energy use will all contribute to improving the utilisation of facilities, infrastructure, and public services or utilities. Satellite and cable television will provide 24-hour entertainment with an increasing number of channels. Alternatives will evolve to replace a bureaucratic welfare state, such as NHS trusts and care in the community. The trend towards shorter hospital stays and increased consultancy and district nursing based on local clinics or medical practices, is part of the emerging pattern for reducing per-capita welfare investment costs. The focus of healthcare is likely to be on improving the quality of life in the context of a realistic life expectancy of say eighty to eighty-five years, before engaging in expensive treatments to extend lifespans.

Economic development and productivity growth have increased the daily output and hourly wages, so that purchasing power for an hour's work has increased over a long period of time. It follows that the value of time has increased for people in employment, and they attempt to pack more consumption into a limited week to achieve a higher return from their free time. The satisfaction of increasing but less essential needs, stimulated by advertising, tends to result in material- and facility-intensive consumption at the expense of time-intensive community, social, or recreational activities. In an ecopolitan society, an increase in the durability, reliability, or lifetime of artefacts, by say twenty-five percent, would generally result in lower resource usage, less consumption and waste, and a qualitative improvement, but downward movement, in the relative purchasing power for an hour's work. This is likely to contribute to the substitution of service- or time-intensive activities for commodity-intensive consumption, which would be environmentally more sustainable.

Through new technology, industries improve their productivity, and if the gains were distributed equitably, they would permit wages to rise and increase consumer incomes so that new businesses could be created and employment generated. At the same time, employment in traditional industries is declining as a result of increasing global competition, because exporting industries tend to be capital-intensive with relatively few workers, and imports are generally in the labour-intensive sectors. Principally, as a result of new production and distribution systems, the amount of work performed over the lifetime of an average person is being reduced, and this is accommodated by a shorter working week, later entry into the workforce, part-time working, and earlier retirement. In recent times, it has generally been accepted that a man had a forty-five-year working life of approximately 80,000 hours. By the year 2025, it is envisaged that a working life will be reduced to thirty-five years of full-time work at, say, 1,200 hours per year to give 42,000 hours of paid employment up to the age of fifty-five years. In order to supplement a reduced pension on early retirement, another ten years will need to be worked part-time to add a further 6,000 hours or a total of 48,000 hours over a lifetime. Similarly, part-time workers may achieve 24,000 hours in thirty-five years. It follows that, in future, the income of an individual over a lifetime may only be sixty percent of the income earned in the past. However, as explained in Section 7.2, it is envisaged that through increased part-time working, some fifty percent of the population will participate in the workforce compared with forty percent currently, so that the available income would be spread more evenly across the population. As a consequence, household incomes may decline in real terms to seventy-five percent of current values and there would be some reduction in the payment of unemployment benefits.

A decline in lifetime incomes in the MDCs will not necessarily result in a corresponding decline in living standards, because, for example, a medium sized family car of the 1980s was replaced at lower real cost in 2000 with a more compact car with extra features, such as higher safety and environmental standards. However, the futuristic vision of a postindustrial society in the advanced economies with more rewarding work, that is intellectual rather than physical, is dependent upon the Washington Consensus view that the benefits of reduced prices in a global economy will compensate for a loss in the real value of incomes. It is only by increasing the standards of living in the LDCs through infrastructure investment that wages will rise there, so that the overall market size can be expanded to utilise excess industrial capacity and generate global employment. In effect, the economic self-interest of the wealthier countries converges with the altruistic aim of raising wages and reducing poverty in the LDCs, which will diminish the gap in GNI per capita.

The study of human time becomes important if a quinary lifespan division is to be defined to include household and residential establishments and both voluntary and salaried or wage employment in human care and life science. The division would include housewives or house-husbands involved in childcare or care of the disabled or elderly and providers of domestic or household services. Publicly-provided health and social services would also be included in this division, together with welfare organisations and charities, which represents some twelve percent of the economically active population. Females not participating full time in the workforce between the ages of twenty and fifty-nine possibly contribute a further eighteen percent, and with an additional six percent from retirees, the voluntary sub-division is double the size of the economically active lifespan sub-division. If the voluntary sub-division were included in the workforce statistics, then the quinary lifespan division would account for thirty percent of the workforce, allowing for a decline in the industrial division and a peaking in the quaternary division before the ecopolitan stage commences. On the assumption that, by the year 2025, the pre-school and student population aged zero to nineteen will comprise twenty-five percent of the population, the working life will be forty years from twenty to fifty-nine years, and those over sixty years of age will comprise twenty-five percent of the population. The potential employment in an ecopolitan society is set out in Table 7 below:

TABLE 7: WORKFORCE ALLOCATION IN AN ECOPOLITAN SOCIETY *(including the voluntary lifespan sub-division)*

Primary resources division	5%
Secondary industrial division	20%
Tertiary commercial division	25%
Quaternary information division	25%
Quinary lifespan division	25%

The potential workforce by division and continental region is given in Appendix 26 for a world population of 12.5 billion in 2150, when the global employment in the primary resources division will have declined from 1.3 billion in 2000 to 540m in 2150. It is important to realise that, in the past, divisional classifications have been based on the perspective that gainful employment lies in the economically active workforce. The choice of divisional classifications influences our image of the future, and it would be better to identify the quinary division as a lifespan division to emphasise its real meaning, rather than to adopt the term social division, which has come to mean funded by the public sector. The important issue is how we use our life cycle rather than how we are employed.

As a result of the information revolution, humans will have greater increments of free time for leisure, life-long learning, and unpaid voluntary work in the lifespan division. However, there will be less opportunity for saving, and in order to maintain a quality of life on some seventy-five percent of past income, there will need to be less consumption, improved durability of goods, longer life assets, higher product quality, minimum materials design, reduced energy use, and more extensive recycling. The shift from wasteful material consumption and planned obsolescence to time-consuming activities, such as human development or care, will result in an increase in social utility. The societal transition will be assisted by mass media communications to create a well-informed society on environmental issues, resource conservation, and family health, such as contraception and dietary matters. The latter will impact on agriculture through a switch from animal protein to vegetable protein, and a demand for less use of inorganic pesticides and fertilisers. The Internet is likely to help people identify employment opportunities and market home-produced services. It is also inevitable that an informal or black economy will develop, and a system of barter is likely to emerge in communities, in which people will exchange services or new and second-hand goods. The informal economy is encouraged by welfare rules that reduce social security payments if they are supplemented by other earnings, and community currencies such as the LETS system (Local Employment and Trade System) are operating in many towns and cities in North America, Australia, and the UK.

At the ecopolitan stage, the suburban dwellers, who enjoy the benefits that the megapolis or metropolitan centre offers for access to jobs, entertainment, and cultural aspects, will have to share the full costs of the city services they use, and not escape the higher levels of municipal expenditure that arise in the city where the residents are poorer. It is also easier to contain crime and violence by the provision of a greater number of smaller cities and fewer megapolises, in which gang warfare and mafia crime flourish. Society needs to take responsibility for reducing income inequality, which is a major contributor to social deprivation. The benefit from ensuring that the lowest forty percent of households receive at least twenty percent of the national income, is that there would be resulting savings in health and welfare services and in the law enforcement system.

To achieve the transition to an ecopolitan society, there will need to be incentives and penalties for corporations to adjust to flexible production and shorter working hours, so that there is a proper sharing of employment. In principle, the motivating factor for organisations will be the realisation that reduced purchasing power and weak consumer demand will result from declining household incomes. But there will also be a high social cost to be paid for protection and security of the wealth-holders from an urban underclass, if the higher income groups are too selfish to adapt to the ecopolitan philosophy. Again, it is likely that a shift in people's values can be achieved with the incentive of more recreation and leisure time, and the opportunity to pursue other fulfilling activities.

The ecopolitan society is likely to influence family values, with families more frequently taking care of elderly parents and with more time to contribute to voluntary work in the community. It is anticipated that cultural activities will flourish, such as concerts, theatre, and live entertainment, together with all forms of art. As people retire, they may move to smaller houses to reduce costs or move to peripheral regions where the cost of living is lower. Alternatively, specialised retirement communities may be developed in warmer southern areas, or rural environments may become attractive, where locally sustainable agriculture would enable communities to become more self-reliant.

7.4 Dynamic investment flows

In *Business Cycles* (1939), Joseph Schumpeter explained that there are discontinuities in economic development in which clusters of innovation change the structure of the economy. He pointed out that economic history has to take account of multiple cycles including the Kitchins inventory cycles that run for three to four years, Juglars investment cycles of seven to eleven years, and Kondratiev long waves of forty-five to sixty years. Nikolai Kondratiev hypothesised that long-waves were an organic part of the capitalist system and that each phase of crisis, recovery, prosperity, and stagnation is a cumulative process arising from the preceding period. The historical record manifests long waves with a periodicity of half a century that displayed growth, expansion, overshoot, and finally collapse in the 1760s, 1820s, the 1870s, the 1920s and again in the 1970s. This is a reminder that investment capital drives the economy away from equilibrium and sets up oscillations that can be explained by chaos theory. Schumpeter described the periodic destructive disruption of the capitalist system and its creative renewal by innovation, which lifts the economy to a higher plane of development.

Simon Kuznets produced an article in 1940, entitled *Schumpeter's Business Cycles,* in which he made the point that the prevalence of fifty-year cycles had not been demonstrated in terms of physical volumes of production or trade. In essence, infrastructure investment and city building cycles result in construction cycles of some twenty to twenty-five years (Kuznet cycles) that are synchronised within the long wave data, and this forms the background to the phenomenon of property cycles. Brian Berry has provided a comprehensive analysis of long-wave explanations in *Long Wave Rhythms in Economic Development and Political Behaviour* (1991), in which he has demonstrated that the evolution of metropolitan America is linked to the life cycles of the dominant transport technology and energy use.

In the UK, residential building cycles of some twenty years' duration have affected the country since 1700, and in the United States, transatlantic migration was the link between inverse building cycles on either side of the Atlantic in the period 1870-1913. The recurrent nature of the long-house building cycle is a direct result of population changes. Construction at home drew on agricultural surplus labour for urban employment, and a wave of capital export to the USA provided an alternative destination so that transatlantic migration was essentially a rural exodus. At the start of the cycle, increased income allowed more people to marry and to marry earlier. An increase in the marriage rate leads to an increase in the birth rate, and some twenty years later these infants become adults seeking marriage and residential accommodation. The resultant increase in household formation leads to an increased demand for housing, furniture, and other consumer durables. It follows that the building cycle contains the seeds of future building cycles.

It is useful to distinguish between construction cycles, which include infrastructure investment, and the term building cycles, which is confined to residential, commercial, industrial, educational, health, and public buildings. The term property is a narrow definition for office, retail, and industrial development held by institutional investors, property investment companies, and developers. Property cycles can then be delineated by means of variables that include rents, rates of return, vacancy rates, property yields, and the flow of investment funds. The recurrent but irregular fluctuations in the rate of all-property total return lie behind the microcycles of the property industry, which have durations of up to a decade, and they define the microstages of urban development. There are two components of the total rate of property return in a period.

The first is the income return, which is the income received in the period as a percentage of the capital value at the start of the period. The second component is the percentage change in the capital value between the start and end of the period. It is variations in capital values that tend to drive the property cycle, and these are affected by rental levels and changes in investment yield.

The longer construction cycles identified by Kuznets are associated with population growth, migration, economic development, and technological evolution. The business and property microcycles are characterised by the boom and bust that results from supply-side production lags, which affect property development. There are also short-term fluctuations in the demand for property of typically four to five years' duration that are affected by changes in political parties and economic variables, such as interest rates, employment levels, consumer expenditure, and manufacturing output. Industrial development is particularly sensitive to interest rates and exchange rates. There are also location-specific variables that will impact upon property investment, including the evolving economic structure of towns and the urban transition.

The pulsations set off by a series of S-curves from energy, transport, and information-infrastructure investments, have stimulated new urban development, the relocation of businesses, and the building of office complexes. However, the process of the urban transition is not smooth, and there are often long lead times between the conception of a property investment opportunity and the full occupation of the completed scheme. Quite frequently, rational assessments are not made in relation to the long-term trend implied by an investment wave, and decisions are incorrectly based on short-term indicators of profitability. There is also a tendency for developers, who have been successful in riding the transitional wave, to think that growth is exponential, and they overlook the decreasing requirement for business space at the upper limits of an S-curve. These characteristics lie behind the microcycles or cyclical fluctuations of the property industry, and the extent of overbuilding at the micro-stage of a town in its transition, determines the length and intensity of the subsequent collapse.

In the office market, stock adjustment models may be derived in which expectations of future demand are sometimes based upon the ratio of the current and the previous period's office employment. Real rents generally decline in relation to vacancy rates that are in excess of the optimal level; however, rents do not fall rapidly to clear the vacant floorspace. Rent free periods or a contribution to fitting out costs are often offered as inducements when take-up rates are slow. The supply of new office space tends to react to reductions in the vacancy rate and rental increases, whereas demand tends to respond to local-office market factors and shortrun improvements in the economy, such as higher growth in GNI. Developers frequently interpret a fall in interest rates as an indicator that output will rise, with additional demand for floorspace, so that speculation in real estate is encouraged since lower yields increase capital values. This contributes to instability in the market and excess supply in property booms.

For a given level of output, technological development is likely to increase the output per unit of input, with savings in energy and the release of labour for alternative employment. However, if one factor of production is increased relative to the other factors, then there are likely to be diminishing returns from the investment. For example, by knowing the employment-to-population ratio (say 0.40), and the average employment per large establishment, the norm for the number of large establishments per 10,000 population may be established. For each category of town (i.e., by economic structure), planning factors may be derived to give the number of large office, retail, and industrial establishments. Any increase in the supply of a property type in excess of that predicated by the planning factors may lead to an imbalance in the commercial or industrial floorspace in relation to the available skilled employment. Consequently, vacant periods or a lower level of occupancy of facilities will lead to reduced returns from property investment. This tendency has already been observed in postindustrial towns.

The central areas of metropolitan cities have significantly more jobs than employed residents and attract a net inflow of work trips. Also, in outer centres, the economic potential or attraction capability of an urban area may be assessed from the number of large workplaces per 10,000 population. However, locational factors, building densities (plot ratios), and space standards also affect land prices, the rental levels, and the total returns from property investment. Planning officers tend to be lenient with planning applications for commercial property in towns with less than the norm of establishments per 10,000 population, to relieve unemployment. On the other hand, in areas where office development exceeds the planning factor or norm, then the planners are understandably tough in an attempt to restrict inward commuting.

An important principle in Ecodynamics is that the rate of investment flow through an ecostructure in a steady state, relative to the capital stock already accumulated there, is equivalent to the reciprocal of the expected lifespan of the stock. It is a measure of the rate at which new investment must be reproduced to replace that which is lost through decommissioning to maintain a steady state. Research in the London Megapolis showed that the average age of the housing stock is sixty to sixty-five years, whereas for industrial and commercial facilities it is forty to sixty years, so that in relation to the total floorspace, the expected lifespan of buildings and infrastructure is around 125 years. The rate of new investment to maintain the status quo will be 1/125 or 0.8% per year. The following relationship connects ecostructure asset ratios and is applicable to artefacts in the economic system:

$$\frac{Investment\ flow}{Capital\ stock} = \frac{Investment}{GNI} \times \frac{GNI}{Capital\ stock}$$

If the infrastructure investment is, say 2.5% of GNI, then the above equation would give the capital stock as 3 x GNI for a city region or state with an infrastructure lifespan of 125 years. If the capital stocks of the economic system are kept constant, then the efficiency of the economy will be improved by a reduction in the investment requirements for a given level of GNI. This may be achieved by lengthening the lifespan of assets and utilising them for longer. The law of diminishing marginal utility applies to the level of stocks, and they will be sufficient at some level.

A background flow of investment capital is required to maintain internal order or coherence in civil ecostructures, and the investment flow parameter is known as the system growth or control parameter, which functions in a similar way to the regulation of the intensity of heat by means of a control knob on a cooker. By controlling the rate of flow of investment into a region or ecostructure, economic decline can be reversed, or a stagnant economy can be made increasingly vibrant. In effect, the investment parameter determines how stagnant or chaotic the system becomes, and increasing the levels of investment capital per capita in due course transforms a region or state to a higher stage of development.

Depending upon the rate of investment flow, the spatio-temporal behaviour of an individual ecostructure will class it as a static, cyclic, vibrant, or chaotic attractor. Cities that receive sufficient investment capital, but not too much, are able to retain their level of complexity. If they do not obtain sufficient investment beyond a minimum threshold, then the level of complexity can no longer be supported. If too much investment is supplied, chaos ensues in the system as the investment overwhelms the diffusive capability of the ecostructure, and speculative property investment or stock-market bubbles destabilise the system. In effect, by increasing the investment parameter, the stable conditions of a static attractor move to the oscillations of a cyclic or periodic attractor, and then through the vibrant phase of 'order out of chaos', and finally to a chaotic condition. In the case of super connectivity, the diffusion by the ecostructure may be insufficient to prevent over activity in the local economy, when the release of accumulated capital held within the system triggers chaotic conditions.

Investment capital is a transformable resource since it can be transformed into financial instruments, facilities, infrastructures, vehicles, equipment, and artefacts. In a complex adaptive system, the investment flows pass through the trophic webs of business networks in which establishments are the nodes. The flows over these networks are variable over time, and the networks are not fixed in time as nodes and linkages may appear or disappear as the transacting entities either adapt and coevolve or malfunction and face extinction. The disturbance resulting from the extinction of establishments opens up new opportunities or niches for further adaptions by other establishments. The effect of new resources or the diversion of resources over a new path may generate significant changes. The recycling of investment with a number of cycles will generally have the overall effect of increasing the resources at each node. The retention of resources can be exploited by their reinvestment in new niches, and establishments that reinvest will tend to capture resources from those that fail to do so. The multiplier effect is a characteristic of networks and flows, and the impact of evolutionary changes is difficult to estimate.

A characteristic of complex adaptive systems is that they have 'critical intervention points' at which strategic investment decisions can become amplified. For example, the long cycles of economic development are caused by sectoral investment waves, which are amplified by the action of the multiplier and the acceleration effects (see Chapter 8). Also, the multi-site allocation of resources by a major corporation to the various business units, may result in a spatial distribution of investment that creates a disproportionate expansion at a specific location.

7.5 Macrolaws of Ecodynamics

At the periphery of the agglomerations of economic activity lie underdeveloped areas that have been unable to build regional economic systems that have the capability of creating an economic growth centre, even if there are pockets of relative prosperity and economic opportunity. Path dependence in location theory is derived from a view that spatial structure is dependent upon a settlement history, rather like geological stratification in which infrastructures are added layer by layer and industries agglomerate sector by sector, to take advantage of economies of locational connectivity or transaction costs. It follows that the major production regions of the world will not only continue to expand, but they will also become more subtly differentiated from one another as national boundaries become less important, and a continental division of labour evolves. A continental economic union such as the USA creates states that are much more specialised than the individual countries of Europe, which are comparable in size. It may be expected that this will lead to heightened forms of differentiation between states, providing there is continental protection and security. The pre-existing natural advantages or built environments that lead to regional specialisation are not permanent, and with changing technologies, advantageous locations change over time so that there will be interstate mobility. Some states will rise and others will decline, with any surplus being accumulated unevenly. A precondition for the reduction in the differentials in social equality between states is to ensure a levelling up of the infrastructural assets in the disadvantaged regions.

In the MDCs (more developed countries) there will be a low population growth, saturation of markets, and the stock of assets will increase slowly as a result of diminishing marginal utility and diminishing returns from productive investments. In the informational economy, increases in economic growth may be reflected in an increase in information and commercial services, which require a relatively small increase in physical stocks. The emphasis in the more developed countries will be on high levels of maintenance efficiency with a focus on protection, stability, and quality. An evolutionary or adaptive strategy will permit the continuing diffusion of investment, until increasing transformation, maintenance, and connectivity costs consume investment capital needed for any further development of diffusive capability. Competing locations in the urban system will attract investment by offering higher investment returns, which produces instability within the geographical landscape. The potential of a state is limited by spatial competition and its ability to attract investment, since additional complexity can only be achieved at the expense of the temporal global investment budget.

A major objective for a state will be to maintain a norm for the number of large establishments per 10,000 population through local economic development initiatives. The aim of an equitable society should be to reduce the high levels of structural unemployment in the transition to a low-growth economy and to provide adequately for the unemployed without impairing the conditions for further economic development and avoiding suffering, degradation, and deterioration in human values. By sharing the work through reduced hours and part-time working, it should be possible to minimise the number of unemployed people. Revenues may be raised from ecotaxes to meet the rising level of social expenditure resulting from an ageing population and earlier retirement.

The evolving structure of the urban economy reflects changes in the functional complexity of the metropolitan region as it adapts to its role within a continental union. Since the locally manifested changes are not local in origin, state governments need to anticipate and facilitate future adjustments. This can be done by inducing inward investment through the provision of investment grants, infrastructure networks, taxation policies, and enterprise zones, business finance initiatives, and training support. A balance has to be struck between providing interim aid to declining localities so that they can be regenerated and the counter-productive efforts of supporting collapsing systems. In the past, expenditure for the amelioration of social costs has sometimes been misallocated so that welfare assistance has been provided rather than investment in regeneration initiatives, or the retraining and relocation of workers to other areas. The long-run interests of citizens are not well served if they are economically, socially, and spatially isolated in declining areas, where they subsist on a combination of welfare assistance and the informal economy. Adaption by both household and corporate establishments through relocation is facilitated by order and regulation that is flexible enough to permit rapid change.

Transactional complexity increases order and stability within a certain spatial and temporal range. If an urban centre is insufficiently developed to take advantage of available investment, then an alternative but better adapted ecostructure will displace it. If there is over development at a particular class of urban centre, the urban system becomes overextended and unstable. Complex dynamic systems exhibit emergent dynamic behaviours in which change does not always arise in a continuous and deterministic way. The evolution of an urban system faces alternative paths for development, and at critical junctures in its history, new pathways for the diffusion of investment are introduced through ecostructure development. Sudden and discontinuous change is known as *punctuated equilibrium* and unforeseen or chance events limit the ability to predict precisely the future shape of an urban system.

The evolution of civil ecostructures in a world urban system remains on a *universal attractor* to increase the overall diffusion of investment capital for the realisation of the full planetary potential. Investment capital has the equivalent properties to 'exergy', which is the free or high-grade energy in an ecosystem, and capital formation drives the system away from equilibrium in the irreversible process of evolution and sets up gradients between locations in the rate of return on investment. This is achieved through the formation of islands of increasing transactional complexity, which result in a reduction in the unit cost of transactions. As populations of households and establishments approach the carrying capacity of a bounded territory, there are diminishing returns on investment so that establishments decentralise and investment capital is diffused to less mature parts of the system. As might be expected, mature ecostructures have different characteristics from less mature ones, in terms of both information and the efficiency of transformation of investment capital. The diffusion of investment flows by ecostructures is subject to a coefficient of connectivity, and in any transformation process there will be a loss of capital.

The behaviour of the civil system is governed by the macrolaws of Ecodynamics, and, as explained in Chapter 2, the evolutionary generating processes are constrained at all levels by the laws of physics, such as the laws of thermodynamics, the law of gravity, the logistical law of atrophy, the instructions encoded in the genome, civil laws, and the laws of economics, such as eventual diminishing returns. The laws of Ecodynamics will need to meet all these constraints, and compliance with the laws of thermodynamics would be based upon the explanation that human societies are themselves dissipative structures that transform food and energy from sources in the regional environment, and the societal wastes are returned to the environmental sinks. However, there are a variety of possibilities for different landscapes in the future, which can be influenced by human vision. The macrolaws for the emergent phenomena are complementary to the microlaws that determine the underlying dynamics, and they are not a substitute for research that provides a coherent explanation for the microsystems. In extreme conditions involving a breakdown in the regime, the macrolaws may no longer provide valid results, and an examination of the microsystems may reveal new patterns in the emergent phenomena so that a reformulation of the macrolaws would be necessary. Civil ecostructures diffuse investment within the world urban system, which is itself transformed endogenously in accordance with the following three macrolaws of Ecodynamics:

First law of Ecodynamics

In an interconnected system of civil ecostructures, investment capital is transferred between establishments so that there is a cumulative transformation of the resource at each trophic position in the vivisystem. This creates irreversible structural change to the system, which evolves through the formation of islands of increasing transactional complexity and connectivity. The potential of an ecostructure is a measure of its capacity to diffuse investment capital within a specific spatio-temporal range. The world urban system is transformed endogenously and remains on a *universal attractor* to increase the overall diffusion of investment capital through the realisation of the full planetary potential. Capital is conserved in so far as a local level of complexity can only be achieved at the expense of the temporal global investment budget.

Second law of Ecodynamics

Ecostructures of households and establishments are complex adaptive systems for the diffusion of investment capital, and they evolve with increasing complexity to economise on the unit cost of transactions. Capital formation drives the urban system away from equilibrium, which sets up gradients between locations in the rate of return on investment. An ecostructure reaches a spatio-temporal peak of diffusive capacity, as less mature ecostructures emerge to provide alternative pathways for the diffusion of investment. The redistribution of investment to less mature parts of the urban system tends to reduce the gradients in the rate of return on investment. Depending upon the rate of investment flow, individual ecostructures may be classed as static, cyclic, vibrant, or chaotic attractors.

Third law of Ecodynamics

Investment capital is dissipated in the transformation of investment, the maintenance of complex civil ecostructures, and by the coefficient of connectivity in transactions between establishments. In mature ecostructures, more investment is dissipated in the maintenance of order and less in extending the system, than for less mature ecostructures. When the planetary system of ecostructures reaches the phase of maturity at which no further evolution or technological change will increase the overall diffusion of investment, a dynamic balance will be achieved if the regenerative capacity of the planetary metasystems (astrophysical, geophysical, physical, biological, and civil) is sufficient to counter the atrophic forces of decline. Investment atrophy is a measure of investment capital deprivation or undernourishment in parts of the urban system, as a consequence of the accumulation of capital in other parts of the system.

Civil Phase Transitions: Chronological references

1. *The Stages of Economic Growth* - Walt W Rostow, 1960.
2. *Territory and Function* - John Friedmann & Clyde Weaver, 1979.
3. *Development in Rich and Poor Countries* - Thorkil Kristensen, 1982.
4. *The Long Wave in Economic Life* - J Jaap Van Duijn, 1983.
5. *Economic Geography* - Brian J Berry, Edgar C Conkling & D Michael Ray, 1987.
6. *The Spatial Impact of Technological Change* - Edited by John F Brotchie, Peter Hall & Peter W Newton, 1987.
7. *Long-Wave Rhythms in Economic Development and Political Behaviour* - Brian J Berry,1991.
8. *Working Time* – Edited by Hugues de Jouvenal, 1993.
9. *Dematerialisation* – Oliviero Bernardini & Riccardo Galli, 1993.
10. *Atlas of World Development* - Edited by Tim Unwin, 1994.
11. *The Age of Diminished Expectations* - Paul Krugman, 1994.
12. *The End of Work* - Jeremy Rifkin, 1995.
13. *The State of Humanity* – Julian L Simon, 1995.
14. *Ecotaxation* - Edited by Timothy O' Riordan, 1997.
15. *Agequake* - Paul Wallace, 1999.
16. *Humanity 3000* – Edited by Sesh Velamoor and Paige Heydon, 2000.
17. *Sprawl Costs* – Robert Burchell, Anthony Downs, Barbara McCann, Sahan Mukherji, 2005.
18. *The Polycentric Metropolis* – Peter Hall & Kathy Pain, 2006.
19. *Spaces of Global Capitalism* – David Harvey, 2006.

8.
TECHNOLOGICAL EVOLUTION OF INFRASTRUCTURE SYSTEMS

8.1 Technological evolution

Within the civil system there are socio-technological systems such as electricity distribution, transportation systems, and communications systems, which are discussed in Chapter 8 as infrastructure systems. The institutional sponsorship and political acceptance of these technological regimes often involves public sector stimulation through government funded research and development contracts, together with a regulatory framework that will motivate and make space for a shift away from the incumbent technology prior to new economies of scale being established. For example, the replacement of horse-drawn transport by motor vehicles created a massive extinction in the network of equestrian orientated establishments. This was subsequently replaced by a new technological system with its own niches in the motor industry, highway construction and maintenance, traffic control systems, car parks, oil companies, petrol stations, car dealerships and garages, motels, and out-of-town retail parks. Societal evolution arises from socio-technological selection of the elements of a technological regime, in which a household's habitual or exploratory behaviour leads to lifestyles and consumer tastes that have been shaped by the experience of earlier technological systems. It follows that, although societal transitions may be technology driven, there is a cultural evaluation process that coevolves with socio-economic trends that determines the envelope of the technological trajectory of any system.

The evolution of 'archetypes' of human-made 'animats' (animals), such as aircraft (birds), submarines (fish), cars (horses), or even robots (humans), contain a unique genetic combination conceived by a multitude of establishments. An archetype corresponds to the biological concept of a phenotype, since the genetic structure of an establishment is embodied within the archetypes it produces. The term 'transpart' is introduced in which a 'transpart' is the smallest complex assembly, component, or compound of an archetype that has the recognisable characteristics of a progeny of the establishment species. For example, complex compounds and assemblies function as organs in the form of television (eyes), telecommunications (voice and ears), radar (sensors) or computers (brains). Establishments transform investment capital and produce transparts that are transferred between establishments in a complex trophic web. Technological evolution is the process of incorporating increasing informational complexity within the transparts and designing transparts that can be combined to form a variety of good animats. Selection within technologies takes place at the level of an animat, and beneficial mutations within the transparts increase the overall performance and survival prospects for the contributing establishments. In the same way that the science of anatomy examines the bodily structure of humans, animals, and plants, then 'archetomy' would mean the examination of the sub-systems of animats.

Technological innovation generally arises from the combination of well used transparts, and complex transparts are generally formed from the combination of simpler transparts. An encompassing hierarchy is formed from a sub-assembly combination of transparts at one level, for component assembly into a mechanical device, for example, at a higher level, which is then incorporated into a complex vehicle or animat at the highest level. The fittest designs are typically built from well-tested and above-average transparts, to give above-average performance for a reasonable lifecycle cost. It is the novel combination of existing transparts that generally produces an innovation, and the discovery of a new transpart may give rise to a variety of different inventions. Technological evolution also involves an increase in sophistication with structural deepening and an increase in complexity, which permits systems to break through previous limits of performance. Additional functions and subsystems allow a technology to operate in a wider or extreme range, to respond to exceptional circumstances, to enhance its reliability, and to adapt to a changing environment.

When a new technology emerges, there is considerable uncertainty as to which of a range of variants will become the dominant design. If, through historical or path-dependent circumstances, a particular technology gains advantage over its competitors, there is additional investment available to make further new technological advances that deprives the rivals of the necessary resources to catch up and seriously compete. The coevolution of complementary systems or products developed to work in conjunction with the dominant design creates an industry 'standard' for the dominant configurations that emerge. This causes a technological 'lock-in' to a system or design that makes it difficult for alternatives to compete, so that incremental product innovation proceeds for what may not in fact be an optimal design. This is an example of a positive feedback loop that amplifies an early success; other types of positive feedback include economies of scale, learning curves, and product branding through advertising.

The location of traditional industrial cities was generally influenced by geographical and transport considerations in relation to the source of energy and raw materials, access to markets, and the availability of a skilled labour force. With high-technology industry, for which the product and material value is high in relation to its size and weight, the knowledge and technology infrastructure is the prime consideration rather than transport costs. *High technology industry* is the term used for all industries based on the most advanced technology. This industry includes micro-electronics (computers, word processors), mechatronics (industrial robots, car, and medical electronics), fibre optics and communication equipment, biotechnology (genetic engineering, drugs, and medicine), new materials (fine ceramics, carbon textiles), aerospace (aircraft, artificial satellites), artificial intelligence, nanotechnology, photoactive materials, and superconducting elements. The locational requirements of high technology industry are often difficult to realise. Firstly, accessibility to academic research institutes is important for R&D activities to constantly collect up-to-date information in order to cope with the rapidly evolving technology. Secondly, specialised skills with deepened labour markets, contractors, and suppliers are required, together with intellectual property lawyers, venture capital firms, social networks, and contacts. Thirdly, living conditions are also important and include urban amenities and a high-quality, natural, and cultural environment.

The evolution of industrial sectors through their life-cycle stages results in a spatial distribution of investments as establishments form sub-regional clusters during the technological development stage, regional agglomerations in the expansion or market penetration stage, and dispersal in the mature stage of market saturation, followed by standardisation and continental rationalisation with plant closures. Silicon Valley in California is a contemporary example of the regional concentration of the US electronics industry, which commenced as early as the 1940s when innovators established their businesses near Stanford University. The availability of engineers, component suppliers, venture capital, and military contracts attracted more electronics firms and resulted in a self-reinforcing concentration of the electronics industry in Santa Clara County. This created a path-dependent advantage for the region with economies of agglomeration that provided advantages in high quality with lower costs, which effectively established an entry barrier to other locations.

Technology substitution, with the replacement of old systems by new ones, has the same characteristics of evolutionary dynamics in which a new species replaces an older one that may eventually become extinct. The issue of *Technological Trajectories* is described by Leonardo Bondi and Riccardo Galli in the journal *Futures* Volume 24 Number 6 July/August 1992. Evolving economic, social, and environmental boundary conditions create obsolescence across the whole spectrum of energy, transport, and communication systems at different capacities and performances of the respective systems. The energy, transport, and communication infrastructures coevolve as they interact, and there are niche envelopes for each technological subsystem. New technologies are adopted over a period of time, and the rate of diffusion of technologies will vary between industrial sectors, regions, and countries.

In the transition to sustainable civil development, new energy and transport technologies will capture market share in relation to future cost efficiencies. These will reflect both technical advances and the sunk-costs in the existing infrastructure, production, and distribution systems. For example, car manufacturers may opt for internal combustion engines modified for hydrogen use. This would accelerate the development of a hydrogen infrastructure, and hydrogen can be produced without greenhouse gas emissions using nuclear energy. These choices will be influenced by government subsidies for research and development, taxation policies, and environmental standards.

8.2 Infrastructure investment waves

The origins of the metropolises of today are the result of historical circumstances. Their subsequent evolution has been dependent upon the ecological context of the city region and a path dependency involving a variety of factors at different points in time, such as the available technology for irrigation, water supply, defence, energy, transport, and communication. Early metropolises were located at seaports or on navigable rivers, which were later supplemented by networks of canals. Increasing urbanisation of the population was accompanied by waves of infrastructure investment that affected the spatial pattern of urban growth. The industrial metropolis was served by the railway system and its converging radial routes, but then the opportunity for dispersal of industry came along with the distribution of electricity. The road system and the growth of car ownership enabled the decentralisation of employment from the metropolis to outer centres, and air travel, telecommunications, and information technologies have connected the global system of cities. Urban systems have evolved from a continental urban hierarchy into an intercontinental urban network in which the linkages are as important as the nodes. An increasing level of connectivity between the nodes accompanies an increase in the level of complexity of the ecostructures.

Infrastructure imposes spatial structure on a region and creates economies of connectivity at locations, which encourages complementary investments that generate additional advantage. This advantage exacerbates the differential between the more developed countries and the less developed countries, and the benefit of raising the standard of infrastructure provision in the LDCs is that it induces a demographic transition with a decline in human population growth. In effect, an increase in the minimum levels of civil artefacts per capita assists in achieving both a new plateau for economic development and a stabilisation in the population. The OECD publication *LINKAGES - OECD and Major Developing Economies* (1995) points out that the MDCs have a productivity advantage derived from a high-quality infrastructure, reliable and low-cost energy supplies, with support from a wide range of specialised service companies and government agencies. The path-dependant advantages of the MDCs creates the mechanism by which a global economy results in unfair competition, and the LDCs would have difficulty in trade competition with the MDCs without low levels of both wages and social charges. It is the high-quality infrastructure in the MDCs that gives them a path-dependent advantage over the LDCs, so that there is no level playing field in a global economy.

Waves of infrastructure investment, such as the railways, electricity distribution, motorways, air transport, satellite communications, and technology-intensive offices arise when there is a convergence of technological innovation. This applies to energy sources (coal, gas, oil, nuclear), materials (steel, chemicals, and synthetic materials), transportation technology (steam engines, electromotors, internal combustion and jet engines), and information technology (electronics, television, telecommunications, and computers). These waves produce a prolonged period of expansion, which stimulates urban development and social infrastructure investments in utilities, housing, health, and education. The new technologies follow an S-shaped logistic curve until saturation of markets is reached. In effect, there have been long waves in economic development that are an historical endogenous process.

Capital investment in infrastructure and the new sectoral technologies will continue as long as the long-term rate of return on the investment at least matches the cost of finance. The economies of scale in major new investment projects are such that construction of the new infrastructure has to be undertaken a considerable time ahead of demand, and little return will be achieved in the early years of operation. As demand eventually rises to utilise the initial capacity created, then profits will increase and it will become attractive to expand the system or the sectoral capacity. If the rate of increase in demand slows down, even though it remains positive, then the level of investment will decline in relation to the expected rate of return. This relationship between the level of investment and expected level of demand is termed the *acceleration principle*.

It is generally accepted that the macrocycles or long cycles of economic development are caused by macrotrends of sectoral transformations and infrastructure investments, which are spurred on through the action of the multiplier and the acceleration principle until eventually there is an excessive accumulation of physical capacity. The levelling off in demand is then accentuated by both the multiplier and the accelerator mechanism, and a long wave downturn commences. It will take the introduction of a cluster of new technologies to create one or more leading sectors capable of absorbing substantial amounts of new investment, before the cumulative effect causes the upward cycle to get under way again. When a sectoral transformation takes place as described above, the performance characteristics of the new technology may be higher, or alternatively, the prices of the output may fall, in relation to existing products. This leads to a shift in demand to the new technologies, and the capital stock of the existing competitors becomes obsolescent. The replacement of equipment may lead to future echoes of the initial investment, where there is a future replacement pattern at, say, ten-year intervals.

Infrastructure may be defined as both civil and social capital, where civil infrastructure includes irrigation, utilities, transport, energy, and telecommunications networks, and social infrastructure includes housing, education, public health, public service facilities, and recreational amenities. There is a close correlation between growth in the stock of a country's infrastructure and its economic output, and as countries develop the composition of the infrastructure stock changes significantly with the level of income (GNI per capita). For low income countries, more basic infrastructure, such as irrigation, water supply, and sanitation is important. As economies mature into the middle-income stage, the agricultural share of the economy shrinks, and more transport infrastructure is provided. The emphasis then shifts to social infrastructure, such as education and health, and the share of investment in power and telecommunications becomes even greater in high-income countries. Good infrastructure has to expand fast enough to cope with population growth and the expansion of trade, whilst diversifying production, lowering transport costs, reducing poverty, and improving environmental conditions. There may be a service provision gap for any infrastructure element where the level of service falls short of the planning standards that the municipal authority proposes as appropriate for the regional stage of development. Also, the long life of urban building stock and infrastructure, in particular water supply, sewerage systems, energy, or transportation assets, does lock in certain development paths. Taking into account the costs of land purchase, compensation, and construction, it is far more expensive to retrofit infrastructure than to ensure that the pre-urbanisation planning standards can accommodate longer-term growth.

The economic impact of infrastructure investment varies according to its sector, the demand and scale of investment, the location, the technology, the time phasing, and the pricing of outputs. The effectiveness of the investment depends upon matching supply to the demand by users and characteristics such as the quality and reliability of services, as well as the quantitative output. Inadequate maintenance shortens the useful life of infrastructure facilities and reduces the capacity to deliver services. Finally, the efficiency with which infrastructure services are provided is a key to realising the potential returns, and because much infrastructure consists of networks, relieving bottlenecks at certain points can result in substantial efficiency gains. The competition for new export markets is especially dependent upon high-quality infrastructure. During the past two decades, increasing globalisation of world trade has arisen not only from the liberalisation of trade policies in many countries, but also from major advances in communications, transport, and storage technologies. These advances centre upon logistics management to achieve cost savings in inventory and working capital and to respond more rapidly to customer demand.

Infrastructure systems technology evolves through improvements in the level of performance as perceived by the user, e.g., in a vehicle the performance characteristics would include weight, speed, size, fuel economy, style, etc., its power (quantity of performance in a given unit of time), the capital cost, the operational costs, and its lifetime. The total life-cycle cost of any process is the sum of the capital cost and the lifetime operating costs, and a new viable technology has to give an equivalent or higher performance than existing technology at a lower or comparable cost. Capital investment is generally made to reduce operating costs so that the total life-cycle costs are less. Better performance is unlikely to cost less than an inferior technology, although cost reductions can often be achieved through economies of scale. Possible trajectories in urban technology can be classified under the following headings:

- Space saving
- Time saving
- Cost reduction
- Resource efficiency
- Length of asset life
- Prevention of system failure
- Market segmentation and customisation
- Economies of scale

The creation, emergence, and adoption of new technologies result in feedback loops, amplification, and nonlinear dynamics. Motor corporations create the blueprints for cars, which become organisms when driven by agents who provide the motoring instructions. In effect, cars function as four-wheeled animats with detachable brains. A technology that is not just conventional or standard exists typically with a number of variants. Jet aircraft, for example, undergo mutations with cumulative airline experience, and designs are constantly modified with improvements for reliability and safety, increased payload, engine and fuel efficiency, and operating systems. An airline becomes locked-in to the new technology as more aircrew and maintenance staff are trained in the systems, and, depending upon the rate of adoption, there can be increasing returns from the investment in research and development. There is diminishing utility from technological advances since the users' appreciation of increasing performance progressively declines beyond a certain level. Technological development is not always devoted to maximising the scale, capacity, or rate of production or speed in the case of transport systems. It is concerned with affordability and increasing utility at reduced cost in an era of strong market competition and environmental constraints. However, when there is a considerable build-up of scientific and technological knowledge in a particular area, then a radical innovation may develop, and the cluster of innovations may generate a new wave of investment.

There are several variants of the life-cycle growth curve, which display the alternative paths of a technology as markets become saturated. The variations include the standard logistic S curve, incremental S curves, multiple S curves, and technology substitution with the emergence of a design innovation. In the case of a standard logistic S curve, a new technology may be adopted in competition with an alternative technology, in which the selected technology could have been influenced by historical events or path dependence, and through the achievement of an early, large market share, derive the benefit of increasing returns and greater market penetration. With large-scale adoption, the capabilities of the technology may be improved with a succession of design changes that result in incremental steps in the logistic curve. Over a period of time, the performance of a technology may consist of multiple S curves when the capabilities are lifted to a new plane and supported by the development of new infrastructures or the achievement of a critical mass in the market. Even in the mature stage, when standardisation and uniformity reduce costs, low-growth rates in a large market can produce large, absolute increases in the number of users of a system. It is difficult to generalise about the length of the various phases of the innovation life cycle, but national infrastructure investments in technologies, such as the telegraph system, railways, city metro systems, motorways, and energy technologies have taken some fifty years to go from ten percent to ninety percent of the saturation level.

Public dominance in infrastructure systems results from the recognition of infrastructure's economic and political importance. Also, systems have to be designed within a unifying structure that offers diversity and redundancy to provide resilience against system failure. Infrastructure investment is not sufficient on its own to generate sustained increases in economic growth. However, public spending on infrastructure construction and maintenance can be a valuable policy tool to provide economic stimulus during recessions. Also, infrastructure investments are often 'lumpy', since new capacity must be created in large increments in advance of the build-up of demand; hence, public ownership has predominated, and a monopoly has been created through economies of scale. In many countries, impressive strides have been made in infrastructure expansion under the earlier stages of this public leadership.

But more recent experience has revealed serious and widespread misallocation of resources, as well as a failure to respond to demand.

Commercial principles need to be applied to infrastructure services, such as water, ports, railways, airports, power, and telecommunications, so that the costs of provision may be recovered through user charges or tariffs. Almost all infrastructures, including roads and sanitation, can be operated as a business in which customers are satisfied, and reasonable returns are achieved on assets through efficient operation and adequate maintenance. Also, competition promotes efficiency and provides users with options that, in turn, make infrastructure providers more accountable. Whilst open competition for users in the market is still not feasible in many infrastructure areas, there are other ways of obtaining the benefits of competition. For activities with high-sunk costs, competing for the right to operate a monopoly can capture many of these benefits. Even where the number of operators is necessarily limited, regulation can compel them to compete against performance benchmarks. An alternative approach to introducing market principles into infrastructure is through privatisation, which transfers assets out of the public sector. The unbundling of services, such as the generation, transmission, and distribution of power, enables them to be operated as separate lines of business, with the opportunity for independent ownership.

The diffusion of socio-technological revolutions that formed investment waves in earlier expansion phases of leader countries will not necessarily form the basis of economic growth of follower countries in later times. To achieve the level of railway construction and private travel appropriate for a country at the distributional stage of development ($6,001-$15,000 per capita), China would have to increase its railway track eight times and its car population from 4m to 310m. The coevolutionary sequence of railways, motorways, and air transport will differ from the previous one in the MDCs with new growth sectors built around information systems, telecommunications, and satellites. Growth in these sectors is intimately related to a new development philosophy emphasising continental integration, economic efficiency, and clean technology. The infrastructural requirements are certain to be less material intensive than the earlier development waves; however, the materials themselves will be qualitatively more sophisticated.

It is generally considered that infrastructure economies of scale exist for medium-sized urban centres of 250,000 population, and this provides an important threshold for a comprehensive range of economic and social activities. The *Economics of Urban Size* (1973), by Harry Richardson, provides a starting point for an analysis of this issue. A population of 1m may be necessary for certain specialised services; however, the average operating costs steadily increase for large cities of up to 2.5m population, and then the costs jump markedly for very large cities at population size classes of 5m and 10m. Planning standards, which are dependent upon the level of GNI per capita, may be established for the sizing of infrastructure facilities per 10,000 population. Appendix 14 shows that, as the GNI per capita increases over sixty times (from $500 to $30,000), the planning standards per capita for housing increase by a factor of four times. Similarly, the changes in planning standards for hospitals, secondary schools, and higher education are given in Appendix 15. In the case of civil infrastructure investments, such as energy, transport, and telecommunications, the correlation with the per capita provision of services is given in Appendix 16. It may be noted that there is a correspondence between energy consumption per capita and GNI per capita. Road investments progress uniformly up to $1,000 GNI per capita, and then a substantial expansion takes place to match the increase in passenger car ownership. There would appear to be a requirement of 1 km of road per forty motor cars at higher levels of national income.

As countries develop, the infrastructure has to adapt to support changing patterns of demand as the shares of power, roads, and telecommunications in the total stock of infrastructure increase relative to those of such basic services as water and irrigation. Infrastructure typically represents about twenty percent of total investment and forty to sixty percent of public investment. In round figures, public infrastructure ranges from two to eight percent (and averages four percent) of GNI. Even these shares understate the social and economic importance of infrastructure, which has strong links to growth, poverty reduction, and environmental sustainability. As countries pass through each development stage, there are sequentially diminishing returns from infrastructure investment in irrigation and drainage, water supply and sewerage, railways, airports, and power.

The OECD publication *Infrastructure to 2030* (2006) assesses the annual world infrastructure investment (additions and renewals) for selected sectors (road, rail, telecommunications, electricity generation and transmission, other energy-related infrastructure investments, water and waste water) at 3.5% of Gross World Product. It does not include airports, sea ports, and storage facilities. The methodology involves looking at these sectors in detail for the OECD countries and also the Big Five economies Brazil, China, India, Indonesia and Russia, which represent around half of the developing economies. It is estimated that the benefit/cost ratio from universal access to improved water and sanitation services would be in the range of 10. Also it is estimated that the global rise in the number of mobile telephone service users will be from 123 million in 2000 to 5 billion in 2020. In Chapter 10, a proposal is made for financing infrastructure investment in the developing countries, because a failure to level the global playing field with infrastructure investment will lead to major disruptions of service in these countries and a slowdown or stagnation in their economic growth.

8.3 Energy

The evolution of the energy system from fuel wood, to coal, to oil, and then to gas is a trajectory of energy decarbonisation, and a continuing reduction in the ratio of carbon to hydrogen atoms with each succeeding energy source. It has been estimated by the International Institute for Applied Systems Analysis in Vienna that the global carbon emissions per unit of primary energy consumed has declined by 0.3 percent annually for the past 140 years, and this trend is likely to continue to 2150. Burning coal releases twice as much carbon dioxide than burning natural gas, whereas burning oil only releases 1.5 times as much as gas. Anthropogenic carbon dioxide emissions result mainly from fossil fuel combustion and cement manufacturing, which releases 0.5 metric tons of carbon dioxide per metric ton of cement produced. The ratio for the mass of carbon dioxide to the mass of carbon is 3.664, and the carbon content of the annual emissions in 2000 was just over 6 billion metric tons.

In 2000, the global energy consumption amounted to 425 Exajoules (EJ), with coal, oil, and natural gas accounting for 80% of energy consumption, nuclear energy (6%), hydro (6.6%), traditional sources such as wood and charcoal (6.4%), and other renewable sources accounting for less than 1%. During the Third Millennium, the remaining fossil fuels will become exhausted and there will need to be a transition to alternative sources of energy such as nuclear, hydropower, tidal power, solar energy, hydrogen, windmills, geothermal, fuel from crops, combustible waste, biogas plants, and fuel wood. The worldwide electricity production in 2000 amounted to 55 EJ, and since the overall efficiency of electricity generation is approximately 35%, the total energy required for electricity production was 160 EJ. A typical, large, modern power station produces an output of 1 GW (1000 MW) from a fuel input of 3 GW, of which 2 GW becomes waste heat. The latest Combined Cycle Gas Turbines (CCGT) can achieve efficiencies of around 50% in the best installations. It is the convention to express the primary energy content of nuclear energy in terms of the energy content of the fossil fuel to generate the same amount of electricity. Since nuclear energy provides 16.9% of electricity production, the energy content is taken as 16.9% of 160 EJ, or 27 EJ, which corresponds to 6% of global energy consumption.

Although the developing countries emit an annual average of 2.5 tons of carbon dioxide per capita compared with 8 tons per capita in Europe (20 tons per capita in the USA), they already account for nearly one half of the global total. The global carbon dioxide emissions in 2000 was 22.5 billion metric tons for a Gross World Product of PPP US$45 trillion, which amounts to 0.5 kg carbon dioxide per PPP $US of GWP. This represents a 55% decline in carbon intensity since the level of 1.1 kg carbon dioxide per PPP $US of GWP in 1980. The developed countries will have to reduce carbon emissions by a further 35% through the more efficient use of energy (a dejouling of energy) and the use of renewable sources so that the developing countries can increase their share of emissions to perhaps 65% of the total by 2025. The aim is to limit the global annual emission of carbon to 5 billion metric tons during the 21st century, with a further reduction to 2.5 billion tons in the next century with the transition to non-fossil fuels.

In view of global warming, it will be beneficial to move away from fossil fuels long before reserves are depleted and to stretch the coal and oil resources as far into the future as possible, in parallel with the use of nuclear energy because it gives time for deep-ocean mixing of carbon dioxide, with a reduction of carbon dioxide in the atmosphere. In 2000, there were approximately 440 nuclear power plants worldwide with just over 100 in the USA. But because of the environmental risks with the use of nuclear power on a large scale, it is likely that this will only get expanded when it becomes evident that a new generation of sustainable nuclear reactors is available and that no other course is possible. It is anticipated that this new generation of reactors will be available by 2025, which will be built to extremely high safety standards and will recover more of the energy contained in uranium with a significant reduction in nuclear waste. Anti-terrorist provisions would include subterranean reactor cores and waste storage facilities, with reinforced auxiliary nuclear plant equipment. Spent nuclear fuel is accumulating in a number of countries, and it contains various radioactive substances, including plutonium, which has a half-life of 24,000 years. This waste is estimated to be harmful to humans and animals for 100,000 years. However, nuclear energy produces electricity without discharging greenhouse gases, and it can be used for the production of hydrogen fuel for transport. In the longer-term, renewable energy sources such as hydropower, tidal power, wind power, and solar energy will be used for hydrogen energy production, once the hydrogen fuel infrastructure has been installed using nuclear technology. In many cases, there will be an optimum technology for specific regions or sites depending upon the available supplies of fossil fuels, the proximity of oceans or rivers, and the annual amount of sunlight.

Renewable energy sources are easier to apply in rural areas than at urban centres, and a city dweller in an industrial developing country consumes 75 GJ per year, whereas a traditional villager may consume less than 15 GJ per year. This does not mean that rural life is less energy intensive than urban life, but demonstrates that villagers often lead miserable lives, consuming little more than is needed to stay alive. In developed countries, rural households are often more energy intensive than urban households because of the increased use of transport in rural areas and higher requirements for heating isolated, badly-insulated farmhouses. On-site sunlight and wind power are insufficient in intensity to support human population densities greater than 2,000 persons/km², with the majority of them consuming the minimum for survival. It follows that cities with population densities in excess of 4,000 persons/km² will be dependent upon non-renewable energy resources (coal or nuclear), with perhaps renewable resources such as solar energy contributing to domestic cooking, space and water heating.

On the assumption that energy efficient vehicles will approach the limits for fuel conservation, liquid fuels will become increasingly expensive and, after the year 2100, conveniently located fossil fuels will have been exploited. Oil shales and tar deposits may be converted into liquid products that may suit existing refineries, and low-grade coal or lignite may be converted to a petroleum-like product. Also, methanol can be produced economically from coal, gas, wood, tar, etc., to give a fuel that is seventy percent as energetic as petroleum, and may be used in existing engines with little modification. Biofuels may be produced from crops by the efficient conversion of biomass, and would require 0.5 hectares per vehicle. This is 2.5 times the amount of land required to feed a human, and it follows that feeding the world's population of vehicles could make excessive demands on the available land. Low-priced natural gas would be the principal source of methanol to the year 2050. Johan Schot, Regma Hoogma, and Boelie Elzen are the authors of an interesting article, *Strategies for shifting technological systems – The case of the automobile system* in *Futures* Volume 26, Number10, December 1994.

Hydrogen may be considered as the ideal transportation fuel from an environmental point of view. Unlike fossil fuels, when it is burned in air it does not produce carbon dioxide, carbon monoxide, sulphur dioxide, or volatile organic compounds. The only by-products of hydrogen combustion are water vapour and a very small amount of nitrogen oxides. The potential of hydrogen as a fuel lies in its production from renewable sources, such as from hydro-power, tidal energy, or wave energy, solar energy, or alternatively, nuclear energy. Hydro-electric power is a form of solar energy as it is generated from the precipitation of rain, whereas tidal energy is principally lunar. The attraction of lunar hydrogen energy is that it provides a means of capturing the intermittent but highly predictable tidal energy, without the high capital cost of a barrage in relation to the useable output that arises with electricity generation. Solar hydrogen energy can be produced in two principal ways: by the electrolysis of water using photovoltaic electricity or by the thermal dissociation of water into hydrogen and oxygen. Depending upon the method of generation, the costs of hydrogen fuel are likely to be three to five times the current price of untaxed petroleum and, with the steady increase in carbon taxes proposed for the MDCs, the substitution of hydrogen for petroleum will commence early in this century. The development of a hydrogen energy infrastructure would enable it to be distributed by regional grids like natural gas to give the cleanest fuel of all. To make the hydrogen burn efficiently, cars will use fuel cells to power an electric motor. In *Tomorrow's Energy – Hydrogen, Fuel Cells and the prospects for a cleaner planet* (2001), Peter Hoffman describes how hydrogen is not an energy source but a carrier that, like electricity, must be manufactured.

8.4 Transport

There is an anthropological constant of about one hour per day for the amount of time humans spend in daily movement about their territory, which is independent of any travel technology adopted. This concept of constant travel time and trip rates is consistent with the notion of travel-time budgets and has implications for land-use patterns and urban scale. In early pre-industrial cities, movement was on foot, and the distance covered limited employment opportunities and the city size. Horse-drawn vehicles and then the tram extended this distance, and in the industrial city, mass rail transit increased it by an additional order of magnitude. In the postindustrial city, employment dispersal has arisen from the wider use of the car and the range of services, and employment opportunities have increased the potential population capacity. Further technologies such as inter-urban motorways and high-speed rail tend to permit the growth of megapolises or megalopolises, and air transport and telecommunications link the global system of cities.

On the basis that much of the daily travel time is allocated to the journey to and from work, it is found that for all modes of transport, walking and the bicycle predominate for journeys up to 2 km, the significance of the bus declines at the 5-15 km range, and the train at above 30-40 km. Cars become progressively more important with greater journey length, and for all journeys car users spend on average two thirds of the amount of time travelling that bus users spend, but travel fifty percent further. For long journeys, rail is often quicker but fares are a severe deterrent when a family is travelling. Changes in average commuting trip times for major European and US cities over recent years show that the impact of job dispersal is to divert trips to less congested suburban destinations and routes, and that journeys are made by car at increased average travel speeds which has the effect of stabilising travel times. In the UK in 2000, the average speed for all journeys and modes was 30 km/hour, and the average distance travelled per person per year over all journeys and all modes was 11,000 km. On average, the total travelling time was 367 hours per year, or one hour per person per day.

Society has become dependent upon the car, and the economic wealth of developed nations is based upon the motor industry with an enormous investment by the oil companies in a hydrocarbon infrastructure with very long payback periods. In 1950, the world population of cars was 22 cars per 1000 population and by 1970 this had increased to 55 cars per 1000 population. By the year 2000, the number of cars reached 540m, or 90 cars per 1,000 population, and annual worldwide travel by car amounts to 16 trillion passenger-km. From the evolutionary assessment in Chapter 10, the number of cars will exceed 1.1 billion by 2025 in relation to a world population of 8.2 billion, or 140 cars per 1,000 population. By the year 2150, the number of cars may exceed 3.1 billion for a world population of 12.5 billion, or 250 cars per 1,000 population. In the period from 2000 to 2025, the increase in the number of cars will come from 230m in East Asia and Pacific, 74m in Latin America, 77m in South Asia, 36m in North America, and 80m in Eastern and South East Europe.

As the number of cars on the roads increase, together with a growth in the annual mileage per vehicle, the annual distances driven in the USA are expected to increase from 2.5 trillion km in 2000 to perhaps 3.6 trillion km by 2020. In the USA, light trucks (Sports Utility Vehicles – SUVs, minivans, and pickup trucks) have increased their share of new car sales to 46% by 2000, although these vehicles are far less fuel-efficient than traditional passenger cars. Consequently, average vehicle fuel-efficiency has declined from 25.9 mpg (miles per gallon) in the early 1980s to 23.8 mpg by 1999, with cars achieving 28.1 mpg and light trucks rated at only 20.3 mpg. The potential gains from fuel-efficiency technologies have been more than offset in the USA by increasing vehicle weight, horsepower, and acceleration performance. European SUVs are smaller and more efficient than US models.

The geopolitics of oil supplies and the carbon emissions problem will require investment by western countries in "clean" cars that run on alternative fuels such as hydrogen or electricity, and this applies to one-third of the vehicle fleet by the year 2025. In addition, some 3.2m km of road will need to be constructed in the less developed countries to accommodate these vehicles, and the future trend will be for smaller cars and long-life vehicles that have standard replacement parts and are fully recyclable. Finally, hybrid electric vehicles (HEVs) based upon a lightweight, high-density battery may capture a share of the market for a family's second car, to cater for local trips within the town for shopping and transporting children to and from school, etc. HEVs are essentially electric cars that also have a small internal combustion engine and an electric generator on board, which can charge the battery en-route to extend the vehicle's range. Full recharging of batteries may take three to eight hours, which can be done overnight at home.

Urban rail systems may be classified into 'suburban' railways (conventional, heavy rail commuter services), 'metros' (traditional, mostly underground, medium weight railways as in London and Paris), 'light rail', or LRT (such as the Docklands Light Railway), and 'tramways' (which include traditional city trams and new systems such as that in Grenoble). The weight of the vehicle affects the acceleration, gradient, and curvature of the track. This in turn affects the average speed and station spacing, the carrying capacity, the headway, and the need for automatic signalling and segregated track. Metro or mass transit systems operate on rights-of-way from which all other traffic is excluded, and they are generally suitable for cities with populations of over 2m and urban densities of over 4,000 inhabitants per km^2, and an annual ridership in excess of 100m passengers. These systems serve central business districts with a minimum of 5m m^2 of office floorspace and are capable of carrying 40,000 to 50,000 passengers per hour. The Paris metro, which is one of the most heavily used and most extensive rapid transit systems in the world, has a route length of 185 km with stations at 0.5 km and carries 1,300m passengers per year, compared with London, which carries 600m passengers over 400 km with stations at 1.5 km average spacing. In 2000, there were some ninety-five heavy-rail urban transit systems with thirty-one percent in Western Europe, eighteen percent in North America, and sixteen percent in Eastern Europe and Central Asia, thirteen percent in East Asia, twelve percent in Latin America, nine percent in Japan, and one percent in Africa.

Typical light rail systems have a capacity of around 20,000 passengers per hour, with station spacing less than 1 km, and an average speed of 22-30 km/hour. Light rail systems are suitable for principal or provincial cities with city populations of over 250,000, and urban densities in excess of 2,500 inhabitants per km^2. Light rail systems include trams, streetcars, and trolley cars that run along tracks at street level that may not be separated from other vehicles. There are some 365 light rail urban transit systems worldwide with forty-five percent of these systems in Eastern Europe and the former Soviet Union, where they have been preserved from the 1930s pre-war era. Western Europe has thirty-three percent of the light rail systems where urban densities reach 3,000 inhabitants per km^2, which is some fifty percent higher than North American cities with densities of say 2,000 inhabitants per km^2 and their share of light rail systems is only nine percent. In contrast to the high-density cities on the eastern coasts of North America and Australia (Sydney), there are low-density auto-dependent 'urban sprawl' cities and towns in west and southern North America and Australia (Perth, Adelaide, and Melbourne) in which urban densities are around 1,500 inhabitants per km^2. In Europe, the planning standard for bus transport is one per 1,000 population, and, in the developing countries, a default planning standard would be 1.5 buses per 1,000 population. It is estimated that there are 8 million buses worldwide, and that annual bus passenger travel amounts 9 trillion passenger-km. In the less developed countries, rights-of-way may be set aside for bus lanes, as at Curitiba in Brazil, on which light rail track can be laid as soon as the means are available for a rail transit system.

For inter-regional transport, a criterion for travel time is the facility to make a return journey within a day, with at least four hours' stay. For intercontinental journeys, the aim is often to reach the destination within a day's travel. The growing network of transport links and modes (road, rail, and air) provides increased mobility, with each mode having a place in the transport market. Motorways and toll-ways dominate for shorter inter-regional trips, fast rail is used for rapid intercity travel, and airlines are facilitating faster international and intercontinental travel. Worldwide there are 1.1m km of rail track, and the world passenger rail travel amounts to 1.9 trillion passenger-km. India accounts for twenty-three percent of rail ridership, China twenty-two percent, Western Europe fifteen percent, Japan thirteen percent, the former Soviet states twelve percent, and the rest of the world fifteen percent. Railways in Western Europe and Japan are orientated towards passenger travel, whereas in North America rail transport is used primarily for freight. In Europe, the High Speed Train (HST) of 200 km/hour is planned to link major cities, and the French TGV has a cruising speed of 300 km/hour. In Australia, the Very Fast Train (VFT) is planned for 350 km/hour, and the Japanese have undertaken to build a new maglev line (magnetic levitation) where train speeds could eventually pass 400-500 km/hour.

It may be noted that the rate of progress of air transport technology has been determined by affordability, and successful developments have offered significant increases in utility at reduced cost. However, once the present standard specification, exemplified by Airbus A300 had been adopted and compatible infrastructure investments made, it became more difficult to generate sufficient improvements in technological efficiency to justify the costs of introduction, except in small increments. There are some 11,000 large commercial aircraft, and the annual number of passengers carried by airlines is 1.5 billion, so that worldwide air travel amounts to 3 trillion passenger-km. It has been suggested that the overall industry growth curve is actually composed of multiple S-curves, each reflecting the impact of major structural change in air transport. Forecasters in the 1970s did not anticipate the conservative path to the present. Most were predicting large numbers of supersonic transports (SSTs) with 1000-seat aircraft, intercity vertical take-off and landing (VTOL) aircraft by the year 2000, and a high probability of hypersonic hydrogen-powered aircraft at Mach 5 or five times the speed of sound. The cruising speed for a wide-bodied jet aircraft such as the 500-passenger Boeing 747 is just under the speed of sound at Mach 0.8 or 880 km/hour, and supersonic aircraft such as Concorde (125 passengers) travel at Mach 2 or 2000 km/hour. In making comparisons between High Speed Trains and air travel, journey times to the station or airport and terminal times for checking in and waiting need to be taken into account. Generally, the breakeven distances for HST/air will be 530 km, and TGV/air 960 km. In summary, within three hours a car will travel over 200 km, the range for VFT is 500 km, and a plane 750 km.

The annual worldwide passenger travel in 2000 was estimated to amount to 30 trillion passenger-km per year, and the modal shares were cars fifty-three percent, buses thirty percent, airlines ten percent, and trains seven percent. It may be shown that per capita travel in terms of passenger-km per year increases at each stage of development from non-motorised transport to public transport, such as buses and low-speed trains that travel between stations at around 20-30 km per hour, and then to cars that offer greater flexibility with door-to-door speeds of 30-60 km per hour. At high incomes, the traffic volumes by car may peak at around 20,000 passenger-km per person, with alternative high-speed long distance journeys by train or air travel increasing their share of the transport market from 10,000 to say 20,000 passenger-km per person. In effect, the share of car journeys will decline from sixty-five percent for annual travel of 30,000 passenger-km to around fifty percent for an annual 40,000 passenger-km travelled at the highest income levels. If the car trips are made at an average speed of 50 km per hour, the daily travel will amount to 1.1 hours. At an average speed of 500 km per hour, the high-speed travel could amount to forty hours per year or an average of less than an hour per week. One consequence of road congestion is that traffic speeds in urban areas may decline to 35 km per hour in high-income countries and, in order to limit the travel-time budget to an hour per day, light-rail systems will be introduced so that the car journeys may decline to 15,000 passenger-km per person, which would represent less than forty percent of the annual distance travelled.

In the MDCs, commercial vehicles account for around twenty percent of the transport on the road, whereas in the LDCs the proportion rises to thirty-five percent at the industrial stage and forty percent at the infrastructural stage. There are various technological innovations that are increasing the effectiveness of transport by road and rail. These developments include in-vehicle communication, vehicle monitoring and control, traffic management systems, electronic road pricing, guided busway systems, driverless trains, transit passenger information, and computer reservation systems. The application of information technology to logistics should improve the scheduling and routing of freight vehicles with an increase in their utilisation. In the UK, approximately one-third of freight vehicles are running empty, and the continued development of new manufacturing techniques with 'just-in-time' delivery promotes more frequent and smaller batch distribution. The growth in road freight is closely related to GDP, and the net transport intensity (i.e., net mass movement) has remained constant in terms of ton-km equivalent per $GDP. However, the gross transport intensity (i.e., the gross mass movement that takes account of under-utilised vehicles) to which transport energy consumption is related has increased. It may be concluded that information technology should be applied to increasing car occupancy and freight vehicle utilisation, because at the postindustrial stage, economies become less intensive in a whole range of inputs including energy, materials, and labour so that transport currently contrasts with other sectors of the economy. Finally, a greater utilisation of the road and other modes of transport can be achieved by spreading peak traffic over time through the use of congestion-charging by means of electronic road pricing, with work-staggering or flexitime, and changes in retail opening hours.

Port container traffic is measured in standard-size container units and amounted to 230 million containers worldwide, of which sixty percent was accounted for by the MDCs. Maritime transport also includes some 80,000 vessels of not less than 100 gross tons, together with a fishing fleet of 1.2 million boats.

8.5 Telecommunications and information technology

Previous technological investment waves have eventually created more jobs than they displaced, and in *Capitalism, Socialism and Democracy*, (1942), the economist Schumpeter pointed out that economic development was a process of creative destruction, with the benefit of raising the standard of living for an increasing population. In the past, technological investment waves such as electric power transmission or the internal combustion engine, whilst displacing existing forms of production, still required labour in sufficient quantity for its own (and attendant) processes, so that its eventual expansion led to the creation of new jobs and industries. Even the earlier development of commercial computers, with its partial displacement of unskilled clerical work, involved so much expansion in other fields such as software development, that the balance of jobs was probably positive. However, in the MDCs where the rate of economic growth is declining, capital-intensive technologies that replace labour are likely to bring a reduction in working hours. In the past, it was the increasing demand that generated the additional employment in absolute terms, which is now true for the LDCs. There are therefore unique properties of the telecommunications and information technology revolution that could inhibit or prevent a repetition of the apparently self-correcting process of investment waves in the more developed countries.

As microelectronics develops, it displays features that make it different from all previous technologies, i.e., its cybernetic properties for control. This technology is a unique metatechnology, which is displacing or transforming all other means of control, notably those using people. As it grows exponentially in power and economy, it displaces, replaces, or improves other means of control with increasing ease. The combination of telecommunications and microcybernetics is causing the shrinking of distance and time so that multinational corporations based in selected cities of the world can control companies and production plants in all parts of the globe. It follows that employment may be destroyed in the more developed countries, but it is likely that more new jobs will be created in the low-cost developing regions of the world.

Appendix 17 gives default planning standards for communications and information technology. Telephone installations increase rapidly with growth of the quaternary information division, with saturation for landline telephones being around 7,000 per 10,000 population. Mobile telephone penetration is currently reaching 7,500 per 10,000 population in Scandinavia. Similarly, the penetration of personal computers is reaching 5,000 per 10,000 population, and in the USA, some sixty percent of personal computers are connected to the internet. In Europe, the penetration of television sets reaches 6,500 per 10,000 population, with newspaper readership at 3,000 per 10,000 population. Not surprisingly, in the USA, the number of television sets reaches 8,500 per 10,000 population, and the newspaper readership declines to 2,000 per 10,000 population.

The shift from analogue technologies to digital technologies has permitted the development of multimedia interactive systems, with the integration of text, images, and sound, which is as revolutionary as the invention of the Greek alphabet in 700 BC. In effect, digital communication is a meta-language that will fundamentally transform society and lead to cultural evolution. The interconnection of both households and business establishments in a global network, with the immediacy of real-time or select-time communications from such a multiplicity of sources, will create a vivisystem of unprecedented complexity. The emergence of new images and visions through the synergy of digital media, computers, and telecommunications will stimulate interactive behaviour between the sender and the receiver. There will be a deepening in the segmentation of consumer markets, with customisation by lifestyle, interests, and values, together with an enhancement in the customer-producer relationships.

Over the past twenty years, there have been dramatic improvements in the cost performance of computing and communications technologies, each of which has grown independently at an annual rate of twenty-five percent. Computers have become so powerful and cost-effective because of their speed of operation, that they have an enormously wide range of applications. For example, computer-aided design (CAD) and computer-aided manufacturing (CAM) have revolutionised the approach to designing and manufacturing complex products. Expert systems (software) consist of a set of decision rules on how to process specific sets of data, which hopefully replicate the best experience of experts. The Internet system and the electronic mail (e-mail) were serving over 250m users in the year 2000, and early in the 21st century the information superhighway is expected to provide a vast range of information, commercial, and financial services to perhaps 2 billion people.

At the regional or local level, telecommunications and information technology can provide an alternative to travel by car by permitting people to work from home (telecommuting) or to shop and obtain services or entertain themselves at any location. These services or entertainment requirements need not be limited by weather conditions or geographic proximity, since the reach can be truly global. Telecommunications technologies are facilitating the globalisation of world financial markets and increasing linkages among principal world cities. Also, the emergence of fibre-optic systems for international and interurban communications is strengthening the telecommunications capacities of large metropolitan centres. Deregulation and private sector competition are reducing the cost of providing advanced telecommunications services in cities. Communications satellites have transformed the Earth into a global village, and the space programme of the second half of the 20th century will have contributed at least as much to the globalisation of the informational economy as to the exploration of space or to intercontinental defence systems.

Surveillance technologies, with spies in the sky, CCTV in city centres, traffic control systems with cameras to enforce congestion charging in urban areas, property security, and medical monitoring services to retirement communities, introduce conflicts between an individual's desire for anonymity and increasing requirements for identifiability. Remote monitoring tasks can be undertaken in low-wage, less developed countries, wherever the skills are available and the time zones are compatible. The electronic tagging of offenders, tracking devices for pet owners, car insurance based upon monitoring of vehicle journeys, child minding systems with chip implants or belt-mounted pagers for the elderly, have set in motion a location-based services industry. The fear for privacy is that by conversing over a cell phone or from surveillance technologies, an individual will leave an archive of electronic trails so that anyone can discover where he has been.

8.6   The exploitation of space

The astrophysical system comprises the solar system with the four inner planets of Mercury, Venus, Earth, and Mars, which are solid globes and comparatively small. Outside the orbit of Mars, there is the main asteroid belt in which move thousands of minor planets. Beyond the asteroid belt come the four giants Jupiter, Saturn, Uranus, and Neptune, which have comparatively small silicate cores surrounded by liquid and deep gaseous atmospheres, and finally Pluto, which is smaller than the moon. The mean distance of the Earth from the Sun is 150m kilometres, and the volume of the Sun is over a million times that of the Earth. Hydrogen is the Sun's fuel with the energy being produced by nuclear transformations deep inside its globe, and the mass loss amounts to 4m tons per second, so that the output is not expected to change markedly for another 5 billion years.

International space exploration has stimulated an appreciation of our planet as a unique and vulnerable biosphere, and as the frontier has been extended by new discoveries in the solar system, possible scenarios have emerged for the exploitation of space. The potential opportunities from the exploitation of space include:

- Space transportation
- Satellite communications
- Defence systems
- Civil space structures
- Environmental research
- Energy sources and treatment of radioactive fuel waste
- Biodome communities
- Planetary terraforming

Space transportation economics is likely to require a convergence of civilian and military applications that provide tangible benefits from investment in space technology. Reusable launch vehicles such as the space shuttle have a substantial cost advantage over expendable or semi-expendable space transportation systems. The development of spaceships that would provide regular and cost-effective transport services to low-Earth orbit would cause a major shift in the design and operating philosophy of spaceports, satellites, space stations, space-based production facilities, and lunar or inter-planetary travel.

Virgin Galactic is planning to become the world's first spaceline with spaceships modelled on SpaceShipOne, which won the $10 million Ansari-X prize for making two manned flights into space within a two-week period at altitudes greater than 100 km, using a reusable spacecraft. The design and construction of SpaceShipTwo is well advanced using the technology developed by Burt Rutan in Mojave, California, with carbon composite materials, a hybrid rocket motor, and a wing-feathering technique for re-entry. The spaceship hitches a ride up to around 15,000 metres attached to a specially designed jet carrier aircraft, 'the mothership'. The spaceship is then released from the mothership and ignites its hybrid rocket, in order to climb to over 100 km in about 90 seconds to reach a speed of around Mach 3, or three times the speed of sound, with around thirty minutes in space. Shortly before the maximum altitude of its flight path, the spaceship feathers (folds its wings) in preparation for re-entry into the earth's atmosphere, drawn by the earth's gravitational pull. As the spaceship meets the resistance of the upper atmosphere, the feathered wings act as air brakes, safely positioning and decelerating the spaceship during re-entry into the earth's atmosphere. At approximately 18,000 km, the spaceship's wings are re-configured to their original position to enable an unpowered (glide) landing back at the spaceport. Commercial flights on the maiden Virgin Galactic craft, the VSS (Virgin SpaceShip) Enterprise, are planned to commence in 2009, from a new $200 million spaceport in the State of New Mexico, USA.

For satellite launches and the space shuttle, the thrust to leave the Earth's atmosphere and enter space is provided by heavy and non-reusable rockets, which require vertical take-off and have relatively low manoeuvrability. An alternative, air-breathing engine technology (i.e., non-rocket), known as ramjets has been able to achieve speeds of up to around Mach 6. These have been used on missiles for which the propulsion is switched to ramjets once the rockets have achieved supersonic velocities. Further developments in the technology involve the supersonic combustion of the rammed airflow in engines that are known as scramjets, which have been tested for speeds of up to Mach 10, and for which the theoretical maximum speed is Mach 20 to 25 so that a vehicle could reach orbit. Scramjet flights into space are unlikely before 2050, and in order for commercial spacelines to use runways at specially designed spaceports, a multimode launching operation would be required as neither ramjets nor scramjets can operate at low speeds. The initial propulsion for a spaceline would be from an advanced turbine engine for speeds up to Mach 2 or 3, then from a ramjet up to around Mach 6, and this would be followed by a scramjet. It is envisaged that spaceports will emerge with the hydrogen-fuel economy, and the pay-off from space flight is that a spaceline could deliver passengers and cargo to any place in the world in less than two hours. A fifteen-hour subsonic flight from New York to Tokyo would be achieved in ninety minutes by using "Fractional Orbital Flight" with take-off, a climb above the atmosphere, a partial orbit, descent, and landing. These orbital flights would be for long-haul intercontinental trips between perhaps ten to fifteen spaceports, which will serve the most densely populated continental zones of a planetary civilisation.

Satellite communications is an area in which the commercialisation of space is already established and from which the pay-off to both developed and less-developed countries has been enormous. Telecommunications and broadcasting are providing traffic growth that has outstripped forecasts, and this is leading to scarcity in the availability of 'slots' for the geostationary orbit of satellites, with respect to a fixed position on the Earth's surface. By 2001, low-orbit communications satellites provided universal telephone access via mobile phones to anywhere on the planet.

Defence systems and the military exploitation of space have largely involved reconnaissance satellites, military surveillance, spacecraft tracking, advanced navigation systems, and command, control, and data acquisition networks. Space warfare would involve ground- or air-based anti-satellite missiles, orbiting hunter-killer satellites, orbiting missile defence systems, laser weapons, electronic jamming, and decoys. The United Nations and arms control treaties prohibit nuclear detonations in outer space, orbital nuclear weapons, and weapons of mass destruction, and there are agreements to prevent the use of Fractional Orbital Missiles.

During the 21st century, civil space structures are likely to be developed to achieve improvements in the technological economics of space exploitation, in the same way as the provision of infrastructure on Earth. As a payload reaches orbit, it must either bring with it all its operational requirements, such as power resources, communications, stabilisation, maintenance; or alternatively, these capabilities could be provided at space stations or platforms. Similarly, remote sensing apparatus can be provided on platforms or spacelabs for scientific research. Manned space stations are already in operation, and following the Salyut (1-7), Skylab and Mir stations, the first International Space Station (ISS) was occupied in 2000 and it will be completed in stages by 2010. In due course, communities of perhaps 1,000 scientists or technologists could inhabit the moon (three days' space flight from Earth) or Mars (nine months' journey away) by living in sealed biodomes. It is anticipated that this will take place at the beginning of the 22nd Century, in the Early Planetary era. Periodical supplies of liquid hydrogen from Earth would be combined with oxygen on the moon (lunar oxides are available), to provide water for the colony, which could be recycled. There are technical difficulties in creating a sustainable environment within a sealed structure, and in particular maintaining a breathable atmosphere. In addition, the psychological problems of humans living together in a container for extended periods of time would need to be overcome. Energy sources in space would be advanced photovoltaic systems, thermal electric systems, and nuclear power. It is possible that radioactive fuel waste could be transported by shuttle from the Earth's biosphere for treatment on another planet.

Environmental research from outer space includes geophysical and geodetic studies, remote sensing, meteorology, atmospheric research, and astronomy. With a worldwide increase in the demand for energy, the hyper-spectral capabilities of remote sensing make it well adapted for space-based oil and gas exploration. In addition to meteorological, oceanographical, and climate-change applications there are numerous uses for agriculture, fisheries, forestry management, and urban planning. Treaty monitoring activities from space would include the environment, such as enforcement of greenhouse gas emissions abatement accords, the verification of EU regional policies (e.g., Common Agricultural Policy), and disarmament agreements. The prevention of natural disasters from earthquakes, volcanic eruptions, tsunamis, storm surges, and flooding are increasingly being monitored to provide early warning systems for the minimisation of injury, loss of life, and other consequences of local and regional disasters.

The Search for Extraterrestrial Intelligence (SETI) using radio astronomy commenced in 1960, when Frank Drake started looking for ETI signals near the hydrogen line frequency from two nearby Sun-like stars – Epsilon Eridani (10.2 light years away) and Tau Ceti (11.9 light years away). In 1982, the International Astronomical Union (IAU) established a Bioastronomy Commission to support a new branch of astronomy for the search for life and intelligence in the cosmos. In 1992, NASA became involved in the search for radio signals with a programme intended to increase from $2 million to $14 million over a period of six years, which would monitor 1,000 target stars out to a distance of 100 light years, using Multi-Channel Spectrum Analysers (MCSAs) with 8 million channels. However, in 1993, Congress voted to terminate all the project's funding and the programme was shut down in 1994. The programme would have been completed by around the year 2000, but instead, the targeted search has been undertaken using private funds by the SETI Institute, whose president is Frank Drake. With the increase in computer power and knowledge, search efforts are being more focused on discovering stars with planetary systems. In 1994, two Swiss astronomers, Michel Major and Didier Queloz, were successful in the search for extrasolar planets with the discovery of 51 Pegasi some forty-five light years away in the constellation Pegasus. The star is similar to the Sun in its surface temperature and mass, but it is older and contains more metals including iron. Around twenty extrasolar planets have been detected since 1995 and, in 1999, a second and third planet was discovered around the star Upsilon Andromedae, which is fifty-four light years from our solar system. These two planets have average distances from their star of 0.8 and 2.5 times the distance between the Earth and the Sun, and they orbit with periods of eight months and four years respectively. It is unlikely that contact will be made with extraterrestrial intelligence until the Near Interplanetary era from 10,000 YF.

In *Space – The First 50 Years* (2007), Patrick Moore and HJP Arnold describe the exploration of space and the various missions so far. Between 1958 and 1968, the Luna, Ranger, Orbiter, and Surveyor spacecraft provided detailed astronomical information and photographs of the moon, prior to the lunar exploration by the Apollo missions (1968-76). Although Venus (110m km from the Sun) is the nearest planet to the Earth, it is over one hundred times as far away as the moon, and the explorations in the period 1961-1995 have indicated that Venus and the Earth began to evolve along similar lines, but as the sun became hotter, Venus was transformed from a potentially life-bearing world into the raging inferno of today. The exploration of Mars (1964 -1993), which is 230m km from the Sun, has revealed a red landscape of volcanic rock, with old river beds, which show that water flowed on Mars in the distant past. The climate is cold, and although at the equator the daytime summer temperature rises to over 22°C (70°F), the night temperature is extremely low as the thin atmosphere retains little heat. The polar caps are made up of a mixture of water ice and carbon dioxide ice, and water exists in buried permafrost in other areas. The main constituent of the atmosphere is carbon dioxide, and whilst life may have existed in the remote past, it is unlikely that there is life on Mars now.

The prospect of terraforming Mars, so that it could be colonised by humans or other species, would appear to be an immense task that would require a wealthy civilisation with a high Gross World Product (GWP) per capita. This is unlikely to arise until the Late Planetary era from the Fourth Millennium. Terraforming is the process of turning a "dead" planet into one where human life would be sustainable, and the approach would be to create the greenhouse effect by manufacturing CFCs (chlorofluorocarbons) on a vast scale over several centuries, and releasing it into the atmosphere. This would warm the planet to near Earth temperatures, which would melt ice at the poles and release carbon dioxide and water vapour from beneath the soil. As both carbon dioxide and water vapour are also greenhouse gases, the warming process would accelerate until temperatures in the Martian tropics rose above the freezing point of water in the summer. In order to create an atmosphere that contained close to the twenty percent oxygen that we have on Earth, genetically engineered plant life and primitive animals or organisms which could survive the abnormal conditions would have to be introduced and allowed to do their work for several centuries before human life could be supported. Within perhaps 100,000 years, a new species of humans could emerge that may possibly be taller and broader chested (as in the Andes), who would be capable of surviving at less than half the air pressure on Earth. Alternatively, genetic engineering, space transportation, planetary terraforming, biodome technology, and civil space infrastructures may provide the convergence of technologies necessary for adapted humans to live on another planet in the second half of the this millennium.

Technological evolution of infrastructure systems: Chronological references

1. *The Economics of Urban Size* - Harry W Richardson, 1973.
2. *Engines of Creation* - K.Eric Drexler, 1989.
3. *Technological Trajectories* – Leonardo Biondi & Riccardo Galli, 1992.
4. *Technology and the Transition to Environmental Sustainability* – Rene Kemp, 1994.
5. *Strategies for Shifting Technological Systems* – J Schot, Hoogma & Elzen, 1994.
6. *Linkages: OECD and Major Developing Economies* – OECD, 1995.
7. *Biotechnology* - Edited by Frederick B Rudolph & Larry V McIntire, 1996.
8. *Renewable Energy* - Edited by Godfrey Boyle, 1996.
9. *Islands in the Sky* - Edited by Stanley Schmidt & Robert Zubrin, 1996.
10. *Transportation* – Scientific American Special Issue, 1997.
11. *Global Energy* – Nebojsa Nakicenovic, Arnulf Grubler & Alan McDonald, 1998.
12. *Communication Futures* – Edited by Sohail Inayatullah & Tony Stevenson, 1998.
13. *Powering the Future* – Tom Koppel, 1999.
14. *Tomorrow's Energy* – Peter Hoffman, 2001.
15. *Hubbert's Peak* – Kenneth S Deffeyes, 2001.
16. *Exploring the Future: Scenarios to 2050* – Shell International, 2001.
17. *The Hydrogen Economy* – Jeremy Rifkin, 2002.
18. *Energy to 2050* – International Energy Agency/OECD, 2003.
19. *Energy at the Crossroads* – Vaclav Smil, 2003.
20. *Space 2030* – OECD, 2004.
21. *Energy's Future Beyond Carbon* – Scientific American Special Issue, 2006.
22. *Energy Technology Perspectives to 2050* – International Energy Agency/OECD, 2006.
23. *Infrastructure to 2030* – OECD, 2006.
24. *Space* – Sir Patrick Moore & H J P Arnold, 2007.
25. *Biofuels for Transport* – Worldwatch Institute, 2007.

9.
TRAJECTORY OF THE WORLD SYSTEM OF CITIES

9.1   World urbanisation

Trade and manufacture flourished under the empires, and cities grew to record sizes with the imperial capitals of Rome (AD 100), Changan (AD 700 – present day Xian) and Baghdad (AD 900) reaching 1m inhabitants and making them the earliest metropolises, although the populations later declined. Much of the bulk transport, including foodstuffs, took place by ship, and most of the major cities of the period were located on coasts or waterways. Long distance land routes were also developed, notably the great Silk Road from Xian in China to Turkestan, which linked East and West Asia 2,000 years ago. Such extensive trade networks brought the different empires into contact with each other and also spread their political and cultural influence far afield.

There is a relationship between the evolution of cities, national life cycles that end in quiescence rather than decline, and the rise and decomposition of empires. The capitals of many former empires have managed to survive for centuries, through the diffusion of knowledge and culture that arises from the transition from one empire to another. After the collapse of Greek power, Athens became attached to the Roman Empire and was subsidised as a centre of learning. The Romans adopted many elements of their civilisation from the Etruscans and the Greeks, and in turn the proliferation of Roman cities in their northward expansion into Western Europe diffused political, administrative, military, and technological skills to other nations. A minor city-state, Byzantium, became the capital of the Eastern Roman Empire as Constantinople and its successor, the Ottoman Empire, and it is still a major city today as Istanbul. The Arab empires adopted Roman technology as a basis for city life with intellectual centres at Baghdad and Damascus in the Middle East, Cairo in North Africa, Spain, and Central Asia. Trade between Byzantium, the Arab empires, and the Mediterranean region played a significant role in the revitalisation of urban life in southern Europe and in the formation of prosperous Italian city-states in the 10th and 11th centuries AD.

From AD 1500, Europe became a dominant power in the world, with primacy of the Renaissance Italians, 16th century Spanish and Portuguese, 17th century Dutch, and 18th and 19th century British, but these colonial powers eventually declined. It was the merchant capitalism or mercantilism from 1500 to 1750 that led to innovations in trade and commerce, such as banking, loan systems, credit transfers, company partnerships, shares in stock, speculation in commodity futures, commercial insurance, and news services. Merchant capitalism was a self-propelling growth system in which the law of diminishing returns provided a continuing impetus for territorial expansion and colonisation. The profits from trading and overseas colonies in pre-industrial Europe flowed into local agriculture, mining, and manufacturing and provided one of the main preconditions for the emergence of industrial capitalism in the 18th century. Outside Europe, maritime gateway towns and trade ports were established along the coasts of North and South America, Africa, South and East Asia, and Australia. Cities of the earliest civilisations corresponded to townships and local centres, but by 3000 BC, cities of 20,000 to 50,000 inhabitants emerged and, from 2250 BC, Tertius Chandler recorded the historical growth of cities in *4000 Years of Urban Growth* (1987). By AD 1000, there were fewer than twenty-five cities of over 100,000 inhabitants, although a few cosmopolitan capitals of the old empires had reached 500,000 at some time in their history. Of the twenty-five largest cities in 1750, ten were seaports, eight riverine ports, and three were located on navigable canals. Table 8 shows the historical growth of the world's population and the city populations by size class.

TABLE 8: HISTORICAL GROWTH OF CITY POPULATIONS

| | | | Number of cities by size | | |
| | | | Cities | Metropolises | Cosmopolises/ Megapolises |
Year	World pop'n	Population in cities of 100,000+	0.1 – 0.99m	1m- 4.99m	5m+
10000BC	10m	-	-	-	-
1000BC	50m	0.3m	3	-	-
200BC	150m	2.5m	15	(1)	-
AD1100	320m	3.6m	25	(1)	-
1700	600m	10.6m	40	-	-
1800	910m	16m	95	2	-
1850	1.2bn	28m	165	4	-
1900	1.6bn	219m	492	20	1
1950	2.5bn	410m	870	71	7
1975	4.1bn	0.9bn	1,720	156	22
2000	6.1bn	1.8bn	3,325	300	50

It was not until the early 1800s that London and Peking reached populations of 1m, and thereby achieved the maximum populations of ancient Rome and China. London was to become the first metropolis to top 2m and, by 1900, had achieved the status of a cosmopolis with a population of 6.5m. By 1925, New York, which has a much larger hinterland, passed London with a population of 8m, and the United States dominated world affairs for much of the second half of the 20th century. Also in 1925, Tokyo became a cosmopolis with a population of 5.3m, and, in 1965, Tokyo became the largest city at that time with a population of 15m. By 1970, the top three world megapolises were Tokyo, New York, and London with populations of 20.5m, 16.9m and 11m respectively.

The main difference between the evolution of cities in Europe and the LDCs is the rate and scale at which they are developing. London grew from a population of 1m to 8m over a period of 130 years. Mexico City did it in 30 years from 1940-1970 and then doubled again in 15 years. The megapolises in the MDCs are now growing slowly if at all and they have built up their housing stocks, civil and social infrastructure, and cultural amenities over many decades of heavy investment, although now the maintenance costs are severe. In the LDCs the investment requirements for extending the cities and megapolises incurs substantial debt that has to be financed, and any economies of location are swallowed up by the high costs of congestion.

Major cities tend to form at a commercial break in transportation such as gateway ports, the junctions between water and railway transport, and in due course all important seaports and river ports have become airports with extensive networks of highways. By 2000, forty of the fifty (i.e., eighty percent) of the cosmopolises and megapolises of the world have been maritime or riverine ports. The twenty-eight seaports include Beijing-Tientsin, Sao Paulo-Santos, and Seoul-Inchon, and the twelve riverine ports include Toronto, Chicago, Paris, the Rhine-Ruhr ports, Baghdad, Cairo, Lahore, Kinshasa, Wuhan, and Santiago. Only Moscow, Mexico City, Bogota, Tehran, Delhi, Bangalore, Hyderabad, and Shenyang are not ports. Also 190 of the 300 metropolises with over 1m inhabitants in 2000 were located on water, so that 230 or sixty-five percent of the total of 350 metropolises, cosmopolises and megapolises are maritime or riverine locations, and this forms the background to the world system of cities. In Appendix 18, metropolises with populations over 2m and capital cities over 1m are listed. Maritime and riverine cities are marked with an M or an R, and it may be noted that nearly seventy percent of the population living in cities of over 2m is located at maritime or riverine ports, and forty percent of the world's population lives along the coastline.

The survival of large cities was dependent upon trade and the ecological context of the city region, which may have a diameter of 200-300 km. A variety of factors will have contributed to the growth of each city, and it is important to take into consideration the available technology at the time for irrigation, water supply, energy, transport, and communication. In looking at the continental regions of the world, it should be noted that Europe is the only continent without an empty interior. All the others have deserts, mountainous areas, frozen interiors, or jungle at their centres such as the Sahara in Africa, the Gobi in Mongolia, the Gibson Desert in Australia, the Rocky Mountains in North America, the frozen interior of Canada, and the tropical forests of the Amazon and the Congo rivers. Within each continent there are countries that have long coastlines in relation to their area. There are also continental countries such as China, India, the former Soviet Union, and the USA with very large populations and relatively limited coastlines in relation to their surface area, and many metropolises began as riverine ports. A number of countries have equilibrium between coastline and land surface, and some ten percent of countries have no coastline, although they may have the benefit of important navigable rivers.

In Western Europe, a high proportion of the largest cities today were in existence by the end of the Roman Empire, and a number were capitals of smaller states before the unification of territory into nation-states. However, the largest proportion of the population in Europe lives in the central part of the continent and not along the coast. Europe has been a continent of colonisers, and although its ports have developed with trade with other continents, it did not have colonial ports, and it was the industrial revolution that created the large number and size of the European cities, particularly where they are close to coalfields. It follows that in Western Europe today many of the cities are predominantly industrial or commercial cities, which predate the American cities, since they have grown out of older urban areas dating to the medieval period or even Greco-Roman times. The economic cycles created spatial adjustments to the growing cities with reconstruction and expansion of commercial districts and disinvestment in other parts of the city. Periods of booming growth caused overloading of the infrastructure in the central city, and, consequently, new tracts of land were assembled for development of a suburban fringe.

9.2 Cities in the continental regions of the world

For the purpose of this research, the world has been divided into five major continental regions and their sub-regions, to give ten regional categories as follows:

- Europe
    Western and Central Europe
    Eastern and South-East Europe

- Africa
    Sub-Saharan Africa
    North Africa

- Asia
    Western and Central Asia
    South Asia
    East Asia and Pacific

- Oceania

- America
    North and Central America
    South America

There are several inconsistencies in the United Nations and the World Bank regional definitions arising from the break-up of the Soviet Union and the inclusion of Turkey within Europe. In this study, the Czech and Slovak Republics, Hungary, and Poland have been included in Central Europe. The Balkan countries of Albania, Bulgaria, Greece, Romania, Turkey, and the five republics of former Yugoslavia have been described as South-East Europe.

The western republics of the former Soviet Union that are contiguous with Europe and Turkey, which include the Baltic states (Estonia, Latvia, and Lithuania), Belarus, Ukraine, Moldova, and Armenia, together with the Russian Federation and Georgia, have been included in Eastern Europe. The Arab countries of the Middle East are taken as Western Asia, and it is proposed that the sub-region of Central Asia will comprise Iran, Afghanistan, and the former Central Asian republics of the Soviet Union. These include Azerbaijan, Turkmenistan, Uzbekistan, Kazakhstan and Tajikistan.

In 2000, the urban population in the MDCs amounted to 0.95 billion and there were 120 cities with over 1m population (metropolises or megapolises). In the less developed regions, the urban population of 1.9 billion was double that of the developed countries, and there are 230 cities with over 1m population. The total number of urban agglomerations of over 2m population amounted to 175 and, if metropolises of under 2m population included urban agglomerations, then the number of metropolises of less than 2m would increase from 175 to 225, and the total number of agglomerations of over 1m would be 400. Metropolitan populations for 2000 are generally based upon the *United Nations World Urbanization Prospects: The 2005 Revision* (2006), for which a concerted effort was made to include the results of the 2000 round of national population censuses.

Oceania has the highest proportion of urban population in metropolises at sixty percent, and this is followed by the United States and Japan with fifty percent of their urban populations in metropolises or megapolises. In Asia, the proportions of the metropolitan populations for China and India are forty-six and thirty-nine percent respectively. The metropolitan populations of Latin America and the Middle East with North Africa are forty-two and thirty-eight percent respectively. Western and Central Europe has thirty percent of its urban population in cities of over 1m inhabitants, and, in the Russian Federation, where Soviet central planners created over 1,000 new cities to limit the size of existing large cities, the proportion is twenty-six percent.

In the year 2000, London was the only megapolitan region in Western Europe with a population of 12m. There were four cosmopolises with over 5m population, which included Paris (9.6m), Essen (6m), Madrid (5m), and the Randstad cosmopolis in the Netherlands, which contained thirty-four percent of the Dutch population with 5m inhabitants as shown in Appendix 18. Randstad ("ring city") in western Netherlands contains the Rotterdam-Hague conurbation (including Dordrecht, Delft, and Leiden) and the Amsterdam-Utrecht conurbation (with Haarlem, Hilversum, and Amersfoort). The Rhine-Ruhr metropolitan region has a population of 9m and is one of the most important urbanised industrial districts in Western Europe, and the region contains thirteen cities of over 200,000 population including Cologne, Essen, Dusseldorf, and Dortmund. The Rhine-Ruhr region lacks a dominant core city and so it has been included as three separate agglomerations, Rhine-Ruhr North, including Essen and Dortmund, Rhine-Ruhr central, including Dusseldorf and Wuppertal, and Rhine-Ruhr South, including Bonn and Cologne. There is also the Rhine-Main metropolitan region based upon the global city Frankurt, which is Germany's second largest agglomeration with a population of nearly 4m. The Rhine River has provided the artery for transport in the region, and the coal from the Ruhr area provided the energy resource, which stimulated the urban growth. There is also a cosmopolis in the Po valley of Italy, where the combined populations of Milan, Turin, Genoa, Novara, Monza, Bergamo, Brescia, Piacenza, and Alessandria amount to over 6m. It is apparent that there is a banana-shaped megalopolitan axis running from central and southern England, through Holland, Belgium, northern France, Germany, Switzerland, and Italy. Other European cities with over 3m population include Berlin, Rome, Naples, and Katowice. There are a further ten cities with over 2m inhabitants and twenty-five cities with over 1m.

In South-East Europe, Istanbul in Turkey is a cosmopolis with a population of nearly 9m, and the other major cities in the Balkans with over 2m inhabitants are Ankara, Izmir, Athens, and Bucharest. In Eastern Europe, the urban population is some seventy percent of the total, and at 2000, Moscow and St Petersburg were the only megapolis and cosmopolis respectively with populations of some 10m and 5m, while Kiev in the Ukraine has 2.6m. There are a further twenty-five cities with over 1m inhabitants in Eastern and South-East Europe.

Africa is one of the least urbanised regions of the world at thirty-six percent, and the major cities of sub-Saharan Africa are on the coast. With the exception of Lusaka in Zambia, no city without access to the sea has a population of over 1m. Cosmopolises include Lagos in Nigeria with over 8m inhabitants and Kinshasa in Zaire with 5m inhabitants. There are two cities with populations over 3m, including Abidjan (Côte d'Ivoire) and

Khartoum (Sudan). There are a further eight cities with over 2m inhabitants, including Addis Ababa in Ethiopia, Cape Town and Johannesburg (South Africa), Nairobi in Kenya, and Dar-es-Salaam in Tanzania and some twenty cities with over 1m inhabitants. This region has one of the world's highest rates of urbanisation at 4.5% per year, which means that the migration to the cities is proceeding at a rate far above that of natural population growth. The major cities are growing much too fast for the governments concerned to deal effectively with the pressures created. Also, the region suffers from the primate city problem so that the few big cities have grown much more rapidly than smaller cities.

In North Africa, the Mediterranean countries have a number of important maritime or riverine cities of over 1m inhabitants. These include Cairo and Alexandria in Egypt with over 10m and 3m inhabitants respectively, Casablanca with over 3m and Algiers with over 2m. The capital cites of Tripoli (Libya) and Rabat (Morocco) have over 1m. In the Middle East, Baghdad has 5m inhabitants, Riyadh has over 3m, and Jeddah, Damascus, Aleppo, and Tel-Aviv-Yafo have over 2m. Amman, Kuwait City, Beirut, and Sana'a have over 1m inhabitants. In Central Asia, Tehran is a cosmopolis with 7m inhabitants, and the major cities with over 2m inhabitants are Esfahan and Mashhad in Iran, Kabul in Afghanistan, and Tashkent in Uzbekistan. In North Africa, the Middle East, and Central Asia, there are a further twelve cities with over 1m population.

South Asia has one of the lowest levels of urbanisation at thirty-two percent, but at the same time with an urban population of nearly 400m, the region has one of the largest urban populations in the world. In India, Mumbai, Calcutta, and Madras are the only major port cities, with populations of 16m, 13m, and 6m respectively. India is not penetrated by rivers, and only the Ganges and Brahmaputra systems can be navigated commercially for any great distance, and so the sub-continent is dependent upon a dense railway network to serve the cities. Delhi has a population of over 12m, and Bangalore and Hyderabad are cosmopolises with populations of over 5m. Ahmedabad and Pune have populations of over 4m and 3m respectively. In Bangladesh, Dhaka is a megapolis with 10m inhabitants, and Chittagong has over 3m. The other megapolis in the region is Karachi in Pakistan with 10m, and Lahore is a cosmopolis with nearly 6m. Yangon in Myanmar has a population of over 3m. There are ten cities with over 2m population and thirty cities with over 1m population in South Asia.

East Asia and the Pacific contain China, which is one of the least urbanised countries in the world with twenty-five percent urban population. Shanghai and Beijing with populations of 13m and 11m, respectively, are the largest cities in China, and there are a further six cosmopolises including Hong Kong with populations of over 5m. The Yangtze delta is a megalopolitan region with Shanghai at the end of a linear megalopolis that includes Suzhou, Wuxi, Changzhou, and Nanjing. In addition, there are twenty-five cities in China with over 2m population, and seventy agglomerations with 1m inhabitants. In Japan, Tokyo-Yokohama has a population of 28m, Osaka-Kobe has 11m inhabitants, and Nagoya 5m; the Tokaido megalopolis includes the six cities along Japan's Pacific coast: Tokyo, Kawasaki, Yokohama, Osaka, Kyoto, Nagoya, and Kobe. Other megacities in East Asia are Jakarta and Manila, all of which have populations of over 10m. Seoul (9.9m), Bangkok and Taipei have over 6m, Singapore and Ho Chi Minh have over 4m, and Bandung, Pyongyang, Pusan, and Hanoi have populations of over 3m. There are nine metropolises with 2m inhabitants and fifteen with over 1m.

The major cities in Oceania are Sydney and Melbourne with populations in excess of 4m and 3m, and the population of Auckland in New Zealand is 1m.

In North America, the four largest cities in the United States are New York, Los Angeles, Chicago, and Philadelphia with populations of 18m, 13m, 8m, and 5m respectively. The north-eastern seaboard of the USA forms a major megalopolis (Boswash) and contains Boston, Hartford, New York, Philadelphia, Baltimore, and Washington. Also a megalopolis is forming along the west coast that will link the San Francisco Bay Area to the Los Angeles/San Diego conurbation. The Great Lakes megalopolis extends from Chicago to Detroit and the southern shores of Lake Erie. Washington DC, Boston, Dallas, and Miami have populations of over 4m, and these are followed by Atlanta, Detroit, Houston, San Diego, and San Francisco with populations of over 3m. The USA has a further seven cities with over 2m inhabitants and twenty metropolises with 1m. In Canada, Toronto has a population of nearly 5m and Montreal, Vancouver, and Ottawa have populations of 3.5m, 2m, and 1m respectively.

Latin America has the highest levels of urbanisation in the developing countries with seventy percent in Central America and the Caribbean, and over eighty percent in South America. In 2000, the megacities of the region were Mexico City and Sao Paulo with populations of 18m and 17m, Rio de Janeiro (10m), Buenos Aires (12m), Lima and Bagota with nearly 7m, and Santiago having over 5m inhabitants. In Brazil, Belo Horizonte (over 4m) and four other cities in Brazil and two in Mexico have over 3m. In Brazil, a megalopolis is emerging in the corridor between Rio de Janeiro and Sao Paulo. There is a polynucleated metropolitan region (or megalopolis) that combines the metropolitan areas of Mexico City, Toluca, Cuernavaca, and Puebla within an area of some 8,200 km². There are a further ten cities in Latin America with populations of over 2m, including Caracas, Havana, and San Juan, and thirty with populations of 1m.

9.3   Projected number of cities

In the year 1500, there were twenty-five cities with over 100,000 inhabitants, of which sixty percent were in South Asia and East Asia, and this share was held until 1800. By 1900, Asia's share had fallen to forty percent and, by 1950 to thirty percent. However, in the year 2000, South and East Asia accounted for thirty-five percent of the world's cities, and this will rise to over forty percent by the year 2025, which reflects a return to the situation in earlier times when Asia had a high concentration of the world's urban population.

By 1825, the world population had passed one billion, of which the urban proportion was ten percent, with over one-half living in cities larger than 100,000 population. A century later in 1925, the total population reached two billion with an urban proportion of twenty percent. A doubling of the world population to four billion was achieved fifty years later in 1975, with nearly forty percent living in urban areas and an urban population of some 730m in the MDCs and 810m in the LDCs. The world's average annual rate of urbanisation reached 3.7% in the late 1950s, after which it began to fall slowly to around 2.4%. However, this figure masks significant regional variations, and in the decade 2000-2010, the growth rates will continue to decline with a range from under one percent per year in Europe and North America, to two percent in the Middle East and Latin America, three percent in Asia and nearly 4.5% in Africa.

The number and size distribution of cities are explored in *Long-Range Futures Research* to the year 2150 with a view to assessing where we are in the life cycle of civilisation and what would be the investment required to complete the world's cities for a global population of 12.5 billion. Non-renewable resources will be severely depleted by the urban development process, and the future operation of the cities will put enormous pressure on natural resources, fossil fuels, the environment, and water resources. Table 9 below gives urban population projections to 2150.

TABLE 9: URBAN POPULATION PROJECTIONS

Year	World Pop'n (m)	Urban population (m)			Urban population as % world pop'n
		MDC	LDC	Total	
1950	2,515	445	290	735	29.2%
1975	4,075	730	810	1,540	37.7%
2000	6,050	950	1,900	2,850	47.0%
2025	8,200	1,130	3,900	5,025	61.3%
2050	9,200	1,100	5,600	6,700	72.0%
2100	10,000	1,170	6,830	8,000	80.0%
2150	12,500	1,440	9,810	11,250	90.0%

In 1950, less than thirty percent of the world's population lived in urban areas and, by 2000, the proportion reached forty-seven percent. By 2025, it may exceed sixty percent. In the year 2000, the world's rural population reached a peak of 3.21 billion, which is expected to decline before 2025. In effect, the net annual addition to the global population will shortly be accommodated entirely in cities and towns, and urban expansion will arise from the natural increase in the urban population rather than migration from rural areas. This will partly be a result of lower infant mortality rates in the cities than in rural villages. Also, rural areas are reclassified as urban by the United Nations Centre for Human Settlements as soon as there are 2,000 inhabitants living in townships with streets and urban utilities. From the year 2000 to 2050, the population of the world will increase from 6.1 billion to 9.2 billion, with the urban proportion rising from nearly fifty percent to over seventy percent, which will require massive capital investment for the construction of the infrastructure and buildings in the expanding cities and towns. Within the twenty-five-year period 2000-2025, there will be a doubling of the urban population in the less developed countries, from just under 2 billion to nearly 4 billion in comparison with a relatively stable urban population of 1 billion in the more developed countries. By the year 2050, the world population is expected to reach 9.2 billion, with an LDC urban population of 5.6 billion, which will be five times the MDC urban population of 1.1 billion.

A universal feature of the historical growth and development of city systems in economically advanced countries is the long-term stability in the national or regional population rank of their leading metropolises. In fact, over two-thirds of the world's metropolises with over 1m population in 2000 were already important cities 200 years ago, and a quarter have been important for at least 500 years. Typically, over long periods of time, the currently most important metropolises of postindustrial countries and the regional city systems have experienced only small upward or downward shifts in ranks. Small countries and larger countries in the early stages of development with central governments, tend to have more primate size distributions of urban areas than sub-continental countries with federal systems of government. For example, the United States, Brazil, and India have federal systems of government that have resulted in an empirical regularity of city sizes that is unexplained by current theory. Zipf (1941) suggested that there is a rank-size rule for integrated systems of cities that can be expressed by a simple formula $Pr = Pi/r$, where Pr is the population of the city ranked r, Pi is the population of the largest city, and r is the rank of city Pr. It follows that the second ranking city has one half of the population of the largest city, the third ranking city one third of the largest, and so on down the scale of regional centres.

World urbanisation increases the number of levels in the urban hierarchy, and megapolises and megalopolises of over 10m inhabitants are expected to accommodate an increasing share of the world urban population in the Third Millennium. From complexity science, the trajectory of the world system of cities will be towards a dynamic balance with an equalisation of populations (equipollence) at each hierarchical level. This is the most likely condition to maintain a stable configuration, since it limits the asymmetry of the urban system, and it will be approached when the global urban population exceeds ninety percent of the total. This permits prognostication for the long-term future that may be approached by 2150, even though the absolute size of the global population may fall within a cone of upper and lower bands. Continental integration within a global economy is likely to result in investment capital flows that maintain dynamic stability with uniform assets per capita across states within a continental union. This would produce an urban pyramid for a nation of 55m urban population, in which a cosmopolis at the top of a pyramid would have two national metropolises (2m+), four sub-national metropolises (1m+), eight provincial cities (500,000+), sixteen principal cities (250,000+), thirty-two regional cities (100,000+), sixty-four urban centres (50,000+), 125 district centres (20,000+), 250 local centres (10,000+), 500 service centres (5,000+), and over 1,000 townships (2,000+). In effect, the national population would be shared between eleven levels in the hierarchy to give 5m urban inhabitants at each level. A national population of 120m would have a hierarchy of twelve levels with a megapolis of 10m+, and urban populations of 10m+ at each level in the hierarchy. A subcontinental population of 250m would have a megalopolis of 20m inhabitants, to correspond to thirteen levels of 20m population. Similarly, a continental population of 700m would have a metapolis of over 50m inhabitants, together with a population of 50m at each of fourteen hierarchical levels. Finally, ten to fifteen supercontinental unions may form with 1 billion inhabitants each, with a metapolis of over 100m and populations of 100m at each of fifteen levels in the urban hierarchy.

The trajectory for the world urban system is derived by examining the historical populations in urban centres of less than 100,000, together with the populations in cities of 100,000 to 0.99m, metropolises of 1m-4.99m, and cosmopolises or megapolises of over 5m. A rule for the current distribution of the urban population is that thirty-eight percent of the urban population lives in urban centres of less than 100,000 population, thirty-six percent in cities of 100,000 to 1.99m population, and twenty-six percent in metropolises and megapolises of over 2m population. In the year 2000, a world urban population of some 2.85 billion was contained in an urban hierarchy of twelve levels, so that each of the five lower levels in classes of less than 100,000 inhabitants contained 7.5% of the urban population or 215m people. The sixty-two percent balance of the urban population was distributed between the seven upper levels, with an average of nearly nine percent or 250m at each upper level.

In 2025, an urban hierarchy of thirteen levels would contain a world urban population of 5.025 billion (sixty percent of the total population), with an average of 350 million population at each of the lower levels and an average of 400m at each of the eight urban size classes larger than 100,000 inhabitants. In 2050, one third of the urban population of 6.75 billion (seventy percent of the total population) would be distributed between the five urban size classes of less than 100,000 inhabitants to give 450m urban population at each level, and the balance of the 4.5 billion urban population would be distributed between the top nine of the fourteen levels in the urban hierarchy to give an average of 500m at each of the upper levels. In 2100, the urban population will reach eighty percent of the world population at 8.0 billion and finally a ninety percent urban population of 11.25 billion is expected to be reached by 2150 in a hierarchy of fifteen levels, with 750m urban population at each level. The default rule for the distribution of the urban population by 2150 is expected to be that one-third of the urban population will occupy the five smaller urban size classes of less than 100,000 population, one-third will be in cities and metropolises of 100,000 to 4.99m, and one-third in cosmopolises, megapolises, megalopolises, or metapolises of over 5m population.

The number of townships, C_1, at the lowest level of the urban hierarchy may be derived by taking one-fifteenth of the urban population (6.67%), and dividing it by 3,000 (the average size of a township of 2,000-4,999 inhabitants). If there are N levels in the urban hierarchy, then a first approximation for a default value for the number of cities, C_N, at the highest level in the urban hierarchy is given by the following equation

$C_N = C_1 /K^{N-1}$. K is the ratio between the number of cities in adjacent levels in the urban hierarchy and lies in the range 2.0-2.5, with an average value of approximately 2.16. For example, in the year 2050 (see Appendix 23), the total urban population is projected to be 6.7 billion with 450m at the lowest level of the urban hierarchy. This gives 150,000 townships and with a hierarchy of 14 levels $K^{N-1} = 2.15^{13} = 21,000$, so that a default value for the number of metapolises at the fourteenth level in the urban hierarchy will be 150,000/21,000 or 7. It should be noted that this hierarchy generates a power law of the same type as Zipf's rule.

In Appendix 19, the urban population is shown for each continental region for the years 2000, 2025, and 2150. Also in Appendix 20, the number of cities is projected for each region by urban-size class for the same years. At the year 2000, there were some 3,325 cities in the size band 100,000 to 0.99m inhabitants, with 300 metropolises of over 1m population, and thirty cosmopolises of over 5m population, and twenty megapolises or megalopolises of over 10m inhabitants. Appendix 21 shows that in the year 2000, thirty-five of the fifty cities of over 5m population were in the less developed countries. It is estimated by extrapolation into the near future from 2000 to 2025 that there will be some fifty-five cosmopolises and over thirty-five megapolises or megalopolises by the year 2025. The new megapolises will include Istanbul in Turkey, Lagos in Nigeria, Kinshasa in the Congo Democratic Republic, Tehran in Iran, and Bangalore, Hyderabad, and Madras in India, Shenzen and Tianjin in China, and Bangkok in Thailand. The figures for city size in 2025 have been derived using annual city population growth rates, or rates of decline, from UN data. These cities were then slotted into the appropriate size band to give the number of cities.

It is prognosticated that, when the global urban population exceeds ninety percent by the year 2150, there will be nearly 10,000 cities of over 100,000 population, of which 750 will be metropolises, together with 100 cosmopolises and eighty-five megapolises or megalopolises with over 10m inhabitants. The majority of these cosmopolises and megapolises already have populations of over 1m, although in sub-Saharan Africa, the major cities of the future may not yet be metropolises in view of the high rate of growth predicted for these countries. In the LDCs the number of cities with populations over 10m will rise to around thirty by the year 2025 and probably seventy in the 22nd century, compared with fifteen megapolises by the year 2150 in the MDCs. A summary of the number of cosmopolises and megapolises at 25- or 50-year intervals since 1925, together with future projections, is given in Table 10 below.

TABLE 10: PROJECTED NUMBER OF COSMOPOLISES AND MEGAPOLISES

		Cosmopolises		Megapolises		Total
Year		MDC	LDC	MDC	LDC	number
1925		3	-	-	-	3
1950		3	2	2	0	7
1975		8	10	2	2	22
2000		9	21	6	14	50
2025		10	45	7	28	90
2050		11	59	11	44	125
2100		12	73	13	57	155
2150		14	86	15	70	185

9.4 The city hierarchy and urban densities

Contiguous megalopolises, megapolises or cosmopolises will form a 'metapolis' of 50-100m population at a density approaching 600-900 inhabitants per km^2 before 2025. In fact, a metapolis is already emerging in the Pearl River (Zhu Jiang) Delta in the Guangdong Province of China, in which a population of 50m will be contained within an area of 50,000 km^2. The development runs 150 km from Hong Kong to Guangzhou (Canton) and includes Shenzhen, Macao, and Zhuhai. Metapolitan belts may form along coastal zones and inland transport axes, and these are likely to extend for some 500-1,500 km and contain a population of 100m in an area of 100,000-180,000 km^2. An early example is the 1,500 km Beijing-Seoul-Tokyo (BESOTO) metapolitan belt in East Asia, which transcends national boundaries and contains some 98m urban inhabitants. The belt encompasses the megalopolises and megapolises of Beijing, Tianjin, Seoul, Tokyo-Yokohama, and Osaka-Kobe to give an overall population density of 500 inhabitants per km^2. It is anticipated that 10-15 planetary metapolises of 50-100m population will be served by very fast trains, a continental network of airlines, and will be interconnected with other planetary metapolises by spacelines.

The sub-continental regions of the world in which metapolises may emerge first are USA, Japan, Northern Europe, China, India, Brazil, South East Asia, East and West Africa. A further Chinese metapolis will form in the Yangtze delta with Shanghai at the coastal end of an industrial corridor extending 300 km to include Suzhou, Wuxi, Changzhou, and Nanjing. Another emerging Asian metapolis will arise from the connection of the Tokyo-Yokohama-Nagoya megalopolis with the Kyoto-Osaka-Kobe megalopolis to form the most extensive MDC metapolis with a population of over 50m. Future metapolises in North America will be the north-eastern seaboard of the USA and the linking of the San Francisco Bay area with the Los Angeles/San Diego complex. In South America, the corridor of development between Rio de Janeiro and Sao Paulo will emerge as a metapolis. In Europe, the Rhine-Ruhr complex will also belong to this class of cities, and in Africa, a metapolis will emerge from the Niger Delta to encompass Lagos and Accra in Ghana. In India, a metapolis is forming along the Ganges Delta from Calcutta to Dhaka in Bangladesh. Finally, an Australian megapolis based upon Sydney may extend towards Melbourne and Brisbane to form a metapolis in the long term.

Since the year 2000, over thirty-five percent of the world's urban population lives in cities of over 1m inhabitants, and some sixty-two percent live in cities of over 100,000 population. In the prospective world city system at the end of the 21st century, around two-thirds will continue to live in cities of over 100,000 population and nearly forty percent of the urban population will live in cities of over 2m inhabitants. An urban population of some 120m that is contained in an urban hierarchy of twelve levels, would comprise 127 cities of over 100,000 population, plus some 3,000 towns of over 2,000 population. Of the 127 cities, there would be one city of over 10m inhabitants, two cities of 5m-9.9m, four cities of 2m-4.9m, eight cities of 1m-1.99m, and 112 cities of over 100,000 but less than 1m.

It is expected that in due course cosmopolises and megapolises with over 5m inhabitants will increase their share of urban population from over fifteen percent to thirty-three percent, but that urban population densities will eventually decline as megapolises, megalopolises, and ecopolitan states emerge. Typical densities by urban size class are indicated in Table 11.

TABLE 11: TYPICAL URBAN DENSITIES BY SIZE CLASS

Urban class	Urban density (inhabitants/km²)	
	MDCs	LDCs
50m+	500	750
20m+	1,000	1,500
10m+	1,250	2,500
1m-9.9m	4,000	8,000
0.1m-0.99m	2,000	4,000
5,000-99,999	1,000	2,000
2,000-4,999	500	1,000
Rural villages	250	250

Densely populated megapolises are around 2,500 inhabitants per km² in the LDCs and, with economic development, the population density declines to around 1,250 inhabitants per km² at the postindustrial stage, as the number of persons per household declines from five in the developing countries to 2.5-3.0 in developed countries. The greater the area under consideration, the less the maximum reasonable density so that the density of metapolises will be lower than megalopolises, and early metapolises with densities of 500 inhabitants per km² will eventually reach 600-900 inhabitants per km². As the size of the spatial unit increases to state or continental levels, the proportion of land covered by urban areas will decline because an increasing proportion will be required to support land uses such as agriculture and forestry. Also, with large-scale areas, a significant proportion may be inhospitable to humans such as arctic regions, deserts, or mountainous areas, although they may support wildlife.

Apart from small islands and city-states, few countries have densities exceeding 400 inhabitants per km², and these include the Netherlands, Bangladesh, South Korea, and Java in Indonesia. However, by the second half of the century, one-third of the world's population will live in countries with these population densities. This would include nearly all of South Asia, the Philippines, and Vietnam, together with a significant number of African countries. Appendix 22 gives the number of cities by urban-size class for 1975, when the land area for urban and rural communities was estimated to be of 11m km², together with the number of cities for 2000 and 2025. Appendix 23 gives default values for the number of cities by urban-size class for 2050, 2100 by interpolation between 2025 and 2150, when the world population is expected to reach 12.5 billion. If the world population were to reach 15 billion, then this could be accommodated at an overall urban density of 1,000 persons per km².

From Appendix 22, as the rural populations become urban at higher densities, the total area of human settlements will reach a peak of 15m km² in 2025 and then decline to 13m km² by 2150 (see Appendix 23), which gives an overall population density for human settlements of 960 inhabitants per km² in 2150. Some ten percent of the world's land surface of 133m km² would be covered by urban areas and villages, to give an overall population density of ninety-five persons per km². If the areas of forest and wildlife are excluded (see Chapter 10.1), the net population density on the habitable areas would be 240 persons per km².

It may be noted that the population density principally affects the average distance between cities of over 100,000 inhabitants, since there is an average of 120m total urban population (i.e., both cities and towns) per 127 cities, or nearly 1m total urban population per city of over 100,000. A net density of over 400 inhabitants per km² will give an average distance of 50 km between cities (i.e., 2,500 km² surrounding each city) and an average distance of 100 km between cities will give a net density of 100 inhabitants per km². In the USA, it should be noted that thirty-two of the fifty states have less than 5m inhabitants, and the seven states with populations of over 10m account for forty-five percent of the national population. If Texas is excluded in view of the relatively low-population density, then the remaining six most populous states contain nearly forty percent of the population at a density of over ninety inhabitants per km², compared with 107 inhabitants per km² in Western Europe. In general, metropolises will be linked by four-lane (or more) interstate highways, and smaller cities will be linked to their respective metropolises, or the interstate highway, by dual carriageway roads. In turn, towns will be linked to their regional centres by roads capable of taking two-way traffic.

Appendix 18 gives gross urban densities for cities in terms of the inhabitants per km², based upon official data for the land area of the Metropolitan or Administrative Area or for the municipality (the city proper). The United Nations Centre for Human Settlements (HABITAT) tracks all cities of over 100,000 inhabitants, and the city-size distribution for 2000 has been derived from named city populations using UN data, in conjunction with a German database that can be accessed on the internet at www.citypopulation.de. For individual cities, the urban densities may be unreliable and should be used with caution because of the variation in the definitions for metropolitan areas and inconsistency in the boundaries used. However, the differences in urban density between cities with populations exceeding 1m inhabitants in the various continental regions of the world vary significantly.

9.5   World cities of the future

World cities or global cities are large and influential cities with financial institutions for directing international investment capital flows, and they may be arranged hierarchically in accordance with the economic power they command. In *The Global City - London, New York, and Tokyo* (1991), Saskia Sassen described the three leading full service world cities, although the actual size of a city does not necessarily make it a global city. It may be a capital city that connects a national economy into the system such as Brussels, Madrid, Paris, Sydney, and Toronto. Other cities such as Zurich, Amsterdam, Frankfurt, Milan, Singapore, and Hong Kong have commanding international roles, whereas Chicago, Los Angeles, and Sao Paulo link important regional or subnational economies. The characteristics of a world city include:

- International orientation and influence in world affairs, such as New York with the United Nations Headquarters.

- A metropolitan city, generally with a population of several million.

- Major international airport and an advanced transport system with multiple modes including mass transit, railway network, and motorways.

- Politically stable country with national policies favouring a market economy.

- International financial institutions for directing international investment capital flows, a stock exchange, and the headquarters of transnational corporations.

- Advanced communications infrastructure with fibre-optics, teleports, satellite communications, and cellular phone services.

- High proportion of quaternary division activities including law firms, international accountants, and management consultants.

- Media corporations with international reach such as Reuters, BBC, Associated Press, *New York Times*, and *The Times, London*.

- World class cultural facilities including theatres, an opera house, concert halls, museums, libraries, and art galleries.

- First class hotels, restaurants with international cuisine, and an entertainment industry with nightclubs and casinos.

- Retail and leisure facilities including health and fitness centres, and sporting facilities for international sporting events, such as the Olympic Games, the Football World Cup, or Tennis at Wimbledon.

World cities have a cosmopolitan culture that attracts the transnational capital class, whose principal concern is uninterrupted flow in the global system of capital accumulation. The predictable stability of nations plays an important part in making a capital city attractive to investors, and, in addition, there needs to be a favourable taxation regime with both financial and personal security. The world cities are ranked by their ability to attract global investment funds, together with their resilience to volatile capital flows and external shocks. Cities may rise to world-city status, they may also lose it, or they may also rise and fall within the world-city hierarchy.

The basic assumption behind the world-city hierarchy is that global control centres are located in the high- and middle-income countries of the world economy. With the decline of centrally planned economies and the reduction in the protective trade barriers of countries weakly linked to international transactions, new centres outside the richer countries are likely to strengthen their status as control points for investment capital in the future. Whilst changes may well occur in the industrial and postindustrial economies, presently at the core of global capitalism, the future expansion of investment-based urban growth is clearly taking place in the lower-middle income LDCs, and to a lesser degree in the low-income countries. It follows that, for each continental region, there will be first, second, third, and fourth order world cities, which may in due course move up the world system hierarchy.

The number of prospective world cities of the future is tabulated in Appendix 24, and the population size categories will refer to the metro-region as follows:

Megapolis (MP)	10m+
Cosmopolis (CM)	5m-9.99m
Metropolis (MT)	1m-4.99m

In summary, there are three prospective first-order world cities, twelve second-order, twenty-seven third-order, and twenty-six fourth-order. The prospective world-city system includes twenty megapolises over

10m inhabitants, eighteen cosmopolises with more than 5m inhabitants, and thirty cities that are currently metropolises. The prospective total of sixty-eight world cities represents the top twenty to twenty-five percent of a total in 2150 of eighty-five megapolises, 100 cosmopolises, and 750 metropolises.

Prospective new entrants to the second-order world cities include Johannesburg, Mumbai, Lagos, and Shanghai. Other rising third-order world cities from the LDCs include Beijing, Buenos Aires, Cairo, Delhi, Karachi, Kinshasa, Mexico City, Nairobi, Rio de Janeiro, Seoul, and Tehran. Fourth-order LDC world cities of the future include Algiers, Baghdad, Bangkok, Bogota, Caracas, Dhaka, Jakarta, Kuala Lumpur, Lima, Manila, Riyadh, and Tel Aviv.

During the 21st century, the world's trade centre is expected to move from the Atlantic to the Pacific, with Tokyo and Shanghai being the principal global cities of East Asia and the Pacific. The prospective future world cities are listed in Appendix 25.

9.6 Megacity hazards and civil disasters

Megacities are becoming increasingly vulnerable to natural and civil disasters due to rapid urbanization in the LDCs, with high population densities and the development of marginal land. Civil disasters create chaotic dynamics at the local and regional level in the short-term, although the consequences and traumas are remembered for many years. A civil hazard is an event that threatens life or property, and potential hazards are described in *Crucibles of Hazard: Megacities and Disasters in Transition* (1999), edited by James Mitchell. A civil disaster is an extreme event, which is, in effect, the realisation of a hazard, where there is a loss of life or damage to the built environment with severe disruption to human activities. Civil disasters include the impact on the built environment of natural hazards such as earthquakes, severe storms, floods and fires, or technological hazards such as transport accidents, toxic spills, and nuclear disasters, and social hazards such as crime, terrorism, and wars. Biohazards include plagues, pandemics, and epidemics of transmittable diseases.

Examples of megacities with multiple hazards include Los Angeles, Tokyo, Mexico City, and Dhaka. Los Angeles is at risk from earthquakes, coastal flooding, wildfires, crime, gang warfare, and smog. Tokyo is susceptible to earthquakes, tsunamis, typhoons, and landslides, but it is relatively free of urban social hazards. Mexico City is vulnerable to earthquakes, volcanic eruptions, subsidence, air pollution, and urban deprivation. Dhaka, the capital of Bangladesh, is located in some of the world's most flood-prone lands, and natural hazards include tropical cyclones and tornadoes. Slums and squatter settlements in Dhaka are vulnerable to environmental, social, and biological hazards with outbreaks of epidemics. Disasters in the megacities of the MDCs tend to result in heavy economic losses, which may disrupt the global finance and trading network, whereas natural disasters in the LDCs often result in thousands of deaths and casualties. These are quantified in *Enhancing Urban Safety & Security: Global Report on Human Settlements* - UN-HABITAT, 2007.

9.6.1 Geophysical hazards

Earthquakes: These occur in mountainous regions that are on average 2000 metres above sea level; an extensive belt several thousand kilometres wide runs through Morocco and Alpine Europe into Greece, Turkey, Iran, and the mountain ranges of central and Southeast Asia. Another wide zone follows the western strip of the Americas especially along the Andes, also the Pacific 'Ring of Fire', and China. It is estimated that fifty-five cities with populations over 1 million lie in high-risk zones, and major cities include Athens, Ankara, Tehran, Jakarta, Manila, Tokyo, Wellington, Santiago, Mexico City, Los Angeles, San Francisco, and Anchorage. Recent earthquakes at megacity locations include Tangshan, Beijing, and Tianjin in 1976 (242,000 deaths), Mexico City in 1985 (9,500 deaths), Los Angeles in 1994 (sixty deaths with over $30 billion economic losses), Kobe, Japan in 1995 (6,400 deaths with $100 billion in losses), and Istanbul/Izmit in 1999 (15,000 deaths). During the past thirty years, civil disasters from earthquakes include Guatemala City in 1976 (23,000 deaths), Naples and Salerno, Southern Italy in 1980 (3,000 deaths), Spitak, Armenia in 1988 (25,000 deaths), Manji, Iran in 1990 (40,000 deaths and 500,000 homeless), Bhuj, India in 2001 (19,700 deaths), Bam,

Iran in 2003 (26,300 deaths), Pakistan in 2005 (87,000 deaths and 106,000 injured) and Sichuan Province, China in 2008 (70,000 deaths, 375,000 injured, and over 1 million homes destroyed).

Volcanoes: These are a hazard to human settlements because more than 500 million people live in the danger zones around active and potentially active volcanoes. The regions and countries at risk include Hawaii, the west coast of the USA, Mexico, Central America, islands of the Caribbean, Italy, the Philippines, Indonesia, and Japan. Megacities at risk from volcanic eruptions include Mexico City, where Popocatepetl erupted in 1920 and became active again in 1995, and Tokyo where the quiescent Mount Fuji presents a low risk. In the past thirty years, civil disasters from volcanoes include Armero, Columbia, in 1985 (22,000 deaths); Nyos, Cameroon, in 1986 (1,750 deaths); Rabaul, Mount Pinatubo, Philippines, in 1991 (1,000 deaths and 40,000 homes destroyed); Papua, New Guinea, in 1994 (inhabitants evacuated but forty percent of buildings, transport, and communications damaged); Nyirgongo, Congo, in 2002 (150 deaths and destruction of one-third of Goma, a town of 200,000 inhabitants).

Tsunamis: These can be generated by impacts from space, earthquakes in the oceanic plates, submarine volcanic eruptions, and landslides. Tsunami waves travel thousands of kilometres across deep oceans without energy loss and cause enormous damage to coastal development. Tsunamis are particularly common in the Pacific 'Ring of Fire'. The megacity of Tokyo is susceptible to tsunamis, and the metropolis of Lima, Peru, faces the hazard of earthquake-generated tsunamis as thirteen tsunamis have hit the low-lying coastal area of Lima (8.4m population) since 1940. In 2004, an undersea earthquake in the Indian Ocean triggered a massive tsunami in which 225,000 people died in a dozen countries around the rim of the Indian Ocean. The most notable volcano-instigated tsunami occurred in 1883 with the eruption of Krakatoa, between the Indonesian islands of Java and Sumatra in which 36,000 people died. Although tsunamis are relatively rare in the Atlantic Ocean, it has been postulated that if the volcano at Las Palmas collapsed into the sea, a megatsunami would sweep across the Atlantic and devastate the eastern seaboard of North America with a wave 20-50 metres high.

Hurricanes and tornados: Extreme events, such as tropical cyclones (hurricanes and typhoons), tornados, and floods are often regional in scale resulting in both loss of life and the destruction of human settlements. Tropical cyclones are violent storms between 300 and 800 km in diameter with wind speeds in excess of 120-240 km/h or 75-150 mph, depending upon their category that ranges from one (weak) to five (devastating). In the Atlantic Ocean, they are termed hurricanes, in the Far East they are called typhoons, and in India and Australia they are known collectively as tropical cyclones. In 1970, a tropical cyclone struck Bangladesh killing 500,000 people and again in 1991 Bangladesh was struck by cyclone 2B with 235 kph winds and a 6-metre-high storm surge, which resulted in 140,000 deaths. The Caribbean Islands, the Gulf of Mexico, and the USA were hit by hurricane Gilbert (Category 5) in 1988, hurricane Andrew (Category 4) in 1992, and the most severe hurricane Mitch in 1998 in which 18,600 people died. Tornados (or twisters) can be spawned by hurricanes, but generally they are caused by violent thunderstorms or triggered by intense heating in continental interiors. Some eighty percent of all tornados occur in the USA, and occasionally they strike in Europe, Africa, and the Indian subcontinent. The paths of tornados are generally short and narrow, although the wind speeds can be double that of a hurricane, so that, from an average tornado, the number of deaths and property damage tends to be limited. In 1999, more than fifty twisters ran across central Oklahoma in a single day with forty deaths in and around Oklahoma City, 1,500 houses and apartments were destroyed, and 4,500 homes damaged at a total cost close to $1 billion.

Flooding: The majority of fatalities and much of the devastation from hurricanes, cyclones, and storms comes in the form of flooding from storm surges, torrential rain, and landslides. In 2005, hurricane Katrina struck the metropolis of New Orleans (1.1 million population), which is 2 metres below sea level and protected by huge levees and drainage channels designed to withstand a Category 3 hurricane. The levees were breached by a storm surge 9 metres high, and hurricane Katrina caused 1,800 deaths and was one of the costliest natural disasters on record at $125 billion. In 1953, a 4-metre-high strong storm-surge struck the coast of the Netherlands, which breached the dykes fronting the Zuider Zee and 1,800 people drowned. The disaster prompted the Dutch government to construct one of the world's largest flood defence systems. The surge also struck the eastern coast of England, in which 307 people died and 24,000 homes were damaged. In August 2002, sustained torrential rainfall caused widespread flooding in central Europe with peak discharges in the Vltava, Danube, and the Elbe rivers that affected the Czech Republic, Austria, Germany, and Italy for a period of a month. Major cities were affected including the Czech capital Prague, where the underground railway system was inundated and regional transport and electricity supplies were seriously disrupted for months. In Mumbai, India, some 60 cm of rainfall fell during a single day in the monsoon of 2005, and 400 people drowned in the city.

Sea level rise: Global warming is expected to cause a rise in sea levels of 6 cm per decade, and the effects will be serious as sixty-five percent of the metropolises and megapolises of the world are located at maritime or riverine ports, and some forty percent of the world's population live in coastal areas. A 1-metre rise in sea level would affect areas such as the Netherlands and Venice in Europe, Florida in the USA, the Nile delta where seventy percent of Egypt's industry is located and 5m people would be seriously affected, and the Ganges delta where over 110m people in Bangladesh live in an area of 144,000 km² of which ten percent would be lost. By 2100, coastal defences will have to be raised and reinforced in a number of cities including New York, Los Angeles, Boston, Miami, Buenos Aires, London, Amsterdam, St Petersburg, Mumbai, Calcutta, Karachi, Hong Kong, Tokyo, and Sydney.

Heat waves: Another consequence of global warming in Western Europe has been a doubling in the duration of heat waves during the 20th century, and the number of unusually hot days has nearly tripled. In megacities, there is a 'heat island effect', which causes a city to be up to 10 degrees centigrade hotter than surrounding areas. Heat islands add to injuries and deaths related to heat stress. In 2003, some 35,000 deaths were linked to the devastating heat wave that hit Europe. It is estimated that any global economic benefits from global warming peak at around 2° C of warming before there are increasingly negative impacts on GWP.

9.6.2 Technological hazards

Transport infrastructures serving metropolitan areas, such as maritime ports, airports with larger aircraft, underground railway systems, railways, and motorways are major technological hazards.

Maritime accidents: Peacetime maritime accidents include the sinking of ferries such as the Herald of Free Enterprise, English Channel in 1987 (193 deaths) or the Estonia in the Baltic Sea in 1994 (852 deaths). Other ship sinkings include MV Joola, Senegal in 2002 (1,863 deaths) and Al-Salam Boccaccio 98, Red Sea in 2006 (1,018 deaths).

Air accidents: Aircraft crashes occur monthly with some 50-100 passengers killed, and, in rare instances, airlines crash into metropolitan areas killing perhaps five to ten civilians on the ground. In 1992, a Boeing 747 cargo flight crashed into high-rise apartment buildings in Amsterdam, killing the crew and forty residents, when two of its engines detached from the wing. One of the worst accidents occurred at Kinshasa, Zaire, in 1996, where an aircraft failed to take-off and crashed into a busy street market killing six passengers and some 260 people on the ground. In 2001, American Airlines Flight 587 crashed into a Queens neighbourhood in New York City, when the plane's vertical tail fin snapped off just after take-off killing 251 passengers and nine crew, as well as five people on the ground.

Metro or subway disasters: These include the Moorgate, London tube crash in 1975 (forty-three killed), the King's Cross, London fire in 1987 (thirty-one killed), and the Baku, Azerbaijan, subway fire disaster in 1995 (337 killed).

Railway disasters: Major train crashes include Bihar, India, in 1981 (286 killed, 300 missing), Ghotki, Pakistan, in 2005 (132 killed), and Amagasaki, Japan, in 2005 (107 killed, 649 injured).

Highway disasters: These include bridge collapses such as a suspension bridge collapse in Nepal in 1974 (148 killed), the Hintze Ribeiro collapse in Northern Portugal in 2001 (70 killed), or the bridge at Minneapolis, USA in 2007 (thirteen killed). Tunnel fires include the Salang, Afghanistan, tunnel fire in 1982 (1,100 deaths), Mont Blanc, in 1999 (thirty-nine killed), and Kaprun, Austria, alpine tunnel fire in 2000 (155 killed). Other highway collisions include bus crashes with thirty to seventy deaths, or petroleum tanker collisions (30-150 killed), and motorway pile-ups in fog (say, ten deaths). Road traffic accidents kill an estimated 1.2 million people worldwide each year and injure about forty times this number. In the USA, annual road accident fatalities are fifteen per 100,000 population, with seventy-five injured per 100m vehicle-km. In the UK, there are annually five fatal road accidents per 100,000 population, of which two are inside urban areas.

Major industrial accidents: These include mining disasters, refinery and chemical plant explosions, nuclear accidents, and toxic spills. Also, exposure to pesticides, noxious chemicals, heavy metal toxins, asbestos, and a variety of air pollutants and land contamination are the consequences of industrial development and need to be limited by environmental protection. Mining disasters may result in 50-150 deaths. For example, the Liaoning mine disaster in Fuxin, China, in 2005 resulted in 210 deaths. In the Aberfan disaster, Wales, a colliery waste-tip collapsed causing a landslide that buried a school in 1966 (144 deaths). An example of a chemical plant explosion was at Flixborough, England, in 1974 (twenty-eight deaths), and the world's worst industrial accident was the Bhopal disaster, India in 1984 (15,000 killed). A major explosion at BP's Texas City Refinery, USA, in 2005, killed fifteen workers and injured 180, and oil platform disasters include the explosion and blaze on the Piper Alpha platform in the North Sea in 1998 (167 killed), and the sinking of the P36 Offshore Platform in the South Atlantic, Brazil in 2001 (11 killed). Oil spills have generally been off-shore with the Torrey Canyon in the English Channel (1967), the Exxon Valdez oil spill in Prince William Sound (1989), and the Prestige spill off the Spanish Coast (2002). As high-quality resources are depleted, more powerful technologies lead to greater risks including radiation leakage from nuclear accidents or waste. Nuclear accidents include the Three Mile Island, USA, accident (1979) with an exposure of 1 millirem to approximately 2 million people, the Chernobyl, Ukraine disaster (1986) with thirty-one immediate deaths and radioactive nuclear fall-out over Europe, and the Tokai-mura, Japan accident (1999) at a uranium reprocessing facility.

9.6.3 Social hazards

Social hazards include the collapse of residential buildings, retail stores, or markets, hotels and accidents at leisure facilities, or sporting events, religious festivals, political demonstrations, and riots in addition to crimes and terrorist attacks.

Residential buildings: The partial collapse of Ronan Point, a twenty-three storey tower block at Newham, East London in 1968, as a result of a natural gas explosion in one of the flats (four deaths) led to major changes in the building regulations. The Highland Towers collapse of a twelve-storey apartment building at Salangor, Malaysia (1993), following a landslide when the car park retaining wall failed, resulted in forty-eight deaths. An overcrowded balcony of a four-story apartment building collapsed during a party in Chicago (2003) with thirteen deaths and fifty-seven seriously injured.

Retail stores and markets: The five-storey Sampoong Department Store in Seoul, South Korea, attracted 40,000 people per day and collapsed in 1995, with 501 deaths and 937 injured.

Snow loadings caused a roof collapse at the Katowice Trade Hall, Poland, in 2006 and resulted in sixty-five deaths and 170 injured. A roof collapse at Basmanny market, Moscow, in 2006 caused the death of sixty-six people.

Hotels and reception halls: The Hyatt Regency hotel walkway collapse in Kansas City, Missouri, in 1981 killed 114 people and injured more than 200 others during a tea dance. Also the six-storey building containing Hotel New World on three floors, collapsed following a gas explosion in Singapore in 1986, killing thirty-three people. Another example involves the collapse of the third floor of the Versailles wedding hall at Talpiot, Jerusalem, during a wedding in 2001, with twenty-three deaths and 380 injured.

Leisure facilities: Fires at leisure facilities account for a number of accidents including the Summerland, Isle of Man fire in 1973 (fifty-one deaths), disco fire at Luyoang, China, in 2000 (309 deaths), and the Republica Cromagnon nightclub fire, Buenos Aires, Argentina, in 2004 (194 deaths). Also the roof of an ice rink collapsed under the weight of heavy snowfall in Bavaria, Germany, in 2006 (fifteen deaths and thirty-two people injured).

Sporting events: Crowd crushes at sporting events generally occur at soccer matches such as Moscow in 1982 (340 deaths) and the Sheffield, England, stadium crush at Hillsborough in 1989 (ninety-six deaths). In the Ellis Park stadium disaster at Johannesburg, South Africa, in 2001, forty-three people were crushed to death.

Religious festivals: Crowds at religious festivals in which participants are trampled to death have occurred frequently at Mecca, Saudi Arabia, in 1990 (1,426 deaths), in 1994 (270 deaths), in 1998 (118 deaths), in 2004 (251 deaths), and in 2006 (362 deaths); another example is a religious festival at Wai, Mahharastra, India, in 2005 (258 deaths).

Political demonstrations: Protests and political demonstrations resulting in high death tolls include Yangon, Myanmar, in 1988 (3,000 deaths), Tiananmen Square, China, in 1989 (1,000 deaths), and Narathiwat province, Thailand, in 2004 (eighty-four deaths).

Riots: These may be race related and can result in perhaps thirty to sixty deaths, such as the Watts Riot, Los Angeles, in 1965 (thirty-four deaths), and Los Angeles riots in 1992 (fifty to sixty deaths).

Crime: In urban areas of more than 100,000 population, over ten percent of the population are victims of crime of some sort each year. In Asia, the proportion is nine percent, in Eastern Europe eleven percent, Western Europe twelve percent, North America thirteen percent, South America fourteen percent, and Africa fifteen percent. There is a correspondence between burglary and theft crimes and levels of unemployment. In the United Kingdom, some seventy percent of offenders are unemployed, and property crime is much more likely where wealth is unevenly distributed. Property crime is used to vent frustration or resentment for the visible material disparity in urban areas, and, in a sense, it is used as a means of redistribution of goods from higher to lower income neighbourhoods. The use of drugs is also a major cause of crime by addicts to obtain the money needed for regular supplies; drug crimes in certain European countries exceed 250 per 100,000 population, with comparable levels in the USA and Canada, and 400 in Australia.

Urban violence: This includes mugging, aggravated theft, grievous bodily harm, and sexual assault is a result of a variety of factors including alcohol and drug abuse, racism and discrimination, the increasing number of single-parent families, an unsupportive and undisciplined home life for youths, and stress from overcrowded housing and poor living conditions. In the USA, the largest eighty cities have a total of 5,000 gangs with an average of fifty members per gang. A consequence of high levels of crime in cities is that people are discouraged from using public transport and public places, and an increasing number of businesses are driven out of urban centres by armed robbery and break-ins, so that buildings deteriorate and property values drop. This leads to further unemployment, falling house values, and the departure of shopping malls and leisure activities to suburban areas, where the higher income groups are tending to live in fortified enclaves. They travel by private car between apartment complexes and office complexes, with secure car parks to reduce the necessity to walk on the streets.

Murders: In most European countries and some wealthy Asian countries, there are less than two murders per year per 100,000 inhabitants with the rates falling below one for some countries. However, the murder rates per 100,000 population for particular cities can be much higher than those for the national averages, and the average for EU capital cities is 2.5 with Belfast at six and Amsterdam at four. In the USA the murder rate is six compared with Washington DC at forty-six per 100,000 inhabitants. In Russia the murder rate is nineteen with a similar rate in Moscow. The murder rate in South Africa is fifty-four per 100,000 population and forty-one in Pretoria. In Latin America, some countries such as Ecuador and Mexico have more than twelve murders per 100,000 inhabitants per year, in comparison with thirty for Venezuela or El Salvador and sixty in Columbia. There are correspondingly high murder rates in some South American cities such as Rio de Janeiro (sixty), Bogota (eighty), and Cali in Colombia (ninety). In a city with a population of 5m, a murder rate of twenty per 100,000 will result in 1,000 murders each year. A number of serial murderers have claimed in excess of twenty victims.

Mass killings: Massacres by lone gunmen have resulted in fifteen to thirty deaths, such as Mc Donald's restaurant, San Ysidro, California, in 1984 (twenty-one deaths), the Dunblane massacre, Scotland, in 1996 (eighteen deaths), Columbine High School Colorado in 1999 (fifteen deaths), and the Virginia Tech shootings, Virginia in 2007 (thirty-two deaths). Arson attacks have resulted in multiple deaths including the Abadan, Iran, theatre in 1978 (400 deaths), the Macedonian disco at Gothenburg, Sweden, in 1998 (sixty-three deaths), and the Daugu, South Korea, subway fire in

2003 (198 deaths). Mass suicides or human sacrifices include the Jonestown, Guyana, murder and suicides in 1978 (913 deaths), Uganda in 2000 (300 deaths), and the Order of the Solar Temple, Switzerland, and Canada in 1994 (fifty-three deaths). Genocide and warfare are extreme social hazards (See Chapter 11).

Terrorism: Incidents of terrorist activity have shifted from attacks on property to indiscriminate violence involving remote-controlled car bombs in streets outside embassies or office complexes, such as the U.S Embassy, Tanzania, in 1998 (225 deaths), or the offices at Oklahoma City in 1995 (168 deaths). Other terrorist attacks in public places include markets such as Baghdad, Iraq, in 2007 (135 deaths), and hotels with Sharm el-Sheikh, Egypt, in 2005 (ninety deaths), theatres with the Moscow, Russia, theatre siege in 2002 (172 deaths), or nightclubs such as Bali, Indonesia, in 2002 (202 deaths), and schools with Beslan, Russia, in 2004 (344 deaths). Attacks on transport systems include railways such as the bombing at the Central Station, Bologna, Italy, in 1980 (eighty-five deaths) or the train bombings in Madrid, Spain, in 2004 (192 deaths). Other transport atrocities include the London Underground and bus bombings in 2005 (fifty-two deaths and 700 injured), marine transport bombing on the SuperFerry 14, Philippines, in 2004 (116 deaths), and the Lockerbie airline bombing in 1988 (270 deaths). Finally, the aerial suicide attacks on the twin towers of the World Trade Centre in New York in 2001 was the most shocking atrocity, in which two airliners destroyed the entire complex and 2,752 people died. A simultaneous attack using another hijacked airliner hit the Pentagon in Washington, D.C., (189 deaths), and a fourth hijacked airliner crashed into a field in Pennsylvania after the passengers attacked the terrorists (forty-four deaths). By 2001, deaths and injuries from international terrorism were in excess of 10,000 per year. State-supported terrorism increases the range of weapons available to achieve political objectives, and chemical, biological, and radiological (CBR) weapons may soon be available to terrorist organisations.

Prison population: The global average prison population is 140 per 100,000 population, with Europe at eighty-seven and the UK at 124 per 100,000 population, which is the highest in Europe due to longer prison sentences. There are higher prison population levels in South Africa (385), some Eastern European countries (up to 465) and the USA (685). The lowest prison populations are in Scandinavia (fifty to sixty-five) and Japan with forty-seven per 100,000 population.

9.6.4 Biological hazards

Disease: Globally, the leading cause of disease and premature death is poverty, malnutrition, inadequate sanitation, and lack of clean water supplies in the LDCs, where the annual number of deaths resulting from poor sanitation and hygiene is estimated to be 1.6 million. The situation is exacerbated by natural disasters and wars, when population movements and refugees transmit outbreaks of diseases such as typhus. Also some 50 million people per year travel between high- and low-income countries, so that air travel is the fastest way of spreading infections long distances around the world. In 2003, there was a scare at airports with incoming flights from the Far East, that a new type of pneumonia, SARS, was being transmitted by passengers from China to Europe and North America. By May 2003, there had been 8,000 cases of SARS and 700 deaths before the virus was identified and a vaccine manufactured.

Epidemics and pandemics: The most commonly contracted diseases that are currently responsible for nearly ninety percent of infection-related deaths under the age of forty-five are AIDS (with over 3 million deaths per year), tuberculosis (3 million deaths per year), malaria (2 million deaths per year), measles (1 million child deaths per year), diarrhoeal/enteric diseases such as typhus (600,000 deaths per year), or cholera and pneumonic or respiratory infections including influenza. An epidemic is generally an outbreak of a disease with over 10,000 cases at a national or regional level, whereas pandemics involve thousands of cases at an international level. For example, a pandemic of Spanish Flu during 1918-19 was responsible for 22 million deaths worldwide. The three diseases that have the potential for pandemic proportions are AIDS, cholera, and influenza. There are also antibiotic-resistant pathogens or 'superbugs' such as MRSA, which are spread by skin contact, with outbreaks in cities in the USA and UK.

Transmission of animal diseases: The growth of a globalised meat trade with factory farms close to congested cities in the developing countries, is likely to lead to the transmission of diseases from animals to humans. Already, intensive livestock farming methods are showing a disregard for biotechnological safeguards, and feeding bone-meal to cattle has led to BSE and the human condition CJD. Similarly, the poultry trade is contributing to the spread of Avian flu', which is a new strain of influenza that can be transmitted from birds or poultry to humans. The consumption of contaminated meat and other agricultural produce that may be infected by the Salmonella bacterium has given rise to fears of Agroterrorism, which is the malicious use of plant or animal pathogens to cause devastating disease in the agricultural sector.

Bioterrorism: A bioterrorism attack is the deliberate release of viruses, bacteria, or other germs to cause fatalities or illness to humans. Examples include a Salmonella attack in ten restaurants in Wasco County, Oregon, by followers of Bhagwan Rajneesh in 1984 (no fatalities but 750 people had gastroenteritis), a Sarin nerve gas attack on the Tokyo Subway in 1995 (twelve deaths), and in the USA, anthrax spores were mailed to targets in 2001 (five deaths). There are concerns that future acts of bioterrorism could involve the anthrax bacillus, the smallpox virus, and the bacterium responsible for bubonic and pneumonic plague (Yersinia Pestis).

Trajectory of the world system of cities: Chronological references

1. *Ecumenopolis: The inevitable city of the future* - C A Doxiadis & J G Papaioannou,1974.
2. *A Global Review of Human Settlements* - United Nations, 1976.
3. *Cities of the World* - Stanley D Brunn & Jack F Williams, 1983.
4. *4000 Years of Urban Growth* - Tertius Chandler, 1987.
5. *Global Report on Human Settlements* - UNCHS (Habitat), 1987.
6. *Global Cities* - Anthony D King, 1990.
7. *World Cities in a World System* - Edited by Paul L Knox & Peter J Taylor, 1995.
8. *States of Disarray* - UNRISD, 1995.
9. *Globalisation and the World of Large Cities* – Eds. Fu-Chen Lo & Yue-Man Yeung, 1998.
10. *The Atlas of the World's Worst Natural Disasters* – Lesley Newson, 1998.
11. *East West Perspectives on 21st Century Urban Development* - Edited by John Brotchie, Peter Newton, Peter Hall & John Dickey, 1999.
12. *Crucibles of Hazard* - Edited by James K Mitchell, 1999.
13. *Global City Regions* - Edited by Roger Simmonds & Gary Hack, 2000.
14. *Cities in a Globalizing World: Global Report on Human Settlements* - UNCHS, 2001.
15. *Compendium of Human Settlement Statistics 2001*- United Nations Department of Economic & Social Affairs, 2001.
16. *The State of the World's Cities 2001* - UNCHS (Habitat), 2001.
17. *Guide to Global Hazards* – Robert Kovach & Bill McGuire, 2003.
18. *The State of the World's Cities 2004/2005* – UN-HABITAT, 2004.
19. *World Urbanization Prospects: The 2005 Revision* - United Nations Department of Economic & Social Affairs, 2006.
20. *The State of the World's Cities 2006/2007* – UN-HABITAT, 2006.
21. *Enhancing Urban Safety & Security: Global Report on Human Settlements* - UN-HABITAT, 2007.

PART 3

CONTEXTUAL MACROSTRUCTURE

10.
SHAPING FUTURE WORLD DEVELOPMENT

10.1 The capacity of the planet

Long-Range Futures Research explains the importance of defining a long-term future state before we can establish a ranking system or set priorities amongst intermediate goals. If the future state achieves human survival and extends the lifetime of an astro-technological civilisation, then intermediate ends are the development of the major systems that will impact on a planetary civilisation, such as the world system of cities, community systems, urban technology, life sciences, and human development. The achievement of the intermediate ends will involve a consumption of resources and the utilisation of physical stocks or assets, which may be viewed as the intermediate means. The problem is to allocate resources amongst competing ends in the knowledge that the ultimate means are scarce. Since the intended future state is survival, then the stock of assets needs to increase to an optimal level of human development, at which point an ecologic transition has to take place or civilisation will be threatened by overuse of resources. By the year 2150, the stock of assets in urban and rural settlements will need to accommodate a population 12.5 billion at planning standards that are appropriate to the stage of development that may be reached by each continental region. If future worldwide development is to be economically efficient, socially equitable, and environmentally sustainable, it is necessary to articulate an acceptable proposal for the future state and to determine the geopolitical, economic, social, technological, and ecological policies for achieving it.

The point at which population growth might outstrip the planet's capacity to supply enough food depends upon the dietary standard and technological developments for raising the efficiency of food production. If 2,500 calories per person is the minimum daily dietary energy supply, in India the standard is 1,900 calories per person per day compared with 3,700 in the USA. Hunger and chronic malnutrition affect some 500m people, and these contribute to stunted physical growth, a lower resistance to disease, higher childhood death rates, and arrested mental development. In Appendix 27, the land use by continental region has been estimated for a world population of 12.5 billion in 2150, and less than ten percent of the world's land surface of 133 m km^2 would be covered by urban areas and villages at a density of 1,000 inhabitants per km^2. Cultivated land and pastures may account for thirty percent of the land, forests may take up to twenty-seven percent, and the remaining uninhabited land for wildlife would be thirty-three percent. If each person were to receive an average of one million calories per year, then one hectare of 'Suitable Land' (United Nations Food and Agriculture Organisation or FAO classification) should support five people, i.e., 0.2 hectares per person. In 2000, the cultivated land and pastures in the developed countries amounted to 1.20 hectares per person compared with 0.59 hectares per person in the developing regions. For a world population of 12.5 billion, the cultivated land and pastures in the MDCs will amount to 0.95 hectares per person compared with 0.23 hectares per person in the LDCs.

By definition, arable land is suitable for growing crops, and the FAO includes land under temporary or permanent crops, land temporarily fallow, temporary meadows for pasture, and market or kitchen gardens. Potential arable land will include cultivated land for crops and pasture for livestock, and some twenty percent of croplands are irrigated, which includes irrigation by controlled flooding. In *Land Resource Potential and Constraints at Regional and Country Levels* (2000), the FAO estimates the suitability of land for rain-fed crop production, and there are some 40m km^2 of potential arable land, which is categorised in terms of quality and weighted with a factor in accordance with its suitability to give an equivalent potential area in terms of 'Very Suitable land.' The suitability classes and weightings are Suitable (x 0.7), Moderately Suitable (x 0.5), and Marginal (x 0.3). Overall, the equivalent potential of arable land corresponds to 0.7 x Very Suitable to give an average of 'Suitable land'` on which 1.0 hectare will support five people. Marginal land and areas for wildlife include inland rivers and lakes, protected natural areas, wetlands, drylands, desert, and steeplands.

One of the limits to increasing agricultural production is that much of the Earth's surface is too mountainous with elevations of 500 metres, or too dry with less than 25 cm annual rainfall, or an annual growing season of less than ninety days, which is too short for normal forms of agriculture. However, irrigation and the use of desalination plants could increase the area of land under agriculture, but these technologies are expensive and the cost of food production would rise. Expanding the cultivated area has the problem that unused land is not in the same location as the population, and there are major political, social, and economic obstacles to allowing the emigration of large populations to these lands. For example, by 2150, there will be severe land pressure in India, whereas North America, South America, and Sub-Saharan Africa have reserves of potential arable land.

The intensification of agriculture entails the technology of mechanisation, cultivation under glass, and also the introduction of improved plant varieties through genetic engineering and increased fertiliser use. Alternative approaches to supplementing future food supplies include increasing the food potential from the seas, which supply some twenty percent of the world's high-quality animal protein, and bypassing animals in the food chain. Both of these would require altering people's eating habits. In addition to the threat of climatic change, it may be expected that additional agricultural yields will become increasingly expensive because of the large capital outlays for irrigation, machinery, and chemicals.

Globally, fresh water is abundant, and each year an average of more than 7,000 m³ per capita enters rivers and aquifers. Most of the countries with limited renewable water resources are in the Middle East, North Africa, and sub-Saharan Africa, the regions where populations are growing fastest. Elsewhere, water scarcity is less of a problem at the national level, but it is nevertheless severe in certain watersheds of northern China, west and south India, and Mexico. The minimum per capita annual water requirement is generally in the range of 800-1200 m³. In a developing country, a typical per capita allocation of the annual water use would be domestic 100 m³, agriculture 450 m³, and industry 250 m³. Water scarcity is considered to be a severe constraint in countries that have renewable water resources of less than 1,000 m³ per capita, and countries with less than 2,000 m³ per capita on average often have dangerously little in years of low rainfall. A measure of water stress is when the use-to-resource ratio exceeds 0.5, and in these regions, water resource management is essential since water can be reused many times. It is estimated that by 2025 a billion people will be living in countries suffering from high levels of water stress.

Water scarcity is frequently both a regional and a political problem because more than 200 river systems, draining over half of the planet's land area, are shared by two or more countries. The cooperative management of shared watersheds and river basins, the control of water pollution, improved maintenance of urban water distribution systems, appropriate water pricing, and the avoidance of waste and development of efficient irrigation systems can extend the availability of scarce supplies. Water sustainability will require an integrated river basin planning framework, a water systems engineering infrastructure, wastewater treatment plants, flood control measures, new irrigation technologies, and sea-water desalination plants. Although desalination is expensive, it is affordable in the oil exporting regions of the Middle East and North Africa. According to the UNEP publication *Global Environmental Outlook 3* (2002), in 2000 Africa's renewable water resources provided an average of 5,000 m³ per capita, but the spatial distribution of water resources on the continent does not match the areas with the highest population densities, so that thirteen African countries suffered water stress or scarcity (less than 1,700 m³ per capita/year or less than 1,000 m³ per capita/year respectively).

The rate at which non-fuel mineral resources are being used is rising steadily throughout the world and especially in the developed regions. Resource use rises steeply during the earlier stages of industrialisation and follows an S-shaped logistic curve until growth levels off at a relatively high level of per capita consumption as shown in Appendix 28. From a World Resources Institute publication, *Resource Flows: The Material Basis of Industrial Economies* (1997), the default values for Total Material Flows (TMF) in the MDCs of 70-80 metric tons per capita may be assessed as Energy (28 tons), Construction (15 tons), Metals and Minerals (10 tons), Agriculture and soil erosion (7 tons), with material variations depending upon the amount of excavation, overburden, and waste. The Direct Material Inputs (DMI), which exclude excavation, overburden, and waste, amount to some 20 metric tons per capita and include Energy (8 tons),

Construction (7.5 tons), Renewables including agriculture (2.5 tons), Metals and Minerals (1.5 tons), and other components (0.5 tons). Some seventy to seventy-five percent of the materials flow in the MDCs in 2000 were derived from the energy and construction sectors, and the lower intensity of per capita energy use, infrastructure provision, and building floorspace per person in the LDCs would indicate default values there for total materials flows at an average of 20 tons per capita to give a world average of 30 tons per capita.

The sectoral use of materials include the building materials used by the construction industry, the metals and plastics used by the motor industry, the chemical and mineral industry products, the textile and clothing manufacturers, and other furniture and equipment for both businesses and households. The expression 'mineral reserves' are a variable quantity, when they are defined as the quantity of mineral that can be mined and extracted at a profit. Julian Simon has made the case in *The Ultimate Resource 2* (1996) that knowledge is the ultimate resource and that technological development and substitution has led historically to a fall in real resource prices. There is a need for caution on this point, because it is necessary to know where we are in the life cycle of planetary development, since the logistic S-curve has a notorious reputation for leading entrepreneurs to believe in increasing returns until they suddenly diminish.

Every mineral resource has an absolute limit, and the aim is to postpone the final day through improved knowledge of regional geology and better exploration techniques. The approaches to conservation of mineral resources include the reduction of waste in extraction and processing, the more efficient use of materials in manufacturing, and the increased durability of manufactured products. Also, there is continuous substitution of one material for another because of scarcities and rising costs. More aluminium is replacing copper in electrical appliances, and aluminium is replacing tin-plated steel in the container industries. Plastics are increasingly substituted for metals in a great many uses, from plumbing pipes to automobile fittings. In spite of these technological changes, the known world reserves of even the most abundant mineral resources would not be sufficient for global consumption at per capita rates of the MDCs.

The oceans, which cover over seventy percent of the Earth's surface to an average depth of nearly 4 km, could provide a new source of minerals since this large volume of salt water contains some 39m tons of solids per km^3. Salt, magnesium, sulphur, calcium, and potassium constitute 99.5% of this. Other more valuable elements in one km^3 of seawater include 11.5 tons of zinc, 3.4 tons of copper, 3.4 tons of tin, 0.25 tons of silver, and 4.5 kg of gold. Sodium, chlorine, magnesium, and bromine are already being extracted electrolytically from the sea. The main problem is that the valuable metals are contained in extremely dilute solutions, which would require a large amount of energy to treat huge quantities of water. The oceans are hydrogen mines, since two out of three atoms in water, H_2O, is hydrogen, and the oceans will be the source of a hydrogen-fuel economy.

The world's proven reserves of oil and gas exceed 300 btoe (billion tons of oil equivalent), and the coal reserves amount to 460 btoe. The current annual (2000) fossil fuel consumption is for coal 2,300 mtoe, oil 3,500 mtoe, and natural gas 2,100 mtoe. At current consumption rates, proven reserves of coal are sufficient for 200 years and oil and gas for fifty years. Estimates of potentially recoverable fossil fuel reserves worldwide are more than 600 times the present annual rate of extraction. In total, fossil fuel resources are probably sufficient to meet world energy demands to 2250 and beyond. Total energy consumption in the developed countries is likely to have peaked in 2000 at 250 EJ (Exajoules) or 115 mbdoe (million barrels a day of oil equivalent), whereas consumption in developing countries will reach 135 mbdoe by the year 2025. Per capita consumption in the MDCs peaked at thirty-five barrels of oil equivalent per year in 2000, and is expected to decline to twenty-five by the year 2025. In developing countries, the per capita consumption is expected to increase from six to seven barrels of oil equivalent per year in the same period.

10.2 Population growth

World population growth estimates are based upon the concept of a demographic transition, in which countries pass through three phases in the course of their development. In the first phase, living standards are low so that a high mortality rate balances a high fertility rate, and the population is relatively stable. In the second phase, infrastructure investment in water supply, sewage systems, and health care result in declining mortality rates and increased life expectancy with a population explosion. In the third phase, equilibrium is re-established through declining birth rates due to social changes as a result of urbanisation, when smaller families are preferred as women become better educated in less developed countries. Although birth rates have fallen, they remain significantly higher than death rates in the developing world and only slightly higher than death rates in the industrial world. In the decade 1960-1969 world population growth reached a peak of 2.1% per year, and this declined to 1.4% by the decade 1990-99. However, in 1965 a 2.1% growth in the world population of 3.3 billion produced a growth in the world population of 70m, compared with a 1.4% increase of 80m in 1995 for a global population of 5.7 billion.

The world average for life expectancy is sixty-six years, and the developing countries have seen an increase from forty-six years to sixty-three years since 1960. Appendix 15 gives the life expectancy for each stage of development, and it rises from fifty-two years in traditional countries to seventy-eight years at the informational stage. In 1900, life expectancy at birth throughout Africa, Asia, and Latin America was twenty-five to twenty-eight years. Since 1960, the world average infant mortality rates have fallen from 128 per 1000 live births to fifty-four. However, in South Asia, there are seventy-three deaths per 1000, and there are still twenty countries where infant mortality rates exceed 100, so that one child in ten dies before their fifth birthday.

Demographic forecasts show the division of the world into slow and fast growing regions. Five regions (Northern America, Western and Central Europe, Eastern and South-East Europe, Australia and New Zealand, and East Asia) with a combined population of 2.4 billion fall into the category of slow-growth rate with an average of 0.9%. The remaining five regions (Africa, Middle East and Central Asia, South-East Asia, Indian Subcontinent, and Latin America) with a total population of 2.8 billion form the high-growth rate group at an average of 2.5% per year, which is nearly three times that of the slow-growth rate group. The group with the low-population growth has an annual increase of 23m people, compared with a population increase of nearly 70m per year in the fast-growth group. It follows that demographic growth in regions is either quick at 2.1% or more, or slow at an average of less than 1%.

Long-term population projections for an ultimate size of a population, when stabilisation is reached with near-zero growth rates, depend enormously upon the polarisation into slow- and fast-growth categories. In particular, the projections for Africa, the fastest growing group, may be unrealistic, simply because the life supporting ecosystems may break down long before these projections become reality. For example, Nigeria, with a current population of 125m, is expected to stabilise at 450m, which would mean that, by the year 2050, Nigeria will have nearly as many people within its borders as there are in the whole of sub-Saharan Africa today. The factors that may change the projections would be a slow down or a reversal of the modernisation process, changes in mortality rates (e.g., AIDS) or interference in the fertility rates. For example, in China, the two-child family policy could not prevent its present population of 1.3 billion growing to 1.8 billion prior to zero growth. China's one-child family plan is an attempt to keep its eventual total to 1.2 billion, although by 2030, one-third of the population may be over the age of sixty-five.

Sustainable development studies should ideally look ahead 250 years in respect of depletion of fossil fuels and climate change, and the United Nations has published *World Population to 2300* (2004). Long-range projections to 2150 were previously based upon the *1998 Revision* (United Nations 2000), but the projections to 2300 are not forecasts but are hypothetical scenarios based upon extrapolations of current trends. We do not know whether fertility rates in the future will be higher or lower than at present and what might be the collective reaction in a given country if fertility rates remained well below replacement for several decades. Mortality exerts some additional influence on the population growth rates, and it largely counteracts that of fertility, since mortality tends to be higher where fertility is higher. Also, we cannot predict the rates of population migration or what new diseases or epidemics will emerge that sharply affect the life statistics used for demographic projections. These factors are dependent upon the stage of development of a country, and so these factors are not independent of future economic growth.

In the UN medium scenario projection, world population peaks at 9 billion in 2050 with minimal growth for the following 250 years. However, the high and low scenarios are considerably different, with a population peaking at 36 billion in 2300 in the high scenario, and the population collapsing to 2.3 billion by 2300 in the low scenario. *Long-Range Futures Research* takes a planning time horizon of 150 years to 2150 to include five future generations, and traces the evolutionary trajectory to 2250. It is not anticipated that there will be a devastating catastrophe on a global scale within this period. World population projections, based upon the United Nations' long-range scenarios for a medium variant, are tabulated in Appendix 19. It is anticipated that the world population will peak at 12.5 billion by 2150 before declining, with a medium population of 10 billion and a high-medium population of 15 billion. If it can be shown that a population of 12.5 billion is sustainable, then there would be no difficulty in supporting a lesser population of 10 billion and modified policies can be formulated for accommodating an increased population of say 15 billion. Alternative population scenarios are considered in Chapter 12.

The United Nations provides some broad regional population estimates that assume that India and China may account for thirty percent of the world's population, the rest of Asia will amount to twenty-five percent, Africa will contain twenty-five percent, Latin America will have ten percent, North America and Oceania will hold five percent, and Europe will have five percent of the world population. It is difficult to assess the impact of immigration on the European population, but it is more likely that the European population will reach eight percent of the total. In the UN population projections, it is assumed that declining mortality worldwide will counterbalance declining fertility and that life expectancy for women will increase from seventy-nine to eighty-eight years between 2050 and 2150, whereas for men it will increase from seventy-four to eighty-three in the 100-year period.

10.3 Measures of economic and social development

Globalisation of the world economy has resulted in the increasing interdependence and transaction linkages between continents, nations, regions, and cities, and changing patterns in production, distribution, and consumption. There are a number of dimensions to the globalisation process, including:

- Population, migration, and employment
- Resources and the environment
- Industry and technology
- Finance and investment
- Marketing and distribution
- Media, telecommunications, and information technology
- Culture and tourism
- Armaments and defence

The global economy is becoming the system within which national economies operate, and capitalist countries are finding that the traditional Keynesian policies with their bias towards investment and growth are no longer sufficient to overcome the global influences. In fact, political, economic, social, and cultural activities are all interrelated in a world system whose trajectory is affected by transnational corporations that are often independent of direct control and regulation by individual nation states. Also, the globalisation of the economic system has resulted in underutilisation of capacity, diminishing returns from investment, closure of inefficient plants, and redundancy for workers. The resulting unemployment and social cost to the community are often ignored, although in the past, the system has been justified by an apparent annual increase in the gross national income (GNI) per capita.

GNI measures the flow of national income arising from transactions in the formal economy. It does not measure services performed within the household or the proportion of agriculture produced in rural communities for consumption by a farmer's family. GNI may also increase with deteriorating social or environmental conditions, since increased crime generates additional law enforcement, and natural disasters or pollution require rehabilitation works or cleaning-up measures. The Stern Review on the economics of climate change points out that some of the negative effects of global warming, such as increasing expenditure on air-conditioning and flood defences, will also increase economic output.

GNI therefore has shortcomings as a measure of well-being or quality of life, and, in order to assess the level of economic development, the stock of assets needs to be taken into account in terms of infrastructure, agricultural, manufacturing, commercial, public, and residential property, equipment and vehicles, and inventories. There is, however, a close relationship between GDP per capita and infrastructure investments per capita, and for practical purposes, it is often used as a proxy for measuring the level of economic development. As a guide, GDP will generally lie in the range twenty-five to forty-five percent of the capital stock, depending upon the composition of the capital stock and the asset life, so that capital stock is often equivalent to three years of GNI.

The International Monetary Fund's World Economic Outlook, 1993, issued an addendum, which converted GNI rankings to international dollars using Purchasing Power Parity (PPP) rates. An international dollar has the same purchasing power over GNI as a US dollar has in the United States. Purchasing Power Parity gives the factor by which the Gross Regional Product in 2000 US$ is increased at each stage of development from up to five times at the agropolitan stage to say 1.5 times at the distributional stage to give the appropriate PPP US$ (see Appendix 28). As the global economy develops in the future, rising per capita standards will enable countries to progress to higher stages of development so that the overall global factor for PPP will reduce from say 1.5 in 2000 (PPP $45 trillion compared to $31 trillion in 2000 US$) towards 1.25 in 2150.

In spite of the limitations of the ratio GNI per capita, it does have the capability of reflecting in a single ratio the life processes of both production and reproduction. Until a less developed country reaches the demographic transition, its rapid population growth dampens the growth in economic development; whereas, in the more developed countries, a high gross national product contributes to a depletion of natural resources and deterioration of the environment, until an ecologic transition is achieved. Another aspect of GNI is that the national income may not be evenly distributed between regions, or between corporations and citizens, or between the various social groups in the community. It follows that there is an enormous loss of intelligence where averages are used in a subcontinental country such as China or India, where specific urban areas may have a per capita GDP that is some ten to fifteen times the average for the country. In more developed continents, the regional disparities may be up to five times the average.

The role of social and environmental indicators became the focus of much attention after the Earth Summit in Rio de Janeiro, 1992. Rio's Agenda 21 commits all 178 signatory countries to expand their national accounts by including environmental costs, benefits, and values. Sustainable development involves improving the well-being of people by raising the standards of living, education, health, civil, and political rights. The number of democracies in the world has risen from thirty percent of all nations in 1975 to some 120 countries or sixty percent by 2000. But in terms of population, only fifty percent of the world's people live in free nations. Economic growth is an essential means for enabling development, but in itself it is a highly imperfect proxy for progress. It has long been recognised that measures of social development such as educational opportunity, infant mortality, and nutritional status are essential complements to GNI.

A range of indicators is necessary to assess social progress and human development, such as health, life expectancy, and civil rights, and the main determinants of well-being such as income, housing quality (including water, sanitation, and drainage within neighbourhoods), and accessibility of schools, health care services, and other social facilities. However, for most nations, there is inadequate data available for most of these constituents and determinants of well being to allow progress to be assessed. There is a tendency for international comparisons to be made on a few indicators that are easily measured, without critical consideration of the extent to which the chosen indicators reflect social progress. The United Nations Development Programme (UNDP) measures social progress using the Human Development Index (HDI). There are three components to the HDI and these are life expectancy at birth (a reduction in this inequality would give all humans the opportunity for a lifespan comparable to that of the MDCs), educational attainment, and GDP per capita.

Appendix 15 shows that the number of doctors per 10,000 population increases from two at the traditional stage to thirty at the informational stage, and the corresponding number of hospital beds increases from twelve to eighty in the most advanced economies. In many cities, there are large differentials in infant mortality rates between the lower income, poor quality areas, and the averages. Secondary school enrolment increases from twenty-five percent in traditional countries to over ninety percent at the informational stage. In the more developed countries, tertiary enrolment increased from fifteen to fifty percent between 1960 and 2000.

An alternative measure to the UNDP Human Development Index is the City Development Index (CDI), which is applied at the urban level and described in *The State of the World's Cities* 2001 (UNCHS-Habitat). Five separate sub-indices are constructed for the City Product per person (analogous to GDP at the city level), Infrastructure, Waste, Health, and Education with values that range from 0-100, which are then combined to create the composite CDI. The city product, infrastructure, and waste components of the index are the key variables for measuring the effectiveness of city governance, and the health, education, and infrastructure components are strong indicators of urban poverty. The infrastructure sub-index is based upon the number of household connections for water, sewerage, electricity, and telephone. The waste sub-index is based upon households with wastewater treatment and formal solid waste disposal. The health sub-index is based upon life expectancy and child mortality, and the education sub-index is based upon literacy levels and school enrolment.

In 2000, one quarter of the urban housing stock in the LDCs consisted of non-permanent structures with a proliferation of slums and squatter settlements. The population living in slums at that time was 910 million, and a millennium goal was set to improve the lives of at least 100 million slum dwellers by 2020. According to *The Challenge of the Slums: Global Report on Human Settlements 2005* (UN-HABITAT, 2005), it is estimated that by 2020 the number of slum dwellers will reach 1.4 billion, with lack of water, lack of sanitation, overcrowding, non-durable housing structures, and lack of security of tenure. Around sixty-five percent of the world's urban population obtain water from a tap, and only thirty percent have piped water within their dwelling. Some ten percent depend upon public taps, eight percent have access to manually pumped water or protected wells, some seven percent purchase water from vendors, and the balance are dependent upon rainwater collection. Some forty percent of the world's population lacks access to sanitation, and over twenty-five percent of the urban population in developing countries lack adequate sanitation. Also around twenty percent of the LDC urban population live in overcrowded conditions with three or more people sharing a bedroom. Low-income settlements are particularly vulnerable to natural disasters and environmental hazards, and mass evictions of slum and squatter settlements increase the vulnerability of the poor. Rural populations represent the largest target group for development interventions, and slum dwellers are a growing population that require investment in infrastructure to reduce the levels of pollution and disease. A worthwhile reference is *Planet of Slums* (2006) by Mike Davis.

Global income growth has tended to exacerbate income inequalities, which is expensive in terms of resources, energy, and the environment. About half the world's population lives in poverty on less than $2 per day, and over 1.1 billion people live in extreme poverty on less than $1 per day. In the least developed nations, the whole output has to be used to maintain society and to keep people alive, so that there is little surplus for the accumulation of either knowledge or artefacts. We need to raise the threshold levels of per-capita consumption in the low-income countries and raise the level of life expectancy, without raising the per capita levels of resource consumption by the high-income countries.

One consequence of ignoring the increasing inequality arising from the global economy is that international terrorists are intent on destroying or levelling down the assets of any states or organisations that they perceive are frustrating a fair distribution of wealth or income. Economic warfare and ethno-religious assassinations or terrorist acts are designed to destroy the status quo, and issue-orientated groups engage in mail bombings, sabotage, and violent protests to secure publicity for their cause. Finally, there is an international terrorist infrastructure with networks of bomb makers, assassins, hostage takers, forgers and document counterfeiters, arms suppliers, training camps, and safe houses, such as the al-Qaeda network, which have the capability for funding multiple attacks that included the strike on the Pentagon. (See Chapter 9, section 9.6).

Society also needs to take responsibility for reducing national income inequality, which is a major contributor to social deprivation. It would appear from national statistics that the range for the ratio of income between the top twenty percent and the bottom twenty percent of the population varies from four to nine, and that, in a decent society, a factor of six would appear to be a realistic level. This may be achieved from progressive income taxes. Democratic governance involves the state in social regulation to avoid the diseconomies of inequality, which are reflected in rising unemployment, reduced purchasing power, and increasing poverty, with rising crime and violence in cities. When the number of people without jobs becomes large in relation to the number of people with jobs, the ensuing resentment and the increase in crime is likely to result in the separation of communities, with the added cost of protection for walled residential areas with security guards, CCTV, and identification cards to gain access.

10.4 Alternative routes to a sustainable future

There are alternative routes to a sustainable future, and the scenarios receiving the most attention involve growth with environmental protection, a technological shift, and societal transformation. The implications of these scenarios can be tested using the formula I=PxAxT, where (I) is the environmental impact of any society, (P) its population, (A) the level of affluence in terms of the capital stock per inhabitant, and the damage done by the particular technologies (T) that support the affluence. This is described in *Beyond the Limits* (1992) by Donella and Dennis Meadows, with Jorgen Randers, who attribute an extension of the formula to Amory Lovins. The IPAT formula may be expanded to examine in more detail the components of affluence (A) and technology (T) as follows:

$$A = \frac{Capital\ stock}{Inhabitants} \ x \ \frac{Material\ throughput}{Capital\ stock}$$

$$\text{and} \quad T = \frac{Energy}{Material\ throughput} \ x \ \frac{Environmental\ impact}{Energy}$$

A reduction in the capital stock per inhabitant reduces the environmental impact of extracting materials, processing them into products, and disposing of them. The material throughput needed to maintain a given capital stock can be reduced through improved durability and longer life assets, more flexible designs that can be adapted to different uses (i.e., buildings), higher total product quality, minimum materials design, better materials choices, and more extensive recycling, reuse, and scrap recovery. By producing energy in more environmentally benign ways, such as a shift to renewable energy resources, the environmental impacts of energy-using technologies can be reduced. Also more efficient use of energy (a dejouling of energy) will lower the energy requirement to support a given level of material throughput. Energy efficiency can be improved by increasing end-use efficiency (i.e., in vehicles, electrical appliances, and lighting), and also through improvements in conversion efficiency, distribution efficiency, system integration, and process design.

In the residential sector of the more developed countries, measures for reductions in energy intensity of some thirty-five percent may be possible by means of double glazing, enhanced building insulation standards, improvements in space heating and cooling systems, fluorescent lighting fittings with sensors, microwave cooking, and technical improvements in appliances. In the industrial sector, improvements in energy intensity of fifteen to twenty-five percent may come from combined heat and power technologies, process improvements, together with recycled material feedstocks. In the transport sector, there is the potential for a reduction in private motoring energy intensities from 1.5 MJ/passenger-kilometre to 0.75 MJ/passenger-kilometre by 2050. Finally, in the electric power generation sector, there is scope for nearly doubling the energy efficiency of power stations from thirty-five to sixty-five percent with the introduction of advances in technology, such as combined-cycle thermal plants, combined heat and power generation, and fuel-cell based technologies.

The scenario of growth with environmental protection may be examined using the assumptions of the Brundtland Commission that there will be a growth in world economic activity of say five times within fifty years. By using energy efficiency as a surrogate for the environmental impacts of technology (T) and setting the terms of the IPAT formula to unity for 2000 as a base case, the environmental impact (I) in 2050 is calculated in Table 12 below.

TABLE 12: ENVIRONMENTAL IMPACT IN 2050 (Brundtland)

Population in year 2050 at 9.5 billion	1.57 times	2000 level
Growth in world product 2000 to 2050	5.00 times	to $150 trillion
Growth in capital stock per inhabitant	2.00 times	$25,000 to $50,000 per capita
Material throughput/capital stock	1.00 times	
Energy efficiency by 2050 Technology factor	1.5 times 0.67 times	
Impact, I = 1.57 x 2.00 x 0.67 =	2.10 times	

It is generally acknowledged that economic growth is beneficial in achieving the demographic transition in less developed countries, and, similarly, it can be argued that economic growth will assist in delivering an ecologic transition in the more developed countries. This would arise because a higher national income will permit increased investment in pollution abatement technologies, including a shift to renewable energy resources, energy efficiency, and longer-life assets. The concern of ecological economists is that growth policies are likely to push economic development beyond the optimal scale of the entire economy relative to the ecosystem. It follows, therefore, that low growth in developed countries is preferable to no growth, and both are better than high growth. However, any physical subsystem of a finite and non-growing planet must eventually become non-growing, and further growth will become unsustainable.

The World Bank data for energy consumption refers to commercial forms of primary energy such as petroleum and natural gas liquids, natural gas, solid fuels (coal, lignite, etc.), and primary electricity (nuclear, hydroelectric, and geothermal power). It also includes combustible renewables and waste, such as firewood, solid biomass, and animal products, gas and liquid from biomass, industrial waste, and municipal waste. A shift to renewable energy would have a major effect in sharply reducing the environmental impacts of every kind of technology that uses energy. Since air pollution is caused principally by energy use, vehicular emissions, and industrial production, city air pollution in industrialising developing countries (which often have relaxed abatement policies) is far worse than in the developed countries.

One technological shift scenario would see a reduction in fossil fuels to sixty-five percent of energy consumption by 2050, with nuclear and hydro reaching twenty percent, and renewable sources rising to a fifteen percent share. Progress in resource efficiency by reducing the material throughput needed to maintain a capital stock would appear to be applicable to the more developed countries, whereas real resource pressures will arise from urban development in the less developed countries. Multi-industry 'industrial ecologies' may be developed in which effluents from industrial processes will not only be minimised but also treated as raw materials for other processes. Rostow's explanation of price rises causing capital to be attracted to developing alternative sources of primary resources, such as energy, materials, and food, would indicate that the leading sectors of the Third Millennium will be new energy sources (nuclear, tidal, solar, wind, etc.), new raw material sources (ocean mining), new foodstuffs industries (biotechnology), and environmental industries. There will also be diminishing marginal utility from minor improvements in the innovations other than cost reduction, the use of less energy, substitute materials, and longer-life artefacts.

The societal transformation scenario rejects the scale of technological change on the basis that it requires too much technology, with the diversion of excessive investment capital, and the ultimate limit is not physical or biological but economic. In fact, growth is more likely to be constrained by shortages of capital than by shortages of energy or raw materials. The mission of the societal transformation scenario is to achieve a shift from quantitative growth to qualitative development; a sustainable scenario with social equity is developed in section 10.5.

10.5 Future gross world product

It is estimated that Gross World Product (GWP) could rise from $31 trillion in 2000 to $60 trillion in 2025 in real terms. For the developing countries as a whole, average incomes could rise by a factor of 2.4 in real terms, from an average of $1,300 today to $3,100 in 2025. Substantial regional differences will persist, although in aggregate, the gap between income levels in developing and industrial countries could narrow. The resulting GNI per capita in 2025 for the various country groups is given in Appendix 29 (Sheet 1 of 2), and this is based upon a redistribution of investment with social equity as outlined in Section 10.6.

By the middle of the 21st century, the developing countries' share of world income is likely to rise from less than twenty percent to almost fifty percent, and if trends continue, it would rise to sixty-five percent by the year 2100. A sustainable Gross World Product (GWP) for a world population of 12.5 billion is estimated to be $200 trillion (2000 US$) in 2150, with a GNI per capita of $40,600 in the more developed countries and $12,400 per capita in the less developed countries. In comparison with GWP per capita projections of $5,150 and $7,300 (2000 US$) in the years 2000 and 2025 respectively, by 2150 the growth in income per capita will increase three times from the 2000 level to give a world average of $16,000 per capita. In testing this scenario with the formula Impact I=PAT (see Section 10.4), Table 13 sets out the assumptions for assessing the environmental impact in year 2025 in relation to the year 2000.

TABLE 13: ENVIRONMENTAL IMPACT IN 2025

Population in year 2025	1.35 times	
Growth in capital stock per inhabitant	1.48 times	(based on investment as Appendix 30)
Material throughput/capital stock	0.85 times	
Energy/material throughput	0.75 times	
Renewable energy factor	0.85 times	
I = 1.35 x (1.48 x 0.85) x (0.75 x 0.85) =	1.10 times	

With a technological shift, it is expected that through longer life assets, recycling, and minimum material design, the material throughput for a given capital stock could be reduced to eighty-five percent by 2025 and sixty-five percent by 2150. Appendix 31 gives the future materials use by continental region; by 2150, the materials use per capita in the MDCs is expected to decline from 75 tons to 45 tons, and converges towards the 35 tons in the LDCs, with a world average of 38 tons per capita. This will require meaningful dematerialisation at the ecopolitan stage of development with an absolute reduction in the per capita use of natural resources by the MDCs. The International Energy Agency estimates that there will be a 1.1% per year increases in energy efficiency, so that energy efficiency could be improved by a factor of 1.25 by 2025 and two times within a timescale of 100-150 years. This would reduce the energy consumption to support a given level of material throughput to two-thirds of its present value by 2025 and to one-third by 2150. Table 14 shows that from 2000 to 2150, a shift to renewable energy resources is expected to reduce the environmental factor for energy used to 0.67, in relation to 2000:

TABLE 14: ENVIRONMENTAL IMPACT IN 2150

Population in year 2150	2.06 times	
Growth in capital stock per inhabitant	2.56 times	($25,000 to $64,000 per capita as Appendix 30)
Material throughput/capital stock	0.65 times	
Energy/material throughput	0.50 times	
Renewable energy factor	0.67 times	
I = 2.06 x (2.56 x 0.65) x (0.50 x 0.67) =	1.15 times	

On the basis of the default planning standards for energy and transport given in Appendix 16 and the technological shift described above, the evolutionary assessment in Appendix 32 (Sheet 1 of 2), gives the world car fleet at over one billion for 2025 and over three billion in 2150, together with the global energy use by continental region. In 2000, the global energy use was 425 EJ (Exajoules) and, on the basis of the assumptions outlined, it is expected that the energy use in 2150 will be 600 EJ. If one-third of commercial energy is produced using renewable sources, then the environmental impact will be equivalent to 400 EJ.

It follows that a sustainable civilisation may be achieved under the assumption of a gross world product of $200 trillion for a world population of 12.5 billion in 2150. However, the MDCs should limit GNI per capita to $40,000-$50,000 (2000 US$) during the urban transition in the LDCs, which is higher than the Gross State Product per capita of California at just under $40,000 in 2000. This would give a world average of $16,000 per capita in 2150. The ratio of per capita income between the wealthiest twenty percent of countries with average GNI/capita of $21,000 and the poorest twenty percent with say $280 is seventy-five times, although using Purchasing Power Parity (PPP) GNI/capita, the ratio would be twenty times ($23,000 compared with $1,150). One consequence of this more equitable scenario of investment transfers from the MDCs is that the GNI/capita ratio could be reduced to thirty times by the year 2025 and twelve times by the middle of the 22nd century. In effect, this development strategy would ensure that the poorest nations achieve the industrial and distributional stage of economic development.

A prognostication of the long-term future was made in Chapter 9 on the basis that the global urban population would reach ninety percent of the total in 2150. It is possible to chart the future evolutionary trajectory of the landscape of the civil system beyond 2150 by making an assessment of the world stock of assets in 2250. Even though the absolute size of the global population beyond 2150 may fall within a cone with an upper level of 12.5 billion and a medium level of say 10 billion, the proportion of the urban population is expected remain constant at ninety percent of the total. In 2250, the Gross World Product is expected to reach $405 trillion (PPP $425 trillion) for a population of 11.5 billion with an asset stock of PPP $1,200 trillion, as shown in Appendices 29 and 30 (Sheet 2 of 2). Table 15 shows that during the years 2000 to 2250, a shift to renewable energy resources is expected to reduce the environmental factor for energy used to 0.50, in relation to the year 2000.

TABLE 15: ENVIRONMENTAL IMPACT IN 2250

Population in year 2250	1.90 times	
Growth in capital stock per inhabitant	4.17 times	($25,000 to $104,000 per capita as Appendix 30)
Material throughput/capital stock	0.60 times	
Energy/material throughput	0.40 times	
Renewable energy factor	0.50 times	
I = 1.90 x (4.17 x 0.60) x (0.40 x 0.50) =	0.95 times	

On the basis of the default planning standards for energy and transport given in Appendix 16 and the technological shift described above, the evolutionary assessment in Appendix 32 (Sheet 2 of 2) gives the world car fleet for 2050 as 1.7 billion, 2100 as 2.7 billion, and over 5 billion for 2250, together with the global energy use by continental region. On the basis of the assumptions outlined, it is expected that the energy use in 2250 will be 640 EJ. If fifty percent of commercial energy is produced using renewable sources, then the environmental impact will be equivalent to 320 EJ.

10.6 Global investment in the planetary infrastructure

From Chapter 5 (section 5.3), the default value for the capital cost of a new city of 100,000 population in Europe has been taken as $6 billion, or $60,000 (2000 US$) per inhabitant, inclusive of housing ($3 bn), other buildings such as industrial, office, retail, hotels, leisure, health, education, civic, and community facilities ($1.5 bn), together with infrastructure and utilities ($1.5 bn). It should be noted that national construction cost indices are calculated from a weighted value of regional construction costs. These are dependent upon local resource costs, variations in housing standards (such as differences between urban and rural housing), or the size of housing units by floor area depending upon the number of occupants, or the technology (i.e., traditional, industrialised, or precast) and single or multi-storey. International construction cost comparisons may be made by adjusting national construction costs by the Purchasing Power Parity factor for a country.

On the basis of the typical default planning standards for the housing, education, health, transport, and energy sectors given in Appendices 14, 15, and 16, it is estimated that at the year 2000, the world stock of investment in both standard and sub-standard urban and rural settlements amounted to some $90 trillion. The world's stock of productive, fixed capital is estimated to be $25 trillion, and taking into account the world investment in primary resources, livestock, transport (vehicles, rolling stock, ships, and aircraft), equipment, communications and information technology, armaments, inventories, and consumer durables, the stock of non-building assets amounts to some $60 trillion. It follows that the replacement value of the total world investment stock amounted to PPP $150 trillion in the year 2000, or PPP $25,000 of assets per capita, of which rural and urban settlements account for sixty percent of total investment.

To calculate a default value for investment stock in the future, an assessment needs to be made of the total settlement costs in 2025 and 2150 on the assumption that the urban and rural housing planning standards set out in Appendix 14 are achieved. The future urban and rural populations are determined by continental region as set out in Appendix 19 and, on the basis of the Gross Regional Product (GRP) per capita from Appendix 29, the cost of the urban and rural housing stock can be estimated. As a rule of thumb, the urban settlement costs may be taken as twice the urban housing costs, whereas a factor of 1.5 has been taken for rural settlement costs. As the proportion of the urban population is expected to increase from say sixty to ninety percent in the period from 2025 to 2150, the rural settlement factor becomes of diminishing importance. The aggregate of the urban and rural housing costs gives the total settlement cost and, as a default value, this has been taken as sixty percent of the total investment cost. These parameters can be adjusted in the light of new information and any changes to the default planning standards in the longer term.

The global economic output in 1950 was $3.8 trillion with a stock of assets of $20 trillion, and by the year 2000, the output amounted to $31 trillion (in 2000 US$) or PPP $45 trillion. Some PPP $5.5 trillion of investment needs to be added annually in the LDCs in the period 2000-2025, to deal with the backlog of investment in human settlements in the LDCs and to increase the world stock of investment to PPP $300 trillion by the year 2025. On the basis that the urban population in the world rises to 11.25 billion in 2150, together with a rural population of 1.25 billion, it is estimated that the stock of investment in urban and rural settlements will amount to PPP $480 trillion, and with productive and other investments amounting to some PPP $320 trillion, the world stock of investment in the year 2150 is estimated to be PPP $800 trillion (PPP $64,000 assets per capita). It follows that the investment required from 2025 to 2150 is PPP $500 trillion, which amounts to $4 trillion per year. For a sustainable civilisation, there will be a global upper limit to the stock of assets, and the stock in 2250 is expected to reach PPP $1,200 trillion (PPP $104,000 per capita). The increase in the capital stock from 2150-2250 will arise from an enhancement in planning standards and the physical quality of life as the less developed countries move to higher stages of development.

The global stock of liquid financial assets, which includes foreign exchange, bonds, equities, and derivatives, exceeded $40 trillion by 1994, which was twice the size of the $20 trillion GDP of the high-income OECD economies. It is expected that the liquid financial stock will grow more rapidly than the world's real economy and, by the year 2025, it will have reached $120 trillion with the less developed countries accounting for twenty percent of the total.

It took fifty years from 1950 to increase the world stock of investment by $130 trillion, and in the twenty-five years from 2000 to 2025, the increase in the stock will amount to a further PPP $150 trillion. This additional investment will need to be shared equally between the more developed and the less developed regions. However, by the year 2000, the world economy was producing sufficient annual investment to accommodate a world population of 12.5 billion by the year 2150, and new investment of $650 trillion over 150 years will deliver the development programme. It follows that at the year 2000 only twenty percent of the world development programme had been undertaken. It should be noted that sixty percent of the total stock of assets is invested in structures with a life of 100-125 years, and the remainder have an asset life of ten to twenty-five years, so that the existing stock will need replacement in addition to new investment.

The net annual addition to the global population will shortly be accommodated entirely in cities and towns, and urban expansion will arise from the natural increase in the urban population rather than migration from rural areas. The additional infrastructure for two billion extra urban residents expected over the next twenty-five years has to be funded on top of the backlog in infrastructure investment that exists in the majority of the world's cities. Alternative sources of funding include cost recovery through user charges and tariffs and the transfer of public-sector infrastructure assets to the private sector, as outlined in chapter 8. Also, private and community Building Societies, Housing Associations, and Friendly Societies can be set up to finance housing development, which accounts for some thirty percent of gross-capital formation in a number of poor countries. In most developing countries, public infrastructure has been funded through domestic taxation and borrowing, savings schemes, donor schemes, Overseas Development

Assistance, and external debt. At the peak of the urban transition, debt financing will generally be essential for 'lumpy' infrastructure investments that have long, useful lives, and the debt can be spread over the lifetime of the assets. Municipal Development Funds and Municipal Bonds are emerging in many developing countries where there is macroeconomic stability, and subnational governments can issue debt where the central government bears the underlying risk for a municipal default.

In the years 2000-2025 the investment capital generated by the more developed countries of $5 trillion per year will yield a surplus of funds by the year 2025, which could be invested in the less developed countries. However, the investment in the less developed countries at twenty percent GRP in the years 2000-2025 will be insufficient to increase the stock of assets from PPP $60 trillion to PPP $180 trillion. To fund the doubling of the urban population in the LDCs in the period 2000-2025, it is essential to transfer $375-500 bn per year from the MDCs to the less developed countries to prevent an appalling deterioration in the conditions of the world's poorest populations. This would correspond to 1.5-2% of the GNI of the MDCs at $25 trillion in 2000, which is two to three times the 0.7% target of GNI by the industrialised countries that would have provided $175 bn. This target was agreed in 1970, following the report of the Pearson Commission on International Development, and this was reaffirmed in Agenda 21 at the Rio Conference in 1992. In reality, the Official Development Assistance and Aid in 2000 amounted to a relatively insignificant $60 bn. After the year 2025, the less developed countries should be able to generate sufficient internal investment for their own growth. By financing an improvement in the quality standards of housing and human settlements in the LDCs, it will be possible to extend the life of the capital stock, which will set them on a path to sustainable development.

Global financial integration (connectivity) weakens the ability of governments to steer their economies, particularly the stimulation of weak economies with high unemployment. If a country lowers its interest rate to encourage investment, there is likely to be an outflow of capital seeking higher interest rates elsewhere. Alternatively, if a government decides to increase government spending or to cut taxes, inflationary fears and the prospect of a fall in the value of the currency may also prompt a capital outflow. For those countries with a debt to GDP ratio in excess of sixty percent, it may become necessary to increase interest rates to prevent capital outflows, and this may result in an increased deficit. A disastrous scenario would arise if a country defaulted on its debts, and with no further finance available, there could be a run on the national banks. If government budget cuts followed, there would probably be an overreaction in the stock markets with an uncontrollable downwards movement in anticipation of a deep recession. There are clearly dangers in removing capital controls and permitting free capital movement, and these are heightened by the super-connectivity of markets. A sensible contingency plan is to create the equivalent of 'firebreaks' as for forest fires and buildings, by the formation of continental economic unions that could prevent global financial collapse and deep, worldwide recessions.

The daily volume of the global foreign exchange market (FX) had reached $1.5 trillion by 2000, in relation to the trading volume of government bonds at some $200 billion a day, and the volume of deals on world's stock exchanges at around $25 billion a day. Some eighty percent of the trading on the FX is currency speculation, and the system is a threat to the international monetary system because raids are mounted on weaker currencies, such as in 1992, when the UK and Italy were forced out of the ERM (Exchange Rate Mechanism) after the combined resources of the European central banks were unable to withstand the pressure.

It has been proposed by James Tobin, the winner of the Nobel Prize for economics in 1981, that a small 0.5% tax should be imposed on foreign exchange transactions to dampen speculative international financial movements, without being a deterrent to genuine international investment commitments or commodity trade. James Tobin had two objectives when he proposed a tax on international currency transactions in 1972, which is described in *The Tobin Tax* (1996), by Ul Haq, Kaul, and Grunberg. His first objective was to make exchange rates reflect long-run fundamentals relative to short-range expectations and risks. His second objective was to preserve and promote autonomy of national macroeconomic and monetary policies. Essentially, the academic support for the Tobin Tax has been concerned with reducing volatility and the volume of transactions, rather than with the generation of revenues. They have advocated that the transactions tax should be the responsibility of the IMF, and each member country should be required as a condition of membership and borrowing privileges to levy the tax. Any transfer of funds to or from tax-free jurisdictions should be included as taxable transactions at penalty rates.

In Scenario S1 – Settlements First (Chapter 12), FX transactions are a critical leverage point and a Tobin tax of 0.5% on transactions of $1.5 trillion per day could provide a United Nations Global Capital Fund, for direct diffusion into infrastructure investments in the LDCs to level the global playing field. This should be treated as a 'user charge' for playing on the global field, in the same way that court fees are payable for using a tennis court. If the imposition of a currency transactions tax resulted in a forty percent reduction in trading volume, and a further fifteen percent was lost through tax-exempt organisations such as governments, central banks, and official international organisations, then collecting governments could retain fifteen percent of the tax for themselves and still leave thirty percent or $450 bn available for a UN Global Capital Fund. By comparison, the Stern Review advocates the expenditure of one percent GWP each year to avoid the worst impacts of climate change, which is equivalent to $450 billion in 2000 and $1 trillion in 2050. The collecting governments could invest their fifteen percent in renewable energy sources and towards a reduction in the infrastructure disparities between their successful and peripheral regions. Estimates of the investment stock requirements in PPP $ (Purchasing Power Parity) are shown in Appendix 30, with default values for Gross Regional Product (GRP) by continental region that are based upon investment transfers from the MDCs to the LDCs. This would aid the poorest LDCs to achieve minimum per capita GDPs of $1,001- $2,500 (2000 US$). By increasing the standards of living in the LDCs through infrastructure investment, a demographic transition will be induced and a rise in wage levels will expand the overall market size, so that excess industrial capacity can be utilised to generate global employment.

Between 1970 and 2000, overall tax revenues for OECD countries rose from thirty to forty percent of GDP, and with the ageing of the developed world's population, the tax revenues would need to approach fifty percent of GDP, unless entitlement cuts are made in the social expenditure budgets (i.e., pensions and unemployment benefits). After the year 2025, there may be a reduction in the international supply of capital from the effect of three factors that include ageing populations and rising dependency ratios on household savings, the age of retirement from the workforce, and the coverage provided by social security benefits. In Asian countries, including India and China, the number of people over sixty-five continues to climb and, in the fifty-year period from 1975-2025, the average age of the Chinese population will have risen from twenty to forty. Also in the period from 1960 to 1995, there has been an increasing trend in Europe for men to take early retirement at age fifty-five, so that the proportion of men in the age group sixty to sixty-four that remain in the workforce has fallen from around seventy-five percent to twenty-five percent.

A long-term historical analysis of the world economy, as described in *The Age of Transition* (1996) by Terence Hopkins and Immanuel Wallerstein et al, shows that there are alternate periods of economic expansion or capital accumulation and periods of stagnation. World supply is generated by the sum of national, corporate, and individual decisions, which are motivated towards increasing total production and investment capital (profits). World demand is dependent upon the political situation in the various nation-states and upon the aggregate of the civil decisions concerning the distribution of the surplus. When the expansion phase ends in overproduction, and a tight world market follows, in order to survive the effects of the recession, the main approaches are cost cutting, capacity reduction (down-sizing), and debt creation. Cost cutting generally involves wage reduction by negotiation, plant relocation to areas with lower labour costs, and erosion of wages via inflation. Capacity reduction involves plant closures, reduced production volumes, and under-utilised capacity with consequential high unemployment. In the downward phase of the economy, there is enormous pressure from civil society for a change in the existing rules that benefit the most privileged within both the global system and within national boundaries. Reforms in the civil system may permit redistribution of income to the poorer groups, which may then lift the world demand curve and, when combined with an investment wave involving technology that has high utility, an upward cycle commences.

Emerging countries offer increasing returns from infrastructure and industrial investments, in the form of decreasing costs per unit of service or output, since expenditure does not increase proportionately with population or demand in the earlier stages of development. Consequently, the lifting of restrictions on foreign investments required from LDCs under loan conditions of the World Bank or the Structural Adjustment Programs of the International Monetary Fund, has resulted in a substantial increase in private investment from the MDCs to wherever yields are highest. There is however a qualitative difference between loans that can lead to negative net transfers between the LDCs and the MDCs if interest rates rise, portfolio investments that may be short-term and unstable, and foreign direct investment with technology transfer that sometimes causes an unwelcome exchange rate appreciation. Therefore, unleashing global capitalism without regional investment regulation could perpetuate a form of colonialism and uneven development in which the LDCs are locked-in to provide cheap sources of labour and raw materials, and exploited as an ever expanding market for technology and services from the MDCs.

Increased investment in the MDCs runs into diminishing returns, which drives down the available return on investment and reduces the capability of the MDCs to provide adequately for retirees. In the slower growing economies of the more developed countries, such as the USA and the UK, pension funds are likely to achieve diminishing returns from domestic investments. For example, between 1955 and 1980, the real profits of the Fortune 500 industrial firms increased by a factor of 2.5 from growing sales, but from 1980 to 1990, economies of scale no longer provided advantages to firms, and the trend was reversed with real profits declining by a factor of 2 on sales of $2.5 trillion. Between 1970 and 1990, profits as a percentage of sales declined from six to three percent. This generates an investment surplus for placement in the global capital markets and for channelling into high return investments in other international centres and the LDCs.

The world cities respond by further property development with large dealing floors and further investment in on-line information systems and continued growth in the quaternary information division. In the global economy, opportunities will exist for portfolio equity investments (i.e., shares of stock purchased as a financial investment, rather than as direct investment) in the transitional or less developed countries. Securities have an advantage for the holder of having a liquid and tradable asset, compared with loans that may be relatively illiquid, such as the physical assets of a manufacturing company.

Shaping future world development: Chronological references

1. *The Future of the World Economy* - Wassily Leontief, 1977.
2. *Ecological Economics* - Edited by Robert Costanza, 1991.
3. *World without End* - David W Pearce & Jeremy Warford, 1993.
4. *Valuing the Earth* - Edited by Herman E Daly and Kenneth N Townsend, 1993.
5. *Visions of Sustainability* – Edited by Doug McKenzie-Mohr & Michael Marien, 1994.
6. *A Survey of Ecological Economics* - Edited by Rajaram Krishnan, Jonathan M Harris & Neva R Goodwin, 1995.
7. *The Economics of Sustainable Development* - Edited by Ian Goldin & Alan Winters,1995.
8. *The United Nations at Fifty* – Harlan Cleveland, Hazel Henderson & Inge Kaul, 1995.
9. *Europe's Environment* - European Environment Agency, 1995.
10. *How Many People Can the Earth Support?* - Joel E Cohen, 1995.
11. *OECD Economies at a Glance* - OECD, 1996.
12. *The Tobin Tax* - Edited by Mahub Ul Haq, Inge Kaul & Isabelle Grunberg, 1996.
13. *The Age of Transition* – Terence Hopkins and Immanuel Wallerstein et al, 1996.
14. *The Ultimate Resource 2* - Julian L Simon, 1996.
15. *Market Unbound* - Lowell Bryan & Diana Farrell, 1996.
16. *Resource Flows* - World Resources Institute (WRI), 1997.
17. *Factor Four* - Ernst Von Weizsacker, Amory B Lovins & L Hunter Lovins, 1997.
18. *The Great Population Spike and After* - Walt W Rostow, 1998.
19. *Land Resource Potential and Constraints at Regional and Country Levels* - United Nations (FAO), 2000.

20. *The Challenge of Slums*: Global Report on Human Settlements 2003 - UNCHS, 2003.

21. *The Real Environmental Crisis* – Jack M Hollander, 2003.

22. *World Population to 2300* – United Nations, 2004.

23. *Financing Urban Shelter: Global Report on Human Settlements 2005* - UN-HABITAT, 2005.

24. *The End of Poverty* – Jeffrey Sachs, 2005.

25. *Planet of Slums* – Mike Davis, 2006.

11.
THE GEOPOLITICAL MACROSTRUCTURE

11.1 Geopolitical perspectives

In the rapidly changing world of the past fifty years, new states have been formed by the break-up of the European colonies in Africa, South and South-East Asia, the Pacific and the Caribbean, and more recently from the break-up of the former Soviet Union, Yugoslavia, Czechoslovakia, and Ethiopia. The United Nations now officially recognises nearly 200 countries or nation-states, compared with only seventy-two states and colonies in 1945, although these states assert sovereignty over some 5,000 ethnic groups whose occupation of their homeland predates the birth of the state. When an ethnic group seeks self-determination, they are generally labelled as rebels, separatists, extremists, terrorists, or insurgents, and some two-thirds of states use military force or intimidation and human rights violations against people they claim as citizens. State-directed genocide has caused over 70m deaths since 1945, high levels of civilian casualties, and an estimated 36m refugees (of which 18m are officially recognised by UNCHR). Since 1991, the General Assembly of the United Nations has accepted the principle that it is the duty of the international community to intervene for humanitarian reasons to protect minorities threatened by starvation, repression, or genocide.

In the 1980s, world military expenditure amounted to $1 trillion every year, which constituted approximately five percent of Gross World Product. Some eighty-five percent of this expenditure was in the more developed countries with the remaining fifteen percent in developing countries. Over one-fifth of the military expenditure in the less developed countries was spent on the import of arms from the MDCs. However, by 2000, defence spending in both the MDCs and LDCs had been reduced to 2.3% of GNI, so that world defence expenditure had declined to 2.3% of GWP at $720 billion, with 21m men under arms as a result of the ending of the Cold War. The number of nuclear warheads has declined from a peak of nearly 70,000 in 1986 to 31,500 in 2000. In the Middle East, military expenditure generally exceeds seven percent of GNI, and this is justified by the events leading to the Gulf War. According to the Stockholm International Peace Research Institute, in 2000, the US defence expenditure amounted to 3.1% of GNI and in the UK it was 2.5%, compared with 1.8% for the European Union. It should be noted that the armed forces of the developing countries amount to over 15m troops compared with 6m in the more developed countries. The more developed countries export $8 billion of conventional weapons to developing countries each year, and a tax on arms sales could have contributed to the cost of UN peacekeeping operations of $2.9 billion in 2000.

Major conflicts since the end of the Second World War include the Korean War (1950-53), the Vietnam War (1957-75), the Arab-Israeli wars (1967, 1973, and 1982), the Iran-Iraq war (1980-89), the First Gulf War (1990-91), and the Second Gulf War or The Iraq War (2003-present). If war is defined as a conflict in which over 1000 people die, since 1950 there have been at least 144 wars, in which 25m people have been killed directly in the fighting. In the fifty years between 1950 and 2000, international wars accounted for 9m deaths (180,000 per year) compared with 16m killed in civil wars (320,000 per year). Since only twenty-five percent of wars result in over 100,000 deaths, it follows that some 5m died in 100 small wars with tens of thousands dead, and 20m died in thirty-nine large wars in which an average of 500,000 were killed in each war. Whereas civilians accounted for fifty-five percent of all war-related deaths in the 1950s, the civilian proportion of deaths had increased to some eighty-five percent in the 1990s, which reflects the changing technology of warfare. Although 25m have been directly killed in the fighting since 1950, some 40m have died in total if those affected by war-related famine or illness are included. UNICEF claims that, during the last decade alone, 2m children have died in civil wars, which are now killing more children than soldiers. Developing countries have carried the burden of over ninety percent of all the conflicts and the casualties. A useful reference is *The State of War and Peace Atlas* (1997) by Dan Smith.

The nations in which over 1m have died in wars since 1945 include Ethiopia, Algeria, Nigeria, Bangladesh, Korea, Vietnam, China, and the Democratic Republic of Congo. The nations with over 100,000 deaths include Angola,

Kenya, Iran, Iraq, Afghanistan, India, Malaysia, and Colombia. Whilst many of the wars have been fought to achieve an end to colonial rule, such as in Algeria, often civil wars have been fuelled by the superpowers as armourers, advisors, and sponsors, and international wars such as Korea and Vietnam have been waged against communism in the north. At the start of the 21st century, the number of wars and armed conflicts are on the decline since peaking in the early 1990s.

During the 21st century, 'Asymmetric Warfare' is the emerging threat to western security. Adaptive enemies do not attempt to put up a symmetrical challenge to advanced conventional technologies, such as precision guided missiles, ground- and space-based navigation systems, surveillance technologies, and communications systems. The objective of these enemies is to inflict sufficient casualties on their adversaries and to extend a conflict indefinitely by avoiding defeat and achieving a stalemate, in order to erode political support for the campaign. These forces counter technological superiority and survive intensive firepower attacks with low-tech counter measures involving dispersal of not only forces, but telecommunications, logistics, and transport infrastructures, together with the use of decoy targets and camouflage. Because of public sensitivity to the human cost of military operations, adversaries threaten to inflict mass casualties by means of chemical, biological, or radiological (CBR) weapons and thereby deter or limit military interventions. CBR weapons can be used against civilian targets by terrorist cells with low-tech delivery systems such as trucks, marine craft, and aircraft to achieve both disruption and destruction.

The geopolitical perspectives for the 21st century indicate that there will be an increasing number of international organisations of all types, and the restructuring of GATT as the World Trade Organisation demonstrates the pressure towards global free trade. As a consequence of this, a new protectionism is developing with multinational economic alliances that have been termed continental unions. Also, it is recognised that the military strength of a continental union is likely to maintain peace in the long term because of the massive deterrence offered by an economy with a gross domestic product exceeding $5 trillion. It is envisaged that it will be the geopolitical considerations and cultural issues that will determine the configuration of borders in the future, and the sovereign state is not facing imminent demise in small and medium sized countries. It is the large and sub-continental economies, particularly in India and China, in which regional differentiation is likely to emerge, and a large number of states will be identified in the future. *The State of the World Atlas* (2003) by Dan Smith provides a unique survey of current events and global trade.

Multi-level governance of the world system is likely to take place at three levels: United Nations, continental union, and the state. The national level may coincide with a sub-continental union level such as the United States, or may correspond to the level of the state for countries with less than 20m population. However, for some fifty countries with larger populations, the possibility exists for formally recognising national regions that may emerge in due course as ecopolitan states. At each level of governance, the aim will be to achieve economic efficiency, social equity, and environmental sustainability, with the upper levels providing the framework of regulations within which the lower levels must operate. In this way, decisions may be made and expenditures balanced at the international, continental, and state levels.

It is envisaged that the United Nations would be responsible for global policies and regulation in respect of international law, the international monetary system, international peacekeeping, and security. It would also continue to supervise aspects such as population control, human rights, refugees, planetary hazards and sustainability, the biosphere, the oceans, and world food and agriculture. UN organisations would continue with the role of monitoring world health, world trade, world communications, and intercontinental transport (maritime and aviation). A possible objective would be to introduce an international system of taxation so that the disparity between the average standards of living of people in the various continental regions of the world can be reduced by the provision of infrastructure assets in the less developed continents.

The continental unions would deal with continental policies and regulation, the continental monetary system, and continental security. The union would be responsible for regional economic infrastructure, environmental regulation, continental borders, immigration and trade, interstate transport and communications, industry and the workforce, and agriculture and rural development. Continental unions would also be responsible for identifying any threats to future societal evolution and for the design of equitable geopolitical, economic, social, technological, and ecological policies to minimise the possibility of a system collapse. The aim of a continental economic union is to achieve sufficient economies of scale so that most economic resources and production conditions can be found within its borders, thereby creating extensive internal trade and reducing the need for international trade and investment. The efficient firms become more willing to invest in new technology to take the share of the market held by the inefficient firms. The cost reduction from intra-union trade allows prices to be lowered, which generates an overall increase in trade. In due course, an economic union can be widened by the inclusion of more countries and deepened by monetary union. In view of the fear of 'social dumping' within an economic union, deepening requires labour market policies with binding norms to protect working conditions, health and safety measures, and welfare payments. These issues are surveyed in *Economic Integration Worldwide* (1997), edited by Ali M. El-Agraa.

The state level would establish state policies and regulations in the context of the continental rules and would be responsible for state finances and taxation. Social infrastructure would be the prime responsibility of the state, and this would include housing, health, education, public services, and urban transport. Urban and rural development, land systems, environmental protection, local economic development, employment, tourism, and social security would also be controlled at the level of the ecopolitan state. It would be a principal objective that public participation and the civil society would be involved in the decision making at the state, municipal, and community levels.

There is an increasing change in the balance of power and distribution of functions between national and local governments. Decentralisation policies of different kinds have been or are being implemented in most countries including those that formally had centralised structures. Some eighty-five percent of countries with populations of 5 million or more have been engaged in some form of transfer of power from national to local levels of government. This involves some devolution of power and functions, with delegation to parastatal organisations and the privatisation of state-controlled organisations.

Simple societies are comparatively small, indigenous communities with a common culture, which occupy homeland territories. These societies may have a sufficient size of population, territory, and resources to remain self-sufficient and stable. However, societies with small populations and limited territory derive additional benefits from increasing their complexity to that of a city state, which retains the primary characteristic of territorial integrity. It also has the advantage of economic and cultural heterogeneity, a sufficient tax base for civil government, a legal system, and a central administration to organise public works and forces for internal order and defence. Civilisation is essentially a learning process that adapts to a political, social, cultural, environmental, and technological stage of development, but with different languages, values, religions, and belief systems. Diversity arises from long-term progressive adaption, and the diversity of civilisations within the continental regions of the world is needed to provide resilience to the external changes and shocks in the evolving urban system. Geographic races include Caucasoid, Negroid, Asiatic or Mongoloid, Indic, Amerindian, Melanesian, Polynesian, Micronesian, and Australoid.

There is a high level of global language diversity with some 6,000 living languages, although there are only 105 major state languages (thirty-five European, ten Central Asian, fifteen South Asian, forty East Asian, and five African). Only 250 languages are spoken by more than 1 million people. Although the official tongue of a country is a key factor in terms of the maintenance and spread of a language, colonialism has resulted in more than half the states having either English, French, or Spanish as an official language. These languages transcend national boundaries to access diverse fields worldwide such as international government, commerce, science and technology, information technology, mass entertainment, and sport. Major regional link languages used by half the world's population comprise English (500m users including as a second language), Chinese (Mandarin 670m), Urdu (630m), Hindi (400m), Spanish (250m), Arabic (170m), Malay-Indonesian (165m), Russian (160m), Bengali (150m), Portuguese (140m), Swahili (130m in the East African states), Japanese (130m), French (100m including second language users) and German (85m).

Religion has provided a major stimulus for language standardisation and, for example, Arabic is an important international language for integration of the Islamic world. The major religions of the world include Roman Catholicism (900m), other western Christians (560m), eastern Christians (160m), Islam (850m), Hinduism (650m), Buddhism (310m), and Chinese religion. There is the prospect that, by the year 2100, the number of spoken languages will decline to 1,000-2,500. By the year 2500, the major civilisations of the world may converge towards perhaps 100 continental cultures, which would reflect the predominant language family and religion.

11.2 Ecopolitan and agropolitan states

Governance systems face two counterbalancing forces, which are globalisation with continental integration of nations and localisation with the devolution of power to states at the subnational level. The formation of the World Trade Organisation (WTO), the European Union, and the devolution of power to the Scottish Assembly in the United Kingdom are examples that operate at the different hierarchical levels and are being closely studied by regional trading blocs throughout the world. The primary objective of localisation is to enhance political stability where a country is divided along geographical or ethnic lines, and decentralisation provides an institutional mechanism for establishing a formal or rule-bound negotiating process between opposing groups. A secondary objective is to provide a standard level of public services on a nationwide or continental basis, and this can be achieved through regional equalisation grants for infrastructure provision at a subnational level. Finally, local government is better placed to target poverty and to achieve social equity through a redistribution process, which involves the provision of education, health, and social safety nets, although central government will be responsible for much of the funding. By decentralising government so that decisions are made at subnational levels, localisation brings increased citizen participation in decision making so that people have more of a chance to shape the context of their own lives.

The emergence of ecopolitan states as basic geopolitical units or bounded landscapes with evolutionary potential is a necessary construct to enable *Long-Range Future Research* to encompass geopolitical, social, and environmental issues that affect the contextual macrostructure. The term *ecopolitan* is formed from its ancient Greek roots 'eco' as used in ecology and economy to mean the sustainable use or, literally, the housekeeping (oikos is the Greek for house) of natural and human resources. The word 'polis' is used with its wider meaning of a city-state with socially equitable policies towards all citizens. An ecopolitan state is also an evolutionary vision to empower civil society in its demands for policies that strike a balance between economic efficiency, social equity, and environmental sustainability. An ecopolitan state is a prospective, sustainable, city region of 5m-20m inhabitants in which human society can evolve to its full potential at a reasonable per capita standard. Ecopolitan states will form the basic units of civilisations, in so far as they have the potential for independence and stability. Nation-states may be expected to accommodate sub-states or smaller states, which are likely to evolve into ecopolitan states. The majority of existing nation-states have been created by the consolidation of occupied territories, when a more powerful nation has formed a colony or an empire. A chapter on "Global Ecopolitics" by Phyllis Mofson is included in an excellent publication *Reordering the World – Geopolitical Perspectives on the 21st Century* (1994), edited by George Demko and William Wood.

The term 'ecopolitan state' has been selected to avoid 'lock-in' to the existing terms for describing territorial entities such as empires, sub-continents, colonies, countries, nations, states, regions, and provinces. Whilst the international importance of nation-states is declining, more countries are being created, and there are on-going negotiations for regional trading blocs and the formation of continental unions. It is envisaged that ecopolitan states will form the building blocks of civilisations, and they will reach a sufficient level of complexity to achieve economic specialisation and economies of scale for a niche within a continental union. Successive generations of governments will modify and rearrange the building blocks in the light of experience, or new alliances may be formed in line with trading agreements or events such as an invasion. History has shown that civilisations rise and decline as natural phases in the evolution of humankind, and continental unions may decompose back to the basic building blocks of ecopolitan states for recombination into different groupings with an emerging planetary civilisation. For the evolution of a complex adaptive system, the building blocks for larger groupings should ideally comprise populations that are socially coherent, with clusters of establishments that are stable and can achieve economies of agglomeration.

Ecopolitan states will emerge by 2025, and a significant number of them will correspond to nation-states, such as Austria, Belgium, Denmark, Finland, Hungary, Norway, Sweden, or Switzerland (during the period 1995-1997 Switzerland achieved a GNI per capita of $40,000, although that has since declined to $38,000). Similarly, Scotland, with a population of over 5m, would be an example of an ecopolitan state that was part of a larger nation-state. In the USA, the fifty individual states would also correspond to ecopolitan states, and in accordance with the slogan "think globally act locally," each state would be responsible for cleaning up its act to meet the economic, social, and environmental challenges of the future. For example, the State of California has had a policy since 2003 that zero emission vehicles should make up ten percent of car sales in the state, and in this way it generates a market for alternative fuels. In the large sub-continental economies such as China and India, regional states could be re-designated into around 300 prospective ecopolitan states.

Intergovernmental action on carbon emissions such as the Kyoto Protocol and the European Union Emission Trading Scheme will assist in a reduction in carbon emissions. Governmental action to encourage a shift to low-carbon energy technologies will involve the introduction of 'ecotaxes', such as a tax on carbon emissions of, say, $100 per metric ton of carbon, which in the case of the UK at 2.5 tons of carbon per capita (9.2 tons of carbon dioxide per capita) would yield $15 billion, or one percent of GNI. In due course, ecotaxes will diminish as they achieve their objective in reducing wasteful energy use and effective decarbonisation to 1.5 tons of carbon per capita during the ecopolitan stage of development, but this saving to the public will be offset by the higher cost of hydrogen energy and renewable sources. A system of tradable 'eco-permits' would be an equitable and efficient way of allocating a share of the permissible carbon dioxide emissions. Trade penalties can be imposed against countries to ensure environmental compliance, and there would be an incentive for corporations to produce clean technologies to gain a share of export markets and to avoid fines at their plant locations. It may be expected that Spot, Futures, and Options markets would develop for eco-permits.

Eco-permits could be issued annually on a continental basis of 150 tons of carbon dioxide per km^2, rather than by GNI, which would not be globally equitable, or by population, which would provide the wrong incentives for unsustainable over-populated countries. One option would be for the allocation to be shared between six super-continental regions, each of 20m-25m km^2, such as North America, South and Central America, Western and Eastern Europe, Pacific (with East Asia and Oceania), South and Central Asia (with the Middle East and North Africa), and Sub-Saharan Africa. Within these super-continental regions, the internal allocation would be by ecopolitan state rather than by population. Regions such as Latin America or Sub-Saharan Africa, with low demand in relation to the quota, could auction their surplus eco-permits to countries with an excess demand with a reserve price of say PPP US$100 per metric ton of carbon.

Ecopolitan policies involve taking action to mitigate global warming through energy efficiency, shifting to low-carbon energy technologies such as hydrogen fuels, biofuels, and renewable sources, and carbon capture at power plants and storage in deep geological formations, deep oceans, or as mineral carbonates. For example, BP plans to develop a 350 MW carbon capture and storage (CCS) plant in Scotland where the carbon from a natural-gas fired power plant is extracted and pumped into the Miller field in the North Sea. Also geo-engineering of the environment could involve large-scale engineering countermeasures or other interventions in climate change, so that humankind has sufficient time to adapt to global warming. Options include increasing carbon sequestration by the oceans using iron fertilisation to increase phytoplankton populations that will draw more carbon dioxide, or by reforestation and afforestion using carbon off-set schemes. The theme of the Human Development Report 2007/2008 is *Fighting Climate Change: Human Solidarity in a Divided World*.

Comparable benefits can be accomplished by extending the lifecycle of fossil fuels by altering the energy mix to reduce the rate of hydrocarbon consumption. If the rate of rise in sea levels can be limited to two metres per millennium, the civil system can adapt by relocating settlements to higher ground, installing flood control structures, and developing irrigation schemes for the dryer regions to ensure the supply of food for the world's population. Geo-engineering to increase the earth's albedo would involve sunlight screening by enhancing low-level clouds or by releasing reflecting micro-balloons into the stratosphere.

One of the most important problems to resolve is the design of an approach for LDCs to achieve the circumstances in which human society can evolve to its full potential with a sufficient per capita standard of living. In particular, it is essential that the LDCs receive investment transfers from the MDCs to achieve the infrastructural stage of development with a per capita GNI of $1,001-$2,500 (2000 US$). The term 'agropolitan' state has been used to differentiate this sub-set of ecopolitan states from the more developed ones. An approach has been outlined in which human society can evolve to its full potential with a sufficient per capita standard of living and without repeating some of the capital intensive and environmentally unsustainable investments made in the MDCs. The essential investment in human capital with high educational and health standards will need to be achieved, with acceptable and rising standards in terms of life expectancy, adult literacy, fertility rates, nutrition, pollution, poverty levels, social equity, and political democracy. Planning standards have been developed for infrastructure investments in the LDCs.

There are sixty-seven countries of over 1m population with a per capita GNI of less than $1,001. In general, these countries have between thirty to thirty-five percent urban population, and sixty percent of their workforces are involved in agriculture or primary resources. At the infrastructural stage, some fifty-five to sixty percent of the population becomes urban, and the proportion of the workforce in agriculture and primary resources declines to thirty percent. A development objective is to avoid the unnecessary process of creating concentrated metropolises with subsequent deconcentration and dispersal. An appropriate urbanisation strategy will be to form urban and district centres and local service centres to serve traditional village communities, with infrastructure investments directed towards the connection of agropolitan districts and states to diminish the advantage of large cities for the attraction of industrial and commercial enterprises. It has already been demonstrated that industry decentralises from metropolitan centres in the MDCs, and capital can be saved by locating them at appropriately-sized centres at the outset. Also, investment needs to be directed towards rural development and areas in which the poor may become land owners and proprietors of small shops and businesses.

The development of compact and sustainable ecostructures, with realistic per capita investment costs, needs to be used effectively by extending the opening hours of facilities and generating additional part-time employment. The priority is for investment in human development to reduce adult illiteracy to below twenty-five percent, to ensure that sixty percent of children of secondary school age attend, and to enable twenty percent of the higher education age group to continue their studies. It is envisaged that there will be four persons per dwelling, and the aim is to ensure that seventy percent of dwellings are connected to water supply and sewerage systems. By improving nutrition to 2500 calories per day per person and reducing infant mortality to thirty per 1000 live births, life expectancy could reach sixty-eight years if there are also twenty doctors and twenty-five hospital beds per 10,000 population.

Energy consumption will rise to 50 GJ per capita, and it is envisaged that the provision of 45 km of road per 10,000 population would permit the growth of a bus transport system, supplemented by seven cars per 100 population. Railways may connect major centres with 2.5 km per 10,000 population, and there will be a need for fifteen percent telephone land-line connections. Encouragement should be given to the necessary institutions to ensure social equity and the availability of finance for small-scale enterprises. There may need to be state regulation of rental levels that can be charged by landlords, interest rates payable to money lenders, and agricultural prices to prevent the rural population from being taken advantage of by the city dwellers. Development of consumer purchasing power and the expansion of domestic markets will enable enterprises to produce an increasing range of artefacts and equipment.

The economy of an agropolitan state will need to be strengthened through the diversification of agricultural production, an increasing variety of industrial enterprises in both urban and rural areas, and a range of commercial and distribution services. Import-substituting industries will permit the economy to evolve. Increasingly, communities will be able to adopt sustainable ways of exploiting local resources, through the development of durable products with appropriate technology. The preservation of both the natural environment and indigenous cultures should enable agropolitan states to flourish within the protection of a continental union.

An Ecopolitan Index (EI) could be developed along the lines of the UNCHS (Habitat) *City Development Index*, to measure the extent to which an ecopolitan state achieves the criteria for efficiency, equity, and sustainability. The norms for an ecopolitan society will aim to reduce social disparities and include the following:

- Peaceful security without an arms race
- Emphasis on collaboration rather than competition
- Provision of justice and equity
- Support of the poorest in other countries
- Enhancement of education and public health
- Caring for future generations
- Protection of natural systems
- Acceptance of a global bioethic
- Preservation of other species and the prevention of suffering

11.3 Continental unions

Since fifty percent of countries have populations of less than 5m, there is a need for regional economic integration and, eventually, the world may comprise some twenty to thirty continental blocs with populations in the range 400m-600m. Continental unions of ecopolitan states will be formed to ensure that fair trade takes place between countries at similar stages of development. Within each continental bloc, there may be perhaps fifty ecopolitan states with populations ranging from 5m to 20m (average 10m inhabitants). Larger nation-states may be expected to accommodate regional sub-states or ecopolitan states, which, in many cases, will correspond to the size of smaller nation-states. Appendix 33 gives a summary of the number of countries by size-band that make up the world's populations, and it is estimated that the number of potential ecopolitan states by the year 2150 will be 1,000.

The spatial dimensions of sustainable-development processes range from studies of the global system, via the continental, to the regional and national scale. It should be noted that international or interregional trade allows one country or region to draw on the ecological carrying capacity of another, and they may be unsustainable in isolation, even though they are sustainable as part of a larger trading bloc. When continental unions face stagnation or diminishing returns, they will have to respond as complex adaptive systems and revise or rearrange their building blocks of ecopolitan states.

The creation of continental unions provides the conditions that make countries invest in other continental unions, to circumvent protective barriers for access to their markets, and to avoid long distance trade with the environmental costs of transport. If the MDCs invest in the LDCs by exporting capital, then the LDCs will have the opportunity to import plant and equipment for major infrastructure investments, such as clean technologies for power generation, airports, transport, and telecommunications networks. The driving force behind a redistribution of investment will be the diminishing returns in the saturated markets of the MDCs. It is envisaged that capital flows will become more important than trade for determining the reciprocity between trading blocs, and developing countries will achieve global competitiveness through the investments in infrastructure and industrial technology. Since continental unions are protective against low-cost imports, investment transfers would compensate less developed countries for the effect of any loss of employment.

National or continental economic unions of ecopolitan states may achieve competitive advantage and further economies of scale through market size and the protection of trade, resource sharing, continental infrastructure, and international security up to the point that there are diminishing returns from the control of immigration over longer frontiers and backward regions to support. Continental unions of 5m-10m km² would seem to be appropriate to the current range and speed of transport and military technology. However, as civilisations or empires exceed a certain size, they become too complex and politically unstable, and parts of the empire break away until a smaller and more stable continental grouping forms. History has shown that civilisations and empires rise and decline as natural phases in the evolution of humankind, and continental unions may decompose back to the basic units of ecopolitan states for recombining into different groups with an emerging planetary civilisation. There is also the possibility that there could be a succession of smaller, efficient ecopolitan states that wish to break away from larger, more stagnant states in the belief that they would be better off as independent or in combination with another sub-state.

An agenda for adapting the global regulatory framework faces a variety of competing proposals for enhancing evolutionary potential. The United States would like to impose free-market principles in accordance with an American ideal of individual rights and liberty. The Washington Consensus promotes the vision that global free trade leads to global economic growth, which creates more jobs and makes companies more competitive, with lower prices for consumers. It also provides LDCs with the opportunity to develop through infusions of foreign capital and technology, and by spreading prosperity it creates the conditions in which democracies and a respect for human rights may flourish. The experience of Latin America reflects the limitations of US policies to achieve both social equity and political stability, and the Mexican financial crises of 1982 and 1995 are an important lesson. Nobel Prizewinner (2001) Joseph Stiglitz has published an important book, *Globalization and its Discontents* (2002), in which he exposes how the IMF and WTO have imposed crippling economic policies on developing nations and how the West has driven the global agenda to further its own financial interests.

The European Union intends to achieve social cohesion as it implements economic and monetary union. The Japanese believe in state intervention to enable it to build up regional and trade networks to achieve long-term development objectives, without regard to short-term efficiency criteria predicated by existing prices. China and India have moved on from state-planned economies by opening sectors to foreign capital, where the emphasis is on the creation of dynamic regional growth zones. The Islamic states are concentrating on national self-reliance and human development, and the oil producers have invested in the region as well as providing assistance to south Asia, which provides an immigrant workforce. The civil wars and turmoil in Africa are explained by complexity science, since the greater the potential difference between the parts of the global system, the greater the atrophy and disorder that will prevail in the territories of least potential. This chronic condition is not curable by a global overdose of free-market economics; instead, planetary cooperation is needed to fund the essential infrastructure so that Africans can also evolve to their full potential.

The creation of continental free trade areas such as the European Union, NAFTA (North American Free Trade Area), and ASEAN (Association of South East Asian Nations) is a necessary intermediate step to constrain global capitalism and world trade, to allow for the wide disparity in labour rates, employment conditions, and environmental standards between the high-income countries and the less developed ones. Continental trading areas will have highly competitive internal markets and will have the option of deciding the nature and type of bilateral agreements with other trading areas to provide sufficient protection for their own civilisations, although political and economic union may follow. A listing of all the regional trade blocs is given in Appendix 34. The higher authority of the continental union will undertake functions that the lower levels could not perform independently. It will take responsibility for the future by ensuring that an adaptable system is designed so that any threats that could cause a system collapse would be identified and appropriate policies could be developed to prevent it. The role of the continental union is to ensure that fair trade is conducted between countries at similar stages of development and to limit the path-dependant advantages of the MDCs in perpetuating the disadvantage of the LDCs. The reciprocal benefit is that wage levels in the MDCs will not be driven down to those of the LDCs (see Chapter 7), and social order can be maintained at the level of the ecopolitan state. This explains the emergence of economic regionalism as a defence against global capitalism. William Greider has written a powerful book: *One world Ready or Not – The Manic Logic of Global Capitalism*.

11.4 Continental futures

In this section, characteristics of each of the continental regions are highlighted for the year 2000 to illustrate some of the demographic, cultural, social, environmental, and economic differences between them, with indications of the significant changes that may be expected by the years 2025 and 2150.

North America accounted for 33% of the Gross World Product in the year 2000, and this may fall to perhaps 24% by the year 2025. The United States is experiencing population growth at an annual rate of 1%, largely as a result of immigration and higher birth rates amongst immigrants. During the 1980s, some 8m legal immigrants and 2m illegal immigrants entered the USA, principally from Asia and Latin America. By the year 2050, it remains possible that the majority of Americans will be non-Latino whites, although a number of major cities have an American black majority. In cities such as Los Angeles, the blacks are increasingly competing with the newer immigrants, both politically and economically, which is a major source of both cultural diversity and tension. The ageing of the population has major implications, and in the year 2000, some 16% of the population was over sixty years old. This is expected to increase to 24% by the year 2025, with consequent implications for social expenditure.

The population density in North America is 15.8 inhabitants per km², taking into account the low densities in Canada and Greenland, whereas in the United States it is 26.7 inhabitants per km². Some 50% of the urban population lives in cities of over 1m inhabitants, and the restructuring of the economy has resulted in economic decline in the nation's older industrial cities and a shift of capital to the high-technology sectors in the sunbelt. Whilst the older states of the north-east (the rust belt) have suffered a long-term population movement to the south and west (the sunbelt), the four states with the highest per capita incomes remain in the north-east (Connecticut, New Jersey, Maryland, and Massachusetts), whereas the four poorest states (Mississippi, Arkansas, Alabama and Louisiana) are in the south. Also, for the first time, California is experiencing a significant population outflow to the neighbouring states of Arizona, Nevada, Oregon, and Washington. The structural changes to an informational economy have resulted in an urban underclass in a society where the lowest 40% of households receive only 16% of the income, and the ratio of the highest 20% to the lowest 20% is a factor of nine. Inevitably, a high crime rate and a cause for concern is the size of the prison population, which is over eight times larger per capita than in Western Europe. In spite of the wide disparity in levels of income in the USA, the average standard of living is one of the highest in the world, which is reflected in the residential space per person, the high level of car ownership at 480 per 1,000 population, and the resulting levels of energy consumption. The environmental impact is that the emissions of carbon dioxide are 20 metric tons per capita compared with an average of 8 tons per capita for Europe.

Since 1945, the United States has been the most powerful nation in the world and, through its generous assistance with the rebuilding of Europe after the war, it has seen the US share of world output decline from fifty to thirty percent as Europe and Japan have developed their economies to challenge the American supremacy. In the post-Cold War world, the opportunity exists for the United States to concentrate on the domestic social issues by controlling excessive consumption and the wasteful use of resources, investing in the regeneration of its cities, and facilitating a redistribution of income to reduce the inequality that exists in order to restore the American dream. In the long run, it is anticipated that it will be tough for the USA to compete in the Asian markets, and that by the year 2025, the USA will have aligned itself much more closely with the Latin American countries. The combined share of world output from the North and South American regions would be 35% by the year 2025 and declining to perhaps 25% by the 22nd century.

Latin America and the Caribbean had a combined population of 515m by the year 2000, with an anticipated annual growth of 1.4%, and the region accounts for nearly 6% of world output. It is envisaged that by the year 2025, the continent will account for 10% of world output, which should increase to perhaps 15% by the 22nd century. The population density is 26.8 persons per km², which is similar to the USA, and some 44% of the urban population live in cities of over 1m. There are wide disparities in income levels in Latin America, with the lowest 40% of households receiving only 12% of the income, and the ratio of the highest 20% to the lowest 20% exceeds a factor of fifteen and indicates the greatest inequality in the world. Crime rates are also one of the highest in the world, second only to Africa, and for example, urban life in Brazil is violent with increasing incidence of armed robbery and drug-related crime, and death-squads target street children. During the 1980s, apart from Jamaica, Colombia, and Chile, all the countries in the Latin American region suffered a decline in per capita income as a result of the structural adjustment programmes to reduce international indebtedness. This resulted in high levels of unemployment, a decline in incomes, and an increase in the informal economy. It is envisaged that the long-term future for Latin America lies in closer links with the USA, which may possibly become more focused on the Americas as the world becomes increasingly multi-polar.

Western and Central Europe accounted for 29% of world output in the year 2000, which is likely to reduce to 24% by the year 2025. With a population growth rate of less than 0.2%, Western Europe may expect to achieve a standard of living approaching that of the USA by the year 2025. The proportion of the population over sixty years old is expected to increase from 20% in the year 2000 to some 27% by the year 2025. In Western Europe, the population density of 107 inhabitants per km² is achieved with only 30% of the urban dwellers living in cities of over 1m population. There would appear to be social advantages in having fewer inhabitants concentrated in metropolitan areas in terms of crime, and the ratio of income between the top 20% and the bottom 20% is a factor of six. Car ownership levels are 425 per 1,000 population, and energy consumption per capita is less than half that of the United States, with a similar proportion for carbon dioxide emissions. The European Union may be expected to rationalise its armed forces to meet the changing requirements of the 21st century, and this will provide the incentive to become independent of the USA for advanced military aircraft and weapons. In due course, it may be expected that European forces will protect Western interests in the Middle East and North Africa; however, in the course of the 21st century, the USA is unlikely to relinquish its interest in the Middle East, particularly with respect to sharing the decreasing petroleum supplies between the continental regions of the world.

Eastern and South-East Europe currently account for 2.5% of world output, but this is expected to increase to 8-10% of world output by the middle of the 21st century. This region had a combined population of over 360m by the year 2000, with an annual growth rate of less than 0.1%. The proportion of the population over sixty years of age is expected to reach 24% by the year 2025. The population density is 185 persons per km², and 27% of the urban population live in cities of over 1m. Crime rates used to be slightly lower than Western Europe, with the lowest 40% of households receiving 24% of income and the ratio between the highest 20% and the lowest 20% being a factor of only four. However, both income disparities and crime rates have increased sharply since 1990. Industry comprises some 40% of employment, compared with 25% for the USA. Car ownership is only 120 cars per 1,000 population. Energy consumption per capita is approximately 75% of that for Western and Central Europe, and carbon dioxide emissions by the Russian Federation are 10 metric tons per capita. The vast natural resources of Russia are a major reason for a desire for closer links by the European Union, in which Germany and the Central European countries are the major trading partners. It is envisaged that by the year 2025, Eastern Europe will have reduced its dependency on industry to perhaps 30% and the service sector will comprise over 60% of the economy. In the second half of the 21st century, Eastern Europe may be expected to reach the present standard of Western Europe, and Eastern and Western Europe combined will account for some 31% of world output by 2025. This would decline to around 19% by the 22nd century.

North Africa, the Middle East, and Central Asia had a combined population of 400m at the year 2000, with a 2.5% share of world output, which is expected to increase to some 5% by the year 2025 and perhaps to 9% by the 22nd century, when the combined populations will reach 1.1 billion. The current proportion of the urban population living in cities of over 1m is 38%. The driving force behind this continental region is Islamic revivalism and the reconstruction of a contemporary Muslim civilisation, based upon the emergence of a newly defined Islamic economics. The ideal behind this intellectual endeavour is a societal transformation that addresses the four basic needs of Muslim communities, namely justice, social welfare, sustainability, and economic activity based upon the spirit and ideals of Islam. In addition, there is an effort to redefine science in Islamic terms and to identify an approach for science to be based upon the values and concepts of Islam. The resurgence of Islam certainly contributed to the break-up of the Soviet Union and the decline of communism. Islamic fundamentalism differs from traditional Islamic beliefs in that the notion of an Islamic state is fundamental to its vision. However, a union or unmah of Islamic nations, whilst retaining the cultural diversity of separate national communities, is consistent with classical Muslim philosophy and would seem to be a basis of a viable future for Muslim civilisation. The Muslim world is now mainly republican rather than royalist; it is fragmented, unstable, and a reconstituted assembly of countries from which a new civilisation will emerge during the 21st century. It may be noted that there are some 60m Chinese Muslims who have been ruthlessly repressed under Chinese communism, but it is evident now that Mohammed has triumphed over Marxism.

Sub-Saharan Africa had a population of 660m by the year 2000, at a density of 29 persons per km², and it accounts for 1% of world output. By the year 2025, the population and production trends will increase total output by a factor of 3.5 under a scenario of investment transfers from the MDCs, although the GNI per capita is unlikely to exceed $850. African countries experience severe difficulties in servicing their high levels of external debt, and economic performance is impeded by long-term structural constraints to development. These include poorly developed institutions, inadequate infrastructure, a low level of development of human resources, and unequal distribution of and access to finance.

During the 1990s, resource flows from Europe dwindled as the European Union diverted funds for the development of depressed areas within Europe, and in addition, the closer relationship between Western and Eastern Europe has resulted in substantial sums in aid being provided to the Commonwealth of Independent States. Sub-Saharan Africa, which comprises some fifty nations, has itself endeavoured to achieve economic integration of its four subregions, West and Central Africa (mainly Francophone states with the exception of Nigeria), East Africa, and Southern Africa (mainly British Commonwealth countries). The economic communities created on the continent of Africa are listed in Appendix 34, and these include ECOWAS, ECAS, COMESA and SADC.

The main hope for Africa is that, by 2025, a Pan-African Economic Community is achieved in response to the global economy, and African countries produce enough food and goods for their own consumption.

An appropriate urbanisation strategy would be more urban centres and local service centres serving traditional village communities, rather than metropolitan centres. Some 35% of the population is urban and nearly 34% of the urban population lives in cities over 1m. Africa has one of the highest crime rates in the world; for example, South Africa is the world's most dangerous country with fifty-four murders per 100,000 population compared with six per 100,000 in the USA. It is envisaged that by 2150, Sub-Saharan Africa will produce some 9% of world output, but it will need substantial assistance with the development of the necessary infrastructure.

South Asia, with a population of 1,375m, currently accounts for 2% of world output, and this is expected to increase to over 4% by the year 2025, and then to perhaps 15% in the 22nd century, when the population is expected to reach 2.8 billion. Population growth rates are expected to exceed 1.5% per year, and the proportion of the population over sixty years of age is expected to increase from 7% to 12% by the year 2025, in a region where life expectancy is sixty-one years. The population density in South Asia is 252 persons per km², and approximately 43% of the urban population lives in cities with over 1m inhabitants. The lowest 40% of households receive 21% of income, and the ratio of the highest 20% to the lowest 20% is a factor of 4.7. Crime rates are slightly lower than in Europe, although violent crime is on the increase in large cities.

India's economy is currently undergoing reform with an opening of markets to foreign competition through a reduction in tariffs and increased opportunity for foreign investors. Privatisation of state enterprises is under way, and international portfolio investors are pumping large sums of money into the local stock market, with a consequential steep rise in market capitalisation. The region has a large, well-educated middle class of 150m and well-developed legal, financial, and other institutions to support a competitive market economy. Like China, it has a large overseas business community, estimated at 20m, that is likely to be an important source of capital. Sustained economic growth in India will help to reduce the absolute level of poverty, although an increase in the number of vehicles and per capita energy consumption will have a significant impact on the environment.

By the year 2000, Japan accounted for 14% of world output, but this is likely to decline to 8% by the year 2025. Zero population growth is expected from the year 2000, and the proportion over age sixty by the year 2025, is expected to rise from 22% to 32%. Japan has the highest level of life expectancy in the world at eighty-two years for women and seventy-six for men, and it is interesting to note that the Japanese consume 2,900 calories per day compared with 3,700 in the USA or 3,300 in Europe. The population density in Japan is 340 inhabitants per km², compared with 238 persons per km² in the UK, and 50% of the population live in cities of over 1m. The lowest 40% of households receive 22% of income, and the ratio of the highest 20% to the lowest 20% is a factor of 4.5. Japan has one of the lowest crime rates in the world and a prison population that is half that of Western Europe.

According to the GNI per capita, in terms of PPP (purchasing power parity), the standard of living in Japan is comparable to the leading countries of Western Europe. However, infrastructure standards are lower than in Europe; for example, housing standards at 16m² per person are less than half that of Western Europe (35m² per person), mainly because of the high cost of housing at over twice the European cost per m². In general, commuters travel 40-50 km each way in overcrowded trains to work in cities, and Japanese railways achieve 2,000 km per person per year, which is the highest ridership in the world. Car ownership is relatively lower at 395 cars per 1,000 population, and this is the result of the high price of car parking and congestion in cities. Industry accounts for 34% of employment, and energy consumption and carbon dioxide emissions are similar to Western Europe. In the long term, Japan is limited by its lack of resources, and currently it is the world's fourth largest generator of nuclear power. By the year 2025, it is envisaged that Japan will have finished the process of catching up with the USA and Europe and that a more relaxed lifestyle may be adopted by the younger generation, with fewer working hours and less commitment to the work ethic. Hostility still lingers towards Japan in the Pacific region from its colonial rule in China and Korea; however, it is anticipated that Japan will increasingly try to develop relations with China to gain a prime position in the world's fastest growing market.

The region of East Asia and the Pacific is currently the fastest growing continental region of the world. Its share of world output is expected to rise from 8% in the year 2000 to over 13% by the year 2025, and to around 19% by the end of the 21st century. The inclusion of Japan and Australia in the Asian-Pacific region would immediately increase the region's share of world output to 23%, and this is likely to remain fairly steady until into the 22nd century. It follows that, in the year 2000, North America was the largest economy in the world with a 33% share of world output, followed by Western and Central Europe at 29%, and then by the Asian-Pacific region at 21%. However, by the 22nd century, North and South America combined will be the largest economy with a 25% share, followed by the Asian-Pacific region at 22%, and with a combined Western, Southern, and Eastern European share of 19%. Under the scenario of investment transfers from the MDCs, the LDCs' share of gross world product (GWP) could be 35% by 2025, 50% by 2050, and by the 22nd century, their share is expected to reach 65%. There is the prospect that by 2150, the LDCs share of GWP will lie in the range 65-75%.

Therefore, there is likely to be a tri-polar concentration of economic strength with the Asian-Pacific region balanced by both Europe and the Americas. In the short term, the Japanese are likely to become the largest foreign investor and aid donor in the Asian-Pacific region, and the other countries are likely to follow the leadership and methodology of Japan because of the obvious commercial advantage. However, by the year 2025, the size of the Chinese economy is likely to approach that of Japan, and by the 22nd century, China is expected to be the undisputed superpower of the Asian-Pacific region. The constraint on China's growth will be the rate at which the physical infrastructure can be put in place.

11.5 Globalisation and transnational corporations

The years 1980-2000 were a revolutionary period for the global economy, with the opening of international markets and great advances in the ease with which goods, capital, and ideas flowed around the world, bringing new opportunities and risks to billions of people. At the start of this period, one-third of the world's workforce lived in countries with centrally planned economies. A further third lived in countries weakly linked to international transactions because of protective barriers to trade and investment. By the year 2000, fewer than ten percent of workers were living in countries that were largely disconnected from world markets.

The trend for increasing international trade and market-based development is encouraging firms to invest in physical capital, efficient new technologies, and workforce skills, with the intention of delivering growth and raising living standards for workers. Inevitably, transnational corporations move production to lower-cost locations by forming alliances with national companies in fast-emerging markets to avoid protectionist import barriers. However, global investment and the transfer of industrial production to developing countries is leading to an accumulation of industrial overcapacity in sectors such as cars, aircraft, steel, chemicals, pharmaceuticals, consumer electronics, and computers. Despite extensive industrial restructuring in the MDCs and the closure of older plants, the increase in capacity is outstripping global demand by some ten to twenty percent. For example, the demand for new cars runs at some ten percent of the total car population, which amounted in 2000 to an annual demand of say fifty-four million cars worldwide in relation to a global car production capacity of eighty million.

The global shift in industrial structure to less mature markets where wages are lower have increased employment insecurity in developed economies with a decline in real wages. In many cases, displaced industrial workers have taken part-time or alternative employment at lower hourly rates of pay. In Appendix 35, the size of the workforce for each of the continental regions has been aligned with the number of large establishments of over fifty employees required to produce the total GRP in PPP (Purchasing Power Parity) by continental region for the years 2000, 2025, and 2150. The total workforce per large establishment is estimated to be 360 in 2150 in comparison with a total employment per large establishment of 300 in the UK in 2000, of which sixty percent work at large establishments of over fifty employees, and forty percent work at small establishments. Allowing 12.5 large establishments per 10,000 population for developed regions in the year 2000, 62.5 large establishments would be required per $bn GRP.

In the developed regions, a growth in GDP of 1.5% per year will increase GRP from PPP $26.7 trillion in 2000 to PPP $40 trillion in 2025, with an increase in the workforce from 640m to 650m. The total workforce per large establishment (over fifty employees) will fall from 380 to 260 in the period 2000-2025, which reflects the increase in productivity required to produce the increase in GRP. Similarly, by the year 2150, the workforce is expected to reach 750m, so that the workforce per large establishment may fall to 185, which would require an increase in worker productivity of two times from the year 2000. In the developing countries, the GRP of PPP $17.8 trillion in 2000 could be produced by 1.1m large establishments, in conjunction with the numerous small establishments that contribute 35-40% of output. The actual workforce is expected to be 2,315m, which gives an average of 2,100 total employment per equivalent large establishment. It follows that there is likely to be low productivity, through lack of capital investment and massive underemployment. By the year 2025, the total employment per large establishment would fall to 1,040 in developing countries, and the workforce by the year 2150 would be 420 per large establishment. In 2150, the national productivity in developing countries would be 70% of that in the UK in 2000. For a world population in the year 2150 of 12.5 billion, there would be a requirement for 15.6m large establishments for a Gross World Product of PPP $250 trillion, which gives 12.5 large establishments per 10,000 population.

In conducting the above analysis in terms of the number of large establishments per PPP $bn GDP, it should be recognised that the use of appropriate technology would reduce unemployment and ameliorate the circumstances of the poor. Conventional development strategies assume that off-the-shelf technologies provide the means for expanding GDP in the LDCs and, generally, transnational corporations will promote their own capital-intensive technologies. Alternative, light-capital technologies designed to use renewable energy resources, recyclable materials, and local resources can be more easily understood and assimilated into the culture without destructive impacts on society or the environment. Low-cost and small-scale technologies that are labour intensive and easy to maintain should not be viewed as obsolete, second-hand, or inefficient. Proper efficiency criteria would not judge increased unemployment, the use of scarce resources, and environmental degradation as signs of economic success.

In view of the Brundtland Commission's prescription for sustainable development, it would take a severe recession for governments to engage in the large scale creation of jobs through public works projects or other methods that would result in increasing the rates of depletion of natural resource stocks. In that context, the reality of the GRP estimates in 2150 for a workforce of 160 per large establishment in the developed regions, is that it would be provided by a workforce of 250 working an annual 1,100 hours in comparison with 1,750 hours today. It follows that, with sixty-five percent of the population in the age group fifteen to sixty-four, the participation rates are likely to increase to ninety percent for males and seventy percent for females, as in the Scandinavian countries, so that over fifty percent of the population participates in the workforce. By increasing the workforce from some forty to fifty percent, which is an increase of one quarter, the available income would be spread more evenly across the population, and the information revolution will provide humans with greater increments of free time for leisure, education, and unpaid voluntary work.

In the context of continental integration within a global economy, strategic analysis involves prescient stage-setting actions to capture an advantage in the dynamics of evolving markets. The aim is to move from a focus on transactions to repeated transactions, then long-term relationships, buyer-seller partnerships, strategic alliances, and the development of a complex business network (trophic web) that will adapt to environmental conditions with innovation, flexibility, and speed as well as being resilient to possible threats. The complexity of the trophic web will be determined by scale as measured by sales, business, and product diversity, technological intensity, geographical scope, and resource or regulatory constraints. The performance trajectory as determined by return on investment (ROI) and earnings per share growth will indicate the stage of evolution of a business within a sector and a geographical market.

The best opportunity for investment will generally be in adjacent territories where capital accumulation is at a relatively lower level than the advanced economies, and for a modest investment there is a substantial increase in the size of the markets. It follows that foreign, direct investment is concentrated in countries or peripheries that are not too disparate in terms of level of complexity from the saturated economies, and the driving force comes from diminishing returns in the former markets. Peripheries are those zones that have lost out in the past in the distribution of surplus to core regions. As a consequence of the acceleration of technological change, research and development costs have to be recouped rapidly with a global sales strategy, which involves the transfer of technology by multinational corporations and the establishment of production facilities in new states in an adjacent continental union. With the protection of the continental union, the multinationals employ an indigenous workforce and a local marketing team to identify the cultural nuances of the new customers.

The regional and local effects of global capitalism would be for world trade to take an increasing share of the remaining eighty-four percent of the world's economy that is locally produced and consumed. This would generally increase the distance between production and consumption, as well as consumption and the environmental consequences. A regime of global free trade requires that countries dismantle any barriers and open up their financial sectors, such as banking, insurance, and brokerage to international competition. Interconnected networks run in both directions, so that local capital can be rapidly transferred to more profitable locations to the detriment of the local economy. Also, services in peripheral regions can be opened up to competition from electronic banking, mail-order houses, and freight transport based in the major centres. Few regions will have any degree of control over their economies, since investment decisions affecting lives of thousands of people will be taken remotely at the headquarters of transnational corporations. A transnational corporation's decision to transfer capital into or out of a country will increase uncertainty with regional rise and decline, ebbs and flows in employment opportunities, and inevitable migration. Foreign direct investment (FDI) by transnational corporations can reach more than twenty percent of the GNP in selected developing countries, and in 2000 FDI at $240 bn was four times the Official Development Assistance and Aid at $60 bn. Multinational investment banks and other financial organisations make most of the $1.5 trillion per day of foreign exchange transactions, to manage their international assets and to profit from the fluctuations between currencies.

According to *Global Inc.*, by Medard Gabel and Henry Bruner, it is estimated that in 2000 there were 820,000 branches, subsidiaries, and affiliates of the 63,000 transnational corporations (TNCs) from the industrialised countries, and these employed directly 90 million people but, if indirect employment is factored in, it is estimated that transnational corporations provided jobs for close to 200 million people. The United Nations Conference on Trade and Development (UNCTAD) tracks the 63,000 parent corporations, and 78% are based in the MDCs, with 63% based in Western Europe, 8% in North America, and 6% in Japan. Nearly 50% come from five countries: Denmark, Germany, Sweden, Switzerland, and the UK. Some 54% of the 820,000 foreign affiliates are based in Asia compared with 12% in the MDCs. Nearly 50% of the export market is held by seven countries, which include the United States (13.2%), Germany (8.2%), Japan (7.7%), China (6.1%), France (4.8%), Canada (4.7%), and UK (4.4%). TNCs are important generators of both tertiary and quaternary division services, including commercial banking, airlines, retail, telecommunications, software, advertising, media and entertainment, consulting, accountancy, and legal services. It is estimated that for metropolises of over 2m population, TNCs may account for 15-20% of employment, whereas in regional centres, the proportion is closer to 5-10%. With increasing world trade, the TNC employment is likely to increase by 4.5 times to 400m by the year 2025 and by 7.5 times to 675m by the year 2150.

The advocates of free trade derive the legitimacy of their ideology from the classical economists Adam Smith and David Ricardo. However, at that time in 1776, the difference in per capita standards between countries or different states was a factor in the range of one to two. From Appendix 29 it may be seen that the inequality between the MDCs and LDCs has since increased to a current ratio of about fifteen. Also, they overlook the classical assumption that there will be a union between capital and territory, whereas corporations can evade social and environmental responsibility at the local level or corporate taxation at the national level. It follows that a borderless world would give undue licence or power to TNCs, and whilst the scale of global capital flows has become beyond the control of nation-states, the regulation of transnational trade will involve protection at the boundaries of the continental trading areas. Foreign corporations will have to invest in production facilities in these countries, employ local people, and contribute to their economies to gain access to their markets. Capital will tend to flow to those cities and localities that offer the maximum opportunity for TNCs to derive additional benefits, and this may include the externalisation of costs onto the community. Since forty percent of all global trade takes place within TNCs, there is plenty of scope for transfer pricing, which diverts declared profits to countries with the lowest tax regimes.

The geopolitical macrostructure: Chronological references

1. *Geography and Trade* - Paul Krugman, 1993.
2. *The Globalisation of Business* - John H. Dunning, 1993.
3. *Reordering the World* - Edited by George J. Demko & William B. Wood, 1994.
4. *Trade Blocs: The Future of Regional Integration* - Edited by Vincent Cable & David Henderson, 1994.
5. *The Cultural Dimension of Development* - UNESCO, 1995.
6. *Global Change, Regional Response* - Edited by Barbara Stallings, 1995.
7. *When Corporations Rule the World* - David C. Korton, 1995.
8. *The Case Against the Global Economy* - Edited by Jerry Mander & Edward Goldsmith, 1996.
9. *One World Ready or Not* - William Greider, 1997.
10. *Economic Integration Worldwide* - Ali M. El-Agraa, 1997.
11. *Sources of Conflict in the 21st Century* – Edited by Zalmay Khalilzad & Ian Lesser, 1998.
12. *Networks and Netwars* – John Arquilla & David Ronfeldt, 2001.
13. *World Investment Report* - United Nations, 2002.
14. *Globalization and its Discontents* – Joseph E Stiglitz, 2002.
15. *Global Inc.,* - Medard Gabel & Henry Bruner, 2003.
16. *The European Dream* – Jeremy Rifkin, 2004.
17. *Futures of Religions* – Edited by William S. Bainbridge, 2004.

12.
LONG-RANGE GLOBAL SCENARIOS

12.1   Emergence of evolutionary futures and landscapes

The Long-Range Futures Research (L-RFR) landscapes emerge from a coherent set of qualitative and quantitative descriptions about the evolutionary parameters that are driving world development, in order to explain the behaviour of the system. Investment capital is the system-growth parameter that creates spatial structure with the evolution of the world's system of cities, and the change in capital stock drives the per capita Gross Regional Product. There is also a qualitative evolutionary driver of the system, which is the increase in planning standards and the physical quality of life that correspond to the stages of development. Sustainable civil development involves intervention in the system to extend life expectancy and the lifetime of our civilisation, through the release of resource constraints and the conservation of our natural environment. The normative drivers of the system would include the following:

- Geopolitical - Planetary governance with continental integration
- Demographic - Sustainable population peak by 2150
- Civil - Urban transition with rising infrastructure standards in LDCs
- Economic - Increasing per capita GRP with an upper limit for MDCs
- Social - International economic convergence and social equity
- Land-use - Allocation for settlements, agriculture, forests, and wildlife
- Technological - Decarbonisation and dejouling of energy with renewables
- Environmental   -   Resource   conservation   through dematerialisation

- Security - Reduction in crime, terrorism, and war
- Hazards - Protection from geo-hazards, bio-hazards, and techno-hazards.

Complexity science unifies the features arising from the evolutionary dynamics of the artificial civil system, and it underlies long-range futures research. The concepts of complexity science have been outlined in descriptive and quantitative terms in the preceding chapters and the relevant appendices as follows:

- Open systems, the system growth parameter, and coevolution
- Endogenous growth, path dependence, civil emergence, and bifurcation
- Transacting entities, complex adaptive system, colonisation, and resilience
- Microstructure, diffusive ecostructures, and gradient reduction
- Information growth, transactional complexity, and network dynamics
- Civil phase transitions, attractors, chaotic dynamics, and macrolaws
- Technological evolution, diversity, and the logistic curve
- Hierarchical trajectory, equipollence, and the power law
- Macrostructure, spatial integration, and decomposition

Data from the base year 2000 are used for the generation of default values for future conditions in years 2025 and 2150 as shown in Appendices 22-23, 26-28, and 31, in order to develop long-range scenario profiles. These are based on a themed combination of possible future configurations for the global macrosystems, to provide an 'identikit picture' that could bear some resemblance to the future. The underlying dynamics of the macrosystems depend upon a number of parameters and parametric variables, which are defined below.

12.1.1 World Development Indicators for the Year 2000

The World Bank publication, *World Development Indicators 2002*, has provided demographic, economic, environmental, rural, and

Long-Range Futures Research

urban data for the base year 2000. Also, statistics on urbanisation trends, cities, population densities, and households have been obtained from the United Nations Centre for Human Settlements. Metropolitan populations for 2000 are generally based upon the *United Nations World Urbanization Prospects: The 2005 Revision* (2006).

12.1.2 Territorial diversity and geopolitical context

The development of evolutionary landscapes for futures research needs to incorporate potential territorial diversity at the outset.

Continental regions: For the purpose of developing evolutionary landscapes, the world has been divided into fifteen continental regions or sub-regions, and eventually the world may comprise some twenty to thirty continental blocs with populations of around 500m. Country data has been aggregated into continental regions.

Ecopolitan states: Within each continental bloc, there may be perhaps fifty ecopolitan states with an average of 10m inhabitants, and by 2150 there may be around 1,000 potential ecopolitan states.

12.1.3 Long-range population projections by continental region

World population projections are based upon the United Nations' long-range scenarios for a medium variant. It is anticipated that the world population will peak at 12.5 billion in 2150 before declining, with a medium population of 10 billion and a high-medium population of 15 billion. The UN provides some broad country and regional population projections that are hypothetical scenarios based upon extrapolations of current trends. In the UN population projections, it is assumed that declining mortality worldwide will counterbalance declining fertility, and that lifespans will increase by one year per decade from the year 2000.

12.1.4 Spatial structure and the future number of cities

Population by urban size class: The basis for the distribution of the cities for 2150 is from complexity science with a dynamic balance and equipollence at each hierarchical level. The city size distributions for 2000 and 2025 have been derived from named city populations using UN data, in conjunction with a German database that can be accessed on the internet at www.citypopulation.de. The city populations for 2025 have been derived using annual population growth rates, or rates of decline, from UN 2000 data. These cities are then slotted into the appropriate size band to give the number of cities.

Rural population and number of villages: In 2000, the world's rural population reached a peak of 3.2 billion, which is expected to decline before 2025. Urban expansion will arise from the natural increase in the urban population rather than migration from rural areas. This will partly be a result of lower infant mortality rates in the cities than in rural villages. Also, rural villages will be reclassified as urban by the United Nations Centre for Human Settlements as soon as there are 2,000 inhabitants living in townships with streets and urban utilities.

Population densities: The area of human settlements reaches a peak of 15m km^2 in 2025 and then declines to 13 m km^2 by 2150, when the low-density rural populations are living at higher urban densities. This will give an overall population density for human settlements of 1,000 inhabitants per km^2 in 2150.

12.1.5 Land-use and natural resources by continental region

The land use by continental region has been estimated for world population of 12.5 billion in 2150, and less than 10% of the world's land surface of 133 m km^2 would be covered by urban areas and villages at a density of 1,000 inhabitants per km^2. Cultivated land and pastures may account for 30% of the land, forests may take up to 27%, and the remaining uninhabited land for wildlife would be 33%. If each person were to receive an average of one million calories per year, then one hectare of Suitable land (United Nations Food and Agriculture Organisation or FAO classification) should support five people, i.e., 0.2 hectares per person.

12.1.6 Civil and societal transitions

For the purpose of futures research, eight stages of development are defined: traditional, agropolitan, infrastructural, industrial, distributional, informational, ecopolitan, and planetary with an indicative Gross National Income (GNI) per capita. The cumulative investment at each of these civil phase transitions gives rise to an increase in the urban proportion of the population, together with the emergence of an ecopolitan society. This will bring slower rates of growth with an ageing population and a decline in lifetime incomes.

12.1.7 Increasing transactional complexity of civil ecostructures

Civil ecostructures behave as complex adaptive systems with changing property profiles and the relocation of establishments. There will be an increase in the proportion of establishments in the quaternary information division, with an eventual decline in property investment returns. The 'Generica' phenomenon results in town centres with identical branches of retailers, leisure facilities, restaurants, hotels, banks, building societies, and insurance companies.

12.1.8 Default planning standards by stage of development

Planning standards, which are dependent upon the level of GNI per capita, may be established for the sizing of infrastructure facilities per 10,000 population. Appendix 14 shows that, as the GNI per capita increases over sixty times (from $500 to $30,000), the number of persons per household halves, and the planning standards per capita for housing increase by a factor of four times. Similarly, the changes in planning standards for hospitals, secondary school places, and higher education are given in Appendix 15, and life expectancy increases by twenty-five percent from sixty-three to seventy-eight years. In the case of civil infrastructure, investments such as energy, transport, and telecommunications, the correlation with the per capita provision of services is given in Appendices 16 and 17.

12.1.9 Assessment of the future capital stock

The MRIS gave the number of large establishments by facility type and size range for each urban classification, so that the total floor area for each type of facility can be assessed by knowing the percentage of the total floorspace that is taken by the proportion of smaller establishments. The area of school and further education facilities are calculated from the planning standards per child for the various age groups. By applying the appropriate construction cost per m^2 for each facility type, the non-residential building costs for a regional centre of 100,000 population amounts to $1.5 bn, in relation to residential building costs of $3 bn. A further $1.5 bn is required for infrastructure and utilities so that as a default value, the capital cost of a new European city of 100,000 population, would amount to $6 bn (2000 US$) or $60,000 per inhabitant.

On the basis of the default planning standards and replacement costs for urban and rural housing, it is estimated that, at the year 2000, the world stock of investment in both standard and sub-standard urban and rural settlements amounted to some $90 trillion, and the stock of non-building assets amounted to some $60 trillion. It follows that the replacement value of the total world investment stock amounted to PPP $150 trillion in the year 2000, or PPP $25,000 of assets per capita, of which rural and urban settlements account for sixty percent of total investment. In 2025, the investment stock in human settlements may double from the 2000 level to PPP $180 trillion. In 2150, the required stock of investment in human settlements is estimated to be PPP $480 trillion, in relation to the global investment stock of PPP $800 trillion or PPP $64,000 per capita.

12.1.10 Future Gross World Product by continental region

In order to fund the doubling of the urban population in the LDCs in the period 2000-2025, it is essential to transfer $375-500 bn per year from the MDCs to the less developed countries or there will be a large-scale urban system failure with an appalling deterioration in the conditions of the world's poorest populations.

Also, to stay within the sustainable GWP of $200 trillion for a world population of 12.5 billion, the MDCs need to limit GNI per capita to $40,000-$50,000 (2000 US$) during the urban transition in the LDCs, which is higher than the Gross State Product per capita of California at just under $40,000 in 2000. This would give a world average of $16,000 (PPP $20,000) per capita in 2150, compared with an average GWP per capita of $5,150 in 2000 and a projection of $7,300 (2000 US$) in 2025.

12.1.11 Future technological trajectories

With a technological shift, it is expected that through longer life assets, recycling, and minimum material design, the material throughput for a given capital stock could be reduced to 85% by 2025 and 65% by 2150. The International Energy Agency estimates that there will be a 1.1% per year increases in energy efficiency, so that energy efficiency could be improved by a factor of 1.2 by 2025 and two times within a timescale of 100-150 years. This would reduce the energy consumption to support a given level of material throughput to two-thirds of its present value by 2025 and to one-third by 2150. In 2000, the global energy use was 425 EJ (Exajoules), and from the evolutionary assessment, the energy use in 2150 is expected to be 600 EJ. If one-third of commercial energy is produced using renewable sources, then the environmental impact will be equivalent to 400 EJ. The technological trajectory will be towards the decarbonisation of energy and a hydrogen fuel economy. The world car fleet for 2025 and 2150 is given in Appendix 32.

12.1.12 Prospective hazards and disastrous scenarios

The macrosystems for an interconnected world will need to coevolve to create a civilisation that is adaptable, robust, and resilient to natural hazards such as floods and famine, bio-hazards such as epidemics or transmittable diseases and social hazards such as currency speculation, riots, or civil wars, and technological hazards such as accidents, explosions, or contamination leading to civil disasters with loss of life, damage to the built environment, and severe disruption to human activities. A hotter and drier climate will require more wells, irrigation systems, and desalination plants, as well as moving settlements away from flood plains or areas that will be submerged by rising sea levels. Improbable events eventually happen on a macro-timescale of a century or more; for example, a 100-year flood gives the magnitude of a flood with a one percent per year probability of occurrence. The probability of it happening in a century is sixty-three percent, and this virtually reaches one-hundred percent within a millennium. A robust strategy will be one that creates sufficient wealth to provide a safety factor against poverty for the least developed countries. Currently, life is far too precarious for billions of people, who are living on the edge of survival.

12.2 Designed intervention in the global macrosystems

Policy decisions taken at the United Nations, continental, and national levels will impact upon the overall behaviour of the civil system, and in due course, planetary governance may provide the means for enhancing planetary potential. As the world population doubles in the period 2000-2150, the existing voting rights in the General Assembly of the United Nations will become increasingly distorted. The sub-continental countries of China and India with a combined population of over 2 billion have only one vote each, and countries with populations of 5 million or less also have a single vote. A world of 200 nations may evolve into 1,000 states, which would be too diverse to deal with at a single global level. Perhaps twenty-five or more continental unions would need to be established to internalise major issues, and each continental union would become a member of the Security Council at the United Nations to provide a forum for the resolution of intercontinental disputes. However, democratic pluralism demands that states with populations of 5m-20m should each be given one vote at the United Nations. The highest evolutionary potential will be achieved by developing a rich diversity within a coherent unifying structure. The incentive for countries to increase their representation could save an enormous amount of bloodshed by giving formal recognition to regional states, minorities, and ethnic groups.

The United Nations will need to respond to a variety of both stabilising and destabilising forces, including multipolar geopolitics, political democratisation, trade liberalisation, financial deregulation, social cohesion, infrastructure investment, human development, and environmental sustainability. The most successful outcome for a planetary civilisation will be an integrated but differentiated planetary economy. Transactions on any scale between continental unions will only arise if they have something different to transact. Continental unions will have to cooperate with each other to produce a suitable regulatory system and tough sanctions to ensure that global stability is increased, with a reduction in the turbulent and atrophic conditions in the least developed countries. Humanitarian considerations have proved insufficient to mobilise adequate international aid for infrastructure, and capital shortages have inevitably resulted in international indebtedness in rapidly urbanising countries. It is a cruel illusion of the affluent MDCs that these LDCs can live within their means and overcome poverty, low calorie diets, and reduced life expectancy.

The analysis has shown that each of the competing approaches to capitalism has made a significant contribution to world development in the 20th century. The conceptual weakness in a transnational liberal economic order is that the disparity in the infrastructure provision between the LDCs and the MDCs results in not merely imperfect but unfair competition, which would exacerbate social inequality and unsustainable development. Institutional changes and planetary governance may be needed to devise new global mechanisms for the collection and distribution of the surplus generated in a global economy. If there was planetary governance, the realisation of full planetary potential would be achieved by transferring investment capital from mature MDCs, which can generate more investment than can be viably reinvested in them, to the less mature LDCs. The LDCs would benefit from a much higher level of investment than they can possibly self-generate.

International policy-making is a design problem with multistage decision processes, and the course of world development can be changed through critical intervention points in the global macrosystems. However, the extent to which the system can be influenced by critical leverage, and the effects and likely outcome of proposed interventions, are often discovered from system experience. *Long-Range Futures Research* provides a basis for exploring the possibilities of enhancing evolutionary potential when investment hypertrophy, diminishing returns, and stagnation arise in the more mature parts of the system. Investment hypertrophy is excessive geographical investment, such as sectoral over-capacity or infrastructural over-development, which leads to investment atrophy in other regions, sectors, or parts of the urban system.

One opportunity that is open to the more developed countries is to raise revenues from ecotaxes to meet an appropriate level of social expenditures to ensure that consumer purchasing power does not decline further than necessary, and to invest in a technological shift in the energy infrastructure. An aim would be to achieve a fifty percent reduction in carbon emissions by the year 2100 and to eliminate the hydrocarbon economy by the year 2200. This approach to investment in ecopolitan states would help raise the level of economic activity whilst reducing environmental pollution. It should also trigger off an upward swing of an investment wave as a result of the cluster of innovations in the transport and energy technologies involved.

The geopolitics of the biosphere is likely to be the ratchet that creates institutional change and planetary governance in the future. For example, China and India's industrial development will be principally fuelled by coal, and by the year 2050, the Asian tiger economies, China and India, will account for some thirty percent of the world's carbon dioxide emissions compared with thirty-five percent for the OECD economies. An environmental strategy would be to reduce the share of fossil fuels from eighty percent of the energy mix in 2000, to perhaps fifty percent by 2100 when nuclear and hydro might account for twenty percent and other renewable sources reach thirty percent of the total. In 2000, fossil fuels accounted for 340 EJ of commercial energy use in relation to a global total of 425 EJ, and the transition to non-fossil fuels could reduce the fossil fuel contribution to 140 EJ towards the end of the fossil fuel era. This would reduce carbon emissions from 6 billion metric tons to 2.5 billion tons, which is assessed as the maximum annual planetary total that can be absorbed, including reabsorption by the oceans, forests, and extraction by photosynthesis and possibly 'geoenvironmental' engineering in the future.

In this research, historical and future continuity has been reinforced by using cities as the foundation for assessing the implications of a population of 12.5 billion humans in 2150, and it has been verified that this state is sustainable in principle, providing the appropriate investment is made in the urban infrastructure and systems are in place for a planetary civilisation. This involves a global urbanisation policy, together with a strategy on a planetary scale, for the industrial, energy, transportation, and telecommunications systems. Early warning systems such as foreign embassies, reconnaissance satellites, remote sensing, or the worldwide seismic network provide a flow of information for the design of secondary adaptive policies as the future unfolds. This is an approach for taking responsibility for the future by designing a complex adaptive system that will be resilient to possible threats. The aim is to minimise the potential for a disastrous scenario in a multipolar world and to determine what policy measures are required to maximise the total number of lives to be lived over time on the planet at a reasonable per capita standard.

The future state of the planet in the period 2025-2150 may be explored by interpolation between the extrapolated near future in 2025 and a prognosticated long-term future state in 2150. In the absence of a model, normative scenarios could be produced using the backcasting approach to explore the prospects for a viable future state some 100-150 years into the future, by examining the extent to which the natural and global macrosystems may be adapted to achieve the desirable outcome. In the long term, discontinuities are bound to occur, but it should be possible to identify a range of threats that could cause a system collapse. Planetary governance, based upon democratic pluralism, will be needed to coordinate and integrate the actions necessary to eliminate the risk of destabilising intercontinental wars, over-population, pandemics, global financial collapse, or eco-catastrophe. This analysis should reveal the combination of different policies that would contribute to the desirable state. The respective policy implications of the alternative scenarios can be identified to determine the scope for freedom of policy action.

In relation to the 250-year horizon for the United Nations population projections, and the 100-year horizon for the Intergovernmental Panel on Climate Change, the appropriate planning horizon for each level of government is tabulated below:

Planetary governance	100–150	years
Continental Unions	50–100	years
National governments	25–50	years
States	20–25	years
Local authorities	10–20	years

12.3 Coevolution of the global macrosystems

In exploring an evolutionary vision for the planet in the period 2000-2150, a combination of different policies were identified that would minimise the possibility of a disastrous scenario in a multi-polar world and achieve a viable future with a desirable end state. A morphological analysis is undertaken for the conception of subsystem parameters that can be combined to produce an overall system design with high performance characteristics. The coevolutionary path for each of the macrosystems will be dependent upon the dynamics of the complex adaptive system, which will have inevitable design defects that may be overcome eventually by trial and error. The future state of the planet in the period 2025-2150 may be explored by interpolation between the extrapolated near future in 2025 and a prognosticated long-term future state in 2150. By the year 2150, each of the following macrosystems and their subsystems will have reached a possible condition or configuration, and a themed combination of the various states for each system will give rise to a scenario.

A   Planetary governance
B   Geopolitical context and military deterrence
C   Continental integration and cultural diversity
D   Population policies and demographic trends

E   Life sciences and medicine
F   Land use and food supplies
G   Resource conservation and the environment
H   High technology and space research
J   World system of cities and the planetary infrastructure
K   Urban technology and telecommunications
L   Community systems and human development
M   Societal transformations and employment
N   The world production system and transnational corporations
P   The world economy and globalisation

Each configuration for a system will be identified by coding; for example, A1, A2, A3, and A4 will represent different conditions for planetary governance. The use of only four configurations for each system rather than five or more, is to illustrate how the methodology can assist in flushing out issues that may threaten planetary stability. Clearly, the approach can be taken to any level of sophistication that is desirable for a specific purpose, and here the analysis will be undertaken to assess the prospects for the realisation of an equitable planetary civilisation with a reasonable standard of living. The objective is to discover how far the world may have evolved towards reaching a sustainable civilisation and the possible time frame for achieving a dynamic balance with the avoidance of a major planetary crisis that would threaten the human species. Exploration of the future by the development of scenarios for the global macrosystems reveals some thirty-five macrotrends.

A   Planetary governance

The United Nations established the Universal Declaration of Human Rights in 1948, and there has been pressure to extend its scope with respect to women's rights, children, unborn children, the dying, future generations, refugees, minorities, and other species. There is also a general consensus that new and improved institutions for planetary governance are essential, because the existing systems are lagging behind the present state of affairs. The shift in global economic development towards the Asian-Pacific region, the Islamic world, and Latin America makes the structure and outlook of the organisation and agencies of the UN seem increasingly out of date. Democratic pluralism demands that continental unions are represented on the Security Council of the UN, and ecopolitan states with 5m-20m population should each have one vote in the General Assembly. Outside the UN, membership of the Group of Ten, the G-10 nations, is still unreasonably restricted, and emerging nations have been refused entry. Similarly, the Bank of International Settlements (the central bankers' bank) and the Organisation for Economic Co-operation and Development (OECD) will need to expand membership.

The configurations that are taken into account for planetary governance are:

A1 Planetary governance lags behind in the changing world.
A2 Institutions are improved and membership widened.
A3 Reform of the UN Security Council with permanent representation by continental unions and new voting rights in the General Assembly.
A4 The role of the UN is expanded and reformed to world government status, with an international taxation system based upon trade.

B Geopolitical context and military deterrence

The United States, Russia, China, France, and the United Kingdom possess nuclear warheads, and Israel, India, Pakistan, and Iran are believed to have the capability of assembling nuclear weapons. However, the military powers are not all economically powerful, and conversely, Japan and Germany are economic powers that are not militarily powerful. The stable bipolar tension of the Cold War is being replaced by an uncertain equilibrium in a multipolar world, in which energy security is a key issue. It is envisaged that Russia will align itself more closely with Western and Central Europe, and that it will protect its territory and mineral reserves against any threats from East Asia. Friction may arise between Russia and the West if the latter tries to increase its influence over the former countries of the Soviet Union; however, it is probable that Central Asia will strengthen its ties with Turkey and South-East Europe, to obtain independent access to ports for its oil. It is likely that wars between the superpowers will be replaced by economic warfare, and military interventions will generally take place against regimes that engage in genocide, terrorism, or the large-scale distribution of drugs.

The configurations considered for the geopolitical system are:

B1   Continuation of the superpower military deterrence.
B2   Demilitarisation, closure of military establishments, and conversion of the military industrial complex to civilian production.
B3   The development of peace-making institutions with weapons inspections and early warning systems, with sanctions by UN Secretariat.
B4   UN peacekeeping force becomes the only legitimate intervention in international, national, and civil wars.

C   Continental integration and cultural diversity

Continental integration is likely to arise to share the cost of security between nations, to protect members of continental unions from unwelcome international trade or dumping, and to control immigration and refugees. At the same time, the recognition of cultural diversity and ethnic and religious groups at the level of the ecopolitan state will reduce the potential for politically dangerous conflicts; however, these will not be eliminated because of wide disparities in the comparative endowments and the economic competitiveness of the various states. The continental union would be responsible for development grants to reduce the inequalities between states.

The configurations taken into account for continental integration and cultural diversity include:

C1   Failure to achieve continental unions and the proliferation of nation-states.
C2   Continental integration of nation-states.
C3   Continental integration with nation-states, whilst recognising cultural diversity between ecopolitan states.
C4   A culturally homogenised world with globalisation, harmonisation, and standardisation, leading ultimately to stagnation and decay.

D   Population policies and demographic trends

The history of civilisation is replete with waves of migrating populations from Asia to Europe, from Europe to the Americas, from India to Africa, from decolonisation in India and Africa, and political and religious refugees. Also, Western and Eastern Europe suffered millions of losses in the two world wars that constrained population growth. With these issues superimposed on demographic trends that indicate slowly growing and ageing populations in the North and rapidly growing young populations in the less developed countries of the South, migration is becoming a very important issue. The drive by populations for higher living standards, better opportunities, and security will create enormous pressure between Eastern Europe and Western Europe, Sub-Saharan Africa and North Africa, North Africa and Southern Europe, and Latin American and North America.

In the United States, immigration and children born to immigrants after arrival account for fifty percent of population growth. Annual growth of 2.5m people per year (or 48,000 per week) is approaching an annual rate of one percent and, by 2050, approximately fifty percent of the population will be non-Latino white.

The predictions for the dates at which zero population growth may be reached in less developed countries, or a rise in birth rates in the western countries, and the potential effects of a major viral epidemic more serious than AIDS, will all contribute to ensuring that there will be a wide range of possibilities for an eventual population on the planet and it is unlikely to remain constant. However, with rising unemployment in the more developed countries, it is inevitable that there will be pressure to control and limit immigration. The three configurations taken for population policies and demographic trends are:

D1   A resumption of population growth in the West with tolerable immigration.

D2   Natural population growth slowing, with migratory flows from eastern and southern countries to the West.

D3   Slower population growth with ageing of the population and immigration control.

D4   Low population projections with a decline in world fertility with higher mortality rates arising from pandemics, such as Ebola, AIDS, and SARS.

E   Life sciences and medicine

The life sciences make significant contributions to the health and welfare of humankind, and they affect some thirty percent of gross world product by way of medicine and genetic engineering, food and beverages, agriculture and forestry, fuel and energy, and environmental biotechnology, such as waste water and sewage treatment and criminal identification. The life sciences provide the means of controlling human fertilisation, extending the lives of humans, controlling disease, and providing warnings on the contamination of the planet and its biosphere. The application of genetic engineering for gene therapy to treat serious, incurable, and generally fatal human diseases is a major step forward in medicine, and its use in the production of novel crops or animal varieties will make a significant contribution to the reduction of hunger. Increasingly, humans will be tested for susceptibility to specific health problems, and in the longer term, we will have choices about how to change ourselves and possibly the evolution of humans. The moral and ethical considerations for the life sciences are profound, and the consequences of biological warfare are unthinkable. In the long term, the alteration of human genetic make-up may increase the potential for life on another planet.

The potential contribution of the life sciences will be the prevention of disease and the promotion of health, so that the high costs of health care may be reduced for an increasing standard of health. The configurations taken into account for life sciences and medicine are:

E1   The moral and ethical issues are uncontrolled, and there is an increasing population of degenerate humans and an increase in immunological diseases or super-viruses that affect vulnerable groups in the population.

E2   Advances in life sciences eliminate many genetic disorders, so that the overall standard of health improves, but life is not unduly extended.

E3   Life sciences make an enormous contribution to the health and the life expectancy of the population.

E4   Genetic manipulation of animals and intensive production, increase the risk of the transmission of animal diseases to humans such as CJD.

F Land use and food supplies

For a world population of 12.5 billion in 2150, less than 10% of the world's land surface would be covered by urban areas and villages. If each person were to receive an average of one million calories per year, then one hectare of suitable land should support five people. It follows that 25m km² of land would be required for food production, whereas the available cultivated land and pastures will exceed 40m km² (30% of the land area). Forests will cover 27% of the land and the remaining uninhabited proportion for wildlife will be 33%. However, there are regional and national limitations to agricultural production in the less developed countries, where the amount of cultivated land and pastures is likely to be 0.23 hectares per person in 2150.

Grains are the staple of world diets, especially wheat, corn, and rice, even though they are relatively low in proteins. However, as the populations of the less developed countries increase, the expansion of world agricultural production may be achieved from both a fifteen percent enlargement of the cultivated area and a significant increase in the yield per hectare. This would arise from mechanisation, improved plant varieties with genetically modified crops, and increased fertiliser use. Alternative approaches would be to increase the food potential from the seas and also to bypass animals in the food chain.

The configurations that have been taken for land use and food supplies include:

F1   Improved rural infrastructure in less developed countries for the transport and storage of crops, together with the mechanisation of agricultural operations.

F2   Irrigation and the use of desalination plants increase the area of land under agriculture, with an increase in the cost of food production.

F3   The introduction of higher-yield varieties of crops with the use of fertilisers and investment in irrigation.

F4   Alteration of people's eating habits with a reduction in calorie intake by the more developed countries, towards the Japanese dietary standard of 2900 calories per day. A shift from meat eating to vegetarianism, with the development of high-protein foods from plant sources such as soyabeans and the development of aquaculture to supplement food supplies from the sea.

G   Resource conservation and the environment

The conservation of natural resources has reached a level of visibility in political consciousness that, for the foreseeable future, scarcity is likely to result in increased prices. International protection of the 'global commons' is essential, and this includes the atmosphere, the oceans outside territorial waters, the biotic resources of the sea such as fish or whales, and the minerals from the oceans or Antarctica. Tradable permits should be purchasable for the use of global commons such as air and sea transportation, fishing, or dangerous or polluting activities. It is expected that the hydrocarbon economy will have run its course within the Third Millennium, and that a diversity of renewable energy sources will be developed, including solar and hydrogen energy. It is envisaged that by the year 2050, half the transport vehicle population will no longer be using petroleum fuels, and this will dramatically change the energy infrastructures on the planet, and governments, utilities, and energy corporations are likely to adapt their strategies accordingly.

In addition to the technological changes, the major concerns are environmental degradation, desertification, deforestation, stratospheric ozone depletion, and a loss of biodiversity. 'Debt for nature' swaps could be made to provide a subsidy for beneficial policies, such as the retention of the tropical rainforests or other natural assets in the interest of humankind. The consequential flooding that would result from global warming and a rise in ocean levels will contribute to an increasing number of environmental refugees that could reach 150m by 2100 for a world population of 10 billion. Other natural hazards could include an earthquake in Tokyo, San Francisco, Mexico, or other major metropolis, or a volcanic eruption as in Pompeii 2,000 years ago, or a cosmic collision. However, the most serious concerns are a major nuclear reactor accident, the problems of nuclear-waste disposal, and the risks arising from proliferation associated with a shift to nuclear energy. Environmental disasters such as major oil spillages at sea or hazards from the disposal of chemical or toxic wastes are increasingly the subject of public concern.

The configurations that are taken into account for resource conservation and the environment are:

G1   An emphasis on dematerialisation with an absolute reduction in the per capita use of natural resources by the MDCs and energy saving strategies.

G2   Continuing reliance on fossil fuels with a decarbonisation trend so that gas becomes the primary fuel, followed by oil and coal and the fossil fuel era ends late in the 22nd century. Oil shales and tar deposits may be converted to liquid products.

G3   Nuclear power generation becomes the primary source of electricity and the production of hydrogen as an energy carrier. The use of nuclear energy for the production of hydrogen is a transitional technology, while the hydrogen energy infrastructure is installed.

G4   Renewable energy becomes the primary source of electricity, with the use of hydrogen as an energy carrier once the infrastructure is installed.

H High technology and space research

Advances in high technology and space research are likely to accelerate the introduction of new materials, space transportation, aerospace developments, automatic control of vehicles, and robotics. Technologies for remote sensing and mapping, geopositioning and navigation, telecommunications, information technology, and microcybernetics, artificial intelligence, nanotechnology, photoactive materials, and superconducting elements will coevolve. There is likely to be a shift from military research to civilian research and commercial applications. A cluster of innovations will include bioreactors and bioelectronic devices for an across-the-board reduction in environmental impacts, which is likely to stimulate an investment wave in the energy and transport sectors.

The configurations that have been taken into account for high-tech and space research are:

H1   The loss of high-tech industries by the more developed countries, which will exacerbate the unemployment problem.

H2   The widespread use of microcybernetics and robotics by corporations that will result in designing people out of the production system, so that their capacity as consumers is reduced.

H3   The convergence of new technologies that will give rise to a new wave of investment in planetary infrastructure such as energy, transportation, telecommunications, and high-tech industries.

H4   Civilian fractional-orbital space flight, a lunar space station, space colonies, and resorts supporting temporary biodome communities.

J   World system of cities and the planetary infrastructure

The long-term trend for the world system of cities and the planetary infrastructure will be for heightened forms of specialisation and differentiation between the states of a continental union, to achieve a world-class production system. These transaction-rich regions will only comprise a limited proportion of continental territory, and other peripheral areas may remain in a relatively underdeveloped state. However, these agglomerations of economic activity attract migrants and low-wage workers alongside high-wage and high-skill employees, which results in congestion and a deterioration of the surrounding area and environment. By providing the preconditions for investment in adjoining states, such as infrastructure and other public services, inward investment will take place in the adjacent states from the push of diminishing returns and the pull of reduced costs, which releases the pressure on the highly developed states. Public sector involvement and a regulatory mechanism for reducing regional inequalities are generally essential for the spatial allocation of infrastructure investments and for funding the civil minimum of amenities in peripheral regions.

The configurations that have been taken into account for the world system of cities and planetary infrastructure are:

J1   The relative decline of the periphery in relation to the core world cities.

J2   The implementation of global and continental regulatory mechanisms to ensure the suitable provision of infrastructure in less developed regions.

J3   The transformation of the less developed continents and nations by creating an increasing number of islands of prosperous metropolitan areas and ecopolitan states, which in due course, will provide a springboard for the development of adjacent states and regions.

J4   A differentiated regional strategy for urban population densities, including a garden-city built environment, high-density megacities and cosmopolises, compact cities with control on suburban development, and low-density urban sprawl.

K   Urban technology and telecommunications

Global development creates an increasing number of nodes and origin-destination city pairs in the world system of cities, and a reduction in the time required to transverse the linkages between them. Time-space convergence can be measured in both travel time and costs, and wide-bodied jets such as the Boeing 747 have reduced air transport costs. Supersonic aircraft such as the Concorde halved the journey time between continents before being taken out of service in 2003, and by the year 2100, hypersonic, hydrogen-powered aircraft or fractional orbital flight will achieve a significant reduction in intercontinental travel times.

The improving speed and cost of transport, together with satellite telecommunication and advanced information technologies, enable transnational corporations to consider the competitiveness of alternative continental locations, as well as alternative states within continents. The impact of this connectivity will be an increasing number of third- and fourth-order world-city regions, which will be served by continental nodes with high-speed intercity links. It follows that a future trend will be the dispersal of global production and the polarisation of superclusters of economic activity through a high-speed transport infrastructure and just-in-time production hinterlands.

Communication technologies and growth of car ownership would seem to be an irreversible trend for an increasingly mobile population. Whilst motoring is likely to be increasingly expensive with alternative fuels, the pattern will be for decentralised urban living in environmentally attractive locations and a steady increase in telecommuting. However, it is the multi-media information technologies and the power of television that will help to transform the culture of society to demand environmentally sustainable policies.

The configurations that have been taken into account for urban technology and telecommunications are:

K1   A failure to develop 'clean' cars and increasing dependence upon public transport systems, with the need for heavy carbon taxes and road tolls to limit vehicle use.

K2   The development of 'clean' cars and incentives to increase vehicle occupancy, together with light railway systems for metropolitan areas.

K3   A combination of new rail technologies and public transit systems for metropolitan and intercity travel, and a combination of 'clean' cars and telecommuting to suit increasing amounts of leisure time and alternative home-working employment.

K4   Dense international and national transport networks served by continental nodes and high-speed intercity links.

L   Community systems and human development

Human communities are under pressure at each stage of development as families move from rural areas to urban centres with the industrial stage, and the distributional economy encourages low-density suburban living with the diffusion of the car. Industry decentralises with the growth of the motorway network and changing production systems, and the informational economy has accelerated the decentralisation of offices. The counterpart of suburbanisation is inner-city decline and disinvestment, which reduces land values and redevelopment eventually becomes profitable again so that opportunities are created for the adaptation of the city with new transport systems and changing land uses.

A marginalised society of the unemployed, the elderly, the migrant workers, and the poor exists in the shadow of new development, where they live in rundown inner-city housing areas and in underutilised transition zones at the urban fringe. Ageing, high-density housing accommodates more low-income urban population than can find jobs.

The tendency to reduce government involvement in social security, public housing, social services, and health care creates an urban underclass that increasingly live by means of the informal economy, crime, drug dealing, prostitution, and barter. Also, banks are reluctant to lend funds for home mortgages or small businesses in the deprived areas, and credit is denied to poorer communities other than by loan sharks.

In the period 1980–2000, median personal income in the USA has remained stagnant and real incomes declined for the lowest four out of five quintiles. The top quintile captured almost all of the increase in GNI over the twenty-year period. Median household income rose by less than five percent in constant US$, and it should be noted that the majority of households in the top quintile had two income earners, compared with an average of one in the mid-quintile and zero-income earners for the lowest quintile. During this period, there has been an increasing female workforce and a considerable increase in the percentage of college graduates. Part of the explanation for declining incomes was that new, low-wage and part-time service jobs do not compensate adequately for the lost manufacturing jobs, and that wages in manufacturing are decreasing as a result of world trade.

The world average life expectancy is sixty-six years, but life expectancy in Sub-Saharan Africa is less than fifty years. In the twenty-year period 1980-2000, living standards in Sub-Saharan Africa fell back to those of the 1960s, and in the same period, the annual growth in per capita consumption in Latin America was only 0.6%. The United Nations admits that during the 1990s, the world's poorest people living on less than $2 per day increased from 2.7 bn to 2.8 bn, which was 58% of the population living in the less developed countries in 2000. There is also a direct correlation between poverty and mounting instability, social unrest, and violence. Development assistance from the MDCs to the LDCs declined from 0.50% of GNI in 1960 to 0.34% of GNI in 1990, and by 2000, Official Development Assistance and Aid at $60 billion had fallen to 0.24% of their GNI of $25 trillion, which is less than half the percentage in 1960.

The provision of community systems for health, education, training, employment, welfare, social, religious, cultural, and recreational activities

Long-Range Futures Research

is essential for human development in both the MDCs and the LDCs. There needs to be an enormous increase in assistance to the LDCs, with conditions attached relating to good governance, civil society, human rights, labour standards, and environmental protection.

The configurations taken into account for the human community systems are:

L1   A continuation of the inequality of income between the MDCs and the LDCs and also within societies, with a failure to achieve social justice.

L2   Increasing urban violence, which includes mugging, aggravated theft, grievous bodily harm, and sexual assault as a result of a variety of factors including alcohol and drug abuse, racism and discrimination, an increasing number of single-parent families, and stress from overcrowded housing and poor living conditions.

L3   A substantial increase in aid to the LDCs and a society that takes more responsibility for supporting social initiatives in deprived areas, particularly in the support of children and young adults who need a break from a family cycle of poor education and poverty.

L4   A fully aware world community with international economic convergence and a closure of the development gap.

M   Societal transitions and employment

A shift from a consumer-driven society with an emphasis on rising standards of living will occur because advances in technology and the productivity of industry will reduce the length of a person's full-time working life and increase the proportion of part-time workers. As a consequence, household income over a lifetime may fall to some seventy-five percent of current values. A transition to an ecopolitan society is expected in which the emphasis will be on quality of life rather than increased consumption, with more civic involvement, the development of the Quinary Lifespan Division of the economy, and greater local self-reliance with long-life artefacts, minimum-material design, resource conservation, and more extensive recycling. Redistribution of income from the employed and the corporate sector will be achieved from progressive income taxation to reduce inequality, consumption taxes such as value-added tax, and ecotaxes on the use of natural resources and carbon emissions.

Mass communications and television will play a major role in education, the promotion of ecological consciousness, and leisure, and it will alleviate the gap between the information-rich in multi-national corporations and the information-poor working on local activities. It is anticipated that there will be an increase in home-working, extension of the informal or black economy, and with a reduction in the size and influence of the middle class, there will be increasing illegal activity and social unrest if appropriate policies are not implemented by government.

The configurations that have been taken into account for social transformation and employment include:

M1   Increasing government debt with an ageing population creates serious financial instability in the MDCs, with large-scale social unrest.

M2   Governments introduce legislation, incentives, and penalties for corporations to adjust to flexible production and shorter working hours, so that there is a proper sharing of employment.

M3   The fully employed proportion of the population welcomes the opportunity for increased leisure time and, at the same time, acknowledges that mortgage repayments and the funding of pension schemes will need to be made over a shorter working life, which will result in a reduction of consumption.

M4   A general acceptance of the quality of life philosophy with increased voluntary work in the lifespan division. Ecotaxes encourage resource conservation and provide the means for income redistribution, particularly towards retired or partially retired groups.

N   The world production system and transnational corporations

World trade accounts for some 16% of GWP, and the remaining 84% of the world's economy is locally produced and consumed. The effect of global capitalism will be for world trade to take an increasing share so that without regulation, transnational corporations would tend to relocate to LDCs with a loss of manufacturing employment in the MDCs. In 2000, some 63,000 transnational corporations and their 820,000 foreign affiliates (in which the transnational corporation would have at least a 20% equity shareholding) account for 25% of GWP and directly employ 90 million people, of which 20 million are in the developing countries. The foreign affiliates account for 54 million employees and, with annual sales of $19 trillion, they account for 10% of GWP. The 1,000 largest transnational corporations account for 80% of world industrial output and are concentrated in four sectors: petrochemicals, automobiles, electronics, and banking. Western Europe accounts for some 40% of world imports; however, if the EFTA intra-trade is excluded on the basis that it is a trading bloc, then Europe's share of world imports is only 16%. Foreign-based transplants are now joining a large set of US-based transnational corporations that have been established for a long time, and competition will greatly intensify with high-volume flexible production.

Up to the year 2000, FDI flows had been increasing at a higher rate than both GWP and the expansion of world trade. However, global, foreign direct investment declined sharply in 2001, as a reflection of a slowdown in the world economy. Inflows fell by 51% to $735 bn and outflows by 55% to $621 bn. FDI inflows to developed countries fell by around half from $1 trillion in 2000 to $503 bn in 2001. Inflows to developing countries decreased by much less from $238 bn to $205 bn, so that some 29% went to the LDCs. The United States received $124 bn in FDI inflows, and inflows and outflows to the EU in 2001 dropped by around 60% to $323 billion and $365 bn respectively.

In view of the comparable salaries between Europe and the USA, high levels of unemployment are more the result of productivity increasing from technology and the production processes than the transfer of production to less developed countries. However, exporting industries tend to be capital-intensive with relatively few workers, and imports, such as clothing and textiles, are generally in labour-intensive sectors. Clearly, the transnational corporations are the engines that drive the global economy, with a number of them having annual sales greater than the GNIs of the countries in which they operate. TNCs require regulation at the level of a continental union or a confederation of states, and it is important that international standards are established in respect of labour conditions, environmental controls, and quality in food processing.

The configurations that have been taken into account for transnational corporations and the world production system include:

N1   Global free trade releases the labour forces of formerly closed economies to compete directly with the workforces of the developed countries. Wages of production workers decline and domestic unemployment rises.

N2   Economic regionalism is considered to be the best defence against unfettered global capitalism, and protectionism and trade wars break out.

N3   Continental unions are formed with highly competitive internal markets, as an intermediate step before global free trade is unleashed. Less developed countries receive an increasing proportion of foreign direct investment.

N4   Continental unions agree on the nature and type of bilateral agreements with other trading areas, and any protective tariffs charged by importing unions are subjected to international taxation for recycling for infrastructure investment in the least developed countries.

P The world economy and globalisation

The Gross World Product is expected to rise from $31 trillion (2000 US$) in the year 2000 to $60 trillion in 2025 and $200 trillion in 2150. The investment requirement is for an increase of $150 trillion in total between 2000 and 2025, and $500 trillion between 2025 and 2150. After the year 2025, the less developed countries should be able to finance their own development, but in the period 2000–2025, a transfer of $450 billion a year from the MDCs to the LDCs could be achieved from a tax on foreign exchange transactions.

It is anticipated that pension funds will find domestic investment returns rather less than in the transitional or developing countries. The LDCs are likely to require increasing capital flows from the MDCs for infrastructure investments, and trade will be balanced by LDC imports of plant equipment and transport. Increasingly, regional trading blocs will be established with multiple international linkages to other countries to control both trade and capital flows. The global economy will discipline governments into adopting deflationary policies, or they will have to pay higher interest rates on debt. A technological shift in the transport and energy infrastructure away from hydrocarbons in the 21st century is likely to initiate an investment wave based upon the convergence of relevant technologies, and this will relieve unemployment.

The configurations taken for the world economy are as follows:

P1   Business-as-usual leads to wide prosperity by mid-century in which developing regions approach standards of living enjoyed in Western Europe in 1980, and GWP reaches 2000 PPP US$500 trillion by 2150. The per capita income in the MDCs in 2150 would be PPP US$125,000 and in the LDCs PPP US$40,000.

P2   Increasing levels of debt in the more developed countries result in low growth and financial instability.

P3   Financial reforms and regulatory changes reduce the risks of the global economy, and ecopolitan policies lead to sustainable development with GWP reaching PPP US$250 trillion by 2150. The per capita income in the MDCs in 2150 would be PPP US$40,000 and in the LDCs PPP US$18,000.

P4   Capital flows to the developing countries are raised to the necessary levels so that LDCs achieve 35% of gross world product in 2025, 50% in 2050, and 65% by the 22nd century.

12.4 Scenario overviews

Based upon the *Long-Range Futures Research* evolutionary landscapes, four scenarios S1, S2, S3, and S4 have been developed with world populations that range from low to high, and with an orientation that is either regional or global, together with varying degrees of international convergence or closure of the development gap. This makes them compatible with the IPCC Special Report on Emissions Scenarios (SRES). The planning time horizon has been taken as 150 years, and it is not anticipated that there will be a devastating catastrophe on a global scale within this period, although there are bound to be local and regional disasters. The aim of futures research is to design intervention strategies that are resilient to possible threats, so that dystopian *Mad Max* and 'Barbarization' or disastrous scenarios are saved for testing the robustness of the selected sets and for 'war-gaming' counter measures. Within the longer term, say 250 years, there will inevitably be discontinuities for any realistic scenario, which could be utopian or dystopian. Complexity science is neutral with respect to political ideology, and single visions cannot encapsulate the enormous complexity of society in a global context. All the scenarios recognise that world energy futures need to follow a decarbonisation path, and each scenario adopts its own emphasis such as human settlements and social equity, cultural diversity, the bio-ethic, and market-driven globalisation. These scenarios are also representative of the concerns of the different schools of futures studies, namely the Systems Science, Social Science, Environmental Science, and Management Science schools. Scenario S1 (Settlements first) has been selected as a base case against which the scenarios below can be evaluated:

S1 - **Settlements first**: A scenario without slums and a more level global infrastructure

S2 - **Continental cultures**: A cultural diversity scenario within a multi-polar world

S3 - **Biotech wave**: An ecological scenario for the world's living resources

S4 - **Generica rules**: A scenario with a worldwide mass market and a global superculture

The characteristics of each scenario are specified and quantified with values for the performance indicators by selection of the expected configuration for each of the systems, for a prognosticated end state in 2150 and intermediate states for 2050 and 2100. The values derived from the exploration of future conditions are used for the preparation of long-range scenario profiles, which are outlined and interpreted in Appendix 36. It is then possible to undertake a multi-criteria evaluation of the various scenarios by comparison with the base case and to identify alternative policies that could deliver the prospective outcome and the future standards of living that may result. In reality, the future outcome is likely to be a hybrid version of all the scenarios, with different elements being played out in the various continental regions of the world.

S1 - Settlements first

The Settlements first (S1) scenario profile describes a world of continental unions in which the economic strategy is to deal with the challenge of slums, squatter settlements, and poverty by direct intervention. This is achieved by investment transfers to the LDCs during the period 2000-2050, to assist with infrastructure and the funding of urban settlements to accommodate an extra 3.5 bn people. Any trade barriers between continental unions would be removed slowly in step with a levelling up of infrastructure in the LDCs, when the global playing field is more even. From 2000, the population of 6.1 bn increases on the lines of the UN median projection and doubles to 12.5 bn by 2150. There will be an increase in GWP in PPP US$ from $44.5 trillion in 2000 to $250 trillion in 2150, and a world increase in per capita GNI from $7,300 to $20,000. This represents a per capita income in the MDCs and LDCs of $40,000 and $17,000 respectively by 2150, to give a rich/poor convergence ratio of 2.3. Medium density human settlements with a garden-city philosophy will result in an overall population density of 1000 persons per km^2, which will take 10% of the world's land surface, with 30% for agriculture and 27% for forests in relation to 30% forest areas in 2000. The remaining uninhabited areas for wildlife will be common for all scenarios at 33%. In view of global warming, it will be beneficial to move away from fossil fuels long before reserves are depleted and to stretch the coal and oil resources as far into the future as possible, in parallel with the use of nuclear energy. A fairly rapid decarbonisation trend will limit global warming to 1° C per century. Technological hazards are the main concern, including transport accidents, toxic spills, and nuclear disasters.

The end-state for scenario **S1-Settlements first**, S1e has the following configuration:

- A4, B4, C2, D2, E3, F1, G3, H3, J2, K3, L4, M2, N4, P4

S2 - Continental cultures

The Continental cultures (S2) scenario profile describes a multi-polar world in which regional cultures may determine the shape of future geopolitical developments rather than globalisation. From 2000, there is a high-medium population increase to 15 bn by 2150, with an increase in GWP in PPP US$ to $320 trillion, and a world increase in per capita GNI to $21,300. This represents a per capita income in the MDCs and LDCs of $45,000 and $19,000 respectively by 2150, to give a rich/poor convergence ratio of 2.4. High density settlements with more persons per household can be achieved with a greater number of megacities and cosmopolises, with high-quality mass-transit and public transport systems, at an overall population density of 2,000 persons per km² will take 5% of the world's land surface. The high population will increase the demand for food with 36% of land for agriculture and 26% for forests. Energy policies will aim to minimise fuel-import dependence, with lower-income resource-rich regions relying on fossil fuels, and higher income, but resource-poor, regions will shift to nuclear and renewable sources. A relatively slow decarbonisation trend will result in global warming at a rate of 2.5° C per century. Geo-hazards such as earthquakes will be of concern in densely populated urban areas, particularly in lower-income regions where building codes may be ignored.

The end-state for scenario **S2-Continental cultures**, S2e has the following configuration:

- A3, B3, C3, D1, E4, F4, G2, H1, J4, K1, L2, M1, N2, P2

S3 - Biotech wave

The Biotech wave (S3) scenario profile describes a globally coherent approach to sustainable development, to ensure that the planet's biological diversity (including human systems) is protected. From 2000, there is low growth in population to 8.5 bn by 2150, with an increase in GWP in PPP US$ to $400 trillion, and a world increase in per capita GNI to $47,000. This represents a per capita income in the MDCs and LDCs of $110,000 and $38,000 respectively by 2150, to give a rich/poor convergence ratio of 2.9.

More compact cities at a high-medium density, with a tight control on suburban development and an emphasis on public and non-motorised transport, will give an overall population density of 1,500 persons per km² and take some 7% of the world's land area. The low population requires less food so that agriculture will take 25% of the land and 35% will be available for forests. Gas becomes a primary fuel for electricity generation using Combined Cycle Gas Turbines in the steady transition to renewable energy sources, which results in global warming at 2° C per century. The main hazards of concern include eco-catastrophes, bio-hazards, and pandemics.

The end-state for scenario **S3 - Biotech wave**, S3e has the following configuration:

- A2, B2, C1, D4, E2, F3, G1, H2, J1, K2, L3, M4, N3, P3

S4 - Generica rules

The Generica rules (S4) scenario profile describes a world with free-market principles, and a worldwide mass market with a global superculture. From 2000, there is low-medium growth in population to 10 bn by 2150, with an increase in GWP in PPP US$ to $500 trillion, and a world increase in per capita GNI to $50,000. This represents a per capita income in the MDCs and LDCs of $125,000 and $40,000 respectively by 2150, to give a rich/poor convergence ratio of 3.1. Lower-density settlements with urban sprawl and high car ownership, will give an overall population density of say 750 persons per km^2 and take up 12% of the world's land surface. A low-medium population with a higher proportion of meat consumption will require 28% of the land for agriculture leaving, 27% for forests. Globally, the energy systems remain predominantly hydrocarbon based to 2100. A mix of energy technologies and supply sources result in a shift from the current shares of fossil fuels and the gradual decarbonisation of energy with global warming at a rate of $3°C$ per century. The main concerns are natural hazards associated with climate change such as flooding, tornados, and tropical cyclones.

The end-state for scenario **S4 - Generica rules**, S4e has the following configuration:

- A1, B1, C4, D3, E1, F2, G4, H4, J3, K4, L1, M3, N1, P1

In making a comparative evaluation of the above 150-year scenarios S1, S2, S3, and S4, it should be noted that the *L-RFR* evolutionary approach gives a low overall rate of annual growth in Gross World Product from 2000 to 2150 of 1.15%, 1.32%, 1.47%, and 1.65%, respectively. In all scenarios there is a convergence between regions in GRP per capita in which the rich/poor ratio ranges from 2.3 to 3.1 by 2150. Scenario S4 has the characteristics of the IPCC SRES scenario AI, S3 has the characteristics of SRES scenario B1, and S2 is not dissimilar from the intention of SRES scenario A2. Since urbanization has not been rigorously modelled in any of the United Nations fifty-year or one-hundred-year reference scenarios described in Chapter 1, there is no direct comparison with S1, although from a population perspective, SI is comparable to the SRES scenario B2. In terms of the UNEP GEO-4 Outlook 2000–2050, S1 would represent a Policy First strategy, S3 would be Sustainability First, and S4 matches the Markets First scenario. As far as the fifty-year scenarios of the Global Scenario Group (GSG) are concerned, S4 is the Conventional World and S3 represents policy reform of the S4 market forces scenario. The GSG Great Transitions scenario lacks any mechanism to deliver the S1-Settlements First scenario, and S2 could be described as a Fortress World scenario but, unlike the GSG Barbarization scenario, it is not dystopian.

Designed interventions and technological development will shape the future by altering the trajectory of the civil system, through the investment-control parameter and the release of resource constraints. Scenario S1 would be the safest and most equitable strategy during the urban transition (2000–2050), and in the longer term (2100–2150) this could evolve into scenario S3 when the global population declines. It is important for future generations that we plan a suitable strategy for the distribution of cities and an energy, transportation, and telecommunications infrastructure on a planetary scale. Civilisation will continuously adapt to visions of the future as the landscape evolves, in accordance with the evolutionary macrolaws of Ecodynamics. In the 21st century, humankind will be ending its period of rapid growth using an excess of natural resources. In the ecopolitan stage of development, societal evolution will slow down with the need to complete the planet's infrastructure using limited natural resources to enhance the effectiveness of both the economic and ecological systems.

Long-range global scenarios: References

1. *The Challenge of Development* - World Development Report, World Bank, 1991.
2. *Development and the Environment* - World Development Report, World Bank, 1992.
3. *Investing in Health* - World Development Report, World Bank, 1993.
4. *Infrastructure for Development* - World Development Report, World Bank, 1994.
5. *Workers in an Integrating World* - World Development Report, World Bank, 1995.
6. *From Plan to Market* - World Development Report, World Bank, 1996.
7. *The State in a Changing World* - World Development Report, World Bank, 1997.
8. *Knowledge for Development* - World Development Report, World Bank, 1998/1999.
9. *Entering the 21st Century* - World Development Report, World Bank, 1999/2000.
10. *Attacking Poverty* - World Development Report, World Bank, 2000/2001.
11. *Building Institutions for Markets* - World Development Report, World Bank, 2002.
12. *Sustainable Development in a Dynamic World* - World Development Report, World Bank, 2003.
13. *Development and the Next Generation* – World Development Report, World Bank, 2007.
14. *World Development Indicators 2002* - World Bank, 2002.
15. *World Development Indicators 2003* - World Bank, 2003.
16. *Human Development Report 2007/2008* – United Nations Development Programme, 2007.
17. *State of the World 2001* - Lester R Brown, 2001.
18. *State of the World 2002* - Worldwatch Institute, 2002
19. *State of the World 2003* - Worldwatch Institute, 2003
20. *State of the World 2004* - Worldwatch Institute, 2004

21. *State of the World 2005* - Worldwatch Institute, 2005

22. *State of the World 2006* - Worldwatch Institute, 2006

23. *State of the World 2007* - Worldwatch Institute, 2007

24. *Vital Signs* - Worldwatch Institute, 2000- 2007.

25. *The State of the World Atlas* - Dan Smith, 2003.

26. *The State of the Environment Atlas* - Joni Seager, 1995.

27. *The State of War and Peace Atlas* - Dan Smith, 1997.

28. *World Desk Reference* - Financial Times, 2002.

13.
EVOLUTIONARY LANDSCAPES FOR A COMPLEX WORLD

Long-Range Futures Research: An Application of Complexity Science sets out the scientific basis for futures studies with an evolutionary vision of the world. Evolutionary generating phenomena are constrained by the transactional dynamics that arise between three distinct tiers, namely the transactional microstructure of the urban system, the global contextual macrostructure, and the planetary metastructure level. The tiers have some degree of independent dynamic pattern, but all the parts interact, and the so-called environment is always a part of the universal system. Futures Researchers will probably find it helpful to have a summary of the planetary metatrends that will influence deep futures beyond 1 million years, the civil megatrends shaping the long-range future to one hundred years and beyond, the global macrotrends affecting the intermediate future within a timespan of twenty-five to fifty years, and the sectoral microtrends impacting on the near future for up to twenty-five years. Since the book encompasses a range of scientific disciplines, the opportunity is taken in Chapter 13 to revisit some of the relevant facts and figures that that have already been covered in the earlier chapters.

13.1 Planetary metatrends and deep futures

The planetary metasystems will induce astrophysical, geophysical, physical, biological, and civil metatrends in a time frame spanning millennia and millions of years that may have both catastrophic and beneficial consequences for life on Earth. These metatrends will be influenced by the following phenomena:

13.1.1 Astrophysical metatrends

Astrophysical metasystem: The galactic spatial structure includes stars, galaxies, groups, clusters, and superclusters. The Milky Way is part of a group that includes the galaxy Andromeda and at least twenty smaller galaxies. Gravity is pulling Andromeda towards our galaxy at 100 km per second. This group lies at the edge of the Virgo cluster, whose centre is fifty million light years away. The clusters are arranged in a still larger filamentary or sheet-like structure of which the most conspicuous is the array of galaxies that form the Great Wall, 200 million light years away. The Great Attractor is another huge concentration of mass that exerts a gravitational force on the entire Virgo cluster of galaxies, and pulls us along at several hundred km per second.

Astrofuturism: Manned space stations are already in operation, and in due course, communities of scientists or technologists could inhabit the moon (three days' space flight from Earth) or Mars (nine months' journey away) by living in sealed biodomes. It is anticipated that this will take place at the beginning of the 22nd Century, in the Early Planetary era. Astrofuturism has visions of planetary terraforming, and the prospect of terraforming Mars so that it could be colonised by humans or other species would appear to be an immense task that could not be undertaken until the Late Planetary era from the Fourth Millennium.

Astrofuturism predicts five to 500 astro-technological civilisations within our galaxy at least 1,000–5,000 light years away. However, it would take 200 millennia to travel to a planet 1,000 light years away at 0.5 percent the speed of light, whereas the fastest spacecraft travels at 0.005 percent light-speed. But it is quite possible that there are civilisations in the older regions of galaxies, which are more advanced than our own. However, it is unlikely that contact will be made with extraterrestrial intelligence until the Near Interplanetary era from 10,000 YF.

Astrophysical timescale: In relation to the meta-timescale shown in Appendix 6, the current Ice Age on Earth is likely to come to an end in the near Galactic Era in 20 MYF (million years into the future), and there will be a clustering of continents in the far Galactic Era in 250 MYF. Also, there will be a collision and distortion of the Milky Way and Andromeda galaxies in the Intergalactic Era in 5 BYF (billion years into the future).

Astrophysical energy sources: The astrophysical system is a source of renewable energy as solar, hydro-power, or tidal energy. Hydro-electric power is a form of solar energy since it is generated from the precipitation of rain, whereas tidal energy is principally lunar. The radiant output of the sun is forty percent hotter than at the time when the Earth was formed, and it will become hotter as the core of the Sun shrinks over the next five billion years. After a further one billion years, it will become a white dwarf and cease to generate energy. The extinction of extraterrestrial life on numerous early planets will have occurred when the energy-producing stars of their planets burned out.

Asteroids: It is probable that within 100 million years, the Earth may be hit by an asteroid larger than 10 km in diameter. Although 200 asteroids have been identified in excess of 1 km wide, it is estimated that there are 2,000 of them and also there are 200,000 larger than 100 metres that could cause an impact similar to the Arizona crater within 10,000 years.

13.1.2 Geophysical metatrends

Geophysical metasystem: The planet's geophysical system involves the evolution of the earth from its molten core to the strata of the earth's crust and terrestrial stocks, to the outer reaches of the atmosphere. The geological evolution of tectonic plates, mountain building, earthquakes, volcanoes, and the geophysical phenomena such as glaciers, oceans, and climate intimately relate to each other.

Sea levels: Milankovitch cycles explain at least sixty percent of climate change during ice ages, and according to the isotope record in deep sea cores, their frequency is around ten complete cycles per million years. There is an alternative body of evidence from tree rings, corals, and ocean sediments that there were at least eight glacial/interglacial cycles in the past one million years. It is therefore possible that other variations may cause the climatic cycle to oscillate between 95,000 years and 125,000 years, and around 125,000 years ago, the sea level was about 5 metres higher than it is today, and temperatures were 1° – 3°C warmer. There is of course the possibility that anthropogenic carbon dioxide emissions may be postponing briefly the descent into the next glacial period, which is expected to commence within the Fourth Millennium.

Gulf Stream Conveyor: By 2500, the warm Gulf Stream conveyor mechanism may cease due to changes in salinity as a result of melting arctic ice, with a sharp decline in north European temperatures to near arctic conditions. This may be the trigger for the ending of the interglacial period at the end of this Millennium with the descent to another glacial period taking 70,000 years.

Tectonic Plates: Tectonic plate motion leads to the clustering of continents twice per billion years. When the continents assemble, there is evidence of a period of continental collision and mountain building, and on fragmentation and dispersal over millions of years, megacontinents eventually drift towards the poles so that the albedo (reflection of solar radiation) of the Earth is increased. Drifting continents at the rate of 40 km per million years change the pattern of ocean circulation and, in combination with glaciation at the poles, an ice age is triggered in which global temperatures drop to 8–12 °C, and sea water becomes locked up in the ice caps so that sea levels fall to 90–120 metres below current levels. In around 250 million years from now, megacontinents will form again and possibly a supercontinent that will eventually fragment.

Earthquakes and Volcanoes: Over 10,000 earthquakes are recorded around the world each year. Most earthquakes and volcanoes occur in narrow zones that follow the mid-ocean ridges and subduction zone boundaries. The magnitude of earthquakes is measured on the Richter scale, and the severity at a location will depend upon the distance from the epicentre and the local geological conditions. The Sichuan Province earthquake in China in 2008 measured 8.0 on the Richter scale, and 70,000 people died, a further 375,000 people were injured, and over one million homes were destroyed. In developing countries, the scale of damage and loss of life are often hugely disproportionate to the size of the quake due to the poor quality of building construction. Volcanic hazards and the level of destruction and loss of life depend upon the scale, the duration, and the timing of the eruption. In any single decade, up to a million people may be detrimentally affected by volcanic activity.

13.1.3 Physical metatrends

Physical metasystem: The planet's physical system is described in terms of physical energy and chemical dynamics. The lowest part of the Earth's atmosphere, the troposphere, extends upwards for approximately 10 km, and between the troposphere and the stratosphere, comes the ozone layer, which filters out the harmful short-wave radiations from space and makes possible advanced life forms on Earth. The energy from the sun is expected to increase by 1% per 100 million years, so that by 1 BYF the heat may become so intense that the biosphere will expire. Cosmic radiation and nuclear fall-out from leakage, accidents, or nuclear weapons represent a short- and long-term threat to humankind.

Climate change: It is anticipated that temperatures will rise by 1°C per century until 2500 when temperatures will peak at 20°C for a fourfold increase in carbon dioxide levels to 0.15% by volume, and then decline by 0.5 °C per century to 17.5 °C by the end of the millennium in 3000, when fossil fuels (coal, oil, and gas resources) become exhausted. The rise in sea levels by 3000 is expected to reach 4 metres. In the absence of policy changes, an eightfold rise in carbon dioxide levels to 0.30% would result in a peak temperature increase of 7.5°C, with an even greater rise in sea levels of 6 metres by the end of the Third Millennium.

Decarbonisation of energy: The evolution of the energy system from fuel wood, to coal, to oil, and then to gas is a trajectory of energy decarbonisation at a rate of 0.3% per year and a continuing reduction in the ratio of carbon to hydrogen atoms with each succeeding energy source. The use of hydrogen as a fuel to power the planet would be environmentally benign, as hydrogen burns in air to emit only water with a relatively low concentration of nitrogen oxides (NO_2). A possible increase in H_2 in the upper atmosphere should not contribute to the greenhouse effect, as the H_2 molecule does not absorb radiation in the same way as CO_2 or CH_4 molecules. In fact, the water vapour would cause clouds to form at high altitude, which would increase the albedo so that there would be a decrease in temperature.

13.1.4 Biological metatrends

Biological metasystem: The biosphere is the region of the Earth's crust and atmosphere in which living matter is found. The biological metasystem involves living organisms, with humans and species of animal and plant populations and their ecosystems. Local climate changes could shift prime food growing regions in a matter of decades, with changes in the length of seasons and rainfall, which could lead to the destruction of many ecosystems and threaten thousands of species. Mass extinctions of biological species have resulted from geophysical changes, physical catastrophes, or economic development, and historically, climate changes have profoundly affected societal evolution. There are also a variety of possible scenarios for eco-catastrophes, bio-hazards, pandemics, and plagues.

Photosynthesis: It has been estimated that humans currently appropriate perhaps thirty percent of the solar energy captured in organic material by land plants. The definition of human appropriation underlying this figure includes direct use by human beings in the form of food, fuel, fibre, and timber, plus losses due to alteration of ecosystems by the conversion of natural systems to agriculture. If the global potential of aquatic forms of life is included, then humans appropriate some twenty percent of the planetary exergy from photosynthesis. It is apparent that, without technological development, a doubling of the human scale would consume forty percent of the exergy, which would leave only sixty percent for all non-human and non-domesticated species. Since humans cannot survive without ecosystems that are made up of other species, it is clear that, in the next generation, the existence of the limits to the biosphere will become more apparent.

Biodiversity: It has been estimated that of the 500 million to one billion species that have existed in evolutionary time, ninety-seven percent are now extinct. Species extinctions are an essential part of the evolutionary process and, in the 600 million years of multicellular life on Earth, it has been estimated that episodic environmental transitions account for sixty percent of all extinctions. Of the estimated twenty to thirty million species living today, some 1.75 million species have been classified. Although species extinction is a natural process, there is concern that natural extinction rates have been increased during the past 400 years (1600–2000) as a result of human activity.

Lifespans: Human lifespans are limited by the number of times that cells can replicate before the body ages; normal, healthy cells in the body divide up to 100 times before becoming senescent. It appears that the DNA strands on the end of every chromosone become shorter and shorter with the division of cells, and this sets a theoretical biological limit of a 120-year lifespan for humans. Longevity is a genetic trait, and scientists are engaged in research in the fields of genetics, neuroscience, and biochemistry in an attempt to understand the multiple causes of ageing. There is the view that there are diminishing returns from advances in the medical battle against ageing, because the body's systems wear out in a sequence predicated by the worst first, so that the treatment of one system will lead to another system failing in its place. Similarly, as a cure is found for one disease, an alternative cause of death will arise.

Human health: The health of human populations will be affected by climate change as a result of natural disasters, cyclones, floods, landslides, and fires. Also, ozone depletion will increase skin-cancer rates, and temperature rises will enlarge the habitat of malarial mosquitoes. A myriad of microbes survive by feeding on and breeding within humans, and contagious infectious diseases, viruses, plagues, and epidemics such as AIDs put humans at risk. Periodical famines, nutritional deficiencies, and diseases of affluent lifestyles such as overnutrition, alcohol, smoking, and drugs are health hazards that affect people through the different stages of development. Exposure to pesticides, noxious chemicals, heavy metal toxins, asbestos, and a variety of air pollutants are the consequences of industrial development and need to be limited by environmental protection.

13.1.5 Civil metatrends

Civil metasystem: The civil system is driven by Ecodynamics with the emergence of complex civil ecostructures (towns and cities), which give long-term stability to civilisations and facilitate the diffusion of investment capital, knowledge, and culture. Civil disasters would include sudden events such as fires, floods, earthquakes, severe storms, transport accidents, toxic spills, nuclear disasters, and terrorism or war.

Civilisation: Arnold Toynbee's view of civilisation was his well known 'challenge and response' thesis in which society encounters a succession of problems, and civilisation develops to surmount them as a fresh dynamic movement. Toynbee saw the collapse of civilisations as a loss of creative power and a failure of vitality. The anthropological perspective on the rise and fall of civilisations is best described in *The Collapse of Complex Societies* by Joseph Tainter in which he studies the reasons for the collapse of past societies. History can provide examples of civilisations, such as the Roman Empire, ancient Egypt, the Indian Mauryan Empire, Chou China, and the Mayan civilisation, in which the increasing costs of social complexity resulted in exhaustion of natural resources and the decline and eventual collapse of these societies.

Evolution of a planetary civilisation: In the 21st Century, the governance systems will face two counterbalancing forces, which are globalisation with continental integration of nations and localisation with the devolution of power to states at the subnational level. Evolution of the civil system involves increasing complexity, which means additional hierarchical levels at the planetary, continental, and subnational levels. The highest evolutionary potential will be achieved by developing a rich diversity within a coherent unifying structure, and this explains the emergence of economic regionalism as a defence against global capitalism. When continental unions face stagnation or diminishing returns, they will have to respond as complex adaptive systems and constantly revise and rearrange their building blocks or subsystems. New institutions will evolve to enhance planetary potential and to prevent military conflict, terrorism, economic and ecological catastrophes, and other civil or natural disasters.

Trajectory of the urban system: World urbanisation progressively adds levels to the urban hierarchy. Towns and cities or civil ecostructures are information-rich structures that evolve through selective changes and movement by individual households and establishments, with immigration and emigration as they respond adaptively to spatial and temporal variations in the environment. The potential of an ecostructure is a measure of its capacity to diffuse investment capital within a specific spatio-temporal range. The world urban system remains on a *universal attractor* to increase the overall diffusion of investment capital through the realisation of potential, and the system will evolve through the formation of islands of increasing transactional complexity and connectivity. The trajectory of the urban system towards equipollence in the urban hierarchy will be approached when the global urban population exceeds ninety percent of the 12.5 billion world total in 2150, with 750m urban population at each of the fifteen levels. The default rule for the distribution of the urban population by the year 2150 is expected to be that one-third of the urban population will occupy the five smaller urban size classes of less than 100,000 population, one-third will be in cities and metropolises of 100,000 to 4.99m, and one-third in cosmopolises, megapolises, megalopolises, or metapolises of over 5m population.

Territorial migration: Increasing populations of both humans and establishments in a bounded territory reach saturation, and diminishing returns ensue as further growth is unsustainable in accordance with the 'atrophy' law or the logistic curve. Increasing returns in less mature states create a rate-of-return gradient and, consequently, streams of investment are diffused to less mature places, which generate establishments and stocks of artefacts that attract populations as the potential is realised. Further investment at these locations will tend to reduce the rate-of-return gradient and, as they become mature, they evolve to their full potential with a reasonable per capita standard. There will be a dynamic balance when there are common levels of stocks of assets per capita within the states or regions of continents.

Geo-engineering and the hydrosphere: Geo-engineering of the environment could involve large-scale engineering countermeasures or other interventions in climate change, so that humankind has sufficient time to adapt to global warming. Options include carbon capture at power plants, increasing carbon sequestration by the oceans using iron fertilisation to increase phytoplankton populations that will draw more carbon dioxide, or by reforestation and afforestation using carbon off-set schemes.

13.2 Civil megatrends shaping the long-range future

The ten most significant civil megatrends that are likely to run for over a century and impact upon the current millennium are as follows:

- Globalisation with increasing interdependence of establishments in a transnational vivisystem of unprecedented complexity.
- Localisation with the emergence of 1,000 ecopolitan states that will lead to a diverse planetary civilisation.
- An ecologic transition in the MDCs to sustainable development.
- The urban transition with a peak rate of expansion of the urban population in the LDCs from two billion in 2000 to four billion in 2025, and then to 5.5 billion in 2050.

- The demographic transition to low fertility in the LDCs and an eventual stabilisation and ageing of the metapopulation.
- Developments in the life sciences, biotechnology, and genetic engineering.
- A technological shift from a hydrocarbon economy to a hydrogen fuel economy.
- Advances in space technology with fractional orbital flight between perhaps ten to fifteen planetary 'metapolises' with 50-100m population by the 22nd century.
- Digital communication in a meta-language that will increase the rate of cultural evolution.
- The continued increase in the Gross World Product per capita from $5,150 in 2000 (US$ 2000) to $7,300 in 2025, and then to $16,000 in 2150.

Sustainable Development: Scenarios for a sustainable future show that a world population of 12.5 billion can be supported providing there is a demographic transition in the LDCs and an ecologic transition in the MDCs. The capacity of the planet could sustain a Gross World Product (GWP) of $200 trillion (PPP $250 trillion) in 2150 in relation to a GWP of $31 trillion (PPP $45 trillion) in the year 2000. The world stock of assets in 2150 is expected to reach PPP $800 trillion, with a per capita investment of PPP $64,000.

Urban transition: In the year 2000, there were some 3,325 cities of over 100,000 inhabitants, 300 metropolises of 1m population, thirty cosmopolises and twenty megapolises. It is estimated that by 2050, there will be seventy cosmopolises and fifty-five megapolises, and the majority of these already have populations of over 1m. By the year 2150, there will be nearly 10,000 cities of over 100,000 population, of which 750 will be metropolises of over 1m inhabitants, and over 100 will be cosmopolises of more than 5m inhabitants, and eighty-five will be megapolises or megalopolises of over 10m inhabitants. In the less developed countries, the number of cities with populations over 10m will rise to over thirty by the year 2025 and probably seventy in the 22nd century, compared with fifteen megapolises or megalopolises by the year 2150 in the MDCs.

Evolution of infrastructure systems: By the year 2000, the world population of cars reached 540m or nine cars per 100 population, and from the evolutionary assessment, the number of cars will exceed 1.1 billion by 2025 in relation to a world population of 8.2 billion or fourteen cars per 100 population. By the year 2150, the number of cars may exceed 3.1 billion for a population of 12.5 billion, or twenty-five cars per 100 population. It is anticipated that by 2050, half the vehicle fleet will no longer be using petroleum fuels and that a shift will have taken place from the hydrocarbon economy to a hydrogen-fuel economy. A hybrid electric car may capture a significant share of the market, and the initial penetration is likely to be for a family's second car for local trips. It is envisaged that space ports will emerge with the hydrogen-fuel economy, and the pay-off from space flight is that a spaceline could deliver passengers and cargo to any place in the world in less than two hours. A fifteen-hour subsonic flight from New York to Tokyo would be achieved in ninety minutes using "Fractional Orbital Flight" with take-off, a climb above the atmosphere, a partial orbit, descent, and landing.

Renewable resource exploitation: The potential global resources of non-fossil fuel (NFF) for the production of hydrogen energy include nuclear power, hydropower, tidal power, wind power, and solar power. Hydrogen provides a means of storing energy from renewable sources and also provides a renewable fuel for use in vehicles. On the basis that hydrogen is produced by its dissociation from the water of the oceans, it is necessary to assess the extent to which the burning of hydrogen will offset rising sea levels. The oceans cover seventy percent of the Earth's surface with an area of 361 million km^2, so that 1 cm depth of the oceans will occupy 3610 billion m^3. If the fuel consumption of a typical car was 20 km per litre, then for 25,000 km per year a car will use 1.25 hectolitres or 12.5 m^3 of hydrogen. On the assumption that a typical car will involve the annual extraction of hydrogen from 20 m^3 of sea water, then one billion cars will use twenty billion m^3 of the oceans each year. It follows that in a century three billion cars would reduce the ocean levels by less than 2 cm, and over 10,000 years this would amount to only 2 metres.

13.3 Global macrotrends and the intermediate future

Within the context of the civil and economic megatrends, scenarios for the future evolution of the global macrosystems (outlined in Chapter 12) reveal some thirty-five macrotrends that will shape the landscape in the following spheres, the politicosphere, the geosphere, the genosphere, the sociosphere, the technosphere, and the infosphere.

13.3.1 Politicosphere

Planetary governance: The introduction of new institutions for planetary governance, with some twenty-five continental unions of 500m population represented on the United Nations Security Council for the resolution of intercontinental disputes.

Globalisation: The creation of a United Nations Global Capital Fund by the imposition of a 0.5% tax on foreign exchange (FX) transactions, for direct diffusion into infrastructure investments in the LDCs to level the global playing field.

Continental integration: There is a need for regional economic integration and, eventually, the world may comprise some twenty to thirty continental blocs with populations in the range 400m–600m. Within each continental bloc there may be perhaps fifty ecopolitan states with average populations of 10m, and these will range from 5m to 20m inhabitants.

Localisation: The United Nations now officially recognises nearly 200 countries or nation-states, although these states assert sovereignty over some 5,000 ethnic groups. It is envisaged that nation-states will accommodate sub-states or smaller states, which are likely to evolve into perhaps 1,000 ecopolitan states by the year 2150.

Democratic reform: As the world population doubles in the period 2000-2150, the existing voting rights in the General Assembly of the United Nations will become increasingly distorted. The sub-continental countries of China and India with a combined population of over two billion have only one vote each, and countries with populations of five million or less also have a single vote. Democratic pluralism may demand that ecopolitan states should each be given one vote in the General Assembly.

13.3.2 Geosphere

Resource scarcity: The long-term goal for civilisations is to achieve a dynamic balance amongst a growing human population, the biosphere of the Earth, the agrarian stocks, which are limited by seasonal units of time and solar flow, and the terrestrial stocks of the geophysical system. The terrestrial stocks of the geophysical system, such as fossil fuels and mineral deposits, are not renewable within a human time span so that there are limits and ultimately absolute shortages.

Land-use: Suitable agricultural land of 0.2 hectares (FAO classification) is required per person to provide one million of calories per year. In the MDCs, the cultivated land and pastures in 2000 amounted to 1.20 hectares per person compared with 0.59 hectares per person in the LDCs. For a world population of 12.5 billion in 2150, the cultivated land and pastures in the MDCs will amount to 0.95 hectares per person compared with 0.23 hectares per person in the LDCs.

Ecologic transition: Economic growth is beneficial in achieving the demographic transition in less developed countries. Through the redistribution of investment rather than the redistribution of wealth, the provision of infrastructure assets will reduce pollution and the vulnerability of the poor. However, there are social limits to economic growth in the more developed countries due to ageing populations, and there is little benefit in the MDCs achieving a higher per capita income than the per capita gross state product of California in 2000, at nearly $40,000. This level of growth would assist in delivering an ecologic transition in the more developed countries, since a higher national income will permit increased investment in pollution abatement technologies and the decarbonisation of fuels, including a shift to renewable energy resources, energy efficiency, and longer-life assets.

Water supplies: Most of the countries with limited renewable water resources are in the Middle East, North Africa, and sub-Saharan Africa, the regions where populations are growing fastest. Elsewhere, water scarcity is less of a problem at the national level, but it is nevertheless severe in certain watersheds of northern China, west and south India, and Mexico. The minimum per capita annual water requirement is generally in the range of 800-1200 m³.

Habitat destruction: The rain forests account for half the species of plants and animals on the planet, say ten to fifteen million species of which the majority are insects, weeds, and fungi, and 0.1-0.2% of these species become extinct annually. Taking into account current extinction rates for mammals and birds at 100-200 times the background rate, current extinction rates for all species lie in the range 100-1,000 times the historical background extinction rate of 2,000 species per century. This reflects a mass extinction event caused by economic development.

13.3.3 Genosphere

Population stabilisation: By the year 2150, the population of the planet will reach a peak of 12.5 billion before declining. This will arise as the standard of infrastructure provision is raised in the LDCs, which induces a demographic transition with a decline in human population growth. In effect, an increase in the minimum levels of civil artefacts per capita assists in achieving both a new plateau for economic development and a stabilisation in the population. It is the growth in the populations of artefacts that creates concerns for environmental sustainability. Many families prefer to have a car than an extra child, and there is a population explosion of 'animats.'

Lifespan: The focus of healthcare is likely to be on improving the quality of life in the context of a realistic life expectancy of say eighty to eighty-five years, before engaging in expensive treatments to extend lifespans by more than just one year per decade after the year 2000. This is in line with United Nations' projections, and an average lifespan of 100 years is probably two centuries away. As stated in paragraph 13.1.4 under biological metatrends, there is a theoretical maximum lifespan for humans of 120 years.

Genome project: The first details released on the human genome in 2001 revealed that the genome contains 30,000–40,000 genes, far fewer than the 100,000 estimate used for most of the last decade. The analysis revealed that humans have ninety-five percent of their genes in common with chimpanzees and eighty-five percent in common with dogs. Some of the genes appear to have come directly from bacteria, and the complete DNA sequencing of other organisms will answer many important questions on biological evolution. The ethical, legal, and social implications of the genome research are a matter of intense debate.

Genetic conservation: Conservation of the genetic material of plant and animal species through biodiversity is necessary if new strains are needed should domesticated strains become vulnerable. This keeps options open for the future, and since some twenty-five percent of prescription drugs are derived from plants, new cures may depend upon the prevention of genetic impoverishment.

Genetic engineering: Animal and plant breeders have for generations selected the features they wanted for domestic varieties in a process of artificial selection. Genomics will accelerate the process of identifying and classifying genetic variations, and the understanding of a human's DNA profile is likely to lead to a more scientific basis for mate selection and an increase in the demand for sperm banks. As the molecular foundation of hereditary diseases emerge, it will be possible to design gene therapies for a specific patient's DNA profile. It is inevitable that people will strive to have offspring that are more intelligent, better looking, more athletic, and have attractive personalities. For example, humans may evolve or genetically engineer themselves into some form of genus intelligens within 100,000 years.

Biotechnology: Biotechnology will change the relationships between agricultural inputs such as land areas under cultivation in different geographic regions and the output in terms of nutritious food or forest products. New strains of crops, vegetables, and fruit that are resistant to pests, frosts, mechanical handling, and refrigeration will have a longer shelf-life for delivery to an expanding population. Already intensive livestock farming methods are showing a disregard for biotechnological safeguards and feeding bonemeal to cattle has led to BSE and the human condition CJD. Bioremediation of contaminated waters and soils is a promising application of biotechnology to environmental problems, and biopharmaceuticals will follow in the wake of genomics.

13.3.4 Sociosphere

Institutional change: Institutions are created to provide the structure that societies impose upon the interactions between humans and the transactions between establishments to prevent chaotic conditions. Institutions are also responsible for the normative enhancement or design of the urban microsystems and also of the global macrosystems. Institutional reforms specify new rules or alter old ones to change the behaviour of the system; however, the difficulty in altering institutional paths means that institutions also act as constraints with respect to societal choices.

Culture: In view of the enormous timescale of anthropogenic evolution over many generations in relation to the shorter time span for societal evolution, it is useful for modelling to partially uncouple the biological metasystem containing individual humans from the civil system in which the household becomes the unit of society. Socio-cultural evolution involves the development of cultural artefacts such as language, communications, mathematics, music, the arts, customs, and religion. Culture embraces traditions, beliefs, value systems, and ways of life; and cultural diversity arises from national, ethnic, and genetic differences that are preserved in nation states, urban quarters, and ghettos.

Languages: Although the official tongue of a country is a key factor in terms of the maintenance and spread of a language, colonialism has resulted in more than half the states having either English, French, or Spanish as an official language. These languages transcend national boundaries to access a wider world in diverse fields such as international government, commerce, science and technology, information technology, mass entertainment and sport. At current rates of language extinction, the number of spoken languages may reduce from 6,000 to fewer than 1,000 by the year 2150.

Religion: At the macro level, continental unions, civil institutions, political ideologies, and religious beliefs provide continuity factors in cultural change and are slow variables. Religion has provided a major stimulus for language standardisation; for example, Arabic is an important international language for integration of the Islamic world. During the period 2000–2050, it is envisaged that conflict between indigenous groups and religions will escalate because of the increasing inequality arising from globalisation. But by 2150, it is likely that cultural and religious diversity will be encouraged by all faiths to promote a shared vision of moral values. By the year 2500, which is a short period on a geological timescale, it is envisaged that the major civilisations of the world will converge towards 100 continental cultures with their own predominant language family and religion. This will be the outcome of the centrifugal force of economic integration acting against the social cohesion of multicultural societies.

Lifestyles: Within the institutional and cultural contexts there are rapid changes in lifestyles, consumer tastes, and fashion. Lifestage and lifestyle profiles are compiled from market and social research surveys that obtain responses on household size (number of persons), marital status (including the length of a couple's relationship); age groups in ten-year bands from 15–24,

25–34, etc., to 65+; children's ages (under 2 years, 2–4, 5–9, and 10–14 years); genders, educational levels, employment status, and occupation; the number of earners and household income (nine bands); years of residence (mobility factor), newspaper readership, and television viewing.

Human development: One of the most important problems to resolve is the design of an approach for LDCs to achieve the circumstances in which human society can evolve to its full potential with a sufficient per capita standard of living. The essential investment in human capital with high educational and health standards will need to be achieved, with acceptable and rising standards in terms of life expectancy, adult literacy, fertility rates, nutrition, pollution, poverty levels, social equity, and political democracy.

Transition to an ecopolitan society: The aim of social policies for an ecopolitan state would be to shift expectations from expansion and growth to human development, ecological consciousness, and quality of life. An ecopolitan society with complexity and diversity would be a complex adaptive system that keeps order and chaos in balance, so that a vibrant and equitable economy can reach higher potential. The opportunity offered by an ecopolitan society will be a shorter working life with fewer working hours, although it will result in a decline in lifetime income. There will be more emphasis on long-life artefacts and self-reliance through local production, with compensation for reduced income from an increase in leisure time and social transfer payments from ecotaxes.

13.3.5 Technosphere

Materials: Steady reductions in energy and resource intensity, which are expressed as the ratio between a country's demand for energy and materials and its GDP (i.e., GDP output per kilogram), have given rise to the concept 'dematerialisation of the economy.' The intensity of use of a given material (or energy) increases with per capita GDP for all economies, reaches a maximum at a similar per capita GDP, and eventually declines. The changing nature of demand arises as economies develop from pre-industrial communities based on agriculture, through the subsequent stages of economic development, with extensive infrastructure build up accompanied by accelerated industrial growth, followed by industrial maturity and saturation of bulk-products demand, and later by the postindustrial era.

Nanotechnology: Nanotechnology involves an extension of existing sciences to the nanoscale where a nanometre (nm) is one billionth of a metre, so that 1 millimetre equals 1m nanometres. A nanometre is the length of 10 hydrogen atoms; the width of a water molecule is 0.3 nm. The branch of physics known as quantum mechanics applies to the scale of atoms and molecules, but nanotechnology cuts across a number of disciplines including colloidal science, chemistry, and materials science as well as electrical and mechanical engineering. New properties of materials can be found at the nanoscale, so that materials can be produced with improved optical, magnetic, thermal, or electrical properties. Nanoscience has been applied to the development of new instruments and apparatus, food and agriculture, electronics and computing, health care, and energy with more efficient solar power and high-efficiency fuel cells. There are a number of potential military applications but there are concerns about the environmental and health risks of manufactured nanomaterials.

Industries: The building of large material-intensive cities and related water supply, sewage facilities, utilities networks, factories, transport, and distribution infrastructures, account for a significant proportion of industrial activity in developing countries. Different waves of infrastructure development tend to follow each other, and this also applies to consumer durables such as the diffusion of the car and household appliances, which have prolonged the phase of the intensity of use of many metals. With saturation of demand, materials use becomes more confined to replacement markets. At the postindustrial stage, recycling acquires a significant role in decreasing the material intensity of economic activities.

Robotics: Robotics encompasses the design, manufacture, and application of robots using the technology of mechanics, electronics, and software. Industrial robots are by far the largest application of robotics, and there are currently around 1m robots with demand increasing as prices fall. There are also agricultural robots that are used for harvesting crops, fruit picking, milking cows, and cutting timber. Also in the United States, billions of dollars are spent on research for military and space applications. Home robotics are at an early stage of development with best-selling robotic toys for children, but it is expected that there will be rapid growth in domestic robots, which are likely to overtake the industrial market.

Energy: The International Energy Agency estimates that there will be a 1.1% per year increase in energy efficiency, so that there will be a 50% 'dejouling' of energy within 100–150 years. During the Third Millennium, the remaining fossil fuels will become exhausted, and there will need to have been a transition to alternative sources of energy such as nuclear, hydropower, tidal power, solar energy, hydrogen, windmills, geothermal, biofuels from crops, combustible waste, biogas plants, and fuel wood. Spent nuclear fuel is accumulating in a number of countries, and it contains various radioactive substances, including plutonium, which has a half-life of 24,000 years. This waste is estimated to be harmful to humans and animals for 100,000 years. In many cases, there will be an optimum technology for specific regions or sites depending upon the available supplies of fossil fuels, the proximity of oceans or rivers, and the annual amount of sunlight.

Transport: In the postindustrial city, employment dispersal has arisen from the wider use of the car and the range of services, and employment opportunities have increased the potential population capacity. Further technologies, such as inter-urban motorways and high-speed rail, tend to permit the growth of megacities or megapolises, and air transport and telecommunications link the global system of cities. The early reduction in the use of hydrocarbon fuels through engine efficiency, and a shift away from hydrocarbon-based transport technology to hydrogen, natural gas, various bio-fuels, or electric motors, lead to the conclusion that high-density compact cities are not the only solution. The freedom for people to use their cars, even at a high price, is highly valued, and it is envisaged that saturation levels for car densities in Europe could be one car per economically active person, i.e. forty-five cars per 100 population.

Military technology: Defence systems and the military exploitation of space have largely related to reconnaissance satellites, military surveillance, spacecraft tracking, advanced navigation systems, and command, control, and data acquisition networks. Space warfare would involve ground- or air-based anti-satellite missiles, orbiting hunter-killer satellites, orbiting missile defence systems, laser weapons, electronic jamming, and decoys. The United Nations and arms control treaties prohibit nuclear detonations in outer space, orbital nuclear weapons, and weapons of mass destruction, and there are agreements to prevent the use of Fractional Orbital Missiles.

13.3.6 Infosphere

Telecommunications: Telephone installations increase rapidly with growth of the quaternary information division, with saturation for mainline telephones being around 6,000 per 10,000 population. Mobile telephone penetration is currently reaching 4,000 per 10,000 population in Scandinavia. Similarly, the penetration of personal computers is reaching 4,000 per 10,000 population, and in the USA the number of internet hosts corresponds to some 25% of personal computers. In Europe, the penetration of television sets reaches 6,000 per 10,000 population, with newspaper readership at 3,000 per 10,000 population. Not surprisingly, in the USA, the number of television sets reaches 8,500 per 10,000 population, and the newspaper readership declines to 2,000 per 10,000 population.

Information technology: Previous technological waves have eventually created more jobs than they displaced, with the benefit of raising the standard of living for an increasing proportion of the population. However, there appear to be unique properties of the communications and information technology revolution that could inhibit or prevent a repetition of this apparently self-correcting process in the MDCs. As this technology grows exponentially in power and economy, it transforms, displaces, or replaces other means of control using people with increasing ease. The combination of telecommunications and microcybernetics is causing the shrinkage of distance and time, so that multinational corporations based in selected cities of the world can control companies and production plants in all parts of the globe. In effect, additional employment will be created in the LDCs with a loss of employment in the MDCs.

Artificial intelligence: The term artificial intelligence was coined in the 1950s by a group of researchers at a conference, which included John McCarthy, Marvin Minsky, Herbert Simon, and Alan Newell. The applications were seen to be speech recognition, medical diagnosis, software, and expert systems. Current uses include analysis of customer behaviour, pattern recognition, analysis of financial transactions, mobile communications management, artificial pets, artificial life, computer gaming, and military applications, including autonomous vehicles.

Multimedia: The shift from analogue technologies to digital technologies has permitted the development of multimedia interactive systems, with the integration of text, images, and sound. The interconnection of both households and business establishments in a global network, with the immediacy of real-time or select-time communications from such a multiplicity of sources, will create a vivisystem of unprecedented complexity. The emergence of new images and visions through the synergy of digital media, computers, and telecommunications will stimulate interactive behaviour between the sender and the receiver.

Infoglut: There is an 'infoglut' of competing problems and messages on cable television with perhaps 200 channels, TV advertising, the internet, and e-mails, faxes, correspondence, newspapers, journals, books, CD-ROMs, videos, radio, audiocassettes, posters, and junk-mail. Even worse are multi-media artefacts and computer games that display violent or pornographic images that result in negative behaviour or emotions. Whilst the best of television provides a window on the world with an unparalleled opportunity for learning, perhaps pay-TV or other forms of user charging will be applied to provide a filter and a rationing of the worst of the excesses.

13.4 Sectoral microtrends and the near future

The spheres of the global macrotrends can be related to the primary, secondary, tertiary, quaternary, and quinary divisions of the economy, and for this purpose the RICI Sectoral Classification is used as a framework for identifying over 150 sectoral microtrends. A number of these microtrends are identified in articles and annual outlooks in the *Futurist*, which is the bimonthly magazine of the World Future Society. These microtrends have been listed below as examples, to enable the identification of specific public sector, institutional, and business problems and opportunities that relate to the overlying macrotrends.

13.4.1 Primary Resources Division

Agribusiness: Changes in eating habits and calorie intake, meat eating and animal feed, genetically modified crops, enlargement of the cultivated land area, and irrigation schemes.

Non-fuel minerals: Some thirty billion tons of non-fuel minerals are extracted annually, including metal ores, chemical minerals, and fertilizers, quarrying, cement, building materials, and overburden. This is twice as much material as the total sediment carried by the world's rivers. Dematerialisation trends and resource substitution will reduce resource extraction and depletion during the urbanisation trajectory of the planet.

Fuels: Hydrocarbon, hydrogen, nuclear

Renewable energy: Hydro, tidal, wind, and solar

Environment: Ocean technology, fishing and aquaculture, forestry and deforestation, climate change, water desalination, wildlife and biodiversity, contaminated waste, geographic information systems (GIS).

13.4.2 Secondary Industrial Division

Property: LDC urbanisation, megacities, artificial worlds, and high-tech architecture and structures.

Construction: Ecodwellings and smart buildings.

Manufacturing: Smart materials, computer-aided manufacturing, robotics, nanotechnology, internet procurement auctions.

Automobile industry: Long-life car, new fuel cars, zero emission vehicles, guidance systems.

Aerospace industry: Superjumbos (800–1000 seat), spaceplanes, space structures.

Recycling: Building materials, domestic waste, vehicles.

13.4.3 Tertiary Commercial Division

Banking and financial services: E-cash, internet banking, direct insurance, and banking services, internationalisation of equity markets.

Transport: Travel ports, integrated transport, high-speed trains, guided busway systems, electronic road pricing.

Distribution and logistics: Just-in-time deliveries, vehicle monitoring and control, computer-load planning.

Retail: Customised design, teleshopping, smart shopping trolleys.

Hotels, tourism, and travel: Themed hotels, high-tech hotels, ecotourism, internet bookings.

Sports and leisure: Extreme sports, techno sports, on-line gaming, virtual reality, themed leisure complexes.

13.4.4 Quaternary Information Division

Media: Multimedia interactive systems, multichannel pay TV, cult TV, internet advertising, virtual advertising.

Business services: Globalisation of business services, virtual meeting places, and virtual organisations.

Telecommunications: Telephone lines and data transmission, internet hosts, mobile telecommunications, low-orbit satellite communications, and real-time/select-time communications.

Information technology: E-mail, speech recognition, supercomputers, web archives.

Design/artificial intelligence: Computer-aided design, expert systems, artificial intellects, artificial life, biomorphic robots.

Institutions: International institutions, institutional design, new social institutions.

Government: Planetary governance, continental unions, international taxation.

Politics: Democratic reform, ecopolitan states, citizen's networks, electronic polls.

Defence: Superpower military deterrence, demilitarisation, missile-defence systems, smart bombs, chemical weapons, biological warfare.

Police and crime: Surveillance, digital crime, electronic terrorism, remote tracking systems, electronic tagging, DNA forensics, smart guns ('Tasars').

13.4.5 Quinary Lifespan Division

Lifestage and demographics: Contraception, population stabilisation, lifespans of ninety to one-hundred years, migration patterns, and active, sheltered, and in-care phases of retirement.

Education: Lifetime learning, virtual schools, internet universities.

Biotechnology: Genetic engineering, gene therapy, disease control.

Health: Telemedicine, non-invasive surgery, animal transplants.

Welfare: Biometric database, personal profiles, welfare records.

Culture: Supercultures, instantaneous language translation, cyber culture, on-line digital jukebox, virtual worlds.

Work: Telecommuting, shorter working week, shorter working life, reduction in lifetime incomes.

Lifestyle: Decline of marriage, single-person households, geodemographics, psychographics, sedentary lifestyles and obesity, use of alcohol, drugs, and tobacco.

13.5   Civilisation in a dynamic balance

The development of *Long-Range Futures Research* is based upon the hypothesis that the civil system is a complex adaptive system that is in far-from-equilibrium stability. In the search for patterns and regularities, it has been observed that diffusive civil ecostructures form as a result of the behaviour of transacting entities, households, and establishments, which have been classified from detailed research in a world-city region. The evolutionary mechanism for the civil system is the capture of investment capital (profit or savings) in the knowledge that returns on investment (ROI) change with the evolution of ecostructures. *Long-*

Range Futures Research provides an evolutionary approach that enables civil-society to foresee the consequences of its actions and to alter the trajectory of the system. In a laissez-faire or free-market economic system, without artificial intervention by government, the law of the jungle would presuppose a natural justice in the selection of the most fit establishments. There is, however, a path-dependent advantage for those entities that have already acquired a stock of assets, so that it is inevitable that an accumulation of capital by the better-off will result in the consumption of an unequal share of scarce natural resources. In a civilised and equitable society, the global playing field should be levelled by using the global surplus to relieve investment atrophy in the LDCs, and to provide voting rights at the United Nations on the basis of uniform and sustainable territorial units.

The ancient Greeks subscribed to an evolutionary life-cycle view of history, and in the 2nd century BC, the historian Polybius predicted the fall of the Roman Empire 600 years into the future. Therefore, the starting point for evolutionary futures research is to establish where human civilisation has reached in history, and the effort so far that has gone into creating the world's infrastructure. The cumulative total of adult humankind to have lived on the planet was 50-60 billion by the year 2000. Human civilisation is still in the growth stage and far from equilibrium, and the *Long-Range Futures Research (L-RFR)* landscapes provide an answer to the question 'where are we now?' in relation to the life cycle of planetary development. The world stock of assets in 2000 was estimated by means of parametric cost models to be PPP $150 trillion (PPP $25,000 of assets per capita) and, in 2150, the stock is expected to reach PPP $800 trillion (PPP $64,000 assets per capita). For a sustainable civilisation, there will be a global upper limit to the stock of assets, which may be approached in 2250 with a PPP $1,200 trillion stock (PPP $104,000 per capita), when the Gross World Product is expected to reach PPP $425 trillion with a per capita GWP of $35,000. It is anticipated that the world population will peak by 2150, and it will decline by one billion before 2250 with the urban proportion staying constant at 90%. The increase in the capital stock from 2150–2250 will arise from an enhancement in planning standards and the physical quality of life as the less developed countries move to higher stages of development, at current European levels.

From a comparison of the PPP $150 trillion global stock of investment in the year 2000 and the possible upper limit of PPP $800 trillion in 2150, or a stock in excess of PPP $1 quadrillion for a planetary civilisation, perhaps no more than twenty percent of the world development programme has been completed. Ontogenesis provides the biogenetic relationship between the life cycle of a human and the evolutionary cycle of humankind, so that a correlation can be made with respect to rates of physical growth, levels of intellectual development, and the stage of maturity for civilisation. In relation to the life cycle of a human who may live to eighty years of age, the stage of development of our civilisation has reached the equivalent stage of a sixteen-year-old, and we are still completing the growth stage to adulthood. The period between 2000 and 2150 will be a period of internal development based upon advanced information systems, and during the period 2250-3000, an early planetary civilisation will form with significantly diminished rates of growth. In 3000, the Gross World Product is expected to reach $600 trillion with an investment stock of PPP $1,500 trillion, for a population of ten billion. By the end of the millennium in 3000, a further shift to renewable energy resources is expected to reduce the environmental factor for energy used to 0.35 in relation to the year 2000 as shown in Table 16.

TABLE 16: ENVIRONMENTAL IMPACT IN 3000

Population in year 3000	1.65 times	
Growth in capital stock per inhabitant	6.00 times	($25,000 to $150,000 per capita.
Material throughput/capital stock	0.60 times	
Energy/material throughput	0.40 times	
Renewable energy factor	0.35 times	
I = 1.65 x (6.00 x 0.60) x (0.40 x 0.35) =	0.85 times	

On the basis of the assumptions outlined above, in 3000 the environmental impact, I, would be 0.85 times that in 2000. It is expected that energy use in 3000 will be 600 EJ, and if two-thirds of commercial energy is produced using renewable sources, then the energy resources will be equivalent to 200 EJ.

From the Drake equation (Chapter 2, section 2.6), the probable number of Astro-technological Civilisations (ATCs) possessing a radio telescope is dominated by the value for the lifespan of a civilisation measured in years. If the duration is short (say 10 millennia), then ATCs are so rare that radio contact would be very unlikely. If the starting point of civilisation was the early settlements, towns, and cities that commenced ten millennia ago, then a case can be made that if only twenty percent of world development has been completed, the potential lifetime of a civilisation is of the order of fifty millennia, providing the biosphere is not destroyed. If the duration of advanced civilisations reaches fifty millennia, then occasional signals are likely to occur. At that point, civilisations may reach their limits before they collapse from economic or catastrophic causes and revert to a more primitive level. An alternative starting point for civilisation was fifty millennia ago (one percent of the age of our species) with the emergence of modern man (*Homo Sapiens Sapiens*), when physiological evolution had created a capability for speech, language, and sophisticated communication. Then, on the basis that twenty percent of world development has been completed, the assessment for the duration of a technological civilisation that can communicate with other astro-technological civilisations could be extended to 250 millennia. This could arise from the emergence of a new species on Earth, *Genus Intelligens*, and consequential multiple 'S-curves' if our capabilities and technological performance are lifted to a higher plane and supported by new infrastructures. If the duration of astro-technological civilisations exceeds 250 millennia, then there is likely to be frequent contact, and it was concluded in Chapter 2 that the number of astro-technological civilisations in our galaxy probably lies in the range of twenty-five to 2,500. If the life of a civilisation is only fifty millennia, then it is unlikely that the number of astro-technological civilisations exceeds the range of five to 500.

With only twenty percent of planetary development complete, the next fifty years provide the greatest opportunity for planning a suitable strategy for the distribution of cities and an energy, transportation, and telecommunications infrastructure on a planetary scale. The urban hierarchy determines the future average population density for human settlements, which is an important issue for sustainable development, and this needs to take into account new city forms at the highest level in the hierarchy for a planetary civilisation. *Long-Range Futures Research* provides a scientific basis for long-range planners to generate default values for the number of cities by urban-size class, planning factors for the number and size of facilities in relation to the stage of development for countries, and estimates of the number of establishments, vehicles, and other artefacts.

It has been shown that human civilisation is still in the growth stage and far from equilibrium, so that for several centuries, chance events and cultural diversity will predominate, although cultures, subcultures, and cults will both coevolve and absorb each other. International travel, telecommunications, television, and cyberspace will encourage a world perspective, a cosmopolitan outlook, and an ecopolitan culture. Futures research has identified a combination of geopolitical, economic, social, technological, and ecological policies that would contribute to achieving a desirable end-state of a sustainable civilisation of 12.5 billion population. Within this millennium, an early planetary civilisation will form with stable continental unions and diminishing rates of growth. As the evolutionary landscape enters the mature phase in the late planetary era with relatively minor changes to the urban pattern, the boundaries between continental unions will be relaxed and a global superculture with a worldwide mass market may permeate everywhere.

Civilisations in decline are characterised by a tendency towards standardisation and uniformity, and a culturally homogenised world could lead to stagnation and decay. However, this is unlikely to be an evolutionary dead end for humans on this planet, because genetic engineering or planetary terraforming may provide the opportunity to initiate interplanetary exploration. An evolutionary civil system can provide a very long organic lifetime, if an intelligent species limits its population so that it can maximise the period of the civilisation by sustainable use of its planetary or interplanetary resources. So an objective for humankind is to extend the lifetime of our civilisation from fifty millennia to perhaps 250-500 millennia, so that we can communicate with other astro-technological civilisations and map the inhabited planets in our universe.

Evolutionary landscapes for a complex world: References

1. *Futures: Volume 22 Number 6,8,10* - 1990
2. *Futures: Volume 23 Number 4* - 1991.
3. *Futures: Volume 24 Numbers 3,6,8,10* - 1992.
4. *Futures: Volume 25 Numbers 1,2,3,4,5,8,10* - 1993.
5. *Futures: Volume 26 Numbers 2,3,5,6,8,10* - 1994.
6. *Futures: Volume 27 Numbers 2,3,4,5,9/10* - 1995.
7. *Futures: Volume 28 Numbers 4,5,6/7,9* - 1996.
8. *Futures: Volume 29 Numbers 4/5,8* - 1997.
9. *Futures: Volume 30 Numbers 2/3* - 1998.
10. *Futures: Volume 31 Numbers 9/10* - 1999.
11. *Futures: Volume 32 Numbers 3/4,6* - 2000.
12. *Futures: Volume 33 Numbers 1,3/4* - 2001.
13. *Futures: Volume 34 Numbers 2,6* - 2002.
14. *Futures: Volume 35 Numbers 6* - 2003.
15. *Futures: Volume 36 Numbers 4* - 2004.
16. *Futures: Volume 37 Numbers 7,8* - 2005.
17. *The Atlas of the Future* - Edited by Ian Pearson, 1998.
18. *Guinness Amazing Future* - Editor Mark Fletcher, 1999.
19. *50 Trends Now Changing the World* – Marvin J Cetron and Owen Davies, 2001.
20. *Trends Shaping Tomorrow's World: Parts 1 & 2* – Marvin J Cetron and Owen Davies, March & May, 2008
21. *Future Survey: Volumes 23-30* – Edited by Michael Marien, 2001-2008.
22. *Futurist Outlook* – 2001
23. *Futurist Outlook* – 2002
24. *Futurist Outlook* – 2003
25. *Futurist Outlook* – 2004
26. *Futurist Outlook* – 2005
27. *Futurist Outlook* – 2006
28. *Futurist Outlook* – 2007
29. *Futurist Outlook* – 2008
30. *Futurist: A Timeline for Technology* – Ian Pearson and Ian Neild, March/April, 2006

PART 4

APPENDICES

APPENDIX 1
A CENTURY OF FUTURES RESEARCH AND SYSTEMS SCIENCE

1902	*Anticipations of the Reaction of Mechanical and Scientific Progress upon Human Life and Thought* - H.G. Wells
1909	Weber's Theory of Industrial Location
1914	*The Astronomical Theory of the Ice Age* - Milutin Milankovitch
1930s	Cost-Benefit Studies for Water Resources by US Corps of Engineers
1940s	Engineering Economics
1940s	Economics of Location - August Losch
1941	Operations Research
1947	General Systems Theory - Ludwig Von Bertalanffy (biologist)
1948	Systems Analysis at RAND
1950s	Military scenario writing and cost-effectiveness studies at RAND
1954	Society for General Systems Research (Social Dynamics) - L. Von Bertalanffy, K. Boulding, R. Gerard, & A. Rapoport
1960	*The Stages of Economic Growth* - Walt Rostow
1960s	Planning, Programming and Budgeting Systems (PPBS)
1960s	Urban Economics
1960s	Transport and Land-use studies
1960s	Management Science and Business Forecasting
1967	*The Art of Conjecture* - Bertrand de Jouvenel
1967	*The Year 2000* - Herman Kahn & Anthony Weiner
1968	Technological forecasting - Delphi Technique and Cross Impact Analysis
1968	Modelling and System Dynamics - Jay W. Forrester
1969	*The Sciences of the Artificial* - Herbert Simon
1972	*The Limits to Growth* - The Club of Rome Report
1974	*Ecumenopolis* - C.A. Doxiadis and J.G. Papaioannou

1974	*The Coming of Post-Industrial Society* - Daniel Bell
1974	Geodemographic Systems - The Clustering of America (PRIZM system)
1976	Self-Organizing Systems and Non-Linear Thermodynamics - Ilya Prigogine
1976	*Global Review of Human Settlements* - United Nations
1977	Application of Systems Analysis to urban development programmes in LDCs
1978	Tenth year of R S. McNamara at World Bank: World Development Indicators
1978	*Ecodynamics: A New Theory of Societal Evolution* - Kenneth Boulding
1970s	Corporate Scenario Planning
1979	*Emergence of the Prospective Approach* - Michel Godet
1980s	Complexity Science & Chaos Theory - Los Alamos National Laboratory
1982	Evolutionary Economics
1987	Complex Adaptive Systems - Santa Fe Institute/John Holland
1987	World Commission on Environment and Development - Gro Harlem Bruntland
1990	*Why the Future is Not What it Was: New models of evolution* - Peter Allen
1994	*Complexity and Thermodynamics: Towards a new ecology* - E. Schneider & J. Kay
1996	*The Knowledge Base of Futures Studies* - Richard Slaughter
1997	*Foundations of Futures Studies* - Wendell Bell
2000	*UN IPPC Special Report on Emissions Scenarios* - UNEP & WMO
2003	International Futures (IF) Global Simulation Model - Barry Hughes
2008	*Long-Range Futures Research: An Application of Complexity Science* - Robert H. Samet

APPENDIX 2

VERHULST LOGISTIC CURVE (S-CURVE)

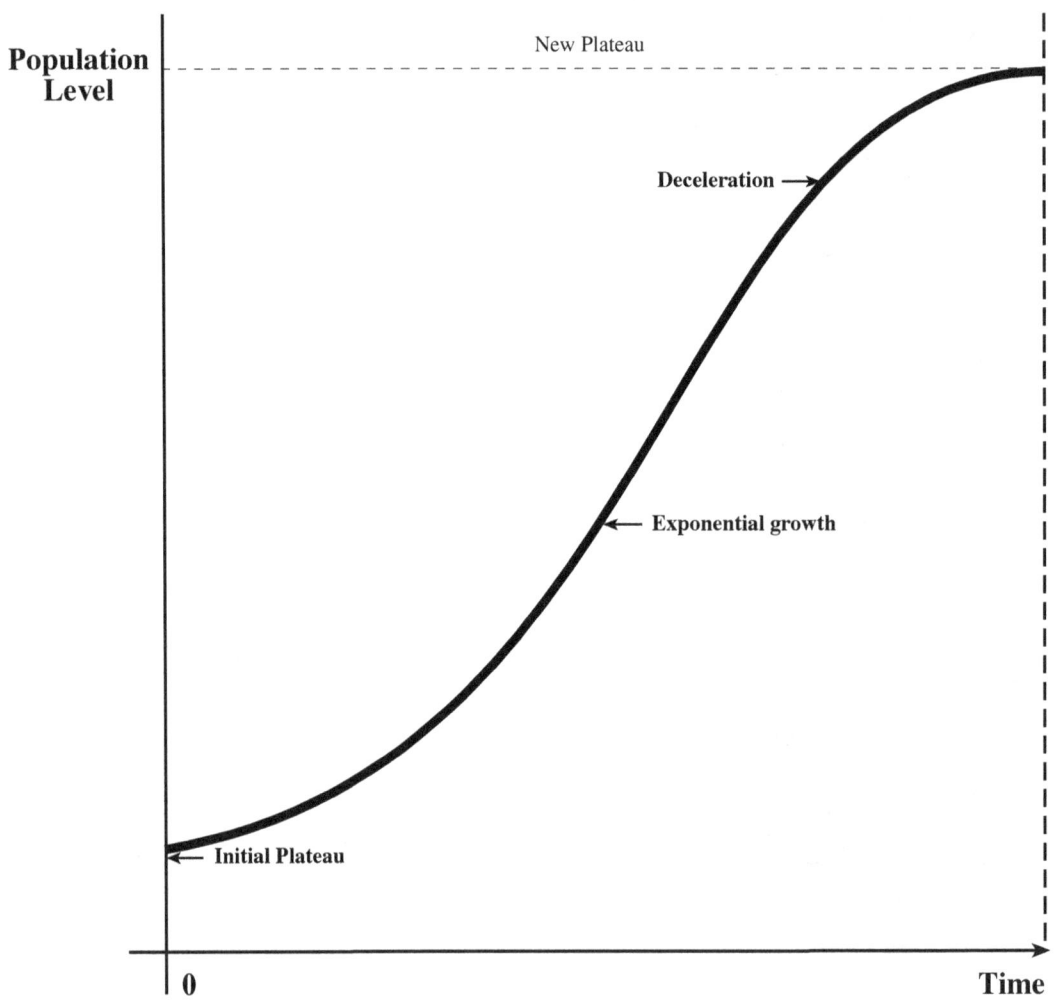

Long-Range Futures Research

APPENDIX 3

PROGNOSTICATION OF A LONG-TERM FUTURE STATE

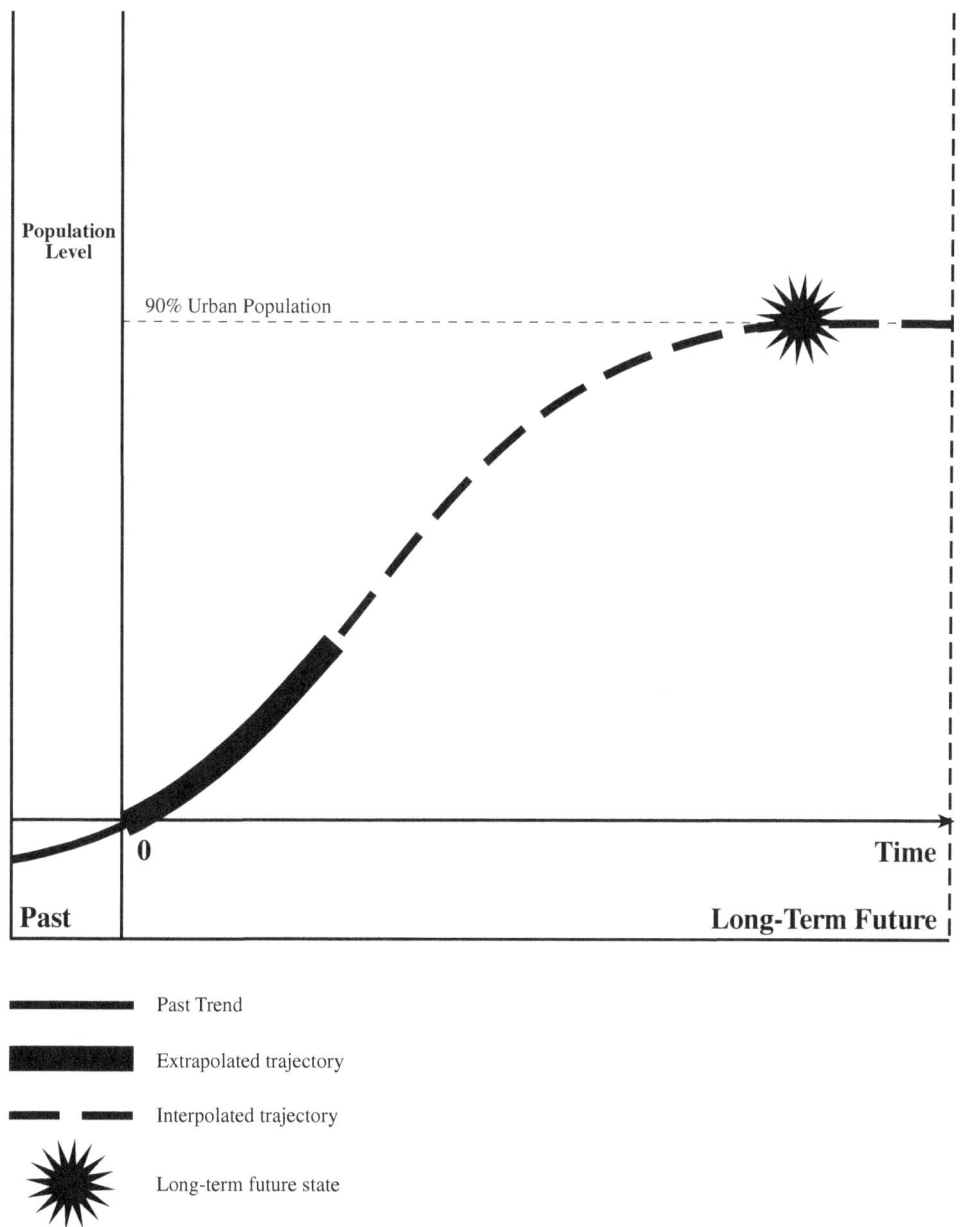

APPENDIX 4
GEOLOGICAL AND ARCHAEOLOGICAL TIMESCALES

Eon	Era	Period & Epoch	Age	Million Years Ago
Hadean	Cryptic & Basin Groups			4500-4000
	Nectarian & Lower Imbrian			4000-3800
Archean	Eoarchean			3800-3600
	Paleoarchean			3600-3200
	Mesoarchean			3200-2800
	Neoarchean			2800-2500
Proterozoic	Paleoproterozoic			2500-1600
	Mesoproterozoic			1600-900
	Neoproterozoic			900-550
Phanerozoic	Paleozoic	Cambrian		550-490
		Ordovician		490-445
		Silurian		445-415
		Devonian		415-355
		Carboniferous		355-290
		Permian		290-245
	Mesozoic	Triassic		245-205
		Jurassic		205-145
		Cretaceous		145-65

	Cenozoic	Paleogene		
		Paleocene		65-55
		Eocene		55-35
		Neogene		
		Oligocene		35-25
		Miocene		25-5
		Pliocene		5-1.75
		Pleistocene		1.75-0.01
			Palaeolithic	
			Lower	1.50-0.25
			Middle	0.25-0.05
			Upper	0.05-0.01
		Holocene		0.010
			Mesolithic	0.010
			Neolithic	0.008
			Chalcolithic	0.006
			Bronze Age	0.004
			Iron Age	0.003
			Roman	0.002
			Medieval	0.001

APPENDIX 5
CIVILISATIONS OF HUMANKIND

Civilisations and empires	Period of existence	Duration in centuries
Western Europe		
• **Minoan**	2000BC - 1400BC	6.00
• **Hellenic**	1300BC - AD558	18.50
• **Celtic**	1000BC - AD200	12.00
• **Spartan**	900BC - AD396	13.00
• **Roman**	31BC - AD378	4.00
• **Byzantine**	AD395 - 1453	10.50
• **Viking**	790 - 1050	2.50
• **Habsburg**	1493 - 1918	4.25
• **Napolenic**	1792 - 1815	0.25
• **British Empire**	1757 - 1931	1.75
Eastern Europe		
• **Siberian (Eskimo)**	1100BC - AD1850	29.50
• **Huns**	AD370 - 500	1.25
• **Slavonic**	500 - 1200	7.00
• **Khazar Empire**	740 - 966	2.25
• **Kievan Empire**	882 - 1245	3.50
• **Muscovite Empire**	1478 - 1917	4.50
Romanov Dynasty (Tsars)	1613 - 1917	3.00
• **Soviet Union (communist)**	1917 - 1989	0.70

Civilisations and empires	Period of existence	Duration in centuries
Middle East and Central Asia		
• **Sumerian**	**Pre-3500BC - 1700BC**	**18.00**
Sumerian Empire	2500BC - 1900BC	6.00
Akkadian Empire	2300BC - 1750BC	5.50
• **Hittite**	**2000BC - 1200BC**	**8.00**
• **Babylonian**	**1500BC - 538BC**	**9.50**
Babylonian Empire	610BC - 539BC	0.75
• **Assyrian**	**1200BC - AD970**	**21.75**
Achaemenian Empire	525BC - 332BC	2.00
• **Persian**	**550BC - AD637**	**12.00**
• **Jewish**	**AD70 - 1497**	**14.25**
• **Arabian**	**975 - 1525**	**5.50**
• **Ottoman**	**1310 - 1919**	**6.00**
• **Islamic**	**1320 - survives**	**7.00 to date**
Africa		
• **Egyptian**	**Pre-3100BC - AD280**	**34.00**
Old Kingdom (Pyramid age)	3300BC - 2150BC	11.50
New Kingdom (Imperial age)	1560BC - 1085BC	4.75
• **Ethiopian Kingdom of Axum**	**AD350 - 800**	**4.50**
• **Ghana**	**700 - 1150**	**4.50**
• **Mali**	**1150 - 1450**	**3.00**
• **Zimbabwe**	**1200 - 1500**	**3.00**
• **Kanuri**	**1350 - 1850**	**5.00**
• **Songhay**	**1460 - 1600**	**1.50**
• **European colonies**	**1750 - 1950**	**2.00**

Civilisations and empires	Period of existence	Duration in centuries
South Asia		
• **Indian**	**Pre-3000BC - AD500**	**35.00**
Mauryan Empire	322BC-185BC	1.25
Gupta Empire	AD390 - 475	0.75
• **Hindu**	**AD775 - survives**	**12.25 to date**
• **Mughal Empire**	**1526 - 1707**	**1.75**
• **British Raj**	**1818 - 1947**	**1.25**
East Asia and Pacific		
• **Chinese**	**1600BC - survives**	**36.00 to date**
Shang Dynasty	1523BC - 1028BC	5.00
Chou China	1100BC - 770BC	3.25
Han China	140BC - AD220	3.50
Sui Dynasty	581 - 687	1.00
Tang Empire	700 - 907	2.00
Sung Dynasty	960 - 1279	3.25
Ming Dynasty	1368 - 1644	2.75
Ching Dynasty	1644 - 1911	2.50
People's Republic	1949 - survives	0.50
• **Khmer Empire**	**AD100 - 1432**	**13.25**
Angkor Kingdom	802 - 1432	6.25
• **Japanese**	**640 - survives**	**13.75 to date**
• **Indonesian - Srivijayan Empire**	**650 - 1250**	**6.00**
• **Korean**	**676 - survives**	**13.25 to date**
• **Burmese - Thai - Cambodian**	**1000 - 1300**	**3.00**
Theravada Buddhism		
• **Mongolian**	**1206 - 1405**	**2.00**

Civilisations and empires	Period of existence	Duration in centuries
America		
• Mayan (Yukatan)	2500BC - AD1550	40.50
• Andean (Incas)	100BC - AD1783	18.75
• American Indian	AD700 - 1600	9.00
• Mexican (Aztec)	1075 - 1821	7.50
• Latin American	1519 - survives	5.00 to date
• Colonial America	1535 - 1783	2.50
• Caribbean	1655 - survives	3.50 to date
• North American (USA)	1776 - survives	2.25 to date
Oceania		
• Aborigine	4000BC - AD1750	57.50
• Micronesian	1550BC - AD1775	33.25
• Maori	900-1750	8.50
• British Commonwealth	1775 - survives	2.25 to date

APPENDIX 6
META-TIMESCALE FOR DEEP FUTURES

Timescale Years Ago	Era	Epoch/ Macro-stage	Age (YA)	Earth & Astro Population
10^{10}	Cosmic		10 billion	
10^{9}	Archean/ Proterozoic		1 billion	
10^{8}	Paleozoic/ Mesozoic		100 million	
10^{7}	Cenozoic	Paleogene/ Neogene	10 million	
10^{6}	Pleistocene	Palaeolithic (Lower)	1 million	Homo Erectus
10^{5}		(Middle)	100,000	Homo Sapiens 1 million
10^{4}	Holocene	Neolithic	10,000	10 million
10^{3}		Agricultural	1,000	
10^{2}		Industrial	150	1 billion
10^{1}		Distributional	50	5 billion
10^{0}	Present (Year 2000)	Informational	25	6 billion

Years Forward			Age (YF)	
10^1	Intermediate future	Informational	25	8 billion
10^2	Long-range future	Ecopolitan	100	10 billion
10^3	Planetary	Early Planetary	1,000	
10^4		Late Planetary (Distant future)	10,000	E.T Intelligence 25 billion
10^5	Interplanetary (Solar system)	Near (Remote future)	100,000	Genus Intelligens 50 billion
10^6		Far (Deep future)	1 million	100 billion
10^7	Galactic	Near	10 million	1 trillion
10^8		Far	100 million	10 trillion
10^9	Intergalactic	Near	1 billion	100 trillion
10^{10}		Far	10 billion	1 quadrillion

Meta-timescale units

10^0	1
10^3	1,000
10^6	1 Million
10^9	1 Billion
10^{12}	1 Trillion
10^{15}	1 Quadrillion

Long-Range Futures Research

APPENDIX 7
RICI/SIC SECTORAL CLASSIFICATION CODING GUIDE

RICI Code © 1996		SIC Code UK 1992	Number of Establishments	% of Total
	TOTAL NUMBER OF LARGE ESTABLISHMENTS		**10897**	**100%**
1	**PRIMARY RESOURCES DIVISION**		**123**	**1.1%**
11	**Agribusiness, mining and energy**		**123**	**1.1%**
11.1	Agribusiness	01, 02, 05	12	
11.2	Mining	10, 12, 13, 14	16	
11.3	Oil	11, 23.2	48	
11.4	Electricity, gas and water supply industries	40, 41	47	
2	**SECONDARY INDUSTRIAL DIVISION**		**2784**	**25.5%**
21	**Construction and property**		**466**	**4.3%**
21.1	Construction	45	180	
21.2	Property	70.1, 70.2, 70.32	108	
21.3	Estate agents	70.3	86	
21.4	Cleaning and sanitary services	74.7, 90	92	
22	**Manufacturing industries**		**576**	**5.3%**
22.1	Metal, mineral products and chemicals	23.1, 23.3, 24, 26, 27, 28.7	126	
22.2	Metal goods, engineering and vehicles	27.5, 28, 29, 30.01, 34, 35, 36.1	116	
22.3	Electrical and electronic engineering	29, 31, 32, 33	77	
22.4	Instruments engineering	30.01, 33	25	
22.5	Food, drink and organic products	15, 16	116	
22.6	Textiles, clothing, footwear and leather products	17, 18, 19	34	

	22.7	Paper, wood, fibres, plastic and rubber products	17.54, 20, 21, 25, 35, 36	61	
	22.8	Recycling	37	5	
	22.9	Industrial holding companies with mixed activities	36, 74.15	16	
23		**Industrial sites**	as above	**1458**	**13.4%**
	23.1	Metal, mineral products and chemicals		195	
	23.2	Metal goods, engineering and vehicles		440	
	23.3	Electrical and electronic engineering		187	
	23.4	Instruments engineering		92	
	23.5	Food, drink and organic products		132	
	23.6	Textiles, clothing, footwear and leather products		76	
	23.7	Paper, wood, fibres, plastic and rubber products		309	
	23.8	Recycling		3	
	23.9	Unclassified manufacturing	(except 74.15)	24	
24		**Industrial sites for the information industry**		**284**	**2.6%**
	24.1	Radio, TV and telecommunications equipment	31, 32, 33	65	
	24.2	Computers and office equipment	30	34	
	24.3	Printing works	22.2	185	
3		**TERTIARY COMMERCIAL DIVISION**		**4724**	**43.4%**
31		**Banking, finance and securities**		**567**	**5.2%**
	31.1	Banks	65.11, 65.12/1	262	
	31.2	Credit granting institutions, except banks	65.12/2, 65.22, 65.23/5	52	
	31.3	Security brokers, money brokers and dealers	65.23/3, 67.11, 67.12/2	74	
	31.4	Commodity brokers and dealers	51.1, 67.11	20	
	31.5	Investment management	65.23/1, 65.23/2, 67.12/1	43	
	31.6	Financial services	65.23/6, 67.13	11	
	31.7	Leasing and renting	65.21, 71	105	

32	**Insurance companies and brokers**		**409**	**3.8%**
32.1	Insurance companies	66	192	
32.2	Insurance agents, brokers and services	67.2	217	

33	**Transport and distribution**		**293**	**2.7%**
33.1	Air transport	62, 63.23	56	
33.2	Water transport	61, 63.11, 63.22	34	
33.3	Road and rail transport	60, 63.21	93	
33.4	Freight brokers and agents	63.4	80	
33.5	Distribution and storage	63.12	30	

34	**Wholesale and retail trade**		**435**	**4.0%**
34.1	Wholesale	50.1, 50.3, 50.4, 51.2 to 51.7	314	
34.2	Retail	50.1, 50.3, 50.4, 50.5, 52.1 to 52.6	121	

35	**Travel, hospitality and consumer services**		**175**	**1.6%**
35.1	Travel and tourism services	63.3	43	
35.2	Hotels and catering	55	58	
35.3	Sports and leisure organisations	91.33, 92.33, 92.34, 92.6, 92.7	24	
35.4	Personal services	74.81, 93, 95	39	
35.5	Consumer goods repairs	50.2, 52.7	11	

36	**Commercial sites**		**2845**	**26.1%**
36.1	Wholesale warehouses	51.2 to 51.7	320	
36.2	Retail distribution centres	52.1 to 52.6	36	
36.3	Retail stores	52	1383	
36.4	Motor services	50.1, 50.2, 50.3, 50.5	106	
36.5	Hotel accommodation	55.1	476	
36.6	Restaurants, pubs and nightclubs	55.3, 55.4	360	
36.7	Sports and leisure facilities	92.6, 92.7	164	

4		QUATERNARY INFORMATION DIVISION		3266	30.0%
	41	**Media**		505	4.6%
	41.1	Publishing of newspapers, periodicals and books	22	273	
	41.2	Advertising agencies	74.4	81	
	41.3	Film production, videos and records	22.14, 22.3, 71.34, 74.8, 92.11, 92.12, 92.32, 71.40/1	85	
	41.4	Radio, television services and performing arts	64.2, 92.2, 92.32	66	
	42	**Business services**		536	4.9%
	42.1	Accountants	74.12	114	
	42.2	Employment agencies	74.5	83	
	42.3	Management consultants	74.14	87	
	42.4	Marketing services	74.13	85	
	42.5	Public relations	74.14	26	
	42.6	Security services	45.31, 74.6	65	
	42.7	Commercial printing services	22.22, 22.23, 22.24, 22.25	18	
	42.8	Miscellaneous business services	74.11, 74.8	58	
	43	**Legal services**		214	2.0%
	43.1	Legal services	74.11	214	
	44	**Telecommunications and IT**		557	5.1%
	44.1	Telecommunications	32.2, 64.2	84	
	44.2	Computer systems and office technology	30, 72.1, 72.2, 72.5, 72.6	384	
	44.3	Information and data services	72.3, 72.4, 74.84, 92.4	50	
	44.4	Post and courier services	64.1	39	
	45	**Designers, consultant engineers and technical services**		350	3.2%
	45.1	Designers	74.14, 74.84	14	
	45.2	Architects, engineers and quantity surveyors	74.2	241	

45.3	Technology, research and development	73	51	
45.4	Miscellaneous technical services	74.2, 74.3	44	
46	**Education and health**		**373**	**3.4%**
46.1	Education	80	130	
46.2	Health	85.1, 85.2	243	
47	**Membership organisations**		**198**	**1.8%**
47.1	Professional and trade associations	74.14, 91.1, 91.2, 91.32, 91.33	112	
47.2	Welfare organisations and charities	85.3, 91.31, 91.33	86	
48	**Government and local authorities**		**344**	**3.2%**
48.1	Government departments and agencies	75.1, 75.3	164	
48.2	Local authorities	75.1, 75.3	89	
48.3	Justice, police, fire services, national defence	75.2	65	
48.4	Foreign embassies	99	26	
49	**Cultural sites**		**189**	**1.7%**
49.1	Cinemas	92.13	43	
49.2	Theatres and concert halls	92.31, 92.32	51	
49.3	Libraries, museums and art galleries	92.51, 92.52	83	
49.4	Zoological and botanical gardens	92.53	12	

LONDON MEGAPOLIS BUSINESS SPACE USERS: LOCATION INDEX ANALYSIS

	Central London		Outer London		London Postcodes		Inner M25		Outer Metropolis		London Metropolis		Outer M25		Megapolis	
	Total	Loc Index	Total	Loc Index	Total	Loc Index	Total	Loc Index	Total	Loc Index	Total	Loc Index	Total	Loc Index	Total	
PRIMARY RESOURCES DIVISION																
11 Agribusiness, mining and energy	38	1.5	18	0.8	56	1.2	23	0.8	41	0.8	79	1.0	44	0.9	123	
PRIMARY RESOURCES DIVISION sub-total	**38**	**1.5**	**18**	**0.8**	**56**	**1.2**	**23**	**0.8**	**41**	**0.8**	**79**	**1.0**	**44**	**0.9**	**123**	
SECONDARY INDUSTRIAL DIVISION																
21 Construction and property	84	0.9	116	1.4	200	1.1	110	1.0	226	1.2	310	1.1	156	0.9	466	
22 Manufacturing industries	55	0.5	88	0.9	143	0.6	162	1.2	250	1.1	305	0.9	271	1.2	576	
23 Industrial sites	20	0.1	231	0.9	251	0.4	365	1.1	596	1.0	616	0.7	842	1.5	1,458	
24 Industrial sites for the information industry	24	0.4	48	0.9	72	0.7	66	1.0	114	1.0	138	0.8	146	1.3	284	
SECONDARY INDUSTRIAL DIVISION sub-total	**183**	**0.3**	**483**	**1.0**	**666**	**0.6**	**703**	**1.1**	**1,186**	**1.0**	**1,369**	**0.8**	**1,415**	**1.3**	**2,784**	
TERTIARY COMMERCIAL DIVISION																
31 Banking, finance and securities	339	2.9	47	0.5	386	1.8	75	0.6	122	0.5	461	1.3	106	0.5	567	
32 Insurance companies and brokers	186	2.2	19	0.3	205	1.3	67	0.7	86	0.5	272	1.1	137	0.9	409	
33 Transport and distribution	53	0.9	38	0.7	91	0.8	108	1.6	146	1.2	199	1.1	94	0.8	293	
34 Wholesale and retail trade	59	0.7	89	1.1	148	0.9	145	1.4	234	1.3	293	1.1	142	0.8	435	
35 Travel, hospitality and consumer services	22	0.6	50	1.6	72	1.1	44	1.1	94	1.3	116	1.1	59	0.9	175	
36 Commercial sites	315	0.5	603	1.2	918	0.8	730	0.8	1,333	1.1	1,648	0.9	1,197	1.1	2,845	
TERTIARY COMMERCIAL DIVISION sub-total	**974**	**1.0**	**846**	**1.0**	**1,820**	**1.0**	**1,169**	**1.0**	**2,015**	**1.0**	**2,989**	**1.0**	**1,735**	**1.0**	**4,724**	

QUATERNARY INFORMATION DIVISION

41	Media	205	2.0	118	1.3	323	1.7	78	0.7	196	0.9	401	1.3	104	0.5	505
42	Business services	184	1.7	94	1.0	278	1.4	93	0.8	187	0.9	371	1.1	165	0.8	536
43	Legal services	143	3.3	7	0.2	150	1.8	17	0.3	24	0.3	167	1.3	47	0.6	214
44	Telecommunications and IT	84	0.7	56	0.6	140	0.7	147	1.1	203	0.9	287	0.8	270	1.3	557
45	Designers, consultant engineers and technical services	81	1.1	54	0.9	135	1.0	81	1.0	135	0.9	216	1.0	134	1.0	350
46	Education and health	58	0.8	108	1.6	166	1.2	94	1.1	202	1.3	260	1.1	113	0.8	373
47	Membership organisations	96	2.4	59	1.7	155	2.0	20	0.4	79	1.0	175	1.4	23	0.3	198
48	Government and local authorities	131	1.9	63	1.0	194	1.5	61	0.8	124	0.9	255	1.2	89	0.7	344
49	Cultural sites	60	1.5	40	1.2	100	1.4	30	0.7	70	0.9	130	1.1	59	0.8	189
	QUATERNARY INFORMATION DIVISION sub-total	**1,042**	**1.6**	**599**	**1.0**	**1,641**	**1.3**	**621**	**0.8**	**1,220**	**0.9**	**2,262**	**1.1**	**1,004**	**0.8**	**3,266**

TOTAL	**2,237**	**1,946**	**4,183**	**2,516**	**4,462**	**6,699**	**4,198**	**10,897**

APPENDIX 9
LONDON MEGAPOLIS BUSINESS SPACE USERS: PROPERTY PROFILES BY URBAN CLASSIFICATION

| Urban Classification and facility type | Unidentified | Size range in square metres | | | | | | | | | | March 1996 |
		200-499	500-999	1000-1999	2000-4999	5000-9999	10000-19999	20000-39999	40000-79999	80000-159999	160000+	Total
Metropolitan Centre												
Depots	11											11
Education	26				2	3	7	2	1	11	2	54
Hospitals						5	16	17	9	1		48
Hotels					100	82	52	29	8			271
Industrial	65				191	51	23	4	1	1		336
Justice	14											14
Leisure	24		1	23	62	21	11					142
Offices		4	803	706	722	262	94	74				2,665
Passenger Transport	13											13
Restaurants		112	33	16	2							163
Retail			49	123	167	20	8			2		369
Trading Units					4	2						6
Warehouses	16		3	3	54	13	1	1				91
Metropolitan Centre Total	**169**	**116**	**889**	**871**	**1,304**	**459**	**212**	**127**	**21**	**13**	**2**	**4,183**
Regional Centres												
Depots	6					1						7
Education	2						1		1		1	5
Hospitals							3	3				6
Hotels				1	8	12	6					27
Industrial	31				114	27	12	9	1			194
Justice	1											1
Leisure	7			3	9	3						22
Offices			160	73	193	41	13	8				488
Restaurants		14	7	6								27
Retail			8	31	75	7	2	2				125
Trading Units					6	1						7
Warehouses	17		5	4	32	11	3					72
Regional Centres Total	**64**	**14**	**180**	**118**	**437**	**103**	**40**	**22**	**2**		**1**	**981**

Urban Classification and facility type	Size range in square metres											March 1996 Total
	Unidentified	200-499	500-999	1000-1999	2000-4999	5000-9999	10000-19999	20000-39999	40000-79999	80000-159999	160000+	
Urban Centres												
Depots	20											20
Education	5					1			1	2	1	10
Hospitals						4	28	17	2			51
Hotels				2	31	36	8	3				80
Industrial	118				304	92	45	24	2	2		587
Justice	2											2
Leisure	17			18	46	18	6					105
Offices			438	264	433	119	38	15				1,307
Passenger Transport	5											5
Restaurants		59	23	9	2							93
Retail			26	165	296	16	10					513
Trading Units	3				34	8	1					46
Warehouses	58	1	5	19	79	22	14	3	3			204
Urban Centres Total	**228**	**60**	**492**	**477**	**1,225**	**316**	**150**	**62**	**8**	**4**	**1**	**3,023**
District Centres												
Depots	9											9
Education	6							1		1	1	9
Hospitals						3	10	5				18
Hotels				1	23	15	1	3	1			44
Industrial	68				249	75	33	11	3			439
Justice	1											1
Leisure	7			8	19	7						41
Offices			285	126	247	36	10	6				710
Passenger Transport	3											3
Restaurants		30	14	3	2							49
Retail			23	73	157	6						259
Trading Units	2				17	6						25
Warehouses	26		2	5	75	17	11	3				139
District Centres Total	**122**	**30**	**324**	**216**	**789**	**165**	**65**	**29**	**4**	**1**	**1**	**1,746**
Local Centres												
Depots	4											4
Education	5											5
Hospitals							4	5				9
Hotels				3	15	12	3	2				35
Industrial	27				87	18	10	6				148
Leisure	7		1	2	10	5	1					26
Offices			92	47	90	13	6					248
Restaurants		11	6	1	1							19
Retail			5	28	46							79
Trading Units	1				1							2
Warehouses	4	1		2	8	4	1					20
Local Centres Total	**48**	**12**	**104**	**83**	**258**	**52**	**25**	**13**				**595**

Urban Classification and facility type	Unidentified	200-499	500-999	1000-1999	2000-4999	5000-9999	10000-19999	20000-39999	40000-79999	80000-159999	160000+	March 1996 Total
Service Centres												
Depots	3				1							4
Education	1											1
Hospitals						1	3	2				6
Hotels				1	4	3	1					9
Industrial	17				54	13	3	1				88
Leisure	4			1	4	3						12
Offices			45	22	38	3						108
Restaurants		6	3									9
Retail			2	15	17							34
Trading Units					2							2
Warehouses	2				6	4	1	1				14
Service Centres Total	**27**	**6**	**50**	**39**	**126**	**27**	**8**	**4**				**287**
Township Communities												
Education	1											**1**
Hospitals						1	1	2	1			**5**
Hotels					7	2	1					**10**
Industrial	5				12	1						**18**
Leisure	5				2							**7**
Offices			8	8	13	3						**32**
Retail			1	2	1							**4**
Trading Units					1							**1**
Warehouses	1				3							**4**
Township Communities Total	**12**	**0**	**9**	**10**	**39**	**7**	**2**	**2**	**1**			**82**

Grand Total	670	238	2,048	1,814	4,178	1,129	502	259	36	18	5	10,897

APPENDIX 10
PLANNING STANDARDS FOR URBAN ESTABLISHMENTS

Facility type	Minimum space per Large estab. (m²)	Large estab. per 10,000 population (number)	Floorspace of large estabs as % of total	Public area of large establishments (m²)	Gross floorspace/ inhabitant (m²)	Gross floorspace/ employee (m²)
Offices	500		65%			
Metropolis		6.75		3,340	3.0	17
Urban centre		3.50-6.00		3,120	2.0	15
Restaurants	500	0.30	30%	620	0.15	8
Retail	1,000		40%			
	(500)		(50%)			
Metropolis		1.25		3,190	1.75	32
Urban centre		1.00-2.25		3,070	1.5	32
Leisure	1,000	0.30	60%	4,600	0.25	50
Hotels	2,000	bedrooms*	65%	bedrooms		per bedroom
Metropolis		135 0.60		200+ 11,420	0.90	65
Urban centre		45 0.25		100+ 7,390	0.30	65
Warehouses	2,000	1.50-3.00	40%	4,810	1.6	200
	1,000		55%	2,830	1.6	200
Industrial sites	2,000		70%			
Metropolis		1.00		5,700	1.0	40
Urban centre		1.00-3.25		6,780	2.0	40
Hospitals	5,000	beds*	80%	beds		per bed
Metropolis		65 0.10		650+ 29,080	0.40	60
Urban centre		50 0.12		450+ 21,180	0.33	60
Universities	5,000	places*	80%	places		per place
Metropolis		665 0.11		6250 100,000	1.00	16
Urban centre		360 0.05		6250 100,000	0.60	16

Note that the number of bedrooms, beds or places is the total per 10,000 population from all establishments (ie, not just large establishments)

APPENDIX 11 A – OFFICE PROPERTY INVESTMENT

Long-Range Futures Research

Urban Area Location Indices for Office Investment

Pattern of long term returns from 1980 to 1995

Note:
All urban areas above the diagonal line have a tertiary location index (TLI) greater than 1.0

Secondary Location Index (SLI)

Quaternary Location Index (QLI)

Secondary Location Index (SLI)	0.0 - 0.2	0.3 - 0.5	0.6 - 0.8	0.9 - 1.1	1.2 - 1.4	1.5 - 1.7	1.8 - 2.0
Q - T OFFICE CENTRES							Westminster (SW1) Oxford Street Southwark (SE1)
						West Ctl London Hammersmith (W6) Richmond	Ldn Docklands (E14)
					East Ctl London Ealing (W5) Kingston	Guildford Maidenhead Redhill	
T - Q COMMERCIAL CENTRES	City of London Kensington				Marlow Rickmansworth		
		Bromley			Croydon Harrow Newbury Reading Staines Woking		
			Chelmsford Haywards Heath Tunbridge Wells		Barnet Basingstoke Camberley St Albans Slough	**Q - S QUATERNARY INDUSTRIAL CENTRES**	
T - S DISTRIBUTION CENTRES				Crawley Farnborough Ilford Maidstone Watford	Aylesbury		
		Hounslow		Southend			
			Barking Grays Hayes	Hitchin			
S - T INDUSTRIAL CENTRES				Dartford Harlow Romford	Hertford High Wycombe		
				Basildon Enfield Stratford (E15)			
		Gillingham		Luton			
			Erith				

7% · 7% · 7% · 8% · 9% · 8% · 8% · 7%

0.0 - 0.2 · 0.3 - 0.5 · 0.6 - 0.8 · 0.9 - 1.1 · 1.2 - 1.4 · 1.5 - 1.7 · 1.8 - 2.0

APPENDIX 11 B – RETAIL PROPERTY INVESTMENT

Long-Range Futures Research

Urban Area Location Indices for Retail Investment

Pattern of long term returns from 1980 to 1995

Tertiary location Index (TLI)	QLI 0.3 - 0.5	QLI 0.6 - 0.8	QLI 0.9 - 1.1	QLI 1.2 - 1.4	QLI 1.5 - 1.7	QLI 1.8 - 2.0	QLI 2.1 - 2.3
0.3 - 0.5							King's Cross/ Marylebone (NW1)
0.6 - 0.8		Hertford, High Wycombe, Hitchin	Aylesbury, Camberley, Slough	Maidenhead, Marlow, Rickmansworth	West Ctl London, Hammersmith (W6), Ldn Docklands (E14)	Westminster (SW1), Oxford Street (W1N, W1P, W1R, W1V)	Mid-town (WC)
0.9 - 1.1	Basildon, Dartford, Enfield, Luton, Stratford (E15)	Crawley, Maidstone, Southend	Barnet, Croydon, Harrow, Reading, St Albans, Sutton	Ealing (W5), Guildford, Redhill	Putney (SW5), Richmond	Southwark (SE1)	
1.2 - 1.4	Grays, Harlow, Hayes, Romford	Chelmsford, Farnborough, Haywards Heath, Ilford, Tunbridge Wells, Watford	City of London, Bromley, Windsor	Kingston			
1.5 - 1.7	Barking, Hounslow		Kensington (W8)				
1.8 - 2.0	Horley						
2.1 - 2.3	Gatwick						

Curve return markers: 9%, 10%, 10%, 11-12%, 11%, 11%, 11%

Quaternary Location Index (QLI)

Note: Regional centres with over 100,000 population such as Watford, Reading, Luton, Southend, Slough and Croydon have become mature for retail, and returns are higher at urban centres with 50,000 - 100,000 population

APPENDIX 11 C – INDUSTRIAL PROPERTY INVESTMENT

APPENDIX 11C

Urban Area Location Indices for Industrial Investment

Category bands (along the diagonal): **S-T INDUSTRIAL CENTRES** · **T-S DISTRIBUTION CENTRES** · **T-Q COMMERCIAL CENTRES** · **Q-T OFFICE CENTRES** · **Q-S QUATERNARY INDUSTRIAL CENTRES**

Secondary Location Index (SLI) ↓ \ Quaternary Location Index (QLI) →	0.0-0.2	0.3-0.5	0.6-0.8	0.9-1.1	1.2-1.4	1.5-1.7	1.8-2.0
0.0-0.2				City of London, Kensington			
0.3-0.5		Hounslow		Bromley	East Ctl London, Ealing (W5), Kingston	West Ctl London, Hammersmith (W6), Richmond	Westminster (SW1), Oxford Street, Southwark (SE1)
0.6-0.8			Chelmsford, Haywards Heath, Tunbridge Wells	Epsom, Sutton, Windsor	Tower Hamlets, Guildford, Maidenhead, Redhill		
0.9-1.1		Barking, Grays, Hayes	Crawley, Farnborough, Ilford, Maidstone, Watford	Croydon, Hemel Hempstead, Newbury, Reading, Staines, Woking	Marlow, Rickmansworth		
1.2-1.4		Dartford, Harlow, Romford	Southend	Barnet, Basingstoke, Camberley, St Albans, Slough			
1.5-1.7	Gillingham	Basildon, Enfield, Stratford (E15)	Hitchin	Aylesbury			
1.8-2.0	Erith	Luton	Hertford, High Wycombe				

Pattern of long term returns from 1980 to 1995

Percentages along the returns curve: 9%, 10%, 11%, 11%, 12%, 11%, 11%

Note:
All urban areas above the diagonal line have a tertiary location index (TLI) greater than 1.0

APPENDIX 12
EXTRA ENERGY CONSUMPTION BY TRANSPORT
FROM URBAN DECENTRALISATION IN THE SOUTH-EAST (1961-1991)

	1991 Population (000s)	Energy consumption MJ per person/week			1991 Consumption	*Extra Consumption
		Car	Other	Total	GJ/week	GJ/week
Greater London	**6,394**	**130.600**	**36.5**	**167.100**	**1,068,265**	**-308,850**
Inner London	2,343	99.6	40.7	140.3	328,957	-186,030
Outer London	4,051	148.3	34.1	182.5	739,308	-122,820
Outer Centres (OMA)						
100,000 - 249,000	730	150.3	30.1	180.5	131,838	+20,247
50,000 - 99,000	2,270	144.5	28.5	173.1	392,937	+60,346
20,000 - 49,000	1,300	142.5	25.5	170.7	222,040	+34,100
10,000 - 19,000	600	170.0	23.0	193.0	115,800	+17,784
5,000 - 9,000	400	185.0	28.0	213.0	85,200	+13,013
2,000 - 4,900	150	205.0	33.3	238.0	35,700	+5,483
Outer Metropolitan Area	5,450	152.8	27.5	180.5	983,515	+150,973
Megapolitan Region	**11,844**	**140.8**	**32.3**	**173.2**	**2,051,780**	**-157,877**
Outer South-East	4,950	150.0	30.0	180.0	891,000	+209,340
South-East Total	**16,794**	**143.5**	**31.6**	**175.1**	**2,942,780**	**+51,463**

* Extra energy consumption by transport from urban decentralisation in the South-East over the 30-year period 1961-1991 is 51,463 GJ/week, which is less than 2% of the total fuel consumption.

APPENDIX 13

EMPLOYMENT STRUCTURE BY STAGE OF DEVELOPMENT

Stage of development		GNI per capita (2000 US$)	Number of countries (1m+ pop'n)	2000 pop'n (m)	% urban pop'n	Workforce estimate* (m)	Default distribution of the workforce							
							Primary Resources (m)	(%)	Secondary Industrial (m)	(%)	Tertiary Commercial (m)	(%)	Quaternary Information (m)	(%)
Ecopolitan	1	40,001 - 60,000	1	7	85	4	-	3.0	1	22.0	1	30.0	2	45.0
Informational	2	25,001 - 40,000	9	540	80	280	10	3.6	73	26.0	85	30.3	112	40.0
	3	15,001 - 25,000	12	261	79	120	5	4.0	31	26.0	37	30.7	47	39.2
Distributional	4	10,001 - 15,000	6	88	79	42	4	10.0	12	28.0	14	32.0	12	30.0
	5	6,001 - 10,000	6	114	79	50	5	10.0	14	28.0	16	32.0	15	30.0
Industrial	6	4,001 - 6,000	9	207	74	91	15	16.5	28	30.8	25	27.5	23	25.2
	7	2,501 - 4,000	15	340	73	156	39	25.0	41	26.3	40	25.6	36	23.1
Infrastructural	8	1,501 - 2,500	18	460	61	213	64	30.0	53	24.9	54	25.4	42	19.7
	9	1,001 - 1,500	10	221	55	92	30	32.6	22	23.8	23	25.0	17	18.5
Agropolitan	10	601 - 1,000	14	1,470	35	846	493	58.3	143	16.9	144	17.0	66	7.8
	11	401 - 600	12	1,432	32	635	378	59.5	105	16.5	107	16.9	45	7.1
Traditional	12	Less than 400	41	900	29	421	258	61.3	66	15.7	68	16.2	29	6.9
			153	6,040	47	2,950	1,301	44.1	589	20.0	614	20.7	446	15.1

*The workforce comprises an average of 63% (55-65% depending upon stage of development) of the population in age group 15-64, and some 75% of this age group are in the workforce (60% male, 40% female), giving 49% of population (range 40-55%).Quaternary division default values are based upon the proportion of urban services (range 30-60%) in Public administration, Education, Health and Social services together with business services that generally account for half the financial services sector. It also includes Post and Telcommunications that are sometimes categorised under Transport & Communications, and Printing & Publishing from the industrial sector.

APPENDIX 14
DEFAULT PLANNING STANDARDS AND COSTS FOR HOUSING

Stage of development		GNI per capita (2000 US$)	Persons per dwelling	Houses per 10,000 population	m² per dwelling	m² dwelling per capita	New urban housing cost per m² (2000 US$) *PPP	New rural housing cost per m² (2000 US$) *PPP
Ecopolitan	1	40,001 - 60,000	2.1	4,750	95	45	825	700
Informational	2	25,001 - 40,000	2.6	3,800	100	39	775	650
	3	15,001 - 25,000	2.5	4,000	90	36	725	600
Distributional	4	10,001 - 15,000	2.9	3,500	85	30	675	550
	5	6,001 - 10,000	3.2	3,100	85	26	625	500
Industrial	6	4,001 - 6,000	3.9	2,550	85	22	575	450
	7	2,501 - 4,000	3.6	2,800	75	20	525	400
Infrastructural	8	1,501 - 2,500	3.8	2,650	65	17	475	350
	9	1,001 - 1,500	4.4	2,250	65	15	425	300
Agropolitan	10	601 - 1,000	4.3**	2,350	55	13	350	250
	11	401 - 600	5.3	1,900	55	10	300	200
Traditional	12	Less than 400	5.7	1,750	45	8	225	150

*Purchasing Power Parity (PPP) data adjust for national cost of living differences by replacing normal exchange rates with PPP international dollars. For example in 2000 India had a GNI per capita of $450 whereas the PPP GNI per capita was $2,340, which differs by a factor of around 5. It follows that the national cost of new urban housing in India would be equivalent to $60 per m² in relation to the PPP rate of $300 per m².

**Including China would reduce the persons per dwelling from 4.3 to 3.5

APPENDIX 15
DEFAULT PLANNING STANDARDS FOR LIFE EXPECTANCY, EDUCATION AND HEALTH

Stage of development		GNI per capita (2000 US$)	Life expectancy at birth in years	Secondary education students as % of age group	Tertiary education students as % of age group	Doctors per 10,000 population	Hospital beds per 10,000 population
Ecopolitan	1	40,001 – 60,000	80	92	65	34	110
Informational	2	25,001 – 40,000	78	92	64	30	80
	3	15,001 – 25,000	78	92	57	30	62
Distributional	4	10,001 – 15,000	77	90	50	27	43
	5	6,001 – 10,000	74	78	46	22	39
Industrial	6	4,001 – 6,000	73	62	25	21	35
	7	2,501 – 4,000	69	62	23	20	33
Infrastructural	8	1,501 – 2,500	68	61	22	20	26
	9	1,001 – 1,500	68	60	21	20	25
Agropolitan	10	601 – 1,000	68	52	10	17	24
	11	401 – 600	63	46	8	4	15
Traditional	12	Less than 400	52	25	5	2	12

APPENDIX 16
DEFAULT PLANNING STANDARDS FOR ENERGY AND TRANSPORT

Stage of development		GNI per capita (2000 US$)	Energy consumption per capita (GJ)	Km railway per 10,000 population	Km road per 10,000 population		Motor car ownership per 10,000 population	Air passengers per 10,000 population
Ecopolitan	1	40,001 - 60,000	175	7	125	n.b	5,000	25
Informational	2	25,001 - 40,000	250*	6	130	n.b	4,530	17
	3	15,001 - 25,000	200	5	125	n.b	4,400	10
Distributional	4	10,001 - 15,000	140	4	110		3,220	9
	5	6,001 - 10,000	140	4	80		1,450	5
Industrial	6	4,001 - 6,000	90	3.5	75		1,440	2
	7	2,501 - 4,000	60	3	60		960	2
Infrastructural	8	1,501 - 2,500	50	2.5	45		720	1.5
	9	1,001 - 1,500	40	2	25		240	0.75
Agropolitan	10	601 - 1,000	40	1	20		90	0.5
	11	401 - 600	25	0.7	20		70	0.25
Traditional	12	Less than 400	15	0.5	15		50	0.17

*175 GJ excluding USA

n.b. Km road per 10,000 population plus/minus 20% depending upon network density (Km per 1,000 km^2.)

APPENDIX 17

DEFAULT PLANNING STANDARDS FOR COMMUNICATIONS AND INFORMATION TECHNOLOGY

Stage of development		GNI per capita (2000 US$)	Telephone landlines per 10,000 pop'n	Mobile phones per 10,000 population	Radios per 10,000 population	Televisions per 10,000 population	Personal computers per 10,000 pop'n
Ecopolitan	1	40,001 - 60,000	7500	6,500	10,000	7,500	5,000
Informational	2	25,001 - 40,000	6500	5,000*	15,000	7,500	4,650
	3	15,001 - 25,000	5500	6,000	12,500	6,000	3,000
Distributional	4	10,001 - 15,000	4500	6,000	5,500	5,750	2,250
	5	6,001 - 10,000	3000	3,000	5,500	3,250	2,250
Industrial	6	4,001 - 6,000	2000	1,800	4,500	3,250	600
	7	2,501 - 4,000	2000	1,800	4,500	3,250	500
Infrastructural	8	1,501 - 2,500	1500	600	3,500	2,750	350
	9	1,001 - 1,500	1000	600	3,500	2,750	200
Agropolitan	10	601 - 1,000	1000	600	3,500	2,750	150
	11	401 - 600	300	60	1,300	1,000	50
Traditional	12	Less than 400	100	30	1,800	500	35

* Note: In the USA there are 4,0000 mobile phones per 10,000 population

APPENDIX 18
METROPOLISES WITH POPULATIONS OVER 2 MILLION AND CAPITAL CITES OF OVER 1 MILLION

M: maritime cities
R: riverine cities

Region and country	National pop'n 2000 (m)	Metropolis		City population (2000)					Gross density inhab'ts/ km²
				5m+	4m+	3m+	2m+	1m+ capital cities	
Western and Central Europe									
Austria	8.1	Vienna	R				2.15		3,500
Belgium	10.3	Brussels						1.00	5,500
Czech Republic	10.3	Prague						1.20	2,400
Denmark	5.3	Copenhagen	M					1.10	3,500
Finland	5.2	Helsinki	M					1.00	2,500
France	58.9	Paris	R	9.70					3,570
Germany	82.2	Berlin	R			3.40			3,800
		Cologne	R			3.05			
		Dusseldorf	R			3.25			
		Essen	R	6.50					
		Frankfurt	R			3.70			4,500
		Hamburg	M				2.65		2,000
		Munich					2.30		4,000
		Rhine-Ruhr	R	(9.15)					1,010
		Stuttgart					2.70		
Hungary	10.0	Budapest	R					1.80	3,400
Ireland	3.8	Dublin	M					1.00	
Italy	57.7	Rome	R			3.40			2,000
		Milan			4.20				1,520
		Naples	M			3.00			10,500
Netherlands	15.9	Amsterdam	M					1.15	3,500
		Randstad	M	(5.10)					970
Norway	4.5	Oslo	M					(0.80)	1,000
Poland	38.7	Warsaw	R				2.20		3,500
		Katowice				3.05			

Region and country	National pop'n 2000 (m)	Metropolis		City population (2000)				1m+ capital cities	Gross density inhab'ts/ km²
				5m+	4m+	3m+	2m+		
Portugal	10.0	Lisbon	M				2.00		7,000
Slovak Republic	5.4	Bratislava	R					(0.45)	1,200
Spain	39.5	Madrid		5.15					3,030
		Barcelona	M				2.80		
Sweden	8.9	Stockholm	M					1.65	3,500
Switzerland	7.2	Bern						(0.35)	1,280
		Zurich						1.05	4,000
United Kingdom	59.7	London	M	12.00					1,100
		Birmingham					2.30		4,500
		Manchester	R				2.25		2,000
South-East Europe									
Bulgaria	8.2	Sofia						1.15	6,500
Croatia	4.4	Zagreb						1.10	3,500
Greece	10.6	Athens	M			3.20			840
Romania	22.4	Bucharest	R				2.00		3,000
Serbia & Montenegro	10.6	Belgrade	R					1.15	2,000
Turkey	65.3	Ankara				3.20			1,280
		Istanbul	M	8.75					1,650
		Izmir	M				2.20		
Eastern Europe									
Armenia	3.8	Yerevan						1.10	
Belarus	10.0	Minsk						1.70	
Georgia	5.0	Tbilisi						1.10	
Latvia	2.4	Riga						(0.80)	3,000
Russian Federation	145.6	Moscow		10.10					4,700
		St Petersburg	M	5.20					4,520
Ukraine	49.5	Kiev	R				2.60		3,000
Sub-Saharan Africa									
Angola	13.1	Luanda	M				2.30		
Benin	6.3	Porto Novo	M					(0.20)	
Cameroon	14.9	Yaounde						1.20	
		Douala	M				(2.00)		
Congo Dem. Rep.	50.9	Kinshasa	R	5.05					
Côte d'Ivoire	16.0	Abidjan	M			3.05			5,550
Ethiopia	64.3	Addis Ababa					2.50		

Region and country	National pop'n 2000 (m)	Metropolis		City population (2000)				1m+ capital cities	Gross density inhab'ts/ km²
				5m+	4m+	3m+	2m+		
Ghana	19.3	Accra	M					1.70	
Guinea	7.4	Conakry	M					1.20	4,200
Kenya	30.1	Nairobi					2.25		1,500
Madagascar	15.5	Antananarivo						1.35	
Mozambique	17.7	Maputo	M					1.10	
Nigeria	126.9	Lagos	M	8.40					13,100
		Ibadan					2.20		
Senegal	9.5	Dakar	M					1.85	
South Africa	42.8	Cape Town	M				2.70		
		Pretoria						1.10	
		Johannesburg	R				2.75		1,680
Sudan	31.1	Khartoum	R			3.95			
Tanzania	33.7	Dar-es-Salaam	M				2.10		
Uganda	22.2	Kampala						1.10	
Zambia	10.1	Lusaka						1.05	
Zimbabwe	12.6	Harare						1.40	
North Africa									
Algeria	30.4	Algiers	M				2.75		10,200
Egypt	64.0	Cairo	R	10.40					7,940
		Alexandria	M			3.50			
Libya	5.3	Tripoli	M					1.90	
Morocco	28.7	Rabat	M					1.50	
		Casablanca	M			3.55			2,190
Tunisia	9.6	Tunis	M					(0.75)	
Western Asia (Middle East)									
	0.7	Al Manamah	M					(0.15)	
Iraq	23.3	Baghdad	R	5.20					7,120
Israel	6.2	Tel-Aviv-Yafo	M				2.20		
Jordan	4.9	Amman						1.45	11,000
Kuwait	2.0	Kuwait City	M					1.20	
Lebanon	4.3	Beirut	M					1.50	
Oman	2.4	Muscat	M					(0.10)	
Qatar	0.6	Doha	M					(0.20)	
Saudi Arabia	20.7	Riyadh				3.60			2,320
		Jeddah	M				2.50		
Syria	16.2	Damascus					2.05		
		Aleppo					2.20		

Region and country	National pop'n 2000 (m)	Metropolis		City population (2000)					Gross density inhab'ts/ km²
				5m+	4m+	3m+	2m+	1m+ capital cities	
United Arab Emirates	2.9	Abu Dhabi	M					(0.95)	
Yemen	17.5	Sana'a						1.35	
Central Asia									
Afghanistan	26.6	Kabul					2.00		
Azerbaijan	8.0	Baku	M					1.80	
Iran	63.7	Tehran		7.00					10,600
		Esfahan					2.00		
		Mashhad					2.00		
Kazakhstan	14.9	Alma-Ata						1.15	
Uzbekistan	24.8	Tashkent					2.15		
South Asia									
Bangladesh	131.1	Dhaka	M	10.15					6,340
		Chittagong	M			3.30			3,330
India	1015.9	Delhi		12.45					8,410
		Ahmedabad	R		4.45				3,420
		Bangalore		5.55					4,340
		Calcutta	M	13.05					7,330
		Hyderabad		5.45					2,930
		Jaipur					2.25		
		Kanpur	R				2.65		
		Kochi (Cochin)	M				(2.00)		
		Lucknow					2.20		
		Madras	M	6.35					5,380
		Mumbai	M	16.10					3,690
		Nagpur					2.10		
		Pune (Poona)				3.65			5,210
		Surat	M				2.70		13,500
		Vadodara					(2.00)		
Myanmar (Burma)	47.7	Yangon	M			3.65			10,430
Nepal	23.0	Kathmandu						(0.50)	
Pakistan	138.1	Islamabad						1.00	
		Faisalabad (Lyallpur)					2.15		
		Gujranwala					(2.00)		
		Karachi	M	10.05					2,850
		Lahore	R	5.45					3,080
		Rawalpindi					(2.00)		
Sri Lanka	19.4	Colombo	M					(0.75)	

Region and country	National pop'n 2000 (m)	Metropolis		City population (2000)					Gross density inhab'ts/ km²
				5m+	4m+	3m+	2m+	1m+ capital cities	
East Asia and Pacific									
Cambodia	12.0	Phnom Penh	M					1.35	
China	1262.5	Beijing		10.85					1,400
		Changchun					2.75		
		Changsha					2.10		
		Chengdu				3.90			1,830
		Chongqing		6.05					
		Dalian	M				2.85		
		Dongguan				3.75			1,520
		Fuzhou	M				2.10		
		Guangzhou	M	7.40					1,020
		Guiyang					2.95		
		Hangzhou	M				2.40		
		Harbin	R			3.45			
		Jinan	R				2.65		
		Jinxi	R				2.00		
		Kunming					2.60		
		Lanzhou					2.05		
		Linyi	R				2.00		
		Nanjing	R			3.50			
		Qingdao	M				2.70		5,000
		Shanghai	M	13.25					2,040
		Shenyang			4.60				1,330
		Shenzhen		6.05					3,100
		Shijiazhuang					2.00		
		Taiyuan					2.50		
		Tangshan					(2.00)		
		Tianjin	M	6.70					940
		Wuhan	R	6.65					
		Xian				3.75			1,060
		Yantai	M				(2.00)		
		Zaozhuang					2.00		
		Zhengzhou	R				2.45		
		Zibo					2.80		
Hong Kong	6.8	Hong Kong	M	6.65					6,050
Indonesia	210.4	Jakarta	M	11.05					8,120
		Bandung				3.45			15,680
		Medan					2.00		
		Palembang	R				(2.00)		

Region and country	National pop'n 2000 (m)	Metropolis		City population (2000)				1m+ capital cities	Gross density inhab'ts/ km²
				5m+	4m+	3m+	2m+		
		Semarang	M				(2.00)		
		Surabaya	M				2.55		15,000
		Tanjung Karang	M				(2.00)		
Japan	126.9	Tokyo-Yokohama	M	27.90					2,000
		Kitakyushu	M				2.75		2,500
		Nagoya	M	5.30					6,500
		Osaka-Kobe	M	11.00					5,320
Malaysia	23.3	Kuala Lumpur						1.30	4,000
North Korea	22.3	Pyongyang	R			3.20			1,520
Philippines	75.6	Manila	M	10.00					6,250
Singapore	4.0	Singapore	M		4.00				5,800
South Korea	47.3	Seoul	M	9.90					4,300
		Inchon	M				2.45		
		Pusan	M			3.65			4,800
		Taegu					2.50		9,000
Taiwan	22.2	Taipei	M	6.65					2,710
		Kaohsiung	M				(2.00)		
Thailand	60.7	Bangkok	M	6.35					4,040
Vietnam	78.5	Hanoi	R			3.75			1,750
		Ho Chi Minh	M		4.60				2,200
Oceania									
Australia	19.2	Canberra						(0.35)	950
		Melbourne	M			3.45			1,500
		Sydney	M		4.10				1,700
New Zealand	3.8	Wellington	M					(0.35)	
North and Central America									
Canada	30.8	Ottawa	R					1.10	3,100
		Montreal	R			3.50			3,200
		Toronto	R		4.75				2,600
		Vancouver	M				2.05		2,000
Costa Rica	3.8	San Jose						1.05	
Cuba	11.2	Havana	M				2.20		
Dominican Republic	8.4	Santo Domingo	M					1.80	
Guatemala	11.4	Guatemala City						1.00	
Haiti	8.0	Port-au-Prince						1.75	

Region and country	National pop'n 2000 (m)	Metropolis		City population (2000)					Gross density inhab'ts/ km²
				5m+	4m+	3m+	2m+	1m+ capital cities	
Jamaica	2.6	Kingston	M					(0.60)	
Mexico	98.0	Mexico City		18.10					2,320
		Guadalajara				3.70			1,360
		Monterrey				3.30			820
Nicaragua	5.1	Managua						1.00	
Panama	2.9	Panama	M					1.10	4,000
Puerto Rico	3.9	San Juan	M				2.25		
United States	281.6	Washington DC	M		4.00				1,330
		Atlanta	R			3.55			750
		Baltimore	M				2.10		1,200
		Boston	M		4.05				1,200
		Chicago	R	8.33					1,510
		Dallas	R		4.15				1,200
		Denver					2.00		1,250
		Detroit	R			3.90			1,190
		Houston				3.85			1,150
		Los Angeles	M	12.70					1,010
		Miami	M		4.95				1,710
		Minneapolis	R				2.40		750
		New York	M	17.80					1,020
		Philadelphia	M	5.15					1,110
		Phoenix	R				2.95		1,040
		Saint Louis	R				2.10		1,030
		San Diego	M			3.00			
		San Francisco	M			3.25			1,600
		Seattle	M				2.75		
		Tampa	M				2.05		1,010
South America									
Argentina	37.0	Buenos Aires	M	11.85					3,220
Bolivia	8.3	La Paz						1.40	6,500
Brazil	170.4	Brasilia					2.75		200
		Belem					2.00		
		Belo Horizonte			4.65				4,500
		Campinas					2.25		
		Curitiba					2.50		
		Fortaleza	M			3.00			2,000
		Porto Alegre	M			3.50			
		Recife	M			3.25			1,170
		Rio de Janeiro	M	10.80					2,160

Region and country	National pop'n 2000 (m)	Metropolis		City population (2000)					Gross density inhab'ts/ km²
				5m+	4m+	3m+	2m+	1m+ capital cities	
		Salvador	M			3.00			1,280
		Sao Paulo	M	17.10					2,120
Chile	15.2	Santiago	R	5.35					2,280
Colombia	42.3	Bogota		6.95					3,500
		Cali					2.70		
		Medellin					2.80		3,000
Ecuador	12.6	Quito						1.40	
		Guayaquil	M				2.10		
Paraguay	5.5	Asuncion						1.50	
Peru	25.7	Lima	M	6.84					2,560
Uruguay	3.3	Montevideo	M					1.30	
Venezuela	24.2	Caracas	M				2.85		
		Maracaibo					2.00		

APPENDIX 19
FUTURE URBAN POPULATION BY CONTINENTAL REGION

Region	Surface Area (m km²)	YEAR 2000 Total pop'n (m)	YEAR 2000 Urban pop'n (m)	YEAR 2000 % urban pop'n	YEAR 2025 Total pop'n (m)	YEAR 2025 Urban pop'n (m)	YEAR 2025 % urban pop'n	YEAR 2150 Total pop'n (m)	YEAR 2150 Urban pop'n (m)	YEAR 2150 % urban pop'n
More developed regions	**52.75**	**1,273**	**954**	**74.9**	**1,365**	**1,125**	**82.4**	**1,600**	**1,440**	**90.0**
North America	19.94	313	241	77.0	370	310	83.8	450	410	91.1
Western & Ctl Europe	4.15	445	346	77.8	445	370	83.1	535	480	89.7
Eastern & S-E Europe	19.72	357	246	68.9	380	315	82.9	455	410	90.1
Japan	0.38	127	100	78.7	129	100	77.5	110	95	86.4
Oceania	8.56	31	21	67.7	41	30	73.2	50	45	90.0
Less developed regions	**81.00**	**4,784**	**1,893**	**39.6**	**6,835**	**3,900**	**57.1**	**10,900**	**9,810**	**90.0**
Africa	30.05									
Sub-Saharan Africa	24.30	660	228	34.5	1,355	733	54.1	2,800	2,520	90.0
North Africa	5.75	138	74	53.6	260	174	66.9	410	370	90.2
West & Central Asia	10.12									
Middle East	3.75	105	72	68.6	180	137	76.1	280	250	89.3
Central Asia	6.37	154	73	47.4	265	172	64.9	410	370	90.2
South Asia	5.17									
India	3.29	1,016	289	28.4	1,230	666	54.1	2,000	1,800	90.0
Other South Asia	1.88	360	104	28.9	490	222	45.3	800	720	90.0
East Asia & Pacific	15.19									
China	9.57	1,270	412	32.4	1,475	653	44.3	1,500	1,350	90.0
Other East Asia	5.62	564	252	44.7	800	479	59.9	1,400	1,260	90.0

Long-Range Futures Research

		YEAR 2000			YEAR 2025			YEAR 2150		
Region	Surface Area (m km²)	Total pop'n (m)	Urban pop'n (m)	% urban pop'n	Total pop'n (m)	Urban pop'n (m)	% urban pop'n	Total pop'n (m)	Urban pop'n (m)	% urban pop'n
Latin America & Caribbean	20.47									
Central America & Caribbean	2.71	171	113	66.1	280	224	80.0	470	425	90.4
South America	17.76	346	276	79.8	500	440	88.0	830	745	89.8
WORLD TOTAL	**133.75**	**6,057**	**2,847**	**47.0**	**8,200**	**5,025**	**61.3**	**12,500**	**11,250**	**90.0**

Central Europe includes the Czech and Slovak Republics, Hungary and Poland.

S-E Europe includes the Balkan countries of Albania, Bulgaria, Greece, Romania, Turkey, & former Yugoslavia.

Central Asia comprises Afghanistan, Azerbaijan, Iran, Kazakhstan, Tajikistan, Turkmenistan, and Uzbekistan.

APPENDIX 20
FUTURE NUMBER OF CITIES BY CONTINENTAL REGION
AND URBAN SIZE CLASS

	Number of cities by urban size class								
	YEAR 2000			**YEAR 2025**			**YEAR 2150**		
Region	0.1m- 0.99m	1m- 4.9m	5m+	0.1m- 0.99m	1m- 4.9m	5m+	0.1m- 0.99m	1m- 4.9m	5m+
More developed regions	**1,100**	**106**	**15**	**1,350**	**140**	**17**	**1,810**	**170**	**29**
North America	240	31	4	320	42	6	475	46	8
Western & Ctl Europe	395	36	5	450	45	5	610	55	9
Eastern & S-E Europe	345	27	3	420	40	3	525	50	7
Japan	100	6	3	130	6	3	150	10	3
Oceania	20	6	-	30	7	-	50	9	2
Less developed regions	**2,225**	**194**	**35**	**4,150**	**360**	**73**	**6,690**	**580**	**156**
Africa									
Sub-Saharan Africa	285	30	2	775	55	12	1,715	145	37
North Africa	105	7	1	175	20	5	215	25	6
West & Central Asia									
Middle East	100	11	1	180	20	4	245	25	6
Central Asia	115	10	1	170	15	3	215	20	5
South Asia									
India	320	27	7	700	55	10	1,230	100	28
Other South Asia	105	10	3	240	20	8	490	40	11
East Asia & Pacific									
China	390	42	8	720	70	15	920	80	22
Other East Asia	315	22	5	520	45	8	860	75	20
Latin America & Caribbean									
Central America & Caribbean	145	13	1	250	20	1	290	25	7
South America	345	22	6	420	40	7	510	45	14
WORLD TOTAL	**3,325**	**300**	**50**	**5,500**	**500**	**90**	**8,500**	**750**	**185**

Note: In 2000 the table includes over 75 agglomerations of over 2m population, so that the number of cities of over 100,000 population and less than 1m are reduced by some 250. If the table included urban agglomerations of 1m population then the number of cities in in the size class 1m-4.9m would increase from 300 to 350, so that the total number of agglomerations of over 1m population would amount to 400.

APPENDIX 21
FUTURE NUMBER OF CITIES IN MDCs AND LCs

Urban area class (population range)	YEAR 2000			YEAR 2025			YEAR 2150		
	No. of MDC cities	No. of LDC cities	Total pop'n (m)	No. of MDC cities	No. of LDC cities	Total pop'n (m)	No. of MDC cities	No. of LDC cities	Total pop'n (m)
100m +	-	-	-	-	-	-	-	5	750
50 - 99.9m	-	-	-	-	1	50	1	9	750
20 - 49.9m	1	-	30	1	9	250	2	18	750
10 - 19.9m	5	14	215	6	18	300	12	38	750
	6	**14**	**245**	**7**	**28**	**600**	**15**	**70**	**3,000**
5 - 9.9m	9	21	195	10	45	325	14	86	750
2 - 4.9m	45	80	300	50	125	525	50	200	750
1 - 1.9m	60	115	270	90	235	475	120	380	750
	120	**230**	**1,010**	**157**	**433**	**1,925**	**199**	**736**	**5,250**
0.5 - 0.99m	90	235	250	175	475	450	235	765	750
0.2 - 0.49m	325	675	250	400	1,200	450	545	1,955	750
0.1 - 0.19m	675	1,325	250	775	2,475	450	1,030	3,970	750
	1,210	**2,465**	**1,760**	**1,507**	**4,583**	**3,275**	**2,009**	**7,426**	**7,500**
50,000 - 99,999	1,150	2,100	220	1,250	3,750	350	1,800	8,200	750
20,000 - 49,999	2,500	5,000	220	3,125	9,375	350	4,500	20,500	750
10,000 - 19,999	5,000	10,000	220	6,250	18,750	350	9,000	41,000	750
5,000 - 9,999	10,000	20,000	220	12,500	37,500	350	18,000	82,000	750
	19,860	**39,565**	**2,640**	**24,632**	**73,958**	**4,675**	**35,309**	**159,126**	**10,500**
2,000 - 4,999	25,000	50,000	207	30,000	95,000	350	45,000	205,000	750
Rural villages	1.5m	13.5m	3,210	1m	11.5m	3,175	0.65m	4.35m	1,250
Total population			**6,057**			**8,200**			**12,500**

Note:

    MDC: More Developed Country
    LDC: Less Develpoed Country

APPENDIX 22
PAST AND FUTURE NUMBER OF CITIES BY URBAN SIZE CLASS

Urban size class (population range)	YEAR 1975			YEAR 2000			YEAR 2025		
	No. of cities	Pop'n (m)	Area (m km²)	No. of cities	Pop'n (m)	Area (m km²)	No. of cities	Pop'n (m)	Area (m km²)
100m +	-	-	-	-	-	-	-	-	-
50m - 99.9m	-	-	-	-	-	-	1	50	0.07
20m - 49.9m	-	-	-	1	30	0.02	10	250	0.20
10m - 19.9m	4	58	0.03	19	215	0.11	24	300	0.15
	1%	**58**	**0.03**	**4%**	**245**	**0.13**	**7%**	**600**	**0.42**
5m - 9.9m	18	136	0.02	30	195	0.03	55	325	0.05
2m - 4.9m	57	168	0.03	125	300	0.05	175	525	0.08
1m - 1.9m	99	136	0.03	175	270	0.05	325	475	0.08
	12%	**498**	**0.11**	**17%**	**1,010**	**0.26**	**24%**	**1,925**	**0.63**
0.5m - 0.99m	152	121	0.04	325	250	0.08	650	450	0.14
0.2m - 0.49m	660	196	0.07	1,000	250	0.08	1,600	450	0.14
0.1m - 0.19m	910	125	0.04	2,000	250	0.08	3,250	450	0.14
	23%	**940**	**0.26**	**29%**	**1,760**	**0.50**	**40%**	**3,275**	**1.05**
50,000 - 99,999	1,800	120	0.09	3,250	220	0.15	5,000	350	0.22
20,000 - 49,999	4,500	120	0.09	7,500	220	0.15	12,500	350	0.22
10,000 - 19,999	9,000	120	0.09	15,000	220	0.15	25,000	350	0.22
5,000 - 9,999	18,000	120	0.09	30,000	220	0.15	50,000	350	0.22
	35%	**1,420**	**0.62**	**44%**	**2,640**	**1.10**	**57%**	**4,675**	**1.93**
2,000 - 4,999	45,000	120	0.18	75,000	207	0.26	125,000	350	0.47
Rural villages	14m	2,535	10.20	15m	3,210	12.64	12.5m	3,175	12.60
	100%	**4,075**	**11.00**	**100%**	**6,057**	**14.00**	**100%**	**8,200**	**15.00**

Note: In 2000 the table includes over 75 agglomerations of over 2m population, so that the number of cities of over 100,000 population and less than 1m are reduced by some 250. The urban densities are as follows:-

Urban class	Population range	Density (inhab/km²)	
		MDCs	LDCs
Metapolis	50m +	500	-750
Megalopolis	20m +	1,000	-1,500
Megapolitan region	10m +	1,250	-2,500
Metropolis	1m - 9.9m	4,000	-8,000
Regional city	0.1m - 0.99m	2,000	-4,000
Urban centre	5,000 - 99,999	1,000	-2,000
Township	2,000 - 4,999	500	-1,000
Rural village	1 - 1,999	250	-250

APPENDIX 23
FUTURE NUMBER OF CITIES BY URBAN SIZE CLASS
(DEFAULT VALUES)

Urban size class (population range)	YEAR 2050			YEAR 2100			YEAR 2150		
	No. of cities	Pop'n (m)	Area (m km²)	No. of cities	Pop'n (m)	Area (m km²)	No. of cities	Pop'n (m)	Area (m km²)
100m +	-	-	-	2	125	0.20	5	750	1.25
50m - 99.9m	7	500	0.69	8	600	0.82	10	750	1.00
20m - 49.9m	14	500	0.37	17	600	0.40	20	750	0.50
10m - 19.9m	35	500	0.27	42	600	0.35	50	750	0.50
	16%	**1,500**	**1.33**	**19%**	**1,925**	**1.77**	**24%**	**3,000**	**3.25**
5m - 9.9m	70	500	0.08	85	600	0.11	100	750	0.15
2m - 4.9m	175	500	0.08	225	600	0.11	250	750	0.15
1m - 1.9m	350	500	0.08	425	600	0.11	500	750	0.15
	33%	**3,000**	**1.57**	**37%**	**3,725**	**2.10**	**42%**	**5,250**	**3.70**
0.5m - 0.99m	700	500	0.16	850	575	0.21	1,000	750	0.30
0.2m - 0.49m	1,750	500	0.16	2,125	575	0.21	2,500	750	0.30
0.1m - 0.19m	3,500	500	0.16	4,250	575	0.21	5,000	750	0.30
	49%	**4,500**	**2.05**	**55%**	**5,450**	**2.73**	**60%**	**7,500**	**4.60**
50,000 - 99,999	6,000	450	0.30	8,000	525	0.41	10,000	750	0.60
20,000 - 49,999	15,000	450	0.30	20,000	525	0.41	25,000	750	0.60
10,000 - 19,999	30,000	450	0.30	40,000	525	0.41	50,000	750	0.60
5,000 - 9,999	60,000	450	0.35	80,000	525	0.41	100,000	750	0.60
	68%	**6,300**	**3.30**	**76%**	**7,550**	**4.37**	**84%**	**10,500**	**7.00**
2,000 - 4,999	150,000	400	0.53	200,000	450	0.70	250,000	750	1.00
Rural villages	10m	2,500	10.00	8m	2,000	8.00	5m	1,250	5.00
	100%	**9,200**	**13.83**	**100%**	**10,000**	**13.07**	**100%**	**12,500**	**13.00**

Long-Range Futures Research

APPENDIX 24
NUMBER OF PROSPECTIVE WORLD CITIES OF THE FUTURE

Region	1st Order	2nd Order			3rd Order			4th Order			Total
	MP	MP	CM	MT	MP	CM	MT	MP	CM	MT	
North America	1	1				1	4			3	10
Western & Cent. Europe	1		2	2		1	4		1	2	13
S-E Europe						1					1
Eastern Europe		1				1					2
Sub-Saharan Africa			1	1		1	1			3	7
North Africa					1					2	3
Middle East & Central Asia						1			1	2	4
South Asia		1			2			2	1	1	7
East Asia & Pacific	1	1			2	2	1	2	2	1	12
Oceania				1			1				2
Latin America		1			3				2	1	7
Total	**3**	**5**	**3**	**4**	**8**	**8**	**11**	**4**	**7**	**15**	**68**

The number of prospective world cities of the future is tabulated above, and the population size categories refer to the metro-region as follows:

Megapolis (MP)   10m+
Cosmopolis (CM)   5m-9.99m
Metropolis (MT)   1m-4.99m

APPENDIX 25
PROSPECTIVE WORLD CITIES OF THE FUTURE

Region and country	1st order	2nd order	3rd order	4th order
Western and Central Europe				
France		Paris		
Germany		Frankfurt	Berlin	Rhine-Ruhr
Italy		Milan	Rome	
Netherlands		Randstad		
Poland			Warsaw	
Spain			Madrid	Barcelona
Switzerland			Zurich	Geneva
United Kingdom	London			
South-East Europe				
Turkey			Istanbul	
Eastern Europe				
Russian Federation			Moscow	St Petersburg
Sub-Saharan Africa				
Ethiopia				Addis Ababa
Kenya			Nairobi	
Nigeria		Lagos		
South Africa		Johannesburg		
Sudan				Khartoum
Tanzania				Dar-es-Salaam
Zaire			Kinshasa	
North Africa				
Algeria				Algiers
Egypt			Cairo	
Morocco				Casablanca
Middle East and Central Asia				
Iran			Tehran	

Iraq			Baghdad
Israel			Tel Aviv
Saudi Arabia			Riyadh
South Asia			
Bangladesh			Dhaka
India		Mumbai	Delhi	Bangalore
				Calcutta
Myanmar (Burma)				Yangon
Pakistan			Karachi
East Asia and Pacific			
China		Shanghai	Beijing	Tianjin
Hong Kong			Hong Kong
Indonesia				Jakarta
Japan	Tokyo-Yokohama		Osaka-Kobe
Malaysia				Kuala Lumpur
Philippines				Manila
Singapore			Singapore
South Korea			Seoul
Thailand				Bangkok
Oceania			
Australia		Sydney	Melbourne
North America			
Canada			Montreal	Vancouver
			Toronto
United States	New York	Los Angeles	Chicago	Houston
			Miami	Seattle
			San Francisco
Latin America			
Argentina			Buenos Aires
Brazil		Sao Paulo	Rio de Janeiro
Colombia				Bogota
Mexico			Mexico City
Peru				Lima
Venezuela				Caracas

Long-Range Futures Research

APPENDIX 26
WORKFORCE BY DIVISION AND CONTINENTAL REGION
FOR A WORLD POPULATION OF 12.5 BILLION IN 2150

Region	Domestic + Workforce (m)	Primary Resources (m)	Secondary Industrial (m)	Tertiary Commercial (m)	Quaternary Information (m)	Quinary Lifespan (m)
More developed regions	**900.00**	**23.76**	**173.40**	**225.84**	**247.00**	**230.00**
North America	255.00	6.40	46.69	63.69	74.40	63.79
Western & Central Europe	300.00	7.50	55.00	75.00	82.50	80.00
Eastern & South-East Europe	250.00	7.56	54.59	63.83	64.00	60.00
Japan	65.00	1.55	11.59	15.79	17.85	18.19
Oceania	30.00	0.75	5.50	7.50	8.25	8.00
	100.0%	**2.6%**	**19.2%**	**25.1%**	**27.4%**	**25.5%**
Less developed regions	**5850.00**	**516.80**	**1376.57**	**1485.27**	**1106.33**	**1370.00**
Africa						
Sub-Saharan Africa	1510.00	207.89	388.07	346.50	237.51	330.00
North Africa	180.00	6.00	39.00	46.19	43.79	45.00
West & Central Asia						
Middle East	110.00	3.60	23.39	27.71	25.27	30.00
Central Asia	180.00	15.00	42.00	48.00	35.00	45.00
South Asia						
India	1100.00	90.00	252.00	288.00	200.00	270.00
Other South Asia	440.00	36.00	100.79	115.19	83.00	105.00
East Asia & Pacific						
China	840.00	70.00	196.00	224.00	160.00	190.00
Other East Asia	770.00	64.50	180.59	206.39	143.50	175.00
Latin America & Caribbean						
Central America & Caribbean	260.00	8.59	55.89	66.21	64.27	65.00
South America	460.00	15.19	98.79	117.03	113.95	115.00
	100.0%	**8.8%**	**23.5%**	**25.3%**	**18.9%**	**23.4%**
WORLD TOTAL	**6750.00**	**540.56**	**1549.98**	**1711.11**	**1353.33**	**1600.00**
	100.0%	**8.0%**	**22.9%**	**25.3%**	**20.0%**	**23.7%**

Note: The workforce has been increased by a factor of 1.2 to include females not participating full time in the workforce between the ages of 20-59, but involved full time in domestic duties in the Quinary Lifespan Division.
Some 25% of the Quaternary Information workforce has been transferred to the Quinary Lifespan Division to include health, social services and charities.

APPENDIX 27
LAND USE BY CONTINENTAL REGION
FOR A WORLD POPULATION OF 12.5 BILLION IN 2150

Region	Total area (m km²)	Urban & villages (m km²)	Cultivated (m km²)	Pastures (m km²)	Forests (m km²)	Wildlife (m km²)
More developed regions	**52.75**	**2.00**	**6.65**	**8.50**	**17.45**	**18.15**
North America	19.94	0.59	2.40	2.44	4.81	9.70
Western & Central Europe	4.15	0.66	1.20	0.60	1.36	0.33
Eastern & South-East Europe	19.72	0.55	2.45	1.95	9.15	5.62
Japan	0.38	0.13	0.05	0.01	0.13	0.06
Oceania	8.56	0.07	0.55	3.50	2.00	2.44
	100.0%	**3.8%**	**12.6%**	**16.1%**	**33.1%**	**34.4%**
Less developed regions	**81.00**	**11.00**	**9.35**	**15.60**	**18.65**	**26.40**
Africa						
Sub-Saharan Africa	24.30	2.93	2.95	6.30	6.07	6.05
North Africa	5.75	0.55	0.20	0.40	0.10	4.50
West & Central Asia						
Middle East	3.75	0.29	0.10	0.20	0.06	3.10
Central Asia	6.37	0.41	0.30	0.50	0.30	4.86
South Asia						
India	3.29	1.85	0.70	0.50	0.14	0.10
Other South Asia	1.88	0.76	0.30	0.40	0.25	0.17
East Asia & Pacific						
China	9.57	1.50	1.25	1.80	1.00	4.02
Other East Asia	5.62	1.41	0.70	1.00	2.00	0.51
Latin America & Caribbean						
Central America & Caribbean	2.71	0.45	0.35	0.65	0.73	0.53
South America	17.76	0.85	2.50	3.85	8.00	2.56
	100.0%	**13.6%**	**11.5%**	**19.3%**	**23.0%**	**32.6%**

Long-Range Futures Research

Region	Total area (m km²)	Urban & villages (m km²)	Cultivated (m km²)	Pastures (m km²)	Forests (m km²)	Wildlife (m km²)
WORLD TOTAL	133.75	13.00	16.00	24.10	36.10	44.55
	100.0%	9.7%	12.0%	18.0%	27.0%	33.3%

Note: The total land area is a region's total area including the land under inland water bodies.

The areas for wildlife include inland rivers and lakes, protected natural areas, wetlands, drylands, desert, and steeplands. The forests, areas for wildlife and grazing land amount to 78% of the total land area and are needed for ecosystem services.

Potential arable land includes cultivated land for crops and pasture for livestock. Some 20% of croplands are irrigated and this includes irrigation by controlled flooding. The UN Food and Agriculture Organization (FAO) estimates the suitability of land for rainfed crop production. The potential arable land is categorised in terms of quality and weighted with a factor in accordance with its suitability to give an equivalent potential area in terms of Very Suitable land. The suitability classes and weightings are as follows: Suitable x 0.7, Moderately Suitable x 0.5, and Marginal x 0.3. Overall the equivalent potential of arable land corresponds to 0.7 x Very Suitable to give an overall average of Suitable land on which 1 hectare will support 5 people.

Long-Range Futures Research

APPENDIX 28
DEFAULT VALUES FOR MATERIAL RESOURCE FLOWS

Stage of development		GNI per capita (2000 US$)	Purchasing Power Parity (PPP$) Factor	2000 Inverse of Energy Intensity PPP$ per EJ	2000 Inverse of Material Intensity PPP$ per metric ton	2000 Material Flows metric tons per capita	2150 Material Flows metric tons per capita
Ecopolitan	1	40,001- 60,000	1.00	200	600	65	40
Informational	2	25,001- 40,000	0.90	150 *	400	75	50
	3	15,001- 25,000	1.20	125	375	65	40
Distributional	4	10,001 - 15,000	1.50	125	350	55	35
	5	6,001 - 10,000	2.00	125	300	55	35
Industrial	6	4,001 - 6,000	1.80	110	250	40	35
	7	2,501 - 4,000	2.30	110	225	30	30
Infrastructural	8	1,501 - 2,500	3.25	80	225	30	30
	9	1,001 - 1,500	3.25	80	200	20	25
Agropolitan	10	601 - 1,000	4.50	95	175	20	25
	11	401 - 600	5.00	95	175	15	20
Traditional	12	Less than 4.00	4.00	70	150	10	15

*150 EJ excluding USA, and 115 EJ inclusive of USA

Note: Energy and Construction material flows represent 70-75% of the Direct Material Inputs in the more developed countries, and 60-65% of the Total Material Flows.

APPENDIX 29
FUTURE GROSS WORLD PRODUCT BY CONTINENTAL REGION
(DEFAULT VALUES FOR GRP BASED UPON INVESTMENT TRANSFERS)

Region	2150 GRP/ m km² US$bn	YEAR 2000			YEAR 2025			YEAR 2150		
		Total pop'n (m)	GRP (2000) US$bn	GRP/cap (2000) US$	Total pop'n (m)	GRP (2000) US$bn	GRP/cap (2000) US$	Total pop'n (m)	GRP (2000) US$bn	GRP/cap (2000) US$
More developed regions	**1,230**	**1,273**	**25,060**	**19,670**	**1,365**	**39,000**	**28,570**	**1,600**	**65,000**	**40,600**
North America	1,020	313	10,250	32,750	370	14,500	39,200	450	20,250	45,000
Western & Ctl Europe	5,180	445	9,080	20,400	445	14,500	32,580	535	21,500	40,190
Eastern & S-E Europe	800	357	760	2,130	380	4,000	10,530	455	15,750	34,620
Japan	13,820	127	4,520	35,590	129	4,900	37,980	110	5,250	47,730
Oceania	260	31	450	14,520	41	1,100	26,830	50	2,250	45,000
Less developed regions	**1,670**	**4,784**	**6,170**	**1,290**	**6,835**	**21,000**	**3,070**	**10,900**	**135,000**	**12,400**
Africa										
Sub-Saharan Africa	700	660	310	470	1,355	1,150	850	2,800	17,000	6,000
North Africa	1,130	138	230	1,670	260	750	2,880	410	6,500	15,850
West & Central Asia										
Middle East	1,800	105	440	4,190	180	1,650	9,170	280	6,750	24,110
Central Asia	860	154	150	970	265	600	2,260	410	5,500	13,410
South Asia										
India	6,610	1,016	450	440	1,230	1,850	1,500	2,000	21,750	10,880
Other South Asia	4,650	360	140	390	490	750	1,530	800	8,750	10,940
East Asia & Pacific										
China	2,040	1,270	1,250	980	1,475	4,250	2,880	1,500	19,500	13,000
Other East Asia	3,290	564	1,280	2,270	800	3,750	4,690	1,400	18,500	13,210
Latin America & Caribbean										
Central America & Caribbean	4,240	171	670	3,920	280	2,250	8,010	470	11,500	24,470
South America	1,080	346	1,250	3,610	500	4,000	8,000	830	19,250	23,190
WORLD TOTAL	**1,500**	**6,057**	**31,230**	**5,150**	**8,200**	**60,000**	**7,300**	**12,500**	**200,000**	**16,000**

Region	2150 GRP/ capita US$	YEAR 2050			YEAR 2100			YEAR 2250		
		Total pop'n (m)	GRP (2000) US$bn	GRP/cap (2000) US$	Total pop'n (m)	GRP (2000) US$bn	GRP/cap (2000) US$	Total pop'n (m)	GRP (2000) US$bn	GRP/cap (2000) US$
More developed regions	**40,600**	**1,250**	**45,000**	**36,000**	**1,300**	**55,000**	**42,300**	**1,500**	**105,000**	**70,000**
North America	45,000	380	17,000	44,740	400	18,500	46,250	490	37,000	75,510
Western & Ctl Europe	40,190	365	16,000	43,840	415	19,000	45,780	470	33,000	70,210
Eastern & S-E Europe	34,620	350	5,700	16,280	340	11,000	32,350	385	23,750	61,700
Japan	47,730	110	5,000	45,450	100	4,750	47,500	100	7,500	75,000
Oceania	45,000	45	1,300	28,890	45	1,750	38,890	55	3,750	68,180
Less developed regions	**12,400**	**7,950**	**45,000**	**5,660**	**8,700**	**85,000**	**9,770**	**10,000**	**300,000**	**30,000**
Africa										
Sub-Saharan Africa	6,000	1,900	2,900	1,530	2,250	8,400	3,730	2,600	49,500	19,000
North Africa	15,850	350	1,570	4,490	385	5,200	13,500	425	17,500	41,200
West & Central Asia										
Middle East	24,110	290	3,370	11,620	265	4,300	16,230	300	14,750	49,170
Central Asia	13,410	430	1,220	2,840	400	4,200	10,500	425	15,500	36,470
South Asia										
India	10,880	1,500	3,650	2,430	1,650	16,000	9,700	1,700	42,000	24,700
Other South Asia	10,940	540	1,100	2,040	650	6,400	9,850	680	13,750	20,220
East Asia & Pacific										
China	13,000	1,400	11,250	8,040	1,350	15,500	11,480	1,350	43,000	31,850
Other East Asia	13,210	620	11,520	18,580	750	9,000	12,000	1,150	40,000	34,780
Latin America & Caribbean										
Central America & Caribbean	24,470	330	2,940	8,910	360	5,500	15,280	495	23,000	46,460
South America	23,190	590	5,480	9,290	640	10,500	16,400	875	41,000	46,860
WORLD TOTAL	**16,000**	**9,200**	**90,000**	**9,800**	**10,000**	**140,000**	**14,000**	**11,500**	**405,000**	**35,000**

APPENDIX 30
FUTURE INVESTMENT STOCK REQUIREMENTS
BY CONTINENTAL REGION
(DEFAULT VALUES FOR GRP BASED UPON INVESTMENT TRANSFERS)

Region	2150 invest./ capita US$ *PPP	YEAR 2000 Urban pop'n (m)	YEAR 2000 GRP (2000) US$bn *PPP	YEAR 2000 Invest. stock US$trn *PPP	YEAR 2025 Urban pop'n (m)	YEAR 2025 GRP (2000) US$bn *PPP	YEAR 2025 Invest. stock US$trn *PPP	YEAR 2150 Urban pop'n (m)	YEAR 2150 GRP (2000) US$bn *PPP	YEAR 2150 Invest. stock US$trn *PPP
More developed regions	125,000	954	26,660	90	1,125	40,000	120	1,440	65,000	200
North America	142,000	241	10,440	32	310	14,500	38	410	20,250	64
Western & Ctl Europe	123,000	346	9,760	36	370	14,000	42	480	21,500	66
Eastern & S-E Europe	103,000	246	2,440	8	315	6,000	24	410	15,750	47
Japan	145,000	100	3,440	12	100	4,500	12	95	5,250	16
Oceania	140,000	21	580	2	30	1,000	4	45	2,250	7
Less developed regions	55,000	1,893	17,840	60	3,900	50,000	180	9,810	185,000	600
Africa										
Sub-Saharan Africa	30,000	228	1,050	3	733	5,250	16	2,520	25,500	84
North Africa	73,000	74	600	3	174	1,750	8	370	9,250	30
West & Central Asia										
Middle East	75,000	72	680	4	137	3,250	9	250	9,250	21
Central Asia	61,000	73	600	2	172	2,000	6	370	7,750	25
South Asia										
India	60,000	289	2,370	7	666	6,000	24	1,800	31,000	120
Other South Asia	60,000	104	590	1	222	2,500	10	720	12,750	48
East Asia & Pacific										
China	60,000	412	5,130	13	653	10,000	39	1,350	28,250	90
Other East Asia	60,000	252	3,140	11	479	6,750	28	1,260	25,750	84
Latin America & Caribbean										
Central America & Caribbean	74,000	113	1,220	5	224	4,500	14	425	12,750	35
South America	76,000	276	2,460	11	440	8,000	26	745	22,750	63
WORLD TOTAL	64,000	2,847	44,500	150	5,025	90,000	300	11,250	250,000	800

Region	2250 invest./ capita US$ *PPP	YEAR 2050			YEAR 2100			YEAR 2250		
		Urban pop'n (m)	GRP (2000) US$bn *PPP	Invest. stock US$trn *PPP	Urban pop'n (m)	GRP (2000) US$bn *PPP	Invest. stock US$trn *PPP	Urban pop'n (m)	GRP (2000) US$bn *PPP	Invest. stock US$trn *PPP
More developed regions	133,000	1,100	46,000	135	1,170	55,000	150	1,350	105,000	200
North America	133,000	340	17,000	45	360	18,500	49	440	37,000	65
Western & Ctl Europe	134,000	330	16,000	44	385	19,000	51	425	33,000	63
Eastern & S-E Europe	133,000	295	6,800	29	300	11,000	34	345	23,750	51
Japan	130,000	97	5,000	13	90	4,750	12	90	7,500	13
Oceania	145,000	38	1,200	4	35	1,750	4	50	3,750	8
Less developed regions	100,000	5,600	84,000	300	6,830	130,000	470	8,900	320,000	1,000
Africa										
Sub-Saharan Africa	83,000	1,155	9,500	42	1,665	18,500	70	2,250	59,500	215
North Africa	120,000	265	2,800	13	310	7,750	25	390	17,500	51
West & Central Asia										
Middle East	133,000	235	5,050	19	215	5,200	22	280	14,750	40
Central Asia	118,000	320	2,800	13	320	6,300	25	375	15,500	50
South Asia										
India	96,000	925	9,250	33	1,320	26,000	82	1,500	50,250	164
Other South Asia	84,000	335	3,300	13	520	10,500	33	620	16,500	57
East Asia & Pacific										
China	96,000	1,125	20,500	70	1,080	23,250	84	1,200	42,750	130
Other East Asia	97,000	495	13,800	50	600	13,500	47	1,035	39,750	112
Latin America & Caribbean										
Central America & Caribbean	131,000	270	6,000	17	290	6,500	30	450	22,500	65
South America	133,000	475	11,000	30	510	12,500	52	800	41,000	116
WORLD TOTAL	104,000	6,700	130,000	435	8,000	185,000	620	10,250	425,000	1,200

Long-Range Futures Research

APPENDIX 31
FUTURE MATERIAL FLOWS BY CONTINENTAL REGION (DEFAULT VALUES)

Region	2150 GRP/ m km² US$bn	YEAR 2000			YEAR 2025			YEAR 2150		
		Total pop'n (m)	Tons per capita	Material Flows tons (bn)	Total pop'n (m)	Tons per capita	Material Flows tons (bn)	Total pop'n (m)	Tons per capita	Material Flows tons (bn)
More developed regions	**1,230**	**1,273**	**75**	**95**	**1,365**	**65**	**90**	**1,600**	**45**	**70**
North America	1,020	313	80	26	370	70	27	450	50	22
Western & Ctl Europe	5,180	445	70	32	445	65	29	535	40	21
Eastern & S-E Europe	800	357	75	27	380	65	25	455	45	21
Japan	13,820	127	50	7	129	45	6	110	40	4
Oceania	260	31	75	3	41	65	3	50	45	2
Less developed regions	**1,670**	**4,784**	**20**	**100**	**6,835**	**30**	**205**	**10,900**	**35**	**400**
Africa										
Sub-Saharan Africa	700	660	15	10	1,355	20	27	2,800	35	98
North Africa	1,130	138	30	4	260	35	9	410	40	17
West & Central Asia										
Middle East	1,800	105	40	4	180	45	8	280	45	13
Central Asia	860	154	20	3	265	30	8	410	35	14
South Asia										
India	6,610	1,016	15	15	1,230	25	33	2,000	35	70
Other South Asia	4,650	360	10	4	490	25	12	800	35	28
East Asia & Pacific										
China	2,040	1,270	20	25	1,475	30	44	1,500	35	53
Other East Asia	3,290	564	30	17	800	35	28	1,400	35	49
Latin America & Caribbean										
Central America & Caribbean	4,240	171	35	6	280	45	13	470	45	21
South America	1,080	346	35	12	500	45	23	830	45	37
WORLD TOTAL	**1,500**	**6,057**	**30**	**195**	**8,200**	**35**	**295**	**12,500**	**38**	**470**

APPENDIX 32
FUTURE WORLD CAR FLEET AND GLOBAL ENERGY USE BY CONTINENTAL REGION (DEFAULT VALUES BASED UPON GRP/CAPITA AND TECHNOLOGICAL SHIFTS)

Region	2150 GRP/cap ('2000) US$	YEAR 2000 Total pop'n (m)	YEAR 2000 Car Fleet (m)	YEAR 2000 Energy Use (EJ)	YEAR 2025 Total pop'n (m)	YEAR 2025 Car Fleet (m)	YEAR 2025 Energy Use (EJ)	YEAR 2150 Total pop'n (m)	YEAR 2150 Car Fleet (m)	YEAR 2150 Energy Use (EJ)
More developed regions	**40,600**	**1,273**	**441**	**250**	**1,365**	**585**	**170**	**1,600**	**675**	**100**
North America	45,000	313	149	107	370	185	61	450	200	32
Western & Ctl Europe	40,190	445	188	72	445	201	53	535	225	31
Eastern & S-E Europe	34,620	357	42	43	380	122	36	455	180	26
Japan	47,730	127	50	22	129	58	15	110	47	7
Oceania	45,000	31	12	6	41	19	5	50	23	4
Less developed regions	**12,400**	**4,784**	**99**	**175**	**6,835**	**545**	**300**	**10,900**	**2,450**	**500**
Africa										
Sub-Saharan Africa	6,000	660	9	16	1,355	16	36	2,800	365	84
North Africa	15,850	138	5	5	260	25	10	410	160	27
West & Central Asia										
Middle East	24,110	105	8	11	180	29	17	280	110	19
Central Asia	13,410	154	4	10	265	21	10	410	120	19
South Asia										
India	10,880	1,016	5	21	1,230	59	33	2,000	410	93
Other South Asia	10,940	360	1	5	490	24	15	800	165	37
East Asia & Pacific										
China	13,000	1,270	4	50	1,475	142	59	1,500	310	70
Other East Asia	13,210	564	23	30	800	115	48	1,400	300	65
Latin America & Caribbean										
Central America & Caribbean	24,470	171	14	10	280	41	26	470	185	31
South America	23,190	346	26	17	500	73	46	830	325	55
WORLD TOTAL	**16,000**	**6,057**	**540**	**425**	**8,200**	**1,130**	**470**	**12,500**	**3,125**	**600**

Note: With a technological shift it is expected that through longer life assets, recycling and minimum material design, the material throughput for a given capital stock could be reduced to 85% by 2025 and 65% by 2150. On the assumption that energy efficiency can be improved by a factor of 1.25 by 2025 and 2 times within a timescale of 100-150 years, this would reduce the energy consumption to support a given level of material throughput to two thirds of it's present value by 2025 and to one third by 2150.

Region	2250 GRP/cap ('2000) US$	YEAR 2050			YEAR 2100			YEAR 2250		
		Total pop'n (m)	Car Fleet (m)	Energy Use (EJ)	Total pop'n (m)	Car Fleet (m)	Energy Use (EJ)	Total pop'n (m)	Car Fleet (m)	Energy Use (EJ)
More developed regions	**70,000**	**1,250**	**600**	**145**	**1,300**	**630**	**120**	**1,500**	**750**	**80**
North America	75,510	380	190	48	400	200	40	490	245	26
Western & Ctl Europe	70,210	365	182	38	415	207	37	470	235	25
Eastern & S-E Europe	61,700	350	153	42	340	153	30	385	193	20
Japan	75,000	110	55	12	100	50	9	100	50	6
Oceania	68,180	45	20	5	45	20	4	55	27	3
Less developed regions	**30,000**	**7,950**	**1,100**	**455**	**8,700**	**2,035**	**555**	**10,000**	**4,560**	**560**
Africa										
Sub-Saharan Africa	19,000	1,900	137	57	2,250	216	67	2,600	1,143	156
North Africa	41,200	350	51	20	385	124	27	425	212	23
West & Central Asia										
Middle East	49,170	290	93	25	265	117	26	300	150	16
Central Asia	36,470	430	62	16	400	129	28	425	192	22
South Asia										
India	24,700	1,500	108	45	1,650	239	115	1,700	748	100
Other South Asia	20,220	540	39	17	650	94	45	680	299	41
East Asia & Pacific										
China	31,850	1,400	203	120	1,350	435	95	1,350	611	70
Other East Asia	34,780	620	273	75	750	241	52	1,150	521	60
Latin America & Caribbean										
Central America & Caribbean	46,460	330	48	30	360	158	36	495	247	26
South America	46,860	590	86	50	640	282	64	875	437	46
WORLD TOTAL	**35,000**	**9,200**	**1,700**	**600**	**10,000**	**2,665**	**675**	**11,500**	**5,310**	**640**

APPENDIX 33
POTENTIAL NUMBER OF ECOPOLITAN STATES

Size band	Population (millions)	Number of Countries Year 2000	Potential No. of states Year 2150
Very small Islands	Less than 1m	40	30
States	Less than 2m	40	40
Small	2 - 4.9m	30	40
	5 - 9.9m	30	40
Medium	10 - 19.9m	30	50
	20 - 49.9m	25	100
Large	50 - 99.9m	14	150
	100 - 249.9m	8	200
	250 - 999.9m	1	50
Sub-continents	1,000m+	2	300
Total		**220**	**1,000**

APPENDIX 34
REGIONAL TRADE BLOCS

INTER-REGIONAL	
APEC	Asia Pacific Economic Cooperation
NAFTA	North American Free Trade Area
INTRA-REGIONAL	
Europe	
CEFTA	Central European Free Trade Area
EU	European Union
Africa	
CEMAC	Economic and Monetary Community of Central Africa
CEPGL	Economic Community of the Great Lakes Countries
COMESA	Common Market of Eastern & Southern Africa Cross Border Initiative
ECCAS	Economic Community of the Central African States
ECOWAS	Economic Community of West African States Indian Ocean Commission
MRU	Mano River Union
SADC	African Development Community
UDEAC	Central African Customs and Economic Union
UEMOA	West African Economic and Monetary Union
Middle East and Asia	
	Arab Common Market
ASEAN	Association of South-East Asia Nations
EAEC	East Asian Economic Caucus
ECO	Economic Cooperation Organisation
ESCAP	Bangkok Agreement for Countries of the Economic and Social Commission for Asia and the Pacific
GCC	Gulf Cooperation Council
SAARC	South Asian Association for Regional Cooperation
UMA	Arab Maghreb Union
Latin America and the Caribbean	
ACS	Association of Caribbean States
	Andean Group
CACM	Central American Common Market
CARICOM	Caribbean Community
	Central American Group of Four
	Group of Three
LAIA	Latin American Integration Association
MERCOSUR	Southern Common Market
OECS	Organisation of Eastern Caribbean States

THE WORK FORCE AND NUMBER OF LARGE ESTABLISHMENTS BY CONTINENTAL REGION

Region	2150 w'force/ large estab	YEAR 2000			YEAR 2025			YEAR 2150		
		Total w'force (m)	GRP (2000) US$bn *PPP	Large estabs '000	Total w'force (m)	GRP (2000) US$bn *PPP	Large ezstabs '000	Tzotal w'force (m)	GRP (2000) US$bn *PPP	Large estabs '000
More developed regions	**185**	**640**	**26,660**	**1,670**	**650**	**40,000**	**2,500**	**750**	**65,000**	**4,060**
North America	170	161	10,440	655	180	14,500	905	215	20,250	1,280
Western & Ctl Europe	185	215	9,760	610	210	14,000	875	250	21,500	1,340
Eastern & S-E Europe	215	181	2,440	155	180	6,000	375	210	15,750	985
Japan	160	68	3,440	215	60	4,500	280	50	5,250	315
Oceania	180	15	580	35	20	1,000	65	25	2,250	140
Less developed regions	**420**	**2,315**	**17,840**	**1,110**	**3,250**	**50,000**	**3,125**	**4,850**	**185,000**	**11,565**
Africa										
Sub-Saharan Africa	790	291	1,050	65	665	5,250	330	1,260	25,500	1,595
North Africa	260	51	600	40	70	1,750	110	150	9,250	580
West & Central Asia										
Middle East	155	34	680	40	50	3,250	200	90	9,250	580
Central Asia	310	59	600	40	70	2,000	125	150	7,750	485
South Asia										
India	465	451	2,370	150	590	6,000	375	900	31,000	1,940
Other South Asia	450	166	590	35	235	2,500	155	360	12,750	795
East Asia & Pacific										
China	390	761	5,130	320	795	10,000	625	700	28,250	1,765
Other East Asia	400	278	3,140	190	440	6,750	425	645	25,750	1,610
Latin America & Caribbean										
Central America & Caribbean	270	72	1,220	75	120	4,500	280	215	12,750	795
South America	270	152	2,460	155	215	8,000	500	380	22,750	1,420
WORLD TOTAL	**360**	**2,955**	**44,500**	**2,780**	**3,900**	**90,000**	**5,625**	**5,600**	**250,000**	**15,625**

All regions:

62.5 large establishments per 2000 US$bn GRP at Purchasing Power Parity (approximately 12.5 large establishments per 10,000 population for developed regions in the year 2000)

Notes:

The total workforce per large establishment in year 2150 compares with 300 total employment per large establishment in the UK of which 60% work at large establishments of over 50 employees, and 40% work at small establishments.

APPENDIX 36
LONG-RANGE SCENARIO PROFILES

Scenario S1: Settlements first

S1 - Settlements First: A scenario without slums and a more level, global infrastructure.

Geopolitical: The shift in global economic development makes the structure and outlook of the organisation and agencies of the UN seem increasingly out of date. The highest evolutionary potential will be achieved by developing a rich diversity within a coherent unifying structure. Democratic pluralism demands that continental unions are represented permanently on the Security Council of the UN to provide a forum for the resolution of intercontinental disputes, with 1,000 ecopolitan states of 5m–20m population having one vote each in the General Assembly by 2150. The political outlook may be described as liberal democratic with human settlements first and strong support for the United Nations Development Programme (UNDP), United Nations Human Settlements Programme (UNHSP), UN Habitat, World Bank Group, and the International Institute for Applied Systems Analysis (IIASA).

Demographic: The population prospects are on the lines of the UN median projections to give a world population of about 9.5 bn in 2050, about 10.2 bn in 2100 and some 12.5 bn by 2150. This reflects a declining mortality rate in conjunction with declining fertility.

Civil: Infrastructure investment with high-quality built-environment and a garden-city philosophy, with investment transfers to LDCs during the period 2000–2050 to assist with funding human settlements to accommodate an extra 3.5 bn people. It is essential that the LDCs achieve the infrastructural stage of development with an annual per capita GNI of 2000 US$1,500, which should be recognised as the basic needs level.

Economic: During the decade 1990–1999 an increasing number of countries in the earlier development stages had negative rates of growth. There is an inherent defect in the equilibrium model of economics, so that there has to be a levelling up of the infrastructure assets in the LDCs to create a level, global playing field before trade barriers can be removed slowly in step with the infrastructure levelling. It is time for financial reforms and regulatory changes, including the introduction of the Tobin tax on foreign exchange transactions for investment transfers to the LDCs.

The economic strategy is to deal with slums, squatter settlements, and poverty by direct intervention, so that global economic growth can increase gradually in the transition away from fossil fuel energy systems. At the same time, the MDCs would need to limit upper levels of GNI per capita to $40,000-$50,000 (2000 US$), which is higher than the Gross State Product per capita of California at just under $40,000 in 2000. From the evolutionary model, the prospects for economic growth under this scenario give a Gross World Product (GWP) of PPP US$130 trillion (2000 US$) in 2050, PPP $180 trillion in 2100, and $250 trillion in 2150. This reflects economic growth from a GWP of PPP $44.5 trillion in 2000 at an annual rate of 2.15% to 2050 and 0.65% from 2050 to 2150, or an overall rate of 1.15% from 2000–2150. The per capita income in 2100 would be MDCs PPP $38,000 and LDCs PPP $15,000, and by 2150, this would be MDCs PPP $40,000 and LDCs PPP $17,000.

Social: One of the most important problems to resolve is the design of an approach for LDCs to achieve the circumstances in which human society can evolve to its full potential. Policies are introduced for infrastructure development in the LDCs and peripheral regions, so that wages will rise there and stimulate a reciprocal demand for capital goods exports from the more developed regions that uses excess industrial capacity and revives employment in the MDCs. A measure of international convergence is the ratio of the GNI per capita in the MDCs to that of the LDCs, which would be a factor of 2.3 in 2150. There will be islands of prosperity on every continent.

Land-use: Medium-density human settlements emerge with an overall population density of 1000 persons per km^2. The intensification of agriculture entails the technology of mechanisation, cultivation under glass, and increased fertiliser use. Irrigation and the use of desalination plants could increase the area of land under agriculture, but these technologies are expensive and the cost of food production would rise. Less than 10% of the world's land surface of 133m km^2 would be covered by human settlements. Cultivated land and pastures would account for some 30% of the land, with forests taking up 27%, and the remaining uninhabited areas for wildlife would account for 33%.

Technological: In view of global warming, it will be beneficial to move away from fossil fuels long before reserves are depleted and to stretch the coal and oil resources as far into the future as possible, in parallel with the use of nuclear energy because it gives time for deep-ocean mixing of carbon dioxide, with a reduction of carbon dioxide in the atmosphere. A new generation of sustainable nuclear technology is accepted by the public because nuclear reactors become the most cost-effective method for the dissociation of hydrogen from water in the hydrogen-fuel economy, with a fairly rapid decarbonisation trend. The convergence of new technologies in energy and transportation gives rise to a new wave of investment in a hydrogen-energy infrastructure.

Environmental: There is a strong emphasis on dematerialisation and on longer life assets with a prolonged replacement of non-permanent structures and slums. Resource conservation and environmental issues remain, in spite of a shift in energy technology, since remotely located nuclear power stations acquire a significant share of the energy market. Improved maintenance of urban water distribution systems, appropriate water pricing, water recycling and the avoidance of waste are part of the strategy for alleviating water shortages. Global warming is assessed at 1° C per century.

Security: Planetary cooperation is needed to fund the essential infrastructure in Africa, to prevent civil wars and disorder in the least developed territories.

Hazards: The main concerns are technological hazards such as transport accidents, toxic spills, and nuclear disasters.

Discontinuities: A failure by the MDCs to impose a currencies transaction tax for investment in infrastructure in the LDCs could result in a large-scale urban system failure, with a severe deterioration in the conditions of the world's poorest populations.

Scenario S2: Continental cultures

S2 - Continental cultures: A cultural diversity scenario within a multi-polar world

Geopolitical: Continental regional cultures may determine the shape of future geopolitical developments rather than globalisation. The European Union intends to achieve social cohesion as it implements economic and monetary union. The Japanese believe in state intervention to enable it to build up regional and trade networks to achieve long-term development objectives, without regard to short-term efficiency criteria predicated by existing prices. China and India have moved on from state-planned economies by opening sectors to foreign capital, where the emphasis is on the creation of dynamic regional-growth zones. The Islamic states are concentrating on national self-reliance and human development, and the oil producers have invested in the region as well as providing assistance to south Asia, which provides an immigrant workforce.

The political outlook may be described as social democratic or culture first, with strong support for United Nations Children's Fund (UNICEF), United Nations Educational, Scientific and Cultural Organization (UNESCO), and the Office of the United Nations High Commissioner for Refugees (UNHCR). Advocates favour the abandonment of the old measure of GDP and replacement with alternative progress indicators such as Human Development Index (HDI) and the World Bank's Wealth Index that reflect qualitative factors such as social and environmental issues.

Demographic: The population prospects are on the lines of the UN high-medium projections to give a world population of about 9.5 bn in 2050, about 12.5 bn in 2100, and some 15 bn by 2150. With an emphasis on family and community life, fertility rates decline relatively slowly, although fertility rates vary in the different continental regions in accordance with differing cultural values and lifestyles.

Civil: Megacities and cosmopolises develop with high-quality mass-transit and public transport systems.

Economic: For the greater part of the 20th Century, the main challenge to the dominant market-driven paradigm came from the Marxist model. However, by 1990, after the break-up of the Soviet Union, the arguments for the capitalist model were reinforced, in spite of its defects. Economic regionalism is considered to be the best defence against unfettered global capitalism, and protectionism and trade wars break out. Economic growth may be slower because of a strengthening of protectionist trade blocs.

The prospects for economic growth under this scenario give a Gross World Product (GWP) of PPP US\$150 trillion (2000 US\$) in 2050, PPP \$225 trillion in 2100 and \$320 trillion in 2150. This reflects economic growth from a GWP of PPP \$44.5 trillion in 2000 at an annual rate of 2.4% to 2050, and 0.8% from 2050 to 2150, or an overall rate of 1.32% from 2000–2150. The per capita income in 2100 would be MDCs PPP \$40,000 and LDCs PPP \$16,000, and by 2150, this would rise to MDCs PPP \$45,000 and LDCs PPP \$19,000.

Social: Increasing government debt with the ageing of the population creates serious financial instability in the MDCs, with large-scale social unrest at the slowness with which governments bring in legislation and incentives to introduce shorter working hours and the sharing of employment. There is an increase in local reliance with investment of local savings in local businesses and small-scale manufacturing. A measure of international convergence is the ratio of the GNI per capita in the MDCs to that of the LDCs, which would be a factor of 2.4 in 2150.

Land-use: High density settlements with more persons per household give an overall population density of 2,000 persons per km². Alternative approaches to supplementing future food supplies include increasing the food potential from the seas, which supply some 20% of the world's high-quality animal protein and bypassing animals in the food chain.

Both of these would require altering people's eating habits. Some 5% of the world's land surface of 133m km² would be covered by human settlements. Cultivated land and pastures would account for some 36% of the land, with forests taking up 26%, and the remaining uninhabited areas for wildlife would account for 33%.

Technological: In many cases, there will be an optimum technology for specific regions or sites depending upon the available supplies of fossil fuels, the proximity of oceans or rivers, and the annual amount of sunlight. The aim is to minimise import dependence, with lower-income resource-rich regions relying on fossil fuels and higher-income but resource-poor regions shifting to nuclear and renewable sources. There will be a relatively slow decarbonisation trend. High-tech industry produces new materials with high insulation values to achieve domestic energy savings, and action to conserve natural resources leads to reduced demand for timber in house construction.

Environmental: There is concern for environmental amenities with an emphasis on pollution control. Water scarcity is frequently both a regional and a political problem because more than 200 river systems, draining over half of the planet's land area, are shared by two or more countries. A measure of water stress is when the use-to-resource ratio exceeds 0.5, and in these regions, water-resource management is essential since water can be reused many times. It is estimated that by 2025, a billion people will be living in countries suffering from high levels of water stress. Global warming is assessed at 2.5° C per century.

Security: The bipolarity of the Cold War becomes replaced by a multipolar world with regional conflicts and a fortress world in relation to trade wars. China emerges as a military superpower in East Asia and the Pacific. Over time, more and more activity takes place in the grey or underground economy, and levels of crime become a major concern

Hazards: The main concerns are geo-hazards such as earthquakes in densely populated urban areas, where building codes may be ignored.

Discontinuities: As a consequence of an increasing population and weak international cooperation for the introduction of ecotaxation, to reduce the levels of carbon dioxide emissions, global warming could result in high levels of soil erosion and desertification with widespread regional famines.

Scenario S3: Biotech wave

S3 - Biotech wave: An ecological scenario for the world's living resources

Geopolitical: A globally coherent approach to sustainable development arises from a high level of environmental consciousness that has been brought about from clear evidence that global and regional pollution, the impacts of natural resource use, such as deforestation, soil depletion, and over-fishing pose a serious threat for humanity. The Cartagena Protocol on Biosafety has set the stage for a regulatory regime for biotechnology and gene transfer, to ensure that the planet's biological diversity (including human systems) are protected. The political outlook may be described as green or environment-first, with strong support for World Health Organisation (WHO), United Nations Environment Programme (UNEP), and UN Intergovernmental Panel on Climate Change (IPCC).

Demographic: The population prospects are on the lines of the UN low projections to give a stable world population of about 9 bn by 2050, which remains constant with about 9 bn in 2100, and then declines slightly to some 8.5 bn by 2150. This reflects a decline in world fertility with higher mortality rates arising from pandemics such as AIDs and mutating lethal viruses.

Civil: Compact cities emerge with a high share of apartments in the housing stock and a tight control on suburban developments, with emphasis on public and non-motorised transport. There will be less than a two-percent saving in transport energy consumption, compared with a decentralised city form. There is increasing regional specialisation and development in the core regions surrounding world cities, with a relative decline in the peripheral regions.

Economic: Ecologists and environmentalists take issue with neoclassical economics insofar as it treats land, the natural environment, as a factor of production. Ecological economics recognises that the scale of the global economy has become large in relation to the carrying capacity of the ecosystems that sustain it. Ecological economics uses a concept of natural capital and recognises that the stocks of nonrenewable natural capital, such as fossil fuels, are finite and also that overuse of renewable natural capital can impair or destroy its ability to regenerate itself. There is a universal awareness that environmental degradation and human overuse of natural ecosystems leads to biodiversity loss.

The prospects for economic growth under this scenario give a Gross World Product (GWP) of PPP US$180 trillion (2000 US$) in 2050, PPP $275 trillion in 2100, and $400 trillion in 2150. This reflects economic growth from a GWP of PPP $44.5 trillion in 2000 at an annual rate of 2.8% to 2050 and 0.8% from 2050 to 2150, or an overall rate of 1.47% from 2000–2150. The per capita income in 2100 would be MDCs PPP $80,000 and LDCs PPP $24,000, and by 2150, this would rise to MDCs PPP $110,000 and LDCs PPP $38,000.

Social: There is rapid progress towards an ecopolitan society and early recognition of a Quinary Lifespan economic division. Unemployment problems are exacerbated in the MDCs by the transfer of high-tech industries to the LDCs. There will be a significant increase in leisure time for the employed proportion of the population, which reduces consumption of material goods but increasing amounts of time are spent on tourism and multi-media entertainment. A measure of international convergence is the ratio of the GNI per capita in the MDCs to that of the LDCs, which would be a factor of 2.9 in 2150.

Land-use: Higher-density urban developments give an overall population density of 1,500 persons per km^2. Agricultural intensification is achieved through genetic engineering with the development of improved animal and plant varieties, together with the maintenance of large areas of wilderness. The globalisation of the production, processing, and handling of food, and changes to breeder stock increase the risk of food-chain diseases such as foot-and-mouth, or bovine spongiform encephalopathy (BSE), and CJD. There will be relatively high food prices with much lower levels of meat consumption. Some 7% of the world's land surface of 133m km^2 would be covered by human settlements. Cultivated land and pastures would account for some 25% of the land, with forests taking up 35%, and the remaining uninhabited areas for wildlife would account for 33%.

Technological: Gas becomes a primary fuel for electricity generation using Combined Cycle Gas Turbines in the steady transition to renewable energy sources, such as hydropower, tidal power, wind power, and solar energy, which will be used in due course for hydrogen-energy production. Technological progress is pushed in the life sciences, artificial intelligence, and nanotechnology.

Environmental: There is an emphasis on high levels of environmental quality with strong ecological resilience. This involves more concentration on the planet's living resources than the mineral resources, and an emphasis on biodiversity. However, there is an increase in forest fires and a drying out of rain forests, with a loss of species. The cooperative management of shared watersheds and river basins, the control of water pollution, and efficient trickle and drip irrigation systems can extend the availability of scarce supplies. Transboundary air pollution (acid rain) is virtually eliminated. Global warming is assessed at 2° C per century.

Security: At the planetary level, the membership of the supranational organisations is widened and demilitarisation continues as governments attempt to reduce expenditures.

Hazards: The main concerns are eco-catastrophes, pandemics, and bio-hazards. Due to global warming, there is an increase in new diseases, with tropical diseases in temperate regions and major epidemics or pandemics. Highly infectious diseases emerge such as Ebola, AIDs, and SARS, and for many there is no treatment, cure or vaccine. Urban health problems reappear after years of decline, and strains of diseases develop resistance to antibiotics. These threats are compounded by the rapid increase in international air travel. Biological weapons become a major bio-hazard.

Discontinuities: Advances in bio-chemistry and gene therapy enable the average life expectancy to reach 100 years by 2200, with the development of genetic techniques to increase intelligence and with the growth and replacement of human organs on demand. However, unforeseen mutations and genetic disorders could be absorbed into the human gene line and passed on to future generations.

Scenario S4: Generica rules

S4 - Generica rules: A scenario with a worldwide mass market and a global superculture

Geopolitical: The United States would like to impose free-market principles in accordance with an American ideal of individual rights and liberty. The Washington Consensus promotes the vision that global free trade leads to global economic growth, which creates more jobs and makes companies more competitive, with lower prices for consumers. It also provides LDCs with the opportunity to develop through infusions of foreign capital and technology, and by spreading prosperity, it creates the conditions in which democracies and a respect for human rights may flourish. The political outlook may be described as hegemonic and laissez-faire or markets first, with support for United Nations Conference on Trade and Development (UNCTAD), World Trade Organisation (WTO), and the International Monetary Fund (IMF).

Demographic: The population prospects for low-medium growth give a world population of about 9 bn in 2050, with 9.5 bn in 2100, and some 10 bn by 2150. Longer lives and smaller families result from a declining mortality rate in conjunction with declining world fertility as the standards of living rise in the LDCs.

Civil: Global cities emerge with an increasing number of third- and fourth-order world city-regions, which will be served by continental nodes with high-speed intercity links. International centres will have identical features with an international airport, financial districts, hotel chains, retailers, leisure facilities, and restaurants. At the local, high-street level, this replication phenomenon is known in the U.S.A. as 'Generica.' This is also referred to as the McDonaldization of society. There is urban sprawl with high car ownership and dense transport networks both nationally and internationally.

Economic: The dominant paradigm of the international community is to allow market forces to drive the global economy, and neoclassical economics and equilibrium theory are claimed to underpin the validity of this model. Transnational corporations make a significant proportion of investments in less developed countries. It may be expected that the world economy becomes more stable as increasing amounts of capital are generated in the less developed countries.

The prospects for economic growth under this scenario give a Gross World Product (GWP) of PPP US$220 trillion (2000 US$) in 2050, PPP $350 trillion in 2100, and $500 trillion in 2150. This reflects economic growth from a GWP of PPP $44.5 trillion in 2000 at an annual rate of 3.3% to 2050 and 0.8% from 2050 to 2150, or an overall rate of 1.65% from 2000–2150. The per capita income in 2100 would be MDCs PPP $110,000 and LDCs PPP $28,000, and by 2150, this would rise to MDCs PPP $125,000 and LDCs PPP $40,000.

Social: This scenario leads to very high per capita incomes in the MDCs, which become absorbed in space exploration, space ports, and fractional orbital flight, artificial worlds on earth, and fine civic and corporate architecture. There will be low-density regions, with a significant proportion of the population owning exclusive trophy houses, second homes, prestigious cars, private planes and boats, racehorses, and robots to do the housework. Investments will be made in the acquisition of scarce positional goods such as well-located property with fine views, fine furnishings, antiques, works of art, and jewellery. A large part of personal expenditure may be spent on education, childcare, medical services, cosmetic surgery, health spas, exotic holidays, and designer clothes. A measure of international convergence is the ratio of the GNI per capita in the MDCs to that of the LDCs, which would be a factor of 3.1 in 2150.

Land-use: With increasing affluence, a higher proportion of wealth is available for urban infrastructure, with low-density urban areas similar to the new frontier towns of western and southern North America and Australia, with overall population densities of 750 persons per km^2. It may be expected that additional agricultural yields will become increasingly expensive because of the large capital outlays for irrigation, machinery, and chemicals in addition to the threat of climatic change. About 12% of the world's land surface of 133m km^2 would be covered by human settlements. Cultivated land and pastures would account for some 28% of the land with a higher level of meat consumption, with forests taking up 27% and the remaining uninhabited areas for wildlife would account for 33%.

Technological: Globally, the energy systems remain predominantly hydrocarbon based to 2100. A mix of energy technologies and supply sources result in a shift from the current shares of fossil fuels and the gradual decarbonisation of energy. The technological emphasis is on defence, air transport, space research, and mechatronics.

Environmental: There is concentration on the management of natural resources rather than the conservation of nature. Water sustainability will require an integrated river basin planning framework, a water systems engineering infrastructure, wastewater treatment plants, flood control measures, new irrigation technologies, and sea-water desalination plants. Most of the countries with limited renewable water resources are in the Middle East, North Africa, and sub-Saharan Africa, the regions where populations are growing fastest. Although desalination is expensive, it is affordable in the oil exporting regions of the Middle East and North Africa. Global warming is assessed at 3° C per century.

Security: Protected enclaves and gated communities with security guards are accepted as a means of keeping out the 'have-nots.' There is also a problem of people from poorer nations trying to migrate to richer nations, often resorting to illegal entry. The oil weapon is seized upon by the Islamic nations as a means to advance their interests.

Hazards: The main concerns are natural hazards associated with climate change such as flooding, tornados, and tropical cyclones.

Discontinuities: By 2500, temperatures in Europe are predicted to decline when the warm Gulf Stream conveyor mechanism slows down due to sea water salinity changes. However, global sea levels rise from 2000 by 1 metre in 2150, and continue at a rate of 0.6 metres per century to reach a peak of 6 metres higher by the end of the millennium in 3000. By the beginning of the Fourth Millennium, the start of the downward decline into another glacial age spurs investment into space travel with space colonies and resorts supporting temporary populations, prior to planetary terraforming becoming a reality.

APPENDIX 37
GLOSSARY OF TERMS

Adaption

Adaption is any change in the structure or habits of an organism, household, or establishment that enables it to survive and reproduce by the capture of energy or investment capital from its transactions within the environment. Adaption is a change in strategy by a transacting entity, based upon system experience.

Agrarian stocks

Agrarian stocks of the biosphere are the sustainable primary resource outputs of the agriculture, forestry, and fisheries sectors, which are limited by gestational or seasonal units of time and solar flow.

Agropolitan state

Agropolitan states are a subset of ecopolitan states in which the per capita GNI is less than $1,001–$2,500 (2000 US$). Some 20%–50% of the population will be urban and 50%–70% of the workforce will be involved in agriculture or primary resources.

Archetype

The genetic structure of an establishment is embodied within the artefacts and transparts that it produces, and the final assembly of an archetype of an animat contains a unique genetic combination conceived by a multitude of establishments. In effect, an archetype is the technological equivalent of the biological concept of a phenotype.

Animat

Animats are artificial equivalents of animals such as aircraft (birds), submarines (fish), cars (horses), tractors (bullocks) or robots (humans).

Atrophy

Atrophy is a measure of undernourishment in a bounded territory or an ecosystem, as an increasing number of species chase a finite supply of resources. Populations reach saturation when resources become limiting in accordance with the logistical Law of Atrophy. See also Investment atrophy.

Attractor

The long-term behaviour of a complex adaptive system follows a universal attractor. The evolution of ecostructures in a world urban system remains on an universal attractor to increase the overall diffusion of investment capital for the realisation of planetary potential. Dynamic stability and far-from-equilibrium stability may correspond to a spectrum of attractor types such as static, cyclic, vibrant, and chaotic attractors.

Backcasting

Backcasting is a methodology for exploring the prospects of a future viable state with a normative scenario for say fifty years into the future and examining the geopolitical, economic, social, technological, and ecological policies that would contribute to achieving the desirable state. It also identifies a range of threats and the manner in which systems would need to be adapted to prevent a disastrous scenario that could cause a system collapse.

Bifurcation

A bifurcation is a branch at which there are two distinct choices available to a system.

Billion

A billion is 1,000 million.

Biosphere

The biosphere is the region of the earth's crust and atmosphere in which living matter is found.

Capital formation

Establishments capture investment capital (profit or savings) in successful transactions so that they can grow and reproduce. This increases the net worth of an establishment, which is the difference between its assets and liabilities. Assets include capital assets, securities, money, and debtors, whereas liabilities include loans and creditors. Business saving is the net increase in net worth after dividends and interest have been paid, and the balance sheet will generally show an increase in both assets and liabilities. If the balance sheets of all establishments in the urban system were consolidated, creditors and debtors would match each other as would securities and loans. Therefore, the consolidated establishment savings represent the increase in the stock of capital assets, plus the growth in investment capital (money stock).

Carrying capacity

The carrying capacity of a territory is the population of biological species, humans, households, establishments, and artefacts that can be supported indefinitely.

Chaotic dynamics

Relatively short-term chaotic dynamics arise in a complex adaptive system when the system shifts between stable states at macrosystem transitions, and so the system will be indeterminate in detail. Natural and civil disasters create chaotic dynamics at the local and regional levels in the short term, as does the closure of a mine in a mining community or the opening of a new superstore that disturbs local traders and customers before conditions settle down and life continues.

Coefficient of connectivity

The coefficient of connectivity is a function of the level of transactional complexity (i.e., the proportion of establishments in the quaternary information division) and the density of large establishments (i.e., the number of large establishments per km²) for an ecopolitan state of 5m–20m population.

Coevolution

Coevolution is a process in which interacting entities and complex systems evolve and adapt to each other.

Coexistence

Coexistence is the presence of potentially competing species in the same habitat.

Colonialism

Colonialism is the exploitation of territories and peoples with less economic power. Geopolitical power is accumulated in the same way as wealth, and inferior power arises from a lack of infrastructure, a lower level of transactional complexity, and inferior military capability.

Competition

Competition is the interaction that arises from the utilisation of a scarce resource by the same or different species. Competition between establishments occurs for the capture of investment capital (profit or savings) in a region. Improved acquisition and accumulation of investment by one sector will reduce the total investment available to the other sectors.

Complex adaptive system

A complex adaptive system or vivisystem is a community of interacting or transacting entities such as organisms, households, or establishments, which change their strategies in accordance with the feedback they receive from their environment. A defining characteristic of a complex adaptive system is that the evolution of the system as a whole survives generations of the transacting entities, which may have relatively short lifespans.

Complexity

Complexity describes the stage of evolution or level of maturity of an evolutionary system, in which there is an elaboration of structure or hierarchy as the system adapts to its environment as a condition of future viability. In the case of an ecosystem or ecostructure, energy or investment capital is diffused with increasing transactional complexity. This involves high levels of diversity within sectors, a range of larger establishments, narrow niche specialisation, trophic webs, rather than chains, higher levels of energy efficiency, an increase in information, and stability or resilience in the presence of external disturbances. Macroscopic properties of a complex adaptive system emerge from the totality of the interactions or transactions within a collective regime.

Complexity science

Complexity science, which is also known as the science of evolution and complexity, explains the ecodynamics of living systems when more energy or investment capital is available than is being used. This results in territorial colonisation through adaption and the formation of complex ecosystems or ecostructures, which diffuse the energy or investment flows in order to eliminate energetic gradients. The rise in the level of complexity corresponds to an increase in the information content relating to the flow of resources necessary to sustain the system. As the energy or investment capital is transformed, the law of atrophy applies, as an increasing number of species chase a finite supply of resources. This causes an evolutionary spiral with natural selection, adaption, coevolution, and migration of species, in which progress is reflected by the centripetal forces of increasing complexity and heterogeneity. However, periodical investment waves integrate or simplify the system to achieve a reduction in transaction costs and more effective resource utilisation. These waves of creative destruction to the environment provide alternative pathways for the diffusion of investment in the centrifugal direction of dispersal, standardisation, and homogeneity. The emergence of macrostructure from the integration and recombination of the constituent parts of the system forms a hierarchy of complexity, which subordinates and regulates the parts of the system and creates the conditions for a subsequent increase in complexity until a climax is reached.

Connectivity
The connectivity of places may be calculated as the number of links with other places or nodes in the network, divided by the maximum possible for a network linking that number of places. More simply, a hierarchy of connectivity for a network as a whole can be determined for places by comparing the actual number of direct connections.

Continental union
A continental union is an alliance of nations or states in which members remove trading barriers for members and maintain a level of protection against trade with non-member states. The arrangement can be extended to economic integration and the unification of economic, social, and environmental policies under a supranational authority.

Cosmopolis
The term cosmopolis is used to define a subset of large metropolises of over 5m population that correspond to a number of cosmopolitan capital cities.

Culture

Culture embraces traditions and beliefs, value systems, ways of life, and forms of art. The cultural heritage of humankind is passed on from generation to generation in the form of memory, language, customs, history, religion, philosophy, moral and aesthetic values, learning, literature, music, dance, and art.

Decomposition

Decomposition of a complex structure into its functional components enables each component to be designed with a degree of independence of the other parts. Similarly, a complex system can be decomposed into subsystems, in order to analyse their behaviour. There may be a number of feasible decompositions of radically different kinds.

Dematerialisation

Dematerialisation of the economy refers to steady reductions in resource intensity in the more developed countries, where the inverse of intensity of use is expressed as the ratio in the form of Gross Domestic Product (GDP) output per kilogram of energy or materials resources.

Determinism

Determinism is a doctrine that claims that events are a consequence of previous causes rather than being generated by free will or random factors.

Developmentalism

Developmentalism or the developmental perspective is used by designers, engineers, and computer scientists to examine how new phenomena can be created from lower-level entities in complex systems and then persist in an evolutionary environment. Transactions between higher-level entities produce macro-level effects that follow macrolaws, which are complementary to the microlaws that explain the micro-level dynamics.

Discontinuity

A discontinuity is a relatively sudden change in the course of events such as a bifurcation point. Within a planning time horizon, it may not be anticipated that there will be a devastating catastrophe on a global scale, although in the longer term there are bound to be discontinuities for any scenario with perhaps local or regional disasters.

Dispersal
Dispersal is the movement of humans, households, establishments, and artefacts away from the urban centres of population.

Ecodiversity
Ecodiversity is the number of species in a specific habitat or community, or it may relate to the pooling of species across habitats within a region.

Ecodynamics
Ecodynamics is the term used to describe the evolution of spatial structure in the civil system, together with the changes in the population of artefacts, which arise from the interactions between households and the transactions with and between establishments. Ecodynamics is concerned with the emergence of complex civil ecostructures through the diffusion of investment capital, in which the urban system is transformed endogenously and undergoes continuous and irreversible change. Capital formation drives the urban system away from equilibrium, which sets up gradients between locations in the rate-of-return on investment. The redistribution of investment to less mature parts of the urban system tends to reduce the gradients, and depending upon the rate of investment flow, individual ecostructures may be classed as static, cyclic, vibrant, or chaotic attractors.

Ecological footprint
The ecological footprint of a city or a state is its environmental impact on the regional, continental, or global ecosystems.

Economic development
Economic development involves the increasing urbanisation of a country's population, with an increase in the per-capita assets and fundamental structural changes in the economy. It is characterised by periods of accelerating and then decelerating population growth, with increasing life expectancy and human development that significantly changes both a nation's age structure and its employment structure. Economic progress implies an increase in the capacity to withstand increasing

levels of environmental fluctuation and the capability to utilise a wider range of resources.

Economic equilibrium

Classical economic theory is dependent upon a notion of market equilibrium. However, the concept of economic equilibrium becomes untenable in the presence of spatial structure and dynamic conditions in which the forces for change are endogenous. Economic development is characterised by progressive, irreversible, and cumulative change, which ensures that the economy is not equilibrating. The concept of economic equilibrium is challenged by complexity science and it's far-from-equilibrium stability.

Ecopolitan State

The emergence of ecopolitan states as basic geopolitical units or bounded landscapes with evolutionary potential, is a necessary construct to enable *Long-Range Futures Research* to encompass geopolitical, social, and environmental issues that affect the contextual macrostructure. The term ecopolitan is formed from its ancient Greek roots 'eco' as used in ecology and economy to mean the sustainable use or literally the housekeeping (oikos is the Greek for house) of natural and human resources. The word 'polis' is used with its wider meaning of a city-state with socially equitable policies towards all citizens. It follows that in an ecopolitan society, a balance would be struck between economic efficiency, social equity, and sustainable civil development. An ecopolitan state is a prospective sustainable city region of 5m-20m inhabitants in which human society can evolve to its full potential at a reasonable per capita standard. An Ecopolitan Index (EI) could be developed along the lines of the UNCHS (Habitat) *City Development Index*, to measure the extent to which a state achieves the criteria for efficiency, equity, and sustainability.

Ecostructure

An ecostructure is a complex adaptive system of households and establishments that utilise a background flow of investment capital to create order and form urban islands of transactional complexity. Investment capital is diffused in the system of connected ecostructures through a series of transactions, and it is transformed into a variety of artefacts or transparts that are transferred between establishments in a complex trophic web. Investment capital is dissipated in sustaining, maintaining, and reproducing internal order and coherence.

Ecosystem
An ecosystem comprises the interacting populations of different species within a biotic community and its abiotic environment.

Ecotaxation
Ecotaxation is the imposition of taxes or charges on activities that have a damaging effect on the environment or the quality of life. Ecotaxation is also adopting a social policy dimension with the objective of shifting taxation policy away from income and labour onto resource depleting and environmentally damaging activities. A system of tradeable ecopermits, which could be allocated annually in relation to continental or contiguous super-continental land areas, could be an equitable way of distributing an upper level of permissible emissions into the atmosphere. Regions with low demand in relation to their quota could auction their surplus ecopermits to countries with excess demand.

Emergence
Emergence is the creation of a complex adaptive or living system from the connection of building blocks in successively more encompassing sets. The building blocks of nature are elementary particles, atoms, molecules, macromolecules, organelles, cells, tissues, organs, organisms, and species within the larger structure that extends to ecosystems and biomes. Spatial structure in the civil system emerges from the combination of households and establishments, districts, towns and cities, regions, states, nations, continents, and the planet.

Emergent property
An emergent property is a feature of a community or a vivisystem that is not deducible from the features of a single household or establishment in a town, or from a single species in an ecosystem.

Endogenous
Endogenous change originates or grows from within or is due to internal causes, whereas exogenous change originates from outside or from external causes.

Equipollence

World urbanisation increases the number of levels in the urban hierarchy, and complexity science indicates that the trajectory of the world system of cities will be towards a dynamic balance with equipollence or an equalisation of populations at each hierarchical level. This is the most likely condition to maintain a stable configuration, since it limits the asymmetry of the urban system, and it will be approached when the global urban population exceeds ninety percent of the total.

Establishment

An establishment is the basic transacting entity of a human settlement, and it may comprise a single or multi-person household or organisation. Although industrial plants, businesses, professional practices, and households may be differentiated functionally as producers, traders, informers, and consumers, respectively, all establishments are transacting entities. Establishments are located at facilities and form part of a complex adaptive system in which the environment is altered when establishments relocate to new or existing facilities. The establishment embraces the stock of assets and artefacts that it uses in the course of its operations. Agencies that are responsible for infrastructure assets, utilities, transport, or mines, for example, are the appropriate establishments for deployment of these assets.

Evolution

Evolution arises from 'quasi-ecological' interactions and transactions between the entities of a complex adaptive system, which involves selection and the diminution of some species to extinction and the expansion of others under constantly changing parameters in the environment. There will also be mutations and an accumulation of genetic changes that lead to diversity. Evolution involves spatial organisation or the formation of hierarchy for information acquisition that results in an increase in complexity to enhance the diffusion of energy or investment flow in the system. Changes in the energetic flow or system growth parameter result in macrosystem transitions with creative destruction and species extinctions, together with a reordering of the hierarchical structure. Evolution combines continuous evolution (gradualism) with discontinuous events (bifurcations) that are generated by irregular instabilities with short-term chaotic dynamics. Although evolution is inherently unpredictable, there are persistent features in far-from-equilibrium systems that may indicate the form or envelope of their evolutionary trajectory.

Evolutionary landscape
An evolutionary landscape provides a silhouette of the survivors of the preceding populations of organisms, households or establishments, facilities, infrastructures, artefacts, and terrestrial stocks.

Evolutionary science
Evolutionary sciences include astronomy, geology, ecology, anthropology, and archaeology, in which the science provides an explanation for the emergence and decline of complex adaptive systems.

Exergy
Exergy is a measure of the capital of energy or the free or high-grade form, which is its available work content. It determines the potential of energy to drive a system away from equilibrium and to set up energetic gradients. Exergy is destroyed in the irreversible process of evolution.

Externalities
Externalities are the unintended costs or benefits of a transaction that are not reflected in the price. Externalities are often imposed upon third parties.

Extinction event
An extinction event is a sharp decrease in the number of species in a relatively short period of time. Five major mass extinctions occurred at the ends of the Ordovician, Devonian, Permian, Triassic, and Cretaceous periods. A sixth Holocene extinction event is being caused by human economic development; the current extinction rates of species lie in the range 100-1,000 times the background rate from the fossil record.

Feedback
Feedback is the mechanism by which the consequences of an ongoing process become factors in modifying or changing that process. The ongoing process is reinforced in positive feedback so that change becomes cumulative, and in negative feedback, the system reacts to change in such a way as to limit or contain it.

Forecast

Forecasts are statements about the probability of an event happening in the future. A forecast implies less certainty than a prediction, which is particularly relevant in the applied sciences such as thermodynamics, hydrodynamics, or aerodynamics, in which fluid flows or structural stresses can be calculated with a degree of confidence.

Foresight

Foresight is the anticipation of future events and involves taking action to avoid harmful situations and to protect ourselves from suffering the consequences of inadequate preparation or errors of judgement. Foresight also implies the ability to plan in order to seize opportunities when they present themselves.

G-7, G-8 and G-10 Countries

The Group of Seven, or G-7 countries is the meeting of the finance ministers from the seven industrialised countries that include Canada, France, Germany, Italy, Japan, the United Kingdom, and the United States of America. This should not be confused with the Group of Eight, or G-8 countries, which is the annual meeting of the heads of government of the Group of Seven plus Russia. The Group of Ten, or G-10 countries is made up of ten industrial countries that include Belgium, Canada, France, Germany, Italy, Japan, the Netherlands, Sweden, the United Kingdom, the United States, and an honorary eleventh member, Switzerland. The G-10 is a forum for discussing international monetary arrangements and it also meets through its central bank, the Bank for International Settlements based in Basle.

Genes

Genes are units of genetic information and are considered to be molecules of DNA and RNA.

Genotype

The genotype is the genetic constitution of an organism or species rather than its observable characteristics.

Gradient Reduction

Nature abhors gradients such as measurable differences in pressure, temperature, or chemical concentration, and ecosystems develop that utilise the exergy from the sun and thereby reduce the ambient gradients. Capital formation drives the urban system away from equilibrium, which sets up gradients between locations in the rate-of-return on investment. The redistribution of investment to less mature parts of the urban system tends to reduce the gradients in the rate-of-return on investment.

Gross National Income (GNI)

In the 1993 System of National Accounts used by the World Bank for Development Indicators from the year 2000, Gross National Income (GNI) replaces Gross National Product (GNP). GNI comprises Gross Domestic Product (GDP) plus net receipts of primary income from foreign sources. GNI per capita is GNI divided by the midyear population.

Habitat

A habitat is the sum of the environmental conditions at the location where an organism lives.

Half-life

Half-life is the time taken for one-half of the atoms of a radioactive isotope to decay into another isotope. It is also used to denote the time taken for chemical compounds, such as pesticides, to lose half their strength.

Hazard

A civil hazard is an event that threatens life or infrastructure and buildings. A civil disaster is an extreme event, which is in effect the realisation of a hazard, where there is a loss of life or damage to the built environment with severe disruption to human activities. Civil disasters include the impact on the built environment of natural hazards such as fires, floods, earthquakes, severe storms, or technological hazards such as transport accidents, toxic spills, and nuclear disasters, and social hazards such as crime, terrorism, and wars.

Hierarchy

The metaconcept of systems science is the existence of hierarchy, such that the different levels in a system are a complex of successively more encompassing sets, with transacting entities such as households and establishments, then residential, business, and industrial districts, with towns and cities, regions, states, nations, continents, and the planet.

Each level is more complex than the level below, and it is characterised by emergent properties that are not apparent at the lower level. This is distinct from the more general use of the term hierarchy, which is the rank order or dominance patterns amongst members of a population such as military rank structures or cities.

Human agents

Interacting human agents are the units of genetic information in a household or establishment, such that the combination of agents contributes to the genetic diversity of the organisation. In effect, the human agents act as genes, and they provide the blueprints or genetic know-how that shapes the organisational form and controls its future development.

Information

The demand for information arises from the uncertainty of dynamic conditions that are changing over time. Information reduces uncertainty and has anti-chaotic properties that create order out of disorder in a far-from-equilibrium system. In the context of complexity science, any movement towards complexity involves an enhancement of the information structures.

Investment atrophy

Investment atrophy is a measure of investment capital deprivation or undernourishment in parts of the urban system, as a consequence of capital accumulation in other parts of the system. Without renewed investment, ecostructures suffer dilapidation and depreciation, and the atrophic forces of deterioration arise from the availability of higher investment returns elsewhere.

Investment hypertrophy

Investment hypertrophy is geographical, sectoral, or infrastructural overdevelopment, or excessive investment such as defence expenditure during the 'cold war,' which leads to investment atrophy in other regions, sectors, or parts of the urban system.

Infrastructure

Infrastructure is defined as civil capital that includes both economic overhead capital and social overhead capital. Economic overhead capital covers utilities, transport, energy, and communications networks, whereas social overhead capital comprises housing, education, public health, and public service facilities. Infrastructure imposes spatial structure on a region and generates locational economies of connectivity, which attracts complementary investments and creates additional advantage.

Intergenerational equity
Intergenerational equity is the term used to express sustainable development in terms of fairness to people living now and in the future. The interval between generations is generally taken as twenty-five years, and a planning time horizon of 125 years will span five future generations.

Investment capital
Investment capital is the sum accumulated for investment in real assets for use by establishments in the course of their operations. Capital investment drives the economic system away from equilibrium in the irreversible process of evolution, and civil ecostructures evolve to diffuse investment capital and to reduce the gradients in return on investment. This property is similar to that of 'exergy,' which is the free or high-grade energy in an ecosystem.

Knowledge
Knowledge is the term used for the stock of information. Society expands its flow of information through education and culture, as organisations or individuals search for alternatives that yield more of the temporarily scarce resources. Information has value as a direct result of the demand from problems to which it can provide solutions.

Law of Atrophy
The law of atrophy states that populations of humans, households, and establishments within a civil ecostructure or a bounded territory reach saturation as resources for growth become limiting, in accordance with the logistic curve first identified by Verhulst (1845). The population dynamics cause the instability that drives evolution. Ecostructure capacity is not constant and evolves over time through investment and technological progress. Once a plateau of capacity has been reached at a particular stage of development, the potential may evolve to successively higher plateaux with further development stages such as industrial, distributional,

informational, ecopolitan, or planetary. Evolutionary changes will be greatest the further the population is from equilibrium, when population growth is unconstrained by resource limits, competition, or selection. The law of atrophy differs from the Malthusian theory of population growth, which predicted that populations would increase geometrically, whereas food would increase arithmetically, with a consequential collapse in population.

Law of Diminishing Returns
The law of diminishing returns refers to the amount of extra output that is obtained from the addition of successively equal extra units of varying input to a fixed amount of some other input.

Laws of thermodynamics
The initial principles of thermodynamics were developed by Carnot in 1824 and later refined by Clausius in 1850. The first law of thermodynamics states that energy can neither be created nor destroyed. The second law of thermodynamics states that in any processes involving energy transformation, free or high-grade energy ('exergy'), which is a measure of the available work content, is degraded to the bound or low-grade form that is equivalent to heat. Clausius adopted the Greek word 'entropy' meaning transformation, as a measure of the energetic cost. Entropy is only strictly defined for equilibrium conditions in closed systems, and in all natural and technological processes the availability of the energy decreases and there is a corresponding increase in entropy. Loosely stated, entropy is a measure of the disorder or chaos in a thermodynamic system. It should be noted that the earth is not a closed system, and that biological evolution has been powered by the sun.

Life cycle
The life cycle is the typical progression through which humans, biological species, and artefacts pass through the phases of inception, growth, internal development, maturity, and senescence. The logistic life cycle curve or S-curve of capability with time is followed by all humans and their artefacts.

LDC (less developed country)
An LDC is a less developed country in which the per capita GNP is relatively low, with lower per capita investment in infrastructure, high

population growth, and lower levels of life expectancy and human development.

Lock-in

Lock-in arises when geopolitical arrangements, institutions, economic processes, or technological solutions accumulate an economic advantage, so that the transformation to a new set of alliances or institutions and changed processes or alternative technologies involves a significant cost barrier. The exit cost from existing situations is a measure of the degree of lock-in.

Macrolaws of Ecodynamics

First law of ecodynamics

In an interconnected system of civil ecostructures, investment capital is transferred between establishments, so that there is a cumulative transformation of the resource at each trophic position in the complex adaptive system. This creates irreversible structural change to the system, which evolves through the formation of islands of increasing transactional complexity and connectivity. The potential of an ecostructure is a measure of its capacity to diffuse investment capital within a specific spatio-temporal range. The world urban system is transformed endogenously and remains on a *universal attractor* to increase the overall diffusion of investment capital through the realisation of the full planetary potential. Capital is conserved in so far as a local level of complexity can only be achieved at the expense of the temporal global investment budget.

Second law of ecodynamics

Ecostructures of households and establishments are complex adaptive systems for the diffusion of investment capital, and they evolve with increasing complexity to economise on transaction unit costs. Capital formation drives the urban system away from equilibrium, which sets up gradients between locations in the rate-of-return on investment. An ecostructure reaches a spatio-temporal peak of diffusive capacity, as less mature ecostructures emerge to provide alternative pathways for the diffusion of investment. The redistribution of investment to less mature parts of the urban system tends to reduce the gradients in the rate-of-return on investment. Depending upon the rate-of-investment flow, individual ecostructures may be classed as static, cyclic, vibrant, or chaotic attractors.

Third law of ecodynamics

Investment capital is dissipated in the transformation of investment, the maintenance of complex civil ecostructures, and by the coefficient of connectivity in transactions between establishments. In mature ecostructures, more investment is dissipated in the maintenance of order and less in extending the system than for less mature ecostructures. When the planetary system of ecostructures reaches the phase of maturity at which no further evolution or technological change will increase the overall diffusion of investment, a dynamic balance will be achieved if the regenerative capacity of the planetary metasystems (astrophysical, geophysical, physical, biological, and civil) is sufficient to counter the atrophic forces of decline. Investment atrophy is a measure of investment capital deprivation or undernourishment in parts of the urban system, as a consequence of the accumulation of capital in other parts of the system.

MDC (more developed country)

An MDC is a more developed country with a relatively high per capita GNP, high per capita levels of investment in infrastructure, low population growth, and high levels of life expectancy and human development.

Megalopolis

The term Megalopolis, which literally means a great city, was introduced by Jean Gottman to delineate major urbanised regions with populations of over 20m, in which several individual metropolitan areas, each in excess of 1m population, had become mutually adjacent.

Megapolis

A megapolis or megacity is defined as a large metropolis with a city population in excess of 10m. A megapolitan city region, in which the metropolitan centre is surrounded by outer centres linked by orbital and radial routes to form a diffused urban system of 10-20m inhabitants, has a density that declines from 2,500 inhabitants per km^2 in the LDCs to 1,250 inhabitants per km^2 in the more developed countries (MDCs). This decline arises from a reduction in the number of persons per household at each stage of economic development and also the extension of the metropolitan region with new perimeter development.

Metapolis

A metapolis is an urban belt of 50-100m population in which contiguous megalopolises, megapolises, cosmopolises, or metropolises form along coastal zones or inland transport axes at a density approaching 600-900 inhabitants per km². Metapolitan belts may extend for some 500-1,500 km and contain a population of 100m within an area of 100,000-180,000 km².

Metasystem

The interacting dynamic metasystems of the planet include the physical, astrophysical, geophysical, biological, and civil systems.

Metropolis

A metropolis is an urban area with over 1m population.

Mineral reserve

Mineral reserves are a variable quantity when defined as the quantity of raw material that can be mined and extracted commercially. Current estimates of a reserve generally assume that the reserves are exploited using existing technology under present economic conditions.

Model

A model is a coherent set of descriptions about the relevant relationships of some aspect of the world, which is intended to clarify our understanding of a problem or the behaviour of a system. A model may take a variety of forms, including a set of statistics or tables, a series of mathematical equations, a computer programme, or a physical simulation. In most futures research, models are the methodology used for predicting a future state in which alternative interventions are proposed, together with an assessment of the corresponding consequences.

Morphological analysis

A morphological analysis of an organism means the study of its organic form; the term is used in systems engineering for the conception of subsystem parameters that can be combined to produce a system design with high performance characteristics. Morphogenises is the development of the complex form of an adult organism from a simple beginning such as an egg or a bud and is the source of emergent evolutionary properties.

Mutation
A mutation is a change in the genetic make-up of an organism. Mutations in the genes produce differences in the form and function of the organism and, conversely, the form and behaviour of the organism will determine the gene changes that have survival value.

Nanotechnology
Nanotechnology is the building of devices on a molecular scale.

Natural resources
Natural resources are resources that can be taken directly from the physical environment.

Natural selection
Natural selection is the process by which organisms best adapted to their environment survive while those less well adapted become extinct. Natural selection within economic sectors applies to the establishment and not directly to the human agents. The principle of natural selection for establishments is inextricably linked to the competition for and the effective utilisation of investment capital.

Network dynamics
Networks that are intended to serve a single coordinated purpose, such as transportation networks, power grids, and telecommunications systems are evolving structures that are built over long periods of time by a variety of independent agents and authorities. Networks have both topological and dynamical properties, which depend upon the traffic and the pattern of connections. Many networks involve the connection and disconnection of linkages and nodes (or vertices), so that the connectivity of the network as a whole depends upon processes operating at the local level, which both constrain and are constrained by the network structure. In a highway network, the nodes are cities and the links are the highways connecting them. In an air traffic system, a large number of airports are connected via a few major hubs (highly connected nodes).

Niche

A niche is the equilibrium population of a species within an ecosystem. Growing populations expand their niches, and if the niche of a species diminishes towards zero, the population is likely to become extinct.

Nonlinear science

Nonlinear science focuses on a specific class of behaviours encountered in many different contexts. In a nonlinear system, the combination of elementary actions can introduce dramatic new effects through positive feedback mechanisms. This can give rise to unexpected structures and events whose properties can be quite different from those of the elementary laws, in the form of abrupt transitions or an irregular and markedly unpredictable evolution in time and space referred to as deterministic chaos.

OECD

The Organisation for Economic Co-operation and Development (OECD) is an international organisation of thirty countries that accept the principles of representative democracy and free-market economy. There are twenty-seven high-income countries in the OECD, including Australia, Austria, Belgium, Canada, Czech Republic, Denmark, Finland, France, Germany, Greece, Hungary, Iceland, Ireland, Italy, Japan, Luxembourg, Netherlands, New Zealand, Norway, Portugal, Slovakia, South Korea, Spain, Sweden, Switzerland, United Kingdom, and the United States. The three upper-middle- income economies in the OECD include Poland, Mexico and Turkey.

Ontogenesis

Ontogenesis provides the biogenetic relationship between the life cycle of a human and the evolutionary cycle of humankind, so that a correlation can be made with respect to rates of physical growth, levels of intellectual development, and the stage of maturity for civilisation.

Open System

An open system exchanges information, energy, and material with its environment. It is able to adapt to its environment, yet may retain a steady-state. The flow of energy or material in the system creates nonlinear dynamic behaviour far-from-equilibrium, and dissipative structures form that oppose the disorder implied by the second law of thermodynamics.

Path dependence

Path dependence in location theory is derived from a view that spatial structure is dependent upon a settlement history, rather like geological stratification in which infrastructures are added layer by layer and industries agglomerate sector by sector, to take advantage of economies of locational connectivity or transaction costs. The urban system is dynamic, and a different pattern of early events or future chance events would cause divergent paths to be followed. Similarly, the early history of technological innovation, which may be the consequence of early events or chance circumstances, can determine its market share rather than the inherent efficiency or superiority of the technology.

Phenotype

The phenotype is the set of observable characteristics of an individual or group of organisms as determined by the interaction of the genotype and the environment.

Policy analysis

Policy analysis is an applied social science in which empirical data is collected on the dynamics of a social system, and policy-relevant information is transformed into knowledge through the conception of new relationships between the variables entering into a policy problem. The results of policy analysis are used to evaluate the options available for the use of present and future resources of a political regime and to predict the cumulative effects so as to determine any necessary changes to the system.

Population density

Population density is the number of individuals per unit area.

Potential

The potential of an ecostructure is a measure of its capacity to diffuse investment capital within a specified spatio-temporal range. The potential of an ecopolitan state is termed the ecopolitan potential, and this will reflect the scale of available natural resources and the absolute number of large establishments, together with the ratio of large establishments per 10,000 population.

Power law
A power law such as Zipf's law or the Pareto distribution is followed when the probability of measuring a particular value of some quantity varies inversely as a power of that value. Power laws arise widely in planetary and earth sciences, physics, demography, economics, and social science.

Prescience
Prescience is the foreknowledge that arises from futures research, and its purpose is both the explanation of past evolution and the exploration of potential futures. There is a continuum for complexity in futures research from 'Projective Futures' involving forecasts, to 'Prospective Futures' based upon scenarios, and finally to 'Evolutionary Futures' with the development of landscapes.

Primary resources division
The primary resources division of the economy includes the industries of agribusiness, mining, energy and water supply.

Prognostication
Prognostication involves an attempt to make an independent assessment of future conditions, and it is used for long-range futures research where there is a scientific basis or phenomenon that limits the number of possible future states of a system. With complex adaptive systems, long-term growth of the system will neither be linear nor exponential as it will generally follow a logistic S-shaped curve. The intermediate or medium-term future may be explored by interpolation between the extrapolated near future and the prognosticated long-term future state.

Punctuated equilibrium
Punctuated equilibrium is a sudden and discontinuous change from unseen or chance events that limit the ability to predict precisely the future evolution of a complex adaptive system.

Purchasing Power Parity
PPP GNI is Gross National Income converted to international dollars using purchasing power parity rates. An international dollar has the

same purchasing power over GNI as a US dollar has in the United States. From 1996, the World Bank has published Purchasing Power Parity (PPP) estimates of GNP per capita in their World Development Indicators.

Quaternary information division

The quaternary information division of the economy encompasses the information service industries such as media, business and legal services, telecommunications and information technology, design, technical services, research and development, education, health, welfare organisations, associations, and government.

Reciprocity

The principle of reciprocity lies behind societal evolution, and it involves a pattern of exchange between individuals, groups, or geopolitical units to achieve cooperation and to discourage the other party from defecting from the arrangement. It is envisaged that capital flows will become more important than trade for determining the reciprocity between trading blocs. Tit-for-tat strategies, in which good behaviour is rewarded by cooperation and uncooperative behaviour is punished by retaliation in the following round, provides a clear message to the other party that coevolution is more survival-positive than competition.

Redistribution of investment

The spatial redistribution of investment refers to the tendency of investment capital to flow away from regions showing diminishing returns on investment, towards less developed regions that offer increasing investment returns.

Reductionism

Reductionism is the scientific approach in which complicated systems are analysed through the interaction of their simpler constituents, which are linked together by relatively simple laws or relationships. Reductionism reveals increasingly microscopic detail of an organism or entity, rather than the emergent macroscopic behaviour of a vivisystem of interacting entities.

Renewable resource

A renewable resource is one that is capable of replenishing itself or of being replenished by human action.

Resilience

Resilience in an evolutionary system relates to conditions far-from-equilibrium and is concerned with the ability to persist and adapt to instabilities that may push the system to a bifurcation point and a new stability domain. Resilience is measured by the magnitude of disturbance that can be absorbed before the system changes structure.

Scenarios

Scenarios are alternative images of the future for policy evaluation, and they are neither predictions nor forecasts but a means of designing more robust strategies. Scenarios generally link qualitative narratives or storylines about the future to quantitative data in the form of tables and figures, which are often generated by computer models. The majority of model-based global studies start with models, and then scenarios are derived on the basis of the models' output. Exploratory scenarios are open-ended paths into the future that could turn out to be utopias or dystopias, whereas normative scenarios are explicitly value based and teleological routes to preferable end states.

Secondary industrial division

The secondary industrial division of the economy includes the industries of construction, property, and manufacturing.

Self-transformation

Self-transformation is the spontaneous emergence of macrostructure, such as with ecostructures or ecosystems, due to the accumulation of the individual micro transactions between a large number of establishments or organisms.

Social equity

Social equity relates to a notion of fairness in the distribution of things like wealth, income, power, position, educational attainment, or access to health facilities. Social equality has come to mean that the rules of the game are equal for all people, such as equality of opportunity or people's rights under the law. In an equitable society, it is necessary to reward real differences in skill, effort, location, danger, or other conditions relating to work. The benefit from ensuring that the lowest forty percent of households receive at least twenty percent of national income is that it reduces the diseconomies of inequality.

Species fitness

Species fitness relates to its ability to survive and reproduce, which is dependant upon past history, the niche it is currently filling, the other organisms in the ecosystem, and the availability of resources. Evolutionary biologists believe that species fitness comes from the dance of coevolution in which each species tries to adapt to all the others.

Substitution

Sustainable development involves conservation of the total stock of capital, in which substitution may take place within the constituent parts, such as infrastructure for natural resources. Substitution of material refers to the replacement of one material by another because of scarcities or rising costs.

Sustainable development

Sustainable development has been defined by the World Commission on Environment and Development (the Brundtland Commission, 1987) as "meeting the needs of the present without compromising the ability of future generations to meet their own needs." However, a more specific definition for sustainable civil development is that it involves extending the lifetime of a planetary civilisation, by leaving viable human settlements for future generations at appropriate planning standards using longer-life assets, clean technologies, and renewable resources to achieve equality in life expectancy for all citizens. Sustainability involves conservation of the overall stock of assets, which includes artefacts, wisdom, natural resources, and genetic material.

System growth parameter

In ecosystems, the system growth parameter is 'exergy,' which drives the system away from equilibrium and is destroyed in the irreversible process of evolution. In the civil system, investment capital is the system growth parameter, and civil ecostructures diffuse investment capital to create endogenous growth with an increase in the asset stock. Civil phase transitions arise with an increase in the system-growth parameter, and at each transition, there is an increase in the urban proportion of the population.

Technological evolution

Technological evolution arises when, for example, horse-drawn transport was replaced by motor vehicles, which created a mass extinction in the network of equestrian-orientated establishments. This was subsequently replaced by a new technological system with its own niches in the motor industry, highway construction and maintenance, traffic control systems, car parks, oil companies, petrol stations, car dealerships and garages, motels, and out-of-town retail parks. Although societal transitions may be technology driven, there is a cultural evaluation process that coevolves with socio-economic trends that determines the envelope of the technological trajectory of any system.

Teleology

Emergent phenomena in the civil system arise from the combined effects of both causality and teleology, which is purposeful human intervention.

Terrestrial stocks

The terrestrial stocks of the biosphere are geological stocks that are not renewable within a human time span.

Territorial colonisation

The Verhulst logistic curve lies behind the waves of territorial colonisation that arise when populations in any given place increase until the limits of environmental resources are approached. From that point in time, the population will stabilise through a reduction in birth rates or an increase in mortality rates, unless the excess population departs to a new region. Colonists tend to settle in the nearest location that offers sufficient resources, and the new regional population will follow the logistic curve until the carrying capacity of the territory is reached and the emigration process is repeated.

Tertiary commercial division

The tertiary commercial division of the economy provides tangible economic services including banking, finance and securities, insurance, transport, wholesale and retail trades, travel, hospitality, and consumer services.

Tons

Tons are metric tons or tonnes, which are equal to 1,000 kilograms.

Transacting or interacting entities

Populations of organisms, households, establishments, and animats within a complex adaptive system or vivisystem are the interacting or transacting entities.

Transpart

A transpart is the smallest complex assembly, component, or compound that has the recognisable characteristics of a progeny of a species of establishments. Archatomy examines how complex assemblies, components or compounds function as organs in the evolution of 'animats' in the same way that the science of anatomy examines the bodily structure of humans, animals, and plants.

Trillion

A trillion is 1,000 billion.

Troposphere

The troposphere is the lowest part of the earth's atmosphere and extends upwards for approximately 10 km. Between the troposphere and the stratosphere comes the ozone layer, which filters out the harmful short-wave radiations from space and makes possible advanced life forms on earth.

Urban System

The term 'urban system' is used in its widest sense to include cities, towns, villages, and isolated establishments, which are connected through infrastructures, even though the coefficient of connectivity may be very low in remote areas.

Vivisystem

A vivisystem or complex adaptive system is a community of interacting organisms and households or transacting entities such as establishments, technological 'animats,' or the animates of artificial life, which change their strategies in accordance with the feedback they receive from their environment. A colony of ants, a swarm of bees, a flock of birds, or a shoal of fish are all vivisystems, and a web of internet users would also be described as a vivisystem.

Wisdom

Civilisation is an evolutionary process in which the stock of wisdom is reflected in the institutions, culture, knowledge, information structures, and socio-technological systems.

APPENDIX 38
CONVERSION OF ENERGY UNITS

Units of energy:

1 Joule	(J)	$= 1$ J
1 Kilojoule	(KJ)	$= 10^3$ J
1 Megajoule	(MJ)	$= 10^6$ J
1 Gigajoule	(GJ)	$= 10^9$ J
1 Terajoule	(TJ)	$= 10^{12}$ J
1 Petajoule	(PJ)	$= 10^{15}$ J
1 Exajoule	(EJ)	$= 10^{18}$ J
1 Zettajoule	(ZJ)	$= 10^{21}$ J

Units of power:

1 Watt (W)	$=$	1 J per second
1 Kilowatt	(KW)	$= 10^3$ W
1 Megawatt	(MW)	$= 10^6$ W
1 Gigawatt	(GW)	$= 10^9$ W
1 Terawatt	(TW)	$= 10^{12}$ W

Note: 1 Terawatt year/year is written as TW

Energy equivalents:

1 BTU = 1.055 KJ

1 Kilocalorie (KCal) = 4.186 KJ

1 Kilowatt hour (KWh) = 3.60 MJ

1 Kilowatt year (KWyear) = 31.5 GJ

1 metric ton of coal equivalent (TCE) = 29.3 GJ

1,000 m^3 of natural gas (NG) = 39.85 GJ

1 metric ton of oil equivalent = 43.1 GJ

1 metric ton of oil equivalent = 7.3 barrels

1 barrel of oil = 5.90 GJ

1 barrel per day = 50 metric tons of oil per year = 2.17 TJ

Energy consumption rates for transport modes are as follows:

Petrol car (over 2.0 litres)	4.65 MJ/passenger km (25% occupancy)
Petrol car (1.4 - 2.0 litres)	1.45 MJ/passenger km (50% occupancy)
Bus	0.58 MJ/passenger km (50% occupancy)
Suburban electric train	0.59 MJ/passenger km (50% occupancy)

APPENDIX 39
BIBLIOGRAPHY

Aburdene, Patricia. *Megatrends 2010 – The rise of conscious capitalism.* Charlottesville, Virginia: Hampton Roads Publishing Company Inc., 2005.

Adouze, Jean and Guy Israel (eds.). *The Cambridge Atlas of Astronomy.* Cambridge: Cambridge University Press, 1988.

Albeverio, Sergio and Denise Andrey, Paolo Giordano, Alberto Vancheri (eds.). *The Dynamics of Complex Urban Systems – An Interdisciplinary Approach.* Heidelberg: Physica-Verlag, 2007.

Allen, Peter M. *Cities and Regions as Self-Organising Systems - Models of Complexity.* Amsterdam: Gordon and Breach Science Publishers, 1997.

Allen, Peter M. *Why the Future is Not What it Was - New models of evolution.* Article in *Futures,* Volume 22, Number 6, July/August 1990.

Allen, Peter M. and Paul Torrens (eds.). *Complexity and the Limits of Knowledge.* Special Issue articles in *Futures,* Volume 37, Number 7, September 2005.

Alley, Richard B. *The Two-Mile Time Machine - Ice Cores, Abrupt Climate Change, and Our Future.* Princeton, New Jersey: Princeton University Press, 2000.

Anderson, Esben Sloth. *Evolutionary Economics - Post-Schumpeterian Contributions.* London: Pinter, 1994.

Arquilla, John and David Ronfeldt. *Networks and Netwars – The future of terror, crime and militancy.* Santa Monica, California: RAND, 2001.

Arthur, W Brian. *Increasing Returns and Path Dependance in the Economy.* Ann Arbor, Michigan: The University of Michigan Press, 1994.

Arthur, W Brian, Steven N Durlauf, and David Lane (eds.). *The Economy as an Evolving Complex System.* Reading, Massachusetts: Perseus Books, 1997.

Bainbridge, William S (ed.). *Futures of Religions.* Special Issue article in *Futures,* Volume 36, Number 9, November 2004.

Batty, Michael and Bruce Hutchinson (eds.). *Systems Analysis in Urban Policy-Making and Planning.* NATO Systems Science Series. New York: Plenum Press, 1983.

Batty, Michael and Sam Cole (eds.). *Time and Space - Geographic Perspectives on the Future*. Special Issue articles in *Futures*, Volume 29, Number 4/5, May/June 1997.

Batty, Michael. *Cities and Complexity – Understanding Cities with Cellular Automata, Agent-Based Models, and Fractals*. Cambridge, Massachusetts: The MIT Press, 2005.

Beinhocker, Eric D. *The Origin of Wealth – Evolution, Complexity and the Radical Remaking of Economics*. McKinsey & Company Inc. London: Random House Business Books, 2006.

Bell, Daniel. *The Coming of Postindustrial Society - A Venture In Social Forecasting*. New York: Basic Books, 1973.

Bell, Wendell. *Foundations of Futures Studies - Human Science for a New Era*. (Volume 1 - *History, Purposes and Knowledge*, Volume 2 - *Values, Objectivity and the Good Society*). New Brunswick, New Jersey: Transaction Publishers, 1997.

Bernardini, Oliviero and Riccardo Galli. *Dematerialization - long-term trends in the intensity of use of materials and energy*. Article in *Futures*, Volume 25, Number 4, May 1993.

Berry, Adrian. *The Next Ten Thousand Years - A Vision of Man's Future in the Universe*. London: The Scientific Book Club, 1975.

Berry, Brian J L, Edgar C. Conkling and D. Michael Ray. *Economic Geography - Resource Use, Locational Choices, and Regional Specialisation in the Global Economy*. Englewood Cliffs, New Jersey: Prentice-Hall Inc., 1987.

Berry, Brian. J. L. *Long-Wave Rhythms in Economic Development and Political Behaviour*. Baltimore, Maryland: The John Hopkins University Press, 1991.

Bertalanffy, Ludwig von. *General System Theory – Foundations, Development, Applications*. New York: George Braziller Inc., 1969.

Biondi, Leonardo and Riccardo Galli. *Technological Trajectories*. Article in *Futures*, Volume 24, Number 6, July/August 1992.

Boulding, Kenneth E. *Ecodynamics - A New Theory of Societal Evolution*. Beverly Hills, California: Sage Publications Inc., 1978.

Boulding, Kenneth E. *The World as a Total System*. Sage Publications Inc., Beverly Hills, California, 1985.

Bova, Ben and Byron Preiss (eds.). *Are We Alone in the Cosmos? - The search for alien contact in the new millennium*. New York: Ibooks Inc., 1999.

Boyle, Godfrey (ed.). *Renewable Energy - Power for a Sustainable Future*. Oxford: Oxford University Press, 1996.

Boyle, Godfrey, Christine Thomas and David Weild (eds.). *Sustainable Futures – Special Issue*. Articles in Futures, Volume 32, Numbers 3/4, April/May 2000.

Bright, James R. *Technological Forecasting for Industry and Government*. Eaglewood Cliffs, New Jersey: Prentice-Hall Inc., 1968.

Broecker, Wallace S. *What if the Conveyor were to Shut Down? – Reflections on the Possible Outcome of the Great Global Experiment*. Article in *GSA Today*, Volume 9, Number 1, January 1999.

Broecker, Wallace S. *The Role of the Ocean in Climate Yesterday, Today, and Tomorrow*. Lamont-Doherty Earth Observatory, Columbia University, New York: Eldigio Press, 2005.

Bronowski, Jacob. *The Ascent of Man*. London: British Broadcasting Corporation, 1973.

Brotchie, John, Peter Newton, Peter Hall, and Peter Nijkamp (eds.). *The Future of Urban Form - The Impact of New Technology*. Beckenham, Kent: Croom Helm Ltd., 1985.

Brotchie, John F, Peter Hall and Peter W Newton (eds.). *The Spatial Impact of Technological Change*. Beckenham, Kent: Croom Helm Ltd., 1987.

Brotchie, John, Michael Batty, Peter Hall, and Peter Newton (eds.). *Cities of the 21st Century -New Technologies and Spatial Systems*. Melbourne: Longman Cheshire Pty Limited,1991.

Brotchie, John, Mike Batty, Ed Blakely, Peter Hall, and Peter Newton (eds.). *Cities in Competition - Productive and Sustainable Cities for the 21st Century*. Melbourne: Longman Australia Pty Ltd, 1995.

Brotchie, John, Peter Newton, Peter Hall, and John Dickey (eds.). *East West Perspectives on 21st Century Urban Development*. Aldershot, Hants: Ashgate Publishing Ltd., 1999.

Brown, Lester R (ed.). *State of the World 2001 - A Worldwatch Institute report on progress toward a sustainable society*. London: Earthscan Publications Ltd., 2001.

Brown, Lester R, Michael Renner, and Brian Halweil. *Vital Signs - The trends that are shaping our future*. London: Worldwatch Institute and Earthscan Publications Ltd., 2000.

Brunn, Stanley D and Jack F Williams. *Cities of the World - World Regional Urban Development*. New York: Harper & Row Publishers Inc., 1983.

Brunn, Stanley D and Thomas R Leinbach. *Collapsing Space and Time - Geographic Aspects of Communication & Information*. London: HarperCollins Academic, 1991.

Bryan, Lowell and Diana Farrell. *Market Unbound - Unleashing Global Capitalism*. New York: John Wiley & Sons, 1996.

Burchell, Robert W., Anthony Downs, Barbara McCann, and Sahan Mukherji. *Sprawl Costs – Economic impacts of unchecked development*. Washington DC: Island Press, 2005.

Cable, Vincent and David Henderson (eds). *Trade Blocs? The Future of Regional Integration*. London: Royal Institute of International Affairs, 1994.

Castells, Manuel. *The Information Age - Economy, Society and Culture*. (Volume 1 - *The Rise of the Network Society*, Volume 2 - *The Power of Identity*, Volume 3 - *End of Millennium*). Oxford: Blackwell Publishers, 1996–1998.

Cathelat, Bernard. *Socio-Styles – The New Lifestyles Classification System for Identifying and Targeting Consumers and Markets*. London: Kogan Page Ltd, 1990.

Cetron, Marvin J and Owen Davies. *50 Trends Now Changing the World*. Bethesda, Maryland: Special Report published by the World Future Society, 2001.

Cetron, Marvin J and Owen Davies. *Trends Shaping Tomorrow's World (Parts 1 & 2) – Forecasts and Implications for Business, Government and Consumers*. Bethesda, Maryland: Published in the *Futurist* by the World Future Society, March & May 2008.

Chandler, Tertius. *4000 Years of Urban Growth*. Lewiston, New York: The Edwin Mellen Press, 1987.

Christian, David. *Maps of Time – An Introduction to Big History*. Berkeley and Los Angeles, California: University of California Press, 2004.

Clark, Colin. *The Conditions of Economic Progress*. London: Macmillan and Co., 1940.

Cleveland, Harlan, Hazel Henderson, and Inge Kaul (eds.). *The United Nations at 50 - Policy and Financing Alternatives*. Articles in *Futures*, Volume 27, Number 2, March 1995.

Cleveland, Harlan and Walter Truett Anderson (eds.). *2000 - The Global Century*. Special Issue. Articles in *Futures*, Volume 31, Numbers 9/10, November/December 1999.

Cline, William R. *The Economics of Global Warming*. Washington, DC: Institute for International Economics, 1992.

Cocks, Doug. *Future Makers, Future Takers – Life in Australia 2050*. Sydney, Australia: University of New South Wales Press, 1999.

Cocks, Doug. *Deep Futures – Our Prospects for Survival*. Sydney, Australia: University of New South Wales Press, 2003.

Cohen, Joel E. *How Many People Can the Earth Support?* New York: W W Norton & Company, 1995.

Cole, Leonard A. *The Eleventh Plague - The Politics of Chemical and Biological Warfare*. New York: W H Freeman and Company, 1997.

Coren, Richard L. *The Evolutionary Trajectory - The Growth of Information in the History and Future of Earth*. Amsterdam: Gordon and Breach Publishers, 1998.

Cornish, Edward. *Futuring - The Exploration of the Future*. Bethesda, Maryland: The World Future Society, 2004.

Costanza, Robert (ed.). *Ecological Economics - The Science and Management of Sustainability*. New York: Columbia University Press, 1991.

Cowan, George A., David Pines, and David Meltzer (eds.). *Complexity - Metaphors, Models and Reality*. Reading, Massachusetts: Addison-Wesley Publishing Company, 1994.

Cox, Kevin R. *An Introduction to Human Geography*. New York: John Wiley & Sons Inc., 1972.

Cronon, William. *Nature's Metropolis - Chicago and the Great West*. New York: W W Norton & Company, 1992.

Curry, David J. *The New Marketing Research Systems*. New York: John Wiley & Sons Inc., 1993.

Daly, Herman E. and Kenneth N. Townsend (eds.). *Valuing the Earth - Economics, Ecology, Ethics*. Cambridge, Massachusetts: The MIT Press, 1993.

Dator, James A. (ed.). *Advancing Futures: Futures Studies in Higher Education*. Westport, Connecticut: Praeger Publishers, 2002.

Davis, Mike. *Planet of Slums*. London: Verso, 2006.

Deffeyes, Kenneth S. *Hubbert's Peak – The impending world oil shortage*. Princeton, New Jersey: Princeton University Press, 2001.

De Jouvenel, Hugues (ed.). *Working Time* - Special Issue. Articles in *Futures*, Volume 25, Number 5, June 1993.

Demko, George J. and William B. Wood (eds.). *Reordering the World – Geopolitical Perspectives on the 21st Century*. Boulder, Colorado: Westview Press Inc., 1994.

Dewar, James A. *Assumption-Based Planning - A Tool for Reducing Avoidable Surprises*. Cambridge: Rand Studies, Cambridge University Press, 2002.

D' Inverno, Mark and Michael Lock. *Understanding Agent Systems*. Heidleberg: Springer-Verlag, 2001.

Dixon, Dougal, Ian Jenkins, Richard Moody, and Andrey Zhuravlev. *Cassell's Atlas of Evolution*. London: Cassell & Co, 2001.

Dogan, Mattei, and John D. Kasarda (eds.). *The Metropolis Era (Volume 1- A World of Giant Cities, Volume 2- Mega-Cities)*. Beverly Hills, California: Sage Publications Inc., 1988.

Doxiadis, Constantinos A. and J. G. Papaioannou. *Ecumenopolis - The Inevitable City of the Future*. New York: W W Norton & Company Inc., 1974.

Dreborg, Karl H. *Essence of Backcasting*. Article in *Futures*, Volume 28, Number 9, November 1996.

Drexler, K. Eric. *Engines of Creation - The Coming Era of Nanotechnology*. London: Fourth Estate Limited, 1990.

Dunning, John H. *The Globalisation of Business*. London: Routledge, 1993.

El-Agraa, Ali M. *Economic Integration Worldwide*. London: Macmillan Press Ltd, 1997.

England, Richard W. *Evolutionary Concepts in Contemporary Economics*. Ann Arbor, Michigan: The University of Michigan Press, 1994.

European Environment Agency, David Stanners and Philippe Bourdeau (eds.). *Europe's Environment - The Dobris Assessment*. Copenhagen: EEA, 1995.

Financial Times. *World Desk Reference*. London: Dorling Kindersley, 2002.

Flanagan, William G. *Contemporary Urban Sociology*. Cambridge: Cambridge University Press, 1993.

Forrester, Jay W. *Urban Dynamics*. Cambridge, Massachusetts: MIT Press, 1969.

Forrester, Jay W. *World Dynamics*. Cambridge, Massachusetts: Wright-Allen Press Inc., 1971.

Friedmann, John and Clyde Weaver. *Territory and Function - The Evolution of Regional Planning*. London: Edward Arnold (Publishers) Ltd., 1979.

Fu-Chen Lo and Yue-Man Yeung (eds.). *Globalisation and the World of Large Cities*. Tokyo: United Nations University Press, 1998.

Fujita, Masahisa, Paul Krugman, and Anthony J. Venables. *The Spatial Economy - Cities, Regions, and International Trade*. Cambridge, Massachusetts: MIT Press, 1999.

Gabel, Medard and Henry Bruner. *Global Inc. – An Atlas of the Multinational Corporation*. New York: The New Press, 2003.

Galli, Riccardo. *Structural and Institutional Adjustments and the New Technological Cycle*. Article in *Futures*, Volume 22, Number 8, October 1992.

Garreau, Joel. *Edge City - Life on the New Frontier*. New York: Doubleday, 1991.

Garrett, Laurie. *The Coming Plague - Newly Emerging Diseases in a World Out of Balance*. London: Penguin Books, 1995.

Gimblett, H. Randy (ed.). *Integrating Geographic Information Systems and Agent-based Modeling Techniques – for Simulating Social and Ecological Processes*. New York: Oxford University Press, 2002.

Glenn, Jerome C., and Theodore J. Gordon. *Millennium Project - 2003 State of the Future*. Washington, D.C.: American Council for The United Nations University, 2003.

Godet, Michel. *The Crisis in Forecasting and the Emergence of the 'Prospective' Approach - With Case Studies in Energy and Air Transport*. New York: Pergamon Press Inc., 1979.

Godet, Michel, Pierre Chapuy and Gerard Comyn. *Global Scenarios - Geopolitical and economic context to the year 2000*. Article in *Futures*, Volume 26, Number 3, April 1994.

Godet, Michel. *Creating Futures - Scenario Planning as a Strategic Management Tool*. London: Economica Ltd., 2001.

Goldin, Ian and L. Alan Winters (eds.). *The Economics of Sustainable Development*. Cambridge: Cambridge University Press, 1995.

Goonatilake, Susantha: *Reconceptualizing the cultural dynamics of the future*. Article in *Futures*, Volume 24, Number 10, December 1992.

Gottmann, Jean. *Megalopolis: The Urbanised Northeastern Seaboard of the US*. New York: The Twentieth Century Fund, 1961.

Gottmann, Jean. *The Coming of the Transactional City*. Maryland: University of Maryland Institute for Urban Studies, 1983.

Gould, Stephen J. *Ontogeny and Phylogeny*. Cambridge, Massachusetts: The Belknap Press of Harvard University Press, 1977.

Gould, Stephen J. *Wonderful Life*. London: Penguin Books Ltd, 1989.

Graham, Stephen and Simon Marvin. *Splintering Urbanism – Networked Infrastructures, Technological Mobilities and the Urban Condition*. London: Routledge, 2001.

Greeenfield, Susan. *Tomorrow's People – How 21st Century Technology is Changing the Way We Think and Feel*. London: Penguin Books Ltd., 2003.

Greider, William. *One World Ready or Not - The Manic Logic of Global Capitalism*. London: Penguin Books Ltd., 1997

Groombridge, Brian and Martin D. Jenkins. *World Atlas of Biodiversity - Earth's Living Resources in the 21st Century*. London: Published in association with UNEP World Conservation Monitoring Centre by University of California Press Ltd., 2002.

Guinness, Mark Fletcher (ed.). *Guinness Amazing Future - The Indispensable Guide to the Next Millennium*. London: Guinness Publishing Limited, 1999.

Halal, William E. *World 2000 - An international planning dialogue to help shape the new global system*. Article in *Futures*, Volume 25, Number 1, January/February 1993.

Hall, Peter and Kathy Pain (eds.). *The Polycentric Metropolis*. London: Earthscan Publications Ltd., 2006.

Harvey, David. *The Urbanization of Capital – Studies in the History and Theory of Capitalist Urbanization*. Oxford: Blackwell Publishers, 1985.

Harvey, David. *Spaces of Global Capitalism – Towards a Theory of Uneven Geographical Development*. London: Verso Books, 2006.

Healey, Michael J (ed.). *Economic Activity & Land Use – The Changing Information Base for Local and Regional Studies*. Harlow, England: Longman Group Ltd., 1991.

Heidmann, Jean. *Extraterrestrial Intelligence*. Cambridge: Cambridge University Press, 1997.

Heijden, Kees Van Der. *Scenarios - The Art of Strategic Conversation*. Chichester: John Wiley & Sons, 1996.

Helmer, Olaf. *Looking Forward. A Guide to Futures Research*. Beverly Hills, California: Sage Publications Inc., 1984.

Hodgson, Geoffrey M. *Economics and Institutions*. Cambridge: Polity Press, 1988.

Hodgson, Geoffrey M. *Economics and Evolution*. Cambridge: Polity Press, 1993.

Hoffmann, Peter. *Tomorrow's Energy - Hydrogen, fuel cells and the prospects for a cleaner planet*. Cambridge, Massachusetts: MIT Press, 2001.

Hoggart, Keith and David R. Green (eds.). *London - A New Metropolitan Geography*. London: Edward Arnold, 1991.

Holland, John H. *Hidden Order - How Adaption Builds Complexity*. Reading, Massachusetts: Addison-Wesley Publishing Company Inc., 1995.

Holland, John H. *Emergence - From Chaos to Order*. Reading, Massachusetts: Addison-Wesley Publishing Company Inc., 1998.

Hollander, Jack M. *The Real Environmental Crisis – Why Poverty, Not Affluence, is the Environment's Number One Enemy*. Berkeley and Los Angeles, California: University of California Press, 2003.

Hopkins, Terence K. and Immanuel Wallerstein. *The Age of Transition – Trajectory of the World-System 1945-2025*. London: Zed Books Ltd., 1996.

Hughes, Barry B. and Peter Johnston. *Sustainable Futures – Policies for global development*. Article in *Futures*, Volume 37, Number 8, October 2005.

Hughes, Barry B. and Evan E. Hillebrand. *Exploring and Shaping International Futures*. Boulder, Colorado: Paradigm Publishers, 2006.

Imbrie, John and Katherine Imbrie. *Ice Ages – Solving the mystery*. Cambridge, Massachusetts: Harvard University Press, 1979.

Inayatullah, Sohail and Tony Stephenson (eds.). *Communication Futures* - Special Issue. Articles in *Futures* Volume 30, Number 2/3, March/April 1998.

Inayatullah, Sohail (ed.). *Layered Methodologies* - Special Issue. Articles in *Futures* Volume 34, Number 6, August 2002.

Intergovernmental Panel on Climate Change (IPCC). *Special Report on Emissions Scenarios*. Cambridge: Cambridge University Press, 2000.

Intergovernmental Panel on Climate Change. *Climate Change 2007 - The Physical Science Basis*. Cambridge: Cambridge University Press, 2007.

International Energy Agency. *Energy to 2050 – Scenarios for a sustainable future*. Paris: Publications Service, OECD/IEA, 2003.

International Energy Agency. *Energy Technology Perspectives – Scenarios and Strategies to 2050*. Paris: Publications Service, OECD/IEA, 2006.

International Labour Office. *World Labour Report*. Geneva: ILO Publications, 1984.

International Labour Office. *World Employment Report 2001:Life at work in the information economy*. Geneva: ILO Publications, 2001.

Jones, Barry. *Sleepers Wake! Technology and the Future of Work*. Brighton, East Sussex: Wheatsheaf Books Ltd., 1982.

Jones, Emrys. *Metropolis - The World's Great Cities*. Oxford: Oxford University Press, 1990.

Jussawalla, Meheroo and D. M. Lamberton (eds). *Communication Economics and Development*. New York: Pergamon Press Inc., 1982.

Jussawalla, Meheroo and Helene Ebenfield. *Communication and Information Economics*. Amsterdam: Elsevier Science Publishers B.V., 1984.

Kahn, Herman and Anthony J Wiener. *The Year 2000 - A Framework for Speculation on the Next Thirty-Three Years*. New York: The Macmillan Company, 1967.

Kahn, Herman, William Brown, and Leon Martel. *The Next 200 Years - A Scenario for America and the World*. New York: William Morrow and Company Inc., 1976.

Kahn, Herman. *World Economic Development - 1979 and Beyond*. Boulder, Colorado: Westview Press Inc., 1979.

Kauffman, Stuart. *At Home in the Universe - The Search for Laws of Complexity*. London: Penguin Books Ltd, 1995.

Kelly, Kevin. *Out of Control - The New Biology of Machines*. London: Fourth Estate Ltd, 1994.

Kemp, Rene. *Technology and the Transition to Environmental Sustainability - The problem of technological regime shifts*. Article in *Futures*, Volume 26, Number 10, December 1994.

Khalilzad, Zalmay and Ian Lesser (eds.). *Sources of Conflict in the 21st Century – Regional futures and U.S. strategy*. Santa Monica, California: RAND, 1998.

Kindleberger, Charles P. *World Economic Primacy 1500-1990*. New York: Oxford University Press Inc., 1996.

King, Anthony D. *Global Cities - Post-Imperialism and the Internationalism of London*. London: Routledge, 1990.

Knox, Paul and John Agnew. *The Geography of the World Economy*. Routledge, New York: Chapman and Hall Inc., 1989.

Knox, Paul L. and Peter J. Taylor (eds.). *World Cities in a World System*. Cambridge: Cambridge University Press, 1995.

Kompanichenko, V. N. *The Cycle and Meaning of the Existence of Humankind*. Article in *Futures*, Volume 26, Number 5, June 1994.

Kooreman, Peter and Sophia Wunderink. *The Economics of Household Behaviour*. London: Macmillan Press Ltd., 1997.

Koppel, Tom. *Powering the Future - The Ballard Fuel Cell and the Race to Change the World*. Toronto: John Wiley & Sons, Canada Ltd., 1999.

Korton, David C. *When Corporations Rule the World*. London: Earthscan Publications Ltd, 1995.

Kovach, Robert and Bill McGuire. *Guide to Global Hazards – A complete reference guide to hazards that endanger life on earth*. London: Philip's, Octopus Publishing Group, 2003.

Krishnan, Rajaram, Jonathan M. Harris, and Neva R Goodwin (eds.). *A Survey of Ecological Economics*. Washington DC: Island Press, 1995.

Kristensen, Thorkil. *Development in Rich and Poor Countries*. New York: Praeger Publishers, 1982.

Krugman, Paul. *Geography and Trade*. Cambridge, Massachusetts: MIT Press, 1993.

Krugman, Paul. *Development, Geography, and Economic Theory*. Cambridge, Massachusetts: MIT Press, 1995.

Krugman, Paul. *The Age of Diminished Expectations*. Cambridge, Massachusetts: MIT Press, 1994.

Kuhn, Thomas S. *The Structure of Scientific Revolutions*. Chicago: The University of Chicago Press, 1962.

Laconte, P., J. Gibson and A. Rapoport. *Human and Energy Factors in Urban Planning: A Systems Approach*. The Hague: Martinus Nijhoff Publishers, 1982.

Lamb, Simon and David Sington. *Earth Story - The Shaping of our World*. London: BBC Worldwide Ltd., 1998.

Laszlo, Ervin. *Evolution: The Grand Synthesis*. Boston and London: New Science Library/ Shambhala, 1987.

Laszlo, Ervin (ed.). *The New Evolutionary Paradigm*. New York: Gordon & Breach, 1991.

Laszlo, Ervin. *The Choice: Evolution or Extinction? - The Thinking Person's Guide to Global Problems*. Los Angeles: Tarcher/Putnam, 1994.

Lempert, Robert J., Steven Popper, and Steven Bankes. *Shaping the Next One Hundred Years*. Santa Monica, California: Rand, 2003.

Leontief, Wassily. *The Future of the World Economy*. New York, Oxford University Press, 1977.

Leydesdorf, Loet and Peter Van den Besselaar (eds.). *Evolutionary Economics and Chaos Theory - New Directions in Technology Studies*. London: Pinter Publishers Ltd., 1994.

Lomberg, Bjorn. *The Skeptical Environmentalist – Measuring the Real State of the World*. Cambridge: Cambridge University Press, 2001.

Mander, Jerry and Edward Goldsmith (eds.). *The Case Against the Global Economy - And for a Turn toward the Local*. San Francisco: Sierra Club Books, 1996.

Mannermaa, Mika. *In Search of an Evolutionary Paradigm for Futures Research*. Article in *Futures*, Volume 23, Number 4, May 1991.

Marien, Michael. *Environmental Problems and Sustainable Futures - Major Literature from WCED to UNCED*. Article in *Futures*, Volume 24, Number 8, October 1992.

Marien, Michael. *World Futures and the United Nations - A guide to recent literature*. Article in *Futures*, Volume 27, Number 3, April 1995.

Marien, Michael (ed). *Future Survey – A monthly abstract of Books, Articles and Reports Concerning Forecasts, Trends and Ideas about the Future*. Bethesda, Maryland: Volumes 23-30, World Future Society Publication, 2001-2008.

Mason, Colin. *The 2030 Spike – Countdown to global catastrophe*. London: Earthscan Publications Ltd., 2003.

McGuire, Bill. *A Guide to the End of the World - Everything You Never Wanted to Know*. Oxford: Oxford University Press, 2002.

McGuire, Bill. *Surviving Armageddon – Solutions for a threatened planet*. Oxford: Oxford University Press, 2005.

McKenzie-Mohr, Doug, and Michael Marien (eds.). *Visions of Sustainability - Special Issue*. Articles in *Futures*, Volume 26, Number 2, March 1994.

Meadows, Donella H, Dennis L Meadows, Jorgen Randers, and William W Behrens. *The Limits to Growth - A Report for the Club of Rome's Project on the Predicament of Mankind*. London: Pan Books Ltd., 1972.

Meadows, Donella H., Dennis L. Meadows, and Jorgen Randers. *Beyond the Limits - Global Collapse or a Sustainable Future*. London: Earthscan Publications Limited, 1992.

Meadows, Donella H, Jorgen Randers, Dennis L Meadows. *Limits to Growth – The 30-year update*. London: Earthscan Publications Limited, 2005.

Meier, Richard L. *A Communications Theory of Urban Growth*. Cambridge, Massachusetts: MIT Press, 1962.

Mesarovic, Mihajlo and Eduard Pestel. *Mankind at the Turning Point*. New York: The New American Library Inc., 1974.

Miser, Hugh J. and Edward S. Quade (eds.). *Handbook of Systems Analysis*. (Volume 1 - *Overview of Uses, Procedures, Applications, and Practice*, Volume 2 - *Craft Issues and Procedural Choices*). Chichester, West Sussex: John Wiley & Sons Ltd., 1985.

Mitchell, James K. (ed.). *Crucibles of Hazard - Mega-cities and Disasters in Transition*. Tokyo: United Nations University Press, 1999.

Mitchell, William J. *e-topia – Urban life, Jim, but not as we know it*. Cambridge, Massachusetts: MIT Press, 2000.

Mitchell, William J. *ME++ - The cyborg self and the networked city*. Cambridge, Massachusetts: MIT Press, 2003.

Monmonier, Mark. *Spying with Maps - Surveillance Technologies and the Future of Privacy*. Chicago: University of Chicago Press, 2002.

Moore, Patrick. *The New Atlas of the Universe*. London: Mitchell Beazley, 1988.

Moore, Patrick and H. J. P. Arnold. *Space – The First 50 years*. Mitchell Beazley, London: The Octopus Publishing Group Ltd., 2007.

Naisbitt, John. *Megatrends - Ten New Directions Transforming Our Lives*. London: Macdonald & Co (Publishers) Ltd., 1982.

Naisbitt, John and Patricia Aburdene. *Megatrends 2000*. London: Sidgwick & Jackson Limited, 1990.

Nakicenovic, Nebojsa, Arnulf Grubler, and Alan McDonald. *Global Energy - Perspectives*. Cambridge: International Institute for Applied Systems Analysis (IIASA) and published by Cambridge University Press, 1998.

National Intelligence Council. *Mapping the Global Future – Report on the National Intelligence Council's 2020 Project*. Pittsburgh, Pennsylvania: Government Printing Office, 2004.

Nelson, Richard R. and Sidney Winter. *An Evolutionary Theory of Economic Change*. Cambridge, Massachusetts: Harvard University Press, 1982.

Nelson, Richard R. *The Sources of Economic Growth*. Cambridge, Massachusetts: Harvard University Press, 1996.

Newman, Mark, Albert-Laszlo Barabasi, and Duncan Watts. *The Structure and Dynamics of Networks*. Princeton, New Jersey: Princeton University Press, 2006.

Newman, Peter and Jeffrey Kenworthy. *Sustainability and Cities - Overcoming Automobile Dependence*. Washington D.C.: Island Press, 1999.

Newson, Lesley. *The Atlas of the World's Worst Natural Disasters*. London: Dorling Kindersley Limited, 1998.

Nicolis, Gregoire and Ilya Prigogine. *Exploring Complexity*. New York: W H Freeman and Company, 1989.

Nicolis, Gregoire. *Introduction to Nonlinear Science*. Cambridge: Cambridge University Press, 1995.

Organisation for Economic Co-operation and Development. *Trends in the Information Economy*. Paris: OECD Publications, 1986.

Organisation for Economic Co-operation and Development. *Linkages - OECD and Major Developing Economies*. Paris: OECD Publications, 1995.

Organisation for Economic Co-operation and Development. *Space 2030 – Exploring the Future of Space Applications*. Paris: OECD Publications, 2004.

Organisation for Economic Co-operation and Development. *Infrastructure to 2030 – Telecom, Land Transport, Water and Electricity*. Paris: OECD Publications, 2006.

O' Riordan, Timothy (ed.). *Ecotaxation*. London: Earthscan Publications Limited, 1997.

Pearce, David W., and Jeremy Warford. *World without End - Economics, Environment, and Sustainable Development*. New York: Oxford University Press Inc., 1993.

Pearson, Ian (ed.). *The Atlas of the Future*. London: Routledge, 1998.

Pearson, Ian and Ian Neild. *A Timeline for Technology: To the Year 2030 and Beyond*. Bethesda, Maryland: Article in *The Futurist*, published by the World Future Society, March 2006.

Perloff, Harvey S. and Lowdon Wingo, Jr. (eds.). *Issues in Urban Economics*. Published for Resources for the Future, Inc. Baltimore, Maryland: The John Hopkins Press, 1968.

Porat, Marc U. *The Information Economy - Definition and Measurement*. Washington D.C.: US Department of Commerce Office of Telecommunications, 1977.

Porter, Michael E. *The Competitive Advantage of Nations*. London: The Macmillan Press Ltd., 1990.

Pred, Allan. *City - Systems in Advanced Economies*. London: Hutchinson & Co (Publishers) Ltd., 1977.

Prestwich, Roger and Peter Taylor. *Regional and Urban Policy in the United Kingdom*. Harlow, Essex: Longman Group UK Limited, 1990.

Raskin, P., G. Gallopin, P. Gutman, A Hammond, and R. Swart. *Bending the Curve - Toward Global Sustainability*. Stockholm, Sweden: Stockholm Environment Institute, 1998.

Raskin, P, T. Banuri, G. Gallopin, P. Gutman, A Hammond, R. Kates, and R Swart. *Great Transition*. Boston, Massachusetts: Stockholm Environment Institute, 2002.

Razak, Victoria and Sam Cole (eds.). *Anthropological Perspectives on the Future of Culture and Society* - Special Issue. Articles in *Futures*, Volume 27, Number 4, May 1995.

Redfern, Ron. *Origins - The Evolution of Continents, Oceans and Life*. London: Cassell & Co, 2000.

Rees, Martin. *Our Final Century - Will the Human Race Survive the Twenty-First Century*? London: William Heinemann, 2003.

Richardson, Harry W. *The Economics of Urban Size*. Farnborough, Hampshire: Saxon House, D C Heath Ltd., 1973.

Rifkin, Jeremy. *The End of Work - The Decline of the Global Labor Force and the Dawn of the Post-Market Era*. New York: G P Putnam's Sons, 1995.

Rifkin, Jeremy. *The Hydrogen Economy*. New York: Penguin Putman Inc., 2002.

Rifkin, Jeremy. *The European Dream*. Cambridge: Polity Press, 2004.

Robinson, John B. *Futures under Glass - A recipe for people who hate to predict*. Article in *Futures*, Volume 22, Number 8, October 1990.

Rostow, Walt W. *The Stages of Economic Growth*. Cambridge: Cambridge University Press, 1960.

Rostow, Walt W. *The World Economy - History and Prospect*. London: The Macmillan Press Ltd., 1978.

Rostow, Walt W. *The Great Population Spike and After - Reflections on the 21st Century*. New York: Oxford University Press Inc., 1998.

Rudolph, Frederick B. and Larry V McIntire (eds.). *Biotechnology - Science, Engineering and Ethical Challenges for the 21st Century*. Washington DC: Joseph Henry Press, 1996.

Sachs, Jeffrey. *The End of Poverty – How we can make it happen in our lifetime*. London: Penguin Books Ltd, 2005.

Salzman, Marian and Ira Matathia. *Next – Trends for the near future*. New York: Overlook Press, 1999.

Salzman, Marian, Ira Matathia, and Ann O'Reilly. *The Future of Men*. New York: Palgrave Macmillan, 2005.

Sardar, Ziauddin and Jerome R. Ravetz (eds.). *Complexity - Fad or Future*? Special Issue articles in *Futures*, Volume 26, Number 6, July/August 1994.

Sassen, Saskia. *The Global City - New York, London, Tokyo*. Princeton, New Jersey: Princeton University Press, 1991.

Sassen, Saskia (ed.). *Global Networks - Linked Cities*. United Nations University, London: Routledge, 2002.

Satterthwaite, David (ed.). *The Earthscan Reader in Sustainable Cities*. London: Earthscan Publications Ltd., 1999.

Schafer, Paul D. *Cultures and Economies - Irresistible forces encounter immovable objects*. Article in *Futures*, Volume 26, Number 8, October 1994.

Schafer, Paul D. *Culture - Beacon of the Future*. Westport, Connecticut: Praeger Studies on the 21st Century. Preager, 1998.

Schmidt, Stanley and Robert Zubrin (eds.). *Islands in the Sky - Bold New Ideas for Colonizing Space*. New York: John Wiley & Sons Inc., 1996.

Schneider, Eric D. and James J. Kay. *Complexity and Thermodynamics - Towards a new ecology*. Article in *Futures*, Volume 26, Number 6, July/August 1994.

Schneider, Eric D. and Dorian Sagan. *Into the Cool – Energy Flow, Thermodynamics, and Life*. Chicago: The University of Chicago Press, 2005.

Schot, Johan, Remco Hoogma and Boelie Elzen. *Strategies for Shifting Technological Systems - The case of the automobile system*. Article in *Futures*, Volume 26, Number 10, December 1994.

Schumpeter, Joseph A. *Capitalism, Socialism and Democracy*. London: Allen & Unwin (Publishers) Ltd, 1942.

Schumpeter, Joseph A. *A History of Economic Analysis*. London: Allen & Unwin (Publishers) Ltd, 1954.

Schwartz, Peter. *The Art of the Long View - Scenario Planning, Protecting Your Company against an Uncertain World*. London: Random Century Group Ltd., 1992.

Schwartz, Peter, Peter Leydon, and Joel Hyatt. *The Long Boom - A Future History of the World 1980-2020*. London: The Orion Publishing Group, Ltd., 2000.

Schwartz, Peter. *Inevitable Surprises - A Survival Guide for the 21st Century*. London: Simon & Schuster UK, Ltd., 2003.

Scientific American. *Transportation – Aerospace, Land and Sea*. Special Issue, Volume 277, Number 4, October 1997.

Scientific American. *Energy's Future Beyond Carbon*. Special Issue, Volume 295, Number 3, September 2006.

Scott, Allen J. *Regional Motors of the Global Economy*. Article in *Futures*, Volume 28, Number 5, June 1996.

Scott, Allen J. (ed.). *Global City-Regions - Trends, Theory, Policy*. Oxford: Oxford University Press, 2001.

Seager, Joni. *The State of the Environment Atlas*. London: Penguin Books Ltd., 1995.

Shell International. *Exploring the Future – Energy Needs, Choices and Possibilities: Scenarios to 2050*. London: Shell International Limited, 2001.

Sherdon, William A. *The Fortune Sellers - The Big Business of Buying and Selling Predictions*. New York: John Wiley & Sons Inc., 1998.

Simmonds, Roger and Gary Hack (eds.). *Global City Regions - Their Emerging Forms*. London: Spon Press, 2000.

Simon, Herbert A. *The Sciences of the Artificial* (3rd edition). Cambridge, Massachusetts: The MIT Press, 1996.

Simon, Julian L. *The State of Humanity*. Oxford: Blackwell Publishers Ltd, 1995.

Simon, Julian L. *The Ultimate Resource 2*. Princeton, New Jersey: Princeton University Press, 1996.

Slaughter, Richard A. *The Foresight Principle*. Article in *Futures*, Volume 22, Number 8, October 1990.

Slaughter, Richard A (ed.). *New Thinking for a New Millennium*. London: Routledge, 1996.

Slaughter, Richard A. (ed.). *The Knowledge Base of Futures Studies* (Volume 1 - Foundations, Volume 2 - Organisations, Practices, Products, Volume 3 - Directions and Outlooks). DDM Media Group, 1996.

Slaughter, Richard A. *Futures Beyond Dystopia – Creating social foresight*. London: RoutledgeFarmer, 2004.

Slee Smith, Paul I. *Think Tanks and Problem Solving*. London: Business Books Limited, 1971.

Smil, Vaclav. *Energy at the Crossroads – Global perspectives and uncertainties*. Cambridge, Massachusetts: The MIT Press, 2003.

Smith, Dan. *The State of War and Peace Atlas*. London: Penguin Books Ltd., 1997.

Smith, Dan. *The State of the World Atlas*. London: Earthscan Publications Ltd., 2003.

Spon, Davis Langdon & Everest (Eds.). *Spon's Architects' and Builders' Price Book 2002*. London: Spon Press, 2002.

Stallings, Barbara (ed.). *Global Change, Regional Response - The New International Context of Development*. Cambridge: Cambridge University Press, 1995.

Stern, Nicholas. *The Economics of Climate Change – The Stern Review*. Cambridge: Cambridge University Press, 2007.

Stiglitz, Joseph E. *Globalization and its Discontents*. London: Penguin Books Ltd, 2002.

Stiling, Peter D. *Ecology - Theories and Applications*. Upper Saddle River, New Jersey: Prentice-Hall Inc., 1996.

Tainter, Joseph A. *The Collapse of Complex Societies*. Cambridge: Cambridge University Press, 1988.

Taylor, Peter J. *World City Network – A global urban analysis*. London: Routledge, 2004.

Times, The. Barraclough, Geoffrey (Editor). *The Times Atlas of World History*. London: Times Books Limited, 1984.

Times, The. *Past Worlds - The Times Atlas of Archaeology*. London: Times Books Limited, 1988.

Toffler, Alvin. *Future Shock*. New York: Random House, 1970.

Toffler, Alvin. *The Third Wave*. London: William Collins Sons & Co Ltd, 1980.

Tonn, Bruce E. *Distant Futures and the Environment*. Article in *Futures*, Volume 34, Number 2, March 2002.

Tough, Allen. *What Future Generations Need from Us*. Article in *Futures*, Volume 25, Number 10, December 1993.

Toynbee, Arnold J. *A Study of History* (twelve volumes). Oxford: Oxford University Press, 1962.

Toynbee, Arnold J. *Mankind and Mother Earth*. Oxford: Oxford University Press, 1976.

Turekian, Karl K. *Global Environmental Change - Past, Present, and Future*. Upper Saddle River, New Jersey: Prentice-Hall Inc., 1996.

Ul Haq, Mahbub, Inge Kaul, and Isabelle Grunberg (eds.). *The Tobin Tax - Coping with Financial Volatility*. Oxford: Oxford University Press Inc., 1996.

United Nations Conference on Human Settlements. *Global Review of Human Settlements*. Oxford: Pergamon Press Ltd., 1976

United Nations Centre for Human Settlements (HABITAT). *Global Report on Human Settlements*. Oxford: Oxford University Press, 1987.

United Nations Centre for Human Settlements (HABITAT). *An Urbanising World - Global Report on Human Settlements 1996*. Oxford: Oxford University Press, 1996.

United Nations Centre for Human Settlements (HABITAT). *Cities in a Globalizing World - Global Report on Human Settlements 2001*. London: Earthscan Publications Ltd., 2001.

United Nations Human Settlements Programme (UN-HABITAT). *The Challenge of Slums - Global Report on Human Settlements 2003*. London: Earthscan, 2004.

United Nations Human Settlements Programme (UN-HABITAT). *Financing Urban Shelter - Global Report on Human Settlements 2005*. London: Earthscan, 2005.

United Nations Human Settlements Programme (UN-HABITAT). *Enhancing Urban Safety and Security - Global Report on Human Settlements 2007*. London: Earthscan, 2007.

United Nations Centre for Human Settlements (HABITAT). *The State of the World's Cities 2001*. Nairobi, Kenya: Publications Unit, UNCHS (Habitat), 2001.

United Nations Human Settlements Programme (UN-HABITAT). *The State of the World's Cities 2004/2005 – Globalisation and Urban Culture*. London: Earthscan, 2004.

United Nations Human Settlements Programme (UN-HABITAT). *The State of the World's Cities 2006/2007 – The Millennium Development Goals and Urban Sustainability*. London: Earthscan, 2006.

United Nations Conference on Trade and Development (UNCTAD). *World Investment Report - Transnational Corporations and Export Competitiveness*. Geneva: United Nations, 2002.

United Nations Department of Economic & Social Affairs. *Compendium of Human Settlement Statistics 2001*. New York: United Nations, 2001.

United Nations Department of Economic & Social Affairs. *World Population Ageing 1950-2050*. New York: United Nations, 2002.

United Nations Department of Economic & Social Affairs. *World Population to 2300*. New York: United Nations, 2004.

United Nations Department of Economic & Social Affairs. *World Urbanization Prospects - The 2005 Revision*. New York: United Nations, 2006.

United Nations Development Programme (UNDP). *Human Development Report 2007/2008*. New York: Palgrave Macmillan Ltd., 2007.

United Nations Educational, Scientific and Cultural Organisation. *The Cultural Dimension of Development - Towards a practical approach*. Paris: UNESCO Publishing, 1995.

United Nations Environment Programme (UNEP). *Global Environment Outlook 3*. London: Earthscan Publications Ltd., 2002.

United Nations Environment Programme (UNEP). *Global Environment Outlook 4*. Nairobi, Kenya: United Nations Environment Programme, 2007.

United Nations Food and Agriculture Organization (FAO). *Land Resource Potential and Constraints at Regional and Country Levels*. Rome: United Nations, 2000.

United Nations Research Institute for Social Development (UNRISD). *States of Disarray - The social effects of globalisation*. Geneva: UNRISD, 1995.

United Nations Research Institute for Social Development (UNRISD). *Social Futures – Global Visions*. Oxford: Blackwell Publishers, 1996.

Unwin, Tim (ed.). *Atlas of World Development*. Chichester, West Sussex: John Wiley & Sons Ltd, 1994.

Van Duijn, J. Jaap. *The Long Wave in Economic Life*. London: George Allen & Unwin, 1983.

Veer, Jeroen Van Der. *Shell Global Scenarios to 2025 – The future business environment: trends, trade-offs and choices*. Shell International Ltd., 2005.

Velamoor, Sesh and Paige Heydon. *Humanity 3000* - Special Issue. Articles in *Futures*, Volume 32, Number 6, August 2000.

Waldrop, Mitchell M. *Complexity - The Emerging Science at the Edge of Order and Chaos*. London: Penguin Books Ltd., 1992.

Wallace, Paul. *Agequake - Riding the Demographic Rollercoaster Shaking Business, Finance and our World*. London: Nicholas Brealey Publishing, 1999.

Ward, Peter and Donald Brownlee. *The Life and Death of Planet Earth – How the new science of astrobiology charts the ultimate fate of our world*. London: Piatkus Publishers, 2002.

Webber, Bruce H, David Depew and James Smith: *Entropy, Information and Evolution – New perspectives on physical and biological evolution*. Cambridge, Massachusetts: The MIT Press, 1988.

Weizsacker, Ernst von, Amory B Lovins and L Hunter Lovins. *Factor Four – Doubling Wealth, Halving Resource Use*. London: Earthscan Publications Ltd., 1997.

Wells, Herbert G. *Anticipations of the Reaction of Mechanical and Scientific Progress upon Human Life and Thought*. London: Harper Brothers, 1902.

Wilson, Edward O. *Sociobiology*. Cambridge, Massachusetts: The Belknap Press of Harvard University Press, 1975.

Wilson, Edward O. *The Diversity of Life*. London: Penguin Books Ltd, 1992.

World Bank. *World Development Reports*. New York: Oxford University Press Inc., 1991– 2002.

World Bank. *World Development Indicators 2002*. Washington D.C. 2002.

World Bank. *World Development Indicators 2003*. Washington D.C. 2003.

World Commission on Environment and Development, Brundtland, Gro Harlem (Chairperson). *Our Common Future*. Oxford: Oxford University Press, 1987.

World Resources Institute (WRI). *Resource Flows - The Material Basis of Industrial Economies*. Washington D.C.: World Resources Institute, 1997.

Worldwatch Institute. *State of the World Reports*. London: Earthscan Publications Ltd, 2001-2006.

Worldwatch Institute. *Vital Signs – The Trends that are Shaping our Future*. London: Earthscan Publications Ltd, 2001-2006.

Worldwatch Institute. *Biofuels for Transport – Global potential and implications for sustaiable energy and agriculture*. London: Earthscan Publications Ltd., 2007.

Zubrin, Robert. *Entering Space - Creating a Spacefaring Civilisation*. New York: Jeremy P/Tarcher Putnam, 1999.

APPENDIX 40
INDEX

Authors cited in the Bibliography are not included in this Index

A

Acceleration principle, 234, 235

Adaption, ii, 13, 15, 22, 23, 26, 29, 88, 96, 99, 105, 114, 122, 178, 184, 211, 222, 340, 516, 520

Aerospace, 67, 231, 384, 425

Afghanistan, 269, 271, 292, 337, 471, 477

Africa
    cities and urbanisation, 263-287, **270-271**, **469-470**, 476, 478, 483, 484
    civilisations, 441
    futures, 32, 355, 356, 476, 478, 487, 488, 497, 498, 502
    GRP and investment, 492-495

Agents, 73, 88, 92, **105**, 111-118, 190, 191, 237, 529, 535

Agglomeration economies, 86, 115

Agrarian stocks, 41, 129, 414, 516

Agribusiness, 144, 424, 447, 452, 538

Agropolitan stage, 125, **200-203**, 313, 341, 369, **462-467**, 491, 516
    See also civil phase transitions

Agropolitan state, 345-347

AIDS, 298, 310, 380, 381, 408, 510, 512

Air transport, 20, 64, 83, 150, 173, 179, 186, 201, 234, 239, 245, 249, 386, 421, 449, 515

Aircraft, 12, 173, 229, 231, 237, **249**, 255, 291, 325, 337, 353, 359, 386, 516

Airports, 8, 143, 149, 155, 179, 182, 239-241, 266, 291, 348, 535

Algeria, 336, 337, 470, 484

American Council for the United Nations University (ACUNU), 19, 30

Amphibians, 45, 48, 58

Anderson, Philip, 91

Angola, 336, 469

Animal, 4, 5, 13, 39, 48-62, 69, 93, 129, 229, 230, 299, 381, 406, 415, 416, 424, 511, 516
    animal protein, 215, 305, 508

Animat, 13, 229, 239, 237, 415, 516, 543

Anthropology, 39, 99, 526

ARC Center for Complex Systems, vi

Archaeological timescales, 61, 438

Archaeology, 39, 99, 526

Archetype, 229, 230, 516

Argentina, 294, 474, 485

Armaments, 312, 325

Arrow, Kenneth, 86

Artefacts, ii, iii, 3-7, 40, 65, 87, 89, 93, 96, 99, 101, 105, 129-132, 161, 516, 525, 526, 541
    civil, ii, 4, 63, 233, 415
    cultural, 4, 417
    length of life, 119, 210, 213, 320, 389, 419, 531
    material, 4, 5, 183, 201, 524
    multi-media, 193, 424

Artificial intelligence, 26, 231, 384, 423, 426, 512

Assets, 111, 113, 117, 119, 127, 159, 303, 313, 316, 323-326, 362, 370, 411, 428, 515, 518

    infrastructural, 224, 236, 239, 327, 338, 414, 505, 525

    length of life, 8, 26, 161, 215, 221, 318-322, 327, 371, 415, 506, 541

    per capita, 25, 99, 131, 132, 199, 276, 325, 326, 370, 410, 428, 522

Association of South East Asian Nations (ASEAN), 350, 501

Asteroid, 20, **44-46**, 50, 68, 69, 254, 403

Astronomy, 3, 66, 258, 526

Astrophysical metasystem. *See under* metasystem

Astro-technological civilisations, **65-69**, 303, 402, 430, 431

Asymmetric warfare, 337

Asymmetry, 92, 123, 276, 525

Atlantic conveyor, 55, 404, 515

    See also Gulf Stream

Atmosphere, 39, 44, 48, 49, 52-56, 65, 163, 243, 254-260, 383, 405, 406, 412, 506, 517, 524, 543

Atrophy, 15, 40, 93, 95, 122, 129, 226, 350, 410, 517, 520, 530, 531

    investment atrophy, 227, 374, 428, 517, 529, 533

Attractor, v, vi, 222, 225, 227, 366, 409, 517, 522, 532, 533

Australia, 47, 62, 216, 248, 249, 264, 267, 280, 289, 295, 309, 357, 473, 485, 514, 536

Automobile, *see* car

    auto dependent, 248

B

Backcasting, 23, 27, 375, 517

Backward linkages, 151, 186

Bacteria, 13, 52, 56, 69, 299, 416

Bangladesh, 206, 271, 280, 282, 287, 289, 290, 336, 471, 485

Banking, finance & securities, 115, 145, 147, 150, 153, 170, 174, 194, 264, 362, 363, 391, 425, 452, 542

Beaumol, William & Binder, Alan, 86

Beijing, 266, 272, 279, 280, 286, 287, 472, 485

Belarus, 268, 469

Belgium, 270, 343, 468, 527, 536

Bertalanffy, Ludwig von, 3, 434

BESOTO metapolitan belt, 279

Bibliography, viii, x, 546

Bifurcation, v, 23, 27, 93, 94, 100, 125, 132, 366, 517, 521, 525, 540

Billion, ix, 517

    billion years ago (BYA), ix, 43, 44, 47, 52, 53, 56

    billion years forward (BYF), ix, 69, 405

Biodiversity, 58-60, 383, 416, 425, 511, 512

Biofuels, 244, 344, 421

Biogas, 242, 421

Biological evolution. *See under* evolution

Biological hazards. *See under* hazards

Biological metasystem. *See under* metasystem

Biomes, 59, 92, 524

Biosphere, 39, 54, 69, 88, 123, 128, 129, 161, 338, 374, 381, 405-407, 414, 516, 517

Biotech wave, 394, 396, 510

Biotechnology, 231, 320, 381, 411, **417**, 510

Birds, 5, 26, 58, 59, 69, 229, 299, 415, 516, 543

Birth rate, 123, 212, 218, 309, 351, 380, 542

Borchert, John, 185

Boston University, Frederick S. Pardee Centre, iv, 19

Boulding, Kenneth, ii, iii, 82-86, 89, 116, 434, 435

Brazil, 14, 87, 241, 248, 273, 275, 280, 293, 352, 474, 485

Breheny, Michael, 164

British Commonwealth, 355, 443

Brookings Institution, iv

Brundtland Commission, 160, 318, 361, 541

Buenos Aires, 273, 286, 290, 294, 474, 485

Building cycles, **217-218**

Buildings, i, ii, 40, 41, 119, 154-158, 178, 218, 221, 275, 288, 291-295, 318, 324, 425, 528

Buses, 248, 250

Business cycle, 114, 157, 217

Business services, 141, 149, 150, 426, 450, 453, 462

C

Cairo, 141, 263, 266, 271, 286, 470, 484

Calcutta, 140, 271, 280, 290, 471, 485

California Institute for the Study of Complex Systems, vi

Calories, 303, 304, 346, 357, 369, 382, 383, 414

Cambodia, 442, 472

Cameroon, 288, 469

Canada, 45, 55, 267, 272, 295, 297, 351, 363, 473, 485, 527, 536

Canals, 83, 185, 201, 233, 264

Capital cities, 138, 266, 296, 468-475, 520

Capital formation, 160, 225, 227, 327, 517, 522, 528, 532

Capital stock, 4, 82, 119, 128-131, 161, 184, 199, 221, 235, 313, 317-326, 328, 365, **370**, 371, 428, 494, 495, 498

Car, ii, 4, 83, 163-167, 179, 186, 211, 214, 245-251, 371, 514,
    accidents, 292
    bombs, 297
    fleet or population, 239, 322, 324, 359, 371, **498-499**
    fuel consumption, 246, 461, 545
    motoring facilities, 155, 173, 229, 293, 296, 542
    ownership, 110, 111, 151, 175, 204, 233, 240, 346, 352-354, 357, 386, 397, 412, 415, 420-422, **465**, 513
    technology, 231, 232, 237, 247, 343, 387, 425, 516
    trips, 246, 249, 250, 254

Carbon, 241, 343, 405
    capture, 344, 410
    dioxide, v, 33, 34, 48, 49, 52-54, 69, 163, 241-244, 260, 405, 506
    emissions, 33, 242, 247, 343, 352-354, 357, 374, 375, 390, 404, 509
    materials, 231, 255
    off-set, 344, 410
    sequestration, 344, 410
    taxes, 164, 212, 245, 343, 387

Caribbean, 273, 288, 289, 335, 352, 443, 477, 478, 487, 488, 492-499, 501, 503

Carnot, Sadi, 89, 531

Carrying capacity, 5, 15, 41, 63, 123, 126, 132, 226, 247, 348, 510, 518, 542

Catastrophe, 8, 14, 20, 376, 512
    astrophysical, 45, 46
    eco-catastrophe, 20, 376, 397, 406, 409
    geophysical, 406
    global, 311, 394, 521
    societal, 100, 132, 162

Census, vii, 107, 112, 117, 140, 148, 149

Center for Complex Systems Research, vi

Center for Naval Analyses (CNA Corporation), vi

Central America
    cities and urbanisation, 268, 273, 473, 474, 476, 478

futures, 344, 476, 478, 487, 488, 497-499, 502

GRP and investment, 492-495

Central Asia

cities and urbanisation, 263, 268, 269, 271, 471, 476-478, 483-485

civilisations, 441

futures, 309, 344, 354, 378, 476-478, 487, 488, 497-499, 502

GRP and investment, 492-495

Central Business District (CBD), 154, 247

Central Europe,

cities and urbanisation, 268, 269, 468, 469, 476, 478, 483, 484

futures, 309, 353, 354, 358, 378, 476-478, 487, 488, 497-499

GRP and investment, 492-495

Centre for Complex Systems (Australia), vi

Chaotic dynamics, vi, 22, 23, 93, 122, 286, 366, 518, 525

Chile, 352, 425

China,

cities and urbanisation, 62, 263, 265, 267, 269, 272, 278-280, 472, 476, 478, 485

civilisations, 64, 65, 408, 442

futures, 31, 32, 239, 310, 311, 338, 343, 349, 358, 372, 378, 413, 476, 487, 488, 497-499, 502, 507, 509

GRP and investment, 241, 313, 363, 492-495

hazards, 287, 288, 292, 294, 295, 298, 336, 404

Cities,

cities by country, **468-475**

cosmopolis, 137, 138, 265, 266, 269-272, 276-279, 281, 285, 286, 386, 396, 409, 411, 468-475, 483, 508, 520, 534

megalopolis, vii, 137, 138, 141, 145, 245, 272-281, 409, 411, 412, 480, 481, 533, 534

megapolis, vii, ix, xi, 88, 113, 137-143, 147-152, 154, 158, 164-167, 175, 185, 196, 216, 221, 245, 265-271, 275-281, 285, 286, 290, 409, 411, 412, 421, 452, 461, 480, 481, 483, 533, 534

metapolis, 138, 276-281, 409, 411, 480, 481, 534

metropolis, ii, 77, 96, 137, 138, 141, 143, 150, 175, 186, 194, 210, 233, 263, 265-267, 269, 272, 275-278, 283, 285-290, 345, 363, 383, 409, 411, 452, 457, 468-475, 480, 481, 520, 533, 534

principal city, 138

provincial city, 138

regional city, 138, 185, 275, 480, 481

world cities, 141, 187, 254, 283-286, 332, 385, 483-485, 510

City Development Index (CDI), 315, 347, 523

Civil disaster, 286-288, 372, 408, 518, 528

Civil emergence, v, 96, 36

Civil engineer, 22, 97, 99, 117

Civil megatrends, 401, 410

Civil phase transitions, vi, 22, 202, 203, 366, 369, 541

See also agropolitan stage

See also distributional stage

See also ecopolitan stage

See also industrial stage

See also informational stage

See also infrastructural stage

See also traditional stage

Civil metasystem. See under metasystem

Civil system, i, viii, xii, 5, 8, 10, 18, 39, 40, 54, 74, 84, 87, 92, 93, 95-99, **116-120**, 170, 176, 191, 226, 229, 323, 331, 344, 366, 372, 398,

408, 417, 427, 431, 522, 524, 534, 541, 542

Civil war, 336, 337, 349, 372, 379, 506

Civilisation

astro-technological, 19, **65-69**, 402, 430, 431

contemporary, ii, x11, 5, 18, 20, 23, 32, 41, 62, 65, 92-96, 130, 188, 205, 209, 348, 350, 429, 431, 544

future evolution, 96, 97, 122, 124, 127-129, 183, 210, 260, 274, 304, 340, 341, 354, 355, 372, 398, 408, 414, 418, 427-431, 536

past civilisations, 4, 41, **62-65**, 99, 100, 180, 263, 264, 380, 408, **440-443**

planetary civilisation, viii, 22, 23, 41, 65, 83, 96, 99, 101, 161, 162, 204, 205, 257, 304, 342, 343, 349, 373, 375, 377, 408, 410, 428-431, 541

sustainable, 323, 326, 342, 365, 377, 428, 431

Classical economics, 81, 82, 86, 88, 510, 513

Classification

establishments, 77, 78, 126, **144-147**, 424-427, 447-451

facility types, 152-156, 159, 454-457

households, **105-112**

land-use, 154, 304, 368-369, 414, 488-489

urban, vii, 107, 138, 147-149, 156, 454-456

Clausius, Rudolf, 90, 531

Climate change, 14, 19, 20, 28, 29, 39, 44-46, 47-51, 52-56, 89, 211, 258, 310, 312, 329, 344, 372, 376, 497, 403-407, 410, 425, 510, 515

Club of Rome, 18, 19, 434

Coal, 49, 54, 122, 128, 201, 234, 241-244, 267, 270, 308, 319, 374, 384, 395, 405, 506, 545

Coevolution

establishments, 99, 114, 122, 125-127, 157, 520

systems, 39, 376, 518

technologies, 25, 182, 231, 239, 366

Coexistence, 518

Colombia, 296, 337, 352, 475, 485

Colonialism, 331, 340, 418, 519

Commonwealth Scientific and Industrial Research Organisation (CSIRO), iv

Comparative advantage, 115, 200

Competition

business, 101, 125, 126, 170, 187, 188, 191, 192, 210, 213, 519, 535, 539

cities, 25, 116, 164, 224

international, 31, 234, 236, 237, 247, 356, 362, 373, 391

technological, 27, 238, 239, 254

Complex adaptive system, v, vi, 1, 10-13, 18, 22-27, 76, 77, 82, 92-98, 112-123, 127-132, 160, 181, 222-227, 366, 409, 427, 435, 517-519, 532, 533, 543

Complex Systems Lab, iv

Complexity

civil phase transitions, 199-204

complexity science, i-xii, 26, 57, 73, 76-78, 89-98, 141, 435, 519-520

diffusive ecostructures, v, 115-120, 128, 141, 156, 160,

dynamic investment flows, 217-223

macrolaws of ecodynamics, 223-227

transactional complexity, 169, 170, 174, 180-185, 187, 225, 227

Computers, 230, 231, 234, 251-254, 359, 422-424, 426, 448, 467

Congestion, 142, 152, 163-167, 179, 250-254, 266, 357, 385

Congo Democratic Republic, 278, 469

Connectivity, 5, 84, 120, 132, 141, 142, 173, 174, 181-184, 209, 222-224, 227, 233, 386, 409, 520, 528, 530, 532, 533, 535, 537

coefficient of connectivity, 117, 226, 227, 518, 533, 543

Conservation, 8, 59, 60, 161-163, 215, 244, 307, 365, 377, 383, 384, 389, 390, 416, 506, 515, 541

Consolidated Metropolitan Statistical Area (CMSA), 139, 142

Construction, 14, 63, **144-147**, 150, 156, 162, 178, 182, 199, **217-219**, 234, 238, 239, 267, 275, 306, 307, 324, 325, 370, 345, 405, 425, 447, 452, 491, 509, 540, 542

Consumer, 7, 99, 101, **105-112**, 118, 145, 147, 175, 184, 188, 191, 204, 210-219, 229, 253, 325, 346, 349, 359, 374, 385, 389, 418, 420, 449, 452, 513, 525, 542

Contextual macrostructure, 9, 13, 22, 26, 40, 144, 301, 342, 410, 523

Continental drift, 49, 51, 80, 81

Continental integration, 170, 239, 276, 341, 347-350, 365, 376, 379, 408, 413

Continental regions, 268, 476, 478, 487, 488, 492, 494, 497, 498, 502

Continental Union, 64, 130, 184, 204, 224, 276, 337-341, 347-350, 372-379, 392, 413, 426, 431, 504, 520

Conurbation, 269, 270, 272

Conversion of energy units, 545

Corporations, 4, 10, 21, 100, 101, 113, 114, 118, 144-152, 156-157, 184, 187-188, 212-216, 229-232, 390-392, 513

See also transnational corporations

Cosmopolis. See under cities

Cote d'Ivoire, 270, 469

Countries, 199-208, 268-273, 287-299, 311-317, 341-345, 413, 527

See also G-7, G-8, G-10

See also OECD

See also LDCs

See also MDCs

Cranfield Complex Systems Management Centre, vi

Crime, 20, 28, 166, 184, 216, 286, 295, 317, 352-357, 388, 426, 509, 528

Cristaller, Walter, 186

Croll, James, 80

Cuba, 473

Cultural diversity, 351, 355, 376, 379, 394, 418, 431, 507

Cultural sites, 451, 453

Culture, 4-6, 28, 63, 97-100, 117, 131, 161, 176, 191, 285, 340, 341, 347, 387, 394-397, 417, 418, 427, 431, 507, 513, 521, 530, 544

Cyberspace, 70, 431

Cyclic attractor. See attractor

Czech Republic, 290, 468, 536

D

Darwin, Charles, ii, 27, 94, 96

Death rate. See mortality rates

Debt, 119, 182, 211, 266, 327-331, 353, 355, 373, 390, 293, 508, 518

Decarbonisation of energy, 241, 343, 365, 371, 384, 394-397, 405, 414, 506, 509, 515

Decomposition, 127, 129, 263, 366, 521

Deep future. See under future

Defence, 6, 12, 63, 169, 179, 184, 233, **254-257**, 312, **335-336**, 340, 422, 426, 451, 515, 529

Deforestation, 383, 425, 510

Degradation,

environmental, 32, 360, 383, 511
    human values, 224

Dejouling of energy, 242, 318, 365, 421

Delhi, 266, 271, 286, 471, 485

Dematerialisation, 34, 204, 322, 365, 384, 419,
    424, 506, 521

Democracy, 28, 83, 169, 251, 345, 419, 536

Democratic reform, 413, 426

Demographic trends, 376, 380

Density,
    urban, 138-141, 164, 167, 248, 279-
    283, 387, 388, 395-397, 422, 430, 468-
    475, 480, 505, 508, 511, 514, 518, 533,
    534
    population, 351-357, 368, 537

Denver, University of, iv, 19, 31

Deprivation, 163, 166, 216, 227, 287, 316, 533

Deregulation, 127, 254, 373

Desalination, 305, 306, 372, 382, 425, 506, 515

Desertification, 583, 510

Designers, consultants & technical services, 98,
    450, 453, 521

Determinism, 521

Deterrence, 130, 337, 376-379, 426

Development,
    economic, 6, 11, 16, 17, 25-28, 59, 82,
    87, 97, 119, 120, 139, 144, 146, 179, 185,
    191, 194, **199-205**, 212, 217-219, 223,
    224, 233-235, 251, 281, 313, 319, 323,
    339, 377, 406, 415, 420, 504, 522. 523,
    526, 534
    property, 175, 184, 194, 219, 332
    sustainable, 19, 160, 161, 310, 314, 328,
    348, 360, 373, 393, 396, 410, 411, 430,
    510, 530, 542
    technological, viii, 4, 6, 15, 63, 127, 188,
    191, 200, 220, 232, 237, 303, 307, 398,
    406

urban, 59, 115, 116, 126, 163, 178, 179,
    183, 218, 219, 234, 274, 320, 386, 397,
    435, 510, 511

Developmentalism, 98, 521

Devolution, 127, 170, 339, 341, 408

Dhaka, 271, 280, 286, 287, 471, 485

Diffusion of investment, 95-97, 118, 122-124,
    132, 142, 160, 224-227, 408, 409, 517,
    520, 522, 525, 532, 533

Diffusive ecostructures. *See under*
    complexity

Diminishing returns, 40, 63, 65, 85, 93, 100, 130,
    160, 183, 189, 190, 220, 224, 226, 264,
    312, 331, 332, 348, 361, 374, 385, 407,
    409, 410, 531, 539

Disarmament, 258

Discontinuity, 76, 94, 181, 521. *See also*
    bifurcation

Disease, 16, 129, 286, 298, 299, 304, 310, 316,
    372, 381, 382, 407, 408, 416, 427, 511,
    512

Disorder, 90, 94, 130, 162, 350, 381, 506, 512,
    529, 531, 537

Dispersal, 122, 127, 128, 130, 160, 173, 200, 232,
    233, 245, 246, 337, 345, 386, 421, 520,
    522

Dissipative structures, 24, 90-92, 95, 117, 226,
    537

Distribution & logistics, 425

Distributional stage, 107, 142, 175, 202, 203,
    205, 239, 313, 323
    See also civil phase transitions

Diversity. *See also* biodiversity
    cultural, 340, 351, 355, 376, 379, 394,
    418, 419, 431, 507
    establishments, 77, 124, 125-127, 132,
    142, 157, 180, 186, 519

territorial, 12, 77, 130, 131, 340, 367, 373, 409, 504

Doxiadis, C.A., 24, 139, 434

Drake, Frank, 65, 67, 258, 259, 430

Drugs, 60, 231, 295, 379, 408, 416, 427

Dynamics,

    chaotic dynamics, 22, 23, 93, 122, 286, 366, 518, 525

    dynamic models, 12, 14, 17, 23, 75, 78, 84, 86, 89, 132, 192

    dynamic stability, 8, 24, 41, 78, 97, 123, 125, 131, 162, 205, 227, 276, 377, 410, 414, 427, 517, 523, 533

    network dynamics, 132, 174, 175, 181-188, 535

    nonlinear dynamics, 12, 22, 59, 78, 87-93, 97, 237, 536, 537

    system dynamics, 12, 13, 17, 18, 86, 96, 101, 120, 132, 434

E

Earth Summit, 314

East Asia and Pacific,

    cities and urbanisation, 264, 268, 272, 273, 279, 280, 286, 472, 473, 476, 478, 483-485

    civilisations, 442

    futures, 246, 344, 357, 476, 478, 487, 488, 497, 498, 499, 502

    GRP and investment, 492-495

Eastern Europe,

    cities and urbanisation, 268, 270, 469, 476, 478, 483, 484

    civilisations, 440

    futures, 344, 353-355, 358, 380, 383, 476, 478, 487, 488, 498, 499, 502

    GRP and investment, 492-495

Ecodiversity. *See* biodiversity

Ecodynamics, 4-7, 25, 40, 83, 87-89, 95-97, 105-132, 144, 221, 223-227, 408, 435, 519, 522

    macrolaws, 532-533

Ecologic transition, 161, 304, 313, 319, 410, 411, 414

Ecological footprint, 162, 522

Economic development. *See under* development

Economic equilibrium, 31, 81-88, 95, 115, 190, 505, 513, 523

Economy, 28, 40, 76, 84, 87, 88, 95, 115, 117, 120, 145, 153, 211, 222, 419, 523, 538-540, 542

    global economy, 25, 147, 151, 171, 175, 184, 205, 209, 214, 234, 276, 311-313, 326, 332, 358, 361, 362, 373, 377, 390-393, 510, 513, 514

    See also world economy

    hydrogen-fuel economy, 256, 308, 412, 506

    informal economy, 216, 225, 353, 388, 390, 509

    informational economy, 145-148, 157-159, 160, 185, 210, 224, 254, 387, 188-196

Eco-permits, 343, 344

Ecopolitan Index (EI), 347, 523

Ecopolitan stage, 200, 203, 204, 214, 216, 322, 343, 398

    See also civil phase transitions,

Ecopolitan State, 13, 26, 77, 132, 162, 281, 338-350, 367, 374, 378, 379, 386, 410, 413, 414, 419, 426, 500, 504, 516, 518, 523, 537

Ecostructure, 137-167, 523-524

    diffusive ecostructures. *See under* complexity

Ecosystem, 28, 33, 39-41, 60, 76, 84, 85, 92-95, 113, 225, 319, 406, 489, 511, 517-520, 522, 524, 527, 530, 536, 541

Ecotaxation, 210, 211, 224, 343, 374, 390, 419, 509, 524

Ecuador, 296, 475

Ecumenopolis, 24, 116, 434

Education, 6, 7, 12, 28, 35, 106, 109, 110, 119, **145-156**, 170-177, 184, 185, 191, 194, 206, 218, 234, 240, 314, 315, 324, 341, 345-347, 363, 370, 389, 390, 419, 451, 453-456, 464, 514, 539, 540

Egypt, 62, 65, 271, 290, 297, 408, 441, 470, 484

Einstein, Albert, 67

Electricity distribution, 83, 201, 229, 234

Electronics, 187, 231-234, 252, 359, 391, 420, 421

Emergence, 26, 78, 90-98, 115, 121, 123, 138, 144, 238, 253, 254, 264, 341, 366, 369, 408-410, 423, 430, 520, 522-524, 526, 540

Emissions. *See under* carbon

Empire, 64, 65, 169, 263, 264, 267, 342, 348, 349, 408, 428, 440-443

Employment, 28, 78, 106, **109-112**, 118, 125, 131, **145-154**, 157, 158, 162, 163, 166, 179, 184-188, **194-196**, 201, 208, **211-221**, 224, 233, 245, 252, 312, 330, 331, 339, 346, 348, 350, 354, **357-363**, 377, 387-393, 419, 421, 425, 450, 502, 505, 508, 522

Endogenous growth, 86, 95, 366, 541

Energy,
geothermal, 242, 319, 421
hydrocarbon, 164, 246, 344, 374, 383, 393, 397, 411, 412, 421, 424, 515
See also coal, oil, and gas
hydroelectric, 319, 320
hydrogen, 164, 167, 188, 232, 241-245, 247, 249, 256, 258, 308, 343, 344, 371,
383, 384, 386, 405, 411, 412, 421-424, 506, 512
nuclear, 20, 173, 232, 234, 241-244, 257, 292, 293, 319, 320, 357, 374, 383, 384, 396, 395, 405, 408, 412, 421, 422, 424, 506, 509
renewable, 28, 30, 33, 34, 129, 161, 211, 242-244, 274, 318-324, 330, 343, 344, 360, 365, 371, 374, 383, 384, 396, 397, 403, 412, 415, 429, 509, 512, 539, 541
solar, 53, 59, 69, 94, 129, 131, 242-244, 320, 383, 403, 406, 412, 414, 420, 421, 424, 512
tidal, 242-244, 320, 403, 412, 421, 424, 512
wind, 242, 243, 289, 320, 412, 421, 424, 512

Energy conversion units, 545

Energy efficiency, 164, 180, 318, 319, 322, 344, 371, 415, 421, 498, 519

Energy use, 35, 54, 132, 212, 215, 218, 307, 320, 322-325, 343, 371, 375, 429

Entity, 76, 112, 192, 193, 516, 525, 539,

Entropy, 90, 93, 531

Environment,
built, 10, 20, 27, 97, 98, 105, 112-117, 122, 126, 144, 152, 164, 167, 178, 223, 286, 372, 386, 504, 528
business, 21, 178, 187, 192, 193, 205
natural, 3, 9, 10, 16, 19, 28, 29, 32-34, 40, 48, 51, 57-60, 92, 93, 97, 100, 118, 123, 128, 160-163, 211, 213-215, 226, 233-237, 240-244, 255, 274, 303, **312-324, 338-352**, 356, 360-365, 371-375, 377, 381-389, 394, 405-410, 415, 417, 429, 435, 506-512, 520-524, 541-543

Environmental,
impact, 317-324, 352, 371, 429, 522

memory, 53

protection, 292, 317, 318, 339, 389, 408

Epidemic, 184, 286, 287, 298, 310, 372, 380, 408, 512

Equality, 7, 16, 161, 171, 224, 540, 541

Equilibrium. *See also* economic equilibrium,

ecosystem, 85, 92-94, 120, 181, 526, 531, 536, 538

population, 93, 129, 132

thermodynamic, 90, 93, 531

Equipollence, 276, 366, 368, 409, 525

Equity,

intergenerational, 160, 530

social, 7, 32, 210, 320, 321, 338, 341, 342, 345-347, 349, 365, 394, 419, 523, 540

Establishment, 3, 4, 13, 15, 22, 26, 40, 41, 76-78, 82, 84, 87, 88, 92, 93, 525

civil emergence, 98-101

complex adaptive system, 112-127,

location, 156-167

planning standards, 152-156

sectoral classification, 144-152

Ethiopia, 271, 335, 336, 441, 469, 484

Ethnic, 5, 64, 109, 172, 184, 335, 341, 373, 379, 413, 418

European Union (EU), 140, 151, 163, 336, 341, 349, 350, 353-355, 501, 507

EU Emission Trading Scheme, 343

Evolution,

biological, 5, 7, 22, 24, 26, 40, 41, 53, 57, 58, 92-94, 121, 381, 407, 416, 430, 526, 530, 531, 535, 541, 544

societal, 3, 4, 6, 7, 10, 18, 23, 24, 27, 40, 58, 61, 76, 89, 94, 99, 129, 174, 190, 191, 209, 210, 229, 339, 406, 411, 417, 429, 435, 536, 539

technological, 8, 25, 27, 29, 58, 84, 123, 124, 173, 181-183, 200, 219, 229, 230, 232, 239, 241, 246, 252

urban, 23, 24, 26, 29, 41, 78, **84-87, 95-99**, 101, **115-119, 121-128**, 131, 142-145, 152, 163, 173, 180-186, 218, 224-227, 233, 264, 266, 311, 323, 342, 343, 349, 365-367, 393, 398, 401, 408, 413, 427, 428, 431, 522, 525, 526, 530, 541

Evolutionary landscape, 12, 22, 77, 101, 367, 393, 431, 526

Evolutionary science, 39, 91, 526

Exergy, 60, 90, 94, 95, 225, 406, 526, 528, 531, 541

Exogenous change, 88, 115, 524

Exploratory scenarios. *See under* scenarios

Exports, 147, 505

Externalities, 526

Extinctions,

artefacts, 418

ecological, 42, 45, 48-50, 56-60, 68, 69, 92, 403, 406, 407, 415, 526

establishments, 76, 114, 120-123, 127, 222, 229, 525, 542

Extraterrestrial intelligence, 68, 70, 258, 259, 402

F

Famine, 100, 336, 372, 408, 510

Far-from-equilibrium, 22, 42, 90, 95, 97, 120, 427, 517, 529, 537, 540

Farm, 62, 87, 162, 243, 299, 312, 417

Feedback, 3, 12, 17, 18, 30, 84, 86, 91, 112, 114, 120, 127, 186, 231, 237, 519, 526, 536, 543

Fertilisers, 216, 305, 382, 506

Fertility rates, 29, 30, 110, 129, 308-311, 345, 367, 381, 411, 419, 504, 507, 510, 513

Fibre optic, 231, 254, 284

Fish, 26, 58, 229, 258, 516, 544

Fishing, 61, 146, 251, 383, 425, 510

Floods, 14, 20, 125, 286, 289, 372, 407, 408, 528

Florida Center for Complex Systems and Brain Sciences, vi

Food & Agriculture Organisation (FAO), 304, 369, 414, 489

Food
    chain, 205, 382, 508, 511
    demand, 28, 106, 396, 397, 406
    prices, 162, 305, 511
    production, 63, 200, 202, 263, 305, 320, 355, 381, 382, 447, 448, 506
    resources, 62, 93, 117, 129, 226, 320, 338, 383, 406, 417, 420, 511
    safety, 178, 392
    supply, 15, 18, 118, 131, 177, 263, 305, 345, 377, 382, 383, 507, 531

Forecast, 8, **10-16**, 18, 20, 21, 41, 117, 195, 249, 257, 309, 434, 527, 538

Foreign exchange transactions, 328, 329, 362, 392, 413, 505

Foresight, 10, 23, 94, 191, 527

Forestry, 59, 61, 146, 258, 267, 281, 282, 304, 344, 365, 368, 381-383, 395-397, 410, 415, 416, 425, 488, 489, 506, 509-512, 514, 516

Fossil fuels, 28, 29, 41, 53, 54, 163, 177, 211, 241-244, 274, 308, 310, 320, 344, 374, 384, 395, 396, 405, 414, 421, 505, 506, 509, 511, 515
    hydrocarbon. *See under* energy
    See also coal
    See also oil
    See also gas

Fractional orbital flight, 257, 386, 411, 412, 514
    fractional orbital missiles, 422

France, 110, 270, 363, 378, 468, 484, 527, 536

Free trade, 337, 349, 350, 362, 363, 392, 501, 513

Freight, 183, 249, 251, 362, 449

Fuel,
    See biofuels
    See energy, hydrogen
    See energy, renewable non-fossil
    See fossil fuels
    See methanol
    See petroleum

Fuel wood, 241, 242, 319, 405, 421

Future,
    deep, 20, 41, 68, 401, 445
    distant, 445
    intermediate, 401, 413, 445
    long-range, 22, 26, 365-398, 401, 410-414, 445, 504-515
    near, 11, 16, 278, 375, 376, 401, 424, 538
    remote, 445

Future global energy use, 322-324, 371, 498, 499

Future Gross World Product, 321, 323, 371, 428, 429, 492, 493, 505, 508, 511, 514

Future investment stock, 326, 370, 428, 494, 495

Future land use, 282, 304, 368, 369, 382, 414, 480, 481, 488, 489, 505, 506, 508, 511, 514

Future material flows, 322, 491, 497

Future number of cities, 278, 282, 478, 479, 480, 481

Future urban population, 278, 476, 477

Future workforce, 215, 359, 360, 487, 502

Future world car fleet, 322, 324, 371, 498, 499

Futures research, vii, 8-13, 17, 22, 27, 28, 435

Futures studies, vii, 3, 9, 10, 89, 99, 401, 435

G

G-7, G-8, G-10 countries, 378, 527

Galactic era, 68, 445

Galaxies, 39, 42, 43, 68-70, 91, 401, 402

Gas, 54, 122, 128, 164, 174, 186, 234, 241-245, 258, 293, 295, 308, 319, 344, 384, 397, 405, 422, 447, 512, 545

Generica, 113, 369, 394, 397, 513

Genetic, 4, 5, 6, 26, 60, 92, 105, 113, 131, 161, 191, 407, 416, 418, 429, 525, 527, 529, 535, 536, 541

Genetic engineering, 40, 129, 188, 229, 231, 260, 305, 381, 382, 411, 416, 417, 424, 427, 431, 511, 512, 516

Genocide, 297, 335, 379

Genome, 40, 226, 416

Genotype, 92, 527, 537

Genus Intelligens, 40, 417, 430, 445

Geodemographic, 107, 108, 427, 435

Geo-engineering, 344, 345, 410

Geological timescales, 22, 40, 47, 61, 91, 418, 438, 439

Geophysical hazards. *See under* hazards

Geophysical metasystem. *See under* metasystem

Geopolitical, 13, 77, 101, 130, 247, 303, 335-342, 363, 367, 373-379, 396, 431, 504, 507, 513, 517, 519, 532, 539
potential, 127-132, 276, 340-343, 349, 361, 367, 394, 408-410, 519, 520

Geothermal energy. *See under* energy

Gerard, Ralph, 3, 434

Germany, 110, 270, 290, 294, 354, 363, 378, 468, 484, 527, 536

Ghana, 280, 441, 470

Global macrosystems, 8, 22, 40, 78, 170, 192, 366, 372-398, 401, 413, 417, 518, 525

Global macrotrends, 401, 413-424

Global Scenario Group (GSG), iv, 19, 32, 398

Global warming, 48, 54, 55, 162-164, 242, 290, 291, 312, 344, 383, 395-397, 410, 506, 509, 512, 515

Globalisation, 30, 34, 141, 170, 171, 236, 254, 311, 312, 341, 358-364, 377, 379, 392-396, 408, 410, 413, 418, 426, 507, 511

Glossary of terms, 516-544

Gould, Stephen Jay, 94

Government, 11, 15, 16, 30, 33, 78, 83, 112, 145-148, 150, **169-175**, 184, 207, 210, 211, 229, 233, **327-330**, 339-343, 376, 378, 383, 388, 390, 393, 418, 426, 428, 451, 453, 508, 512, 527, 539

Gradients,
energetic, 53, 60, 94, 520, 526, 527
informational, 92, 115
return on investment, 122, 131, 156, 160, 225, 227, 366, 410, 522, 528, 530, 532
socio-economic, 183

Greece, 169, 268, 287, 469, 477, 536

Greenhouse effect, 48, 49, 52-55, 163, 260, 406

Gross Domestic Product (GDP), 337, 521, 528

Gross National Income (GNI), 83, 194, 202, 312, 369, 528, 538,

Gross Regional Product (GRP), 28, 313, 325, 327, 330, 359-361, 365, 398, 492-495, 498, 499, 502, 503

Gross World Product (GWP), 28, 30-35, 54, 203, 241, 242, 260, 321, 323-324, 329, 351, 358, 360, 371, 390-398, 417, 428, 429, 505, 508, 511, 514

Growth rate,
demographic, 274, 309, 310, 353, 356
economic, 30, 33, 200-204
technological development, 15, 238
urban, 126, 141, 278, 368

Guatemala, 287, 473

Gulf Stream, 55, 404, 515

See also Atlantic Conveyor

H

Habitat destruction, 59

Half-life, 243, 421, 528

Hardware, 10, 99

Hawaii Research Center for Futures Studies, iv,

Hazard,

  biological, 287, 298, 299, 366, 372, 397, 406, 408, 512

  geophysical, 20, 29, 286-291, 316, 319, 338, 366, 372, 383, 396, 404, 405, 509, 515, 528

  social, 20, 286, 287, 293-299, 372, 528

  technological, 20, 29, 162, 286, 291-293, 366, 372, 384, 395, 506, 528

  See also civil disaster

Health care, 118, 309, 314, 381, 388, 420

Hierarchical trajectory, **275-279**, 280, 281, 366

Hierarchy,

  encompassing sets, 9, 98, 121, 524, 528

  rank order, 107, 275, 313, 529

  social, 107, 209

  spatial, 87, 91, 122, 123, 132, 137, 186, 275, 525, 528

  technological, 230

  See also urban hierarchy

High-tech, 112, 151, 201, 231, 351, 384, 385, 425, 426, 509, 511

Highways, 173-175, 182, 229, 266, 283, 292, 535, 542

Homo Erectus, 57, 61, 444

Homo Sapiens, 57, 61, 444

Homo Sapiens Sapiens, 5, 41, 57, 61, 430

Hospitals, 8, 83, 147-150, 154, 212, 240, 314, 346, 369, 454-457, 464

Hotels, 113, 147-150, 153, 154, 284, 293, 294, 297, 324, 369, 426, 449, 454-457, 513

Household,

  behaviour, 107, 111, 112, 172, 177, 188, 192, 193, 204, 218, 229, 420

  classification, 105-112, 145, 214

  income, 35, 83, **106-110**, 118, 119, 193, **213-216**, 243, 312, 315, 330, 351-357, 388, 389, 419, 541

  movement, 116, 118, 225, 226, 409, 518, 522, 530

  size, 106, 139, 141, 281, 369, 396, 418, 427, 508, 534

  transacting entity, 3, 26, 76, 87, **88**, 98, **118-123**, 160, 175, 177, 189, 190, 227, 253, 307, 423, 427, 516, 519, 522-525, 528, 532, 543

  unit of society, 23, 40, 88, 92, 99, 105-112, 116, 117, 147, 170, 193, 367, 417, 524, 526, 529

Housing,

  costs, 324, 325, 357, 370

  finance, 79, 118, 157, 175, 234, 235, 327, 328, 339, 530

  quality, 163, 295, 314, 315, 324, 328, 388, 389

  planning standards, 240, 324, 325, 357, 369, 463

  stock, 109, 119, 218, 221, 266, 315, 325, 510

Houston, University of, iv,

Hudson Institute, 17

Hughes, Barry, 19, 31, 435

Human agents, 88, 105, 111, 118, 529, 535

Human Development Index (HDI), 314, 315, 507

Human rights, 169, 171, 335, 338, 349, 377, 389, 513

Humankind, 4-6, 40, 41, 63, 64, 92, 96, 128, 169, 191, 209, 342, 344, 349, 381, 383, 398, 405, 410, 428-431, 521, 536

Hungary, 268, 343, 468, 477, 536

Hybrid electric vehicles (HEVs), 247, 412

Hydrocarbon. *See under* energy

Hydrogen. *See under* energy

Hydrogen-fuel economy. *See under* economy

Hydropower. *See under* energy, hydroelectric

I

Ice age, 44, 47-50, 53, 57, 59, 68, 80, 81, 402-404, 434

Illinois Center for Complex Systems Research, vi

Image, 5, 6 16, 27, 75, 94, 193, 205, 210, 215, 253, 423, 424, 540

Immigration, 116, 195, 206, 311, 338, 348, 351, 379, 380, 409

Imports, 213, 348, 391, 393

Income,
  disparity, 32-34, 216, 316, 317, 350-357, 389, 541
  lifetime income, 210, 213, 215, 369, 389, 419, 427
  per capita, 28, 30-33, 119, 154, 158, 161, 312, 321, 323, 351, 353, 369, 393-397, 414, 505, 508, 511, 514
  redistribution, 33, 211, 295, 331, 352, 361, 390
  tax, 7, 211, 316, 317, 390, 524
  See also gross national income
  See also household income

India,
  cities and urbanisation, 87, 269, 271, 275, 278, 280, 471, 476, 478, 485
  civilisations, 64, 65, 380, 408, 442
  futures, 343, 356, 372, 374, 413, 476, 478, 487, 488, 497-499, 502

GRP and investment, 313, 338, 492-495

hazards, 287-290, 292, 294, 337

Indigenous, 161, 340, 347, 362, 418

Individuals, 6, 23, 92, 94, 96, 191, 530, 537, 539

Indonesia, 62, 206, 241, 282, 288, 297, 340, 442, 472, 473, 485

Industrial sites, 146, 147, 150, 157, 159, 448, 452, 454-457

Industrial stage. 201, 203-205, 250, 251, 387
  See also civil phase transitions

Inequality, 7, 119, 160, 216, 314-317, 352, 363, 373, 389, 390, 418, 541

Infant mortality, 274, 309, 314, 315, 346, 368

Infoglut. *See* information overload

Informal economy. *See under* economy

Information,
  overload, 183, 192, 193, 423
  network, 130, 141, 187
  services, 119, 141, 146, 150, 152
  society, 144, 146, 193
  superhighway, 253
  system, 13, 24, 60, 88, 114, 147, 149, 170, 172, 176, 210, 239, 332, 425, 429
  technology, 12, 20, 145, 151, 167, 172, 179, 201, 234, 250-253, 312, 340, 384, 418, 422, 423, 426, 539
  See also quaternary information division

Informational economy. *See under* economy

Informational stage, 18, 142, 202-204, 309, 314, 315
  See also civil phase transitions

Infrastructural stage, 201-203, 250, 345, 504
  See also civil phase transitions

Infrastructure,
  investment, 27, 83, 117, 132, 142, 156, 200, 211, 214, **217-225, 233-241**, 246,

249, 266, 309, 313, 316, **324-332**, 345-348, 370, 373, 385, 392-395, 413, 504, 505, 531, 533

civil, 25, 119, 199, 200, 235, 240, 266, 504, 529, 530

See also irrigation, utilities, highways, railways, ports, airports, energy, telecommunications

industrial, 211, 331, 348

See also industrial sites, warehousing

social, 119, 179, 199, 200, 234, 235, 266, 529, 530

See also housing, hospitals, schools, universities, public service facilities, offices, retail, hotels, recreational amenities

Innovation, 4, 27, 83, 101, 112, 127, 129, 146, 191, 192, 199, 200, 217, 230, 231, 234, 237, 238, 250, 320, 361, 374, 384

Insects, 5, 58-60, 415

Institut des Systemes Complexes Rhone Alpes, vi

Institute for Alternative Futures, iv

Institute of International Economics, 54

Institutions, 4, 5, 13, 32, 99, 127, 161, 169-176, 178, 193, 346, 355, 356, 377-379, 409, 413, 417, 418, 426, 448, 532, 544

Insurance companies & brokers, 113, 369, 449, 452

Interdependence, 86, 181, 311, 410

Interest rates, 157, 219, 220, 328, 331, 346, 393

Intergalactic era, 68-70, 402, 445

Intergenerational equity, 160, 530

Interglacial period, 48-50, 55, 81, 404

Intergovernmental Panel on Climate Change (IPCC), 19, 29, 32, 54, 89, 376, 393, 398, 510

Intermediate future. *See under* future

International economic convergence, 29, 364, 389

International Energy Agency (IEA), iv, 322, 371, 421

International Futures (IF) Global Simulation Model, ix, 19, 31, 34, 89, 435

International Futures Programme (OECD), iv. *See also* OECD

International Institute for Applied Systems Analysis (IIASA), iv, vi, 19, 241, 504

International Monetary Fund (IMF), 313, 331, 513

International Society for Systems Science, 3

Internet, 108, 172, 174, 188, 193, 216, 252, 253, 422-427, 544

Interplanetary, 68, 69, 259, 402, 431, 445

Interregional, 210, 348

Interurban, 185, 186, 254

Intervention point, 27, 101, 223, 374

Investment,

atrophy, 227, 374, 428, 517, 529, 533

capital, 18, 25, 28, 63, 75, 86, 95-97, 100, 116-119, 122, 131, 132, 180, 191, 234, 236, 320, 365, 517, 530, 532, 541

capture, 114, 115, 119, 131, 156, 192, 427, 516, 519

diffusion, 122-124, 132, 142, 186, 224-227, 408-410, 413, 517-523, 532, 533, 537

flows, 26, 96, 116, 122, 124, 131, 191, 202, **217-222**, 276, 283, 284, 520, 522, 523, 525, 533, 539

hypertrophy, 374, 529

returns, 88, 131, 156, 219, 220, **224-227**, 234, 237, 312, **331-332**, 348, 361, 392, 410, 528, 530, 532

stock, 325-327, 330, 370, 428, 429

transfers, 201, 323, 330, 345, 348, 355, 358, 395, 504, 505

wave, 16, 21, 25, 27, 114, 122, 184, 211, 219, 220, **233-237**, 331, 393, 520

See also infrastructure investment

See also property investment

Investment Property Databank (IPD), xi, 158

Iowa Complex Adaptive Systems Group, vi,

Iran, 62, 269, 271, 278, 287-288, 337, 378, 471, 477, 484

Iraq, 62, 297, 336, 337, 470, 485

Irreversible, 4, 5, 8, 22, 60, 76, 82, 94, 97, 128, 141, 225, 226, 386, 522, 523, 526, 530, 532, 541

Irrigation, 63, 99, 179, 183, 233, 235, 240, 267, 304-306, 345, 372, 382, 424, 489, 506, 512, 514, 515

Isard, Walter, 115

Islam, 206, 341, 349, 354, 355, 377, 418, 441, 507, 515

Italy, 110, 140, 208, 270, 287, 288, 290, 297, 329, 468, 484, 527, 536

J

Jakarta, 272, 286, 287, 472, 485

Japan,
cities and urbanisation, 269, 272, 280, 473, 476, 478, 485
civilisations, 442
futures, 357, 358, 476, 478, 487, 488, 497-499, 502
GRP and investment, 492-495
hazards, 207, 288. 292, 293, 299

Jeffreys, Sir Harold, 80

Jobs, 216, 220, 251, 252, 317, 349, 361, 363, 388, 422, 513

Justice, 6, 155, 169, 210, 347, 354, 389, 428, 451, 454, 455

Just-in-time, 180, 251, 386, 425

K

Kahn, Herman, 17, 202, 434

Karachi, 271, 286, 290, 471, 485

Kazakhstan, 269, 471, 477

Kenya, 271, 337, 470, 484

Keynes, John Maynard, 312

Knowledge, 4, 5, 9, 10, 58, 84, 86, 96, 117, 120, 128-130, 151, 172, 181, 183, 189-196, 408, 435, 530, 537

Kondratiev long-waves, 217

Kuhn, Thomas, 80, 87

Kuznet cycles, 217

Kyoto Protocol, 343

L

Land degradation, 383, 510, 511
See also degradation

Landscape,
artificial, 10, 78, 96
bounded, 341, 523
evolutionary, 12, 22, 365, 367, 393, 401, 538

Land-use, 24, 28, 29, 113, 148, 154, 176, 178, 211, 245, 365, 368, 414, 434, 488, 505, 508, 511, 514
See also classification

Language diversity, 340, 341, 418

Larger Urban Zone (LUZ), 140

Latin America,
cities and urbanisation, 269, 273, 274, 474, 475, 477, 478, 483, 485
civilisations, 442
futures, 246, 311, 344, 349, 351-353, 377, 477, 478, 487, 488, 497-499, 502
GRP and investment, 492-495

Law of Atrophy, 15, 40, 93, 95, 226, 517, 520, 530, 531

Less developed countries (LDCs),
    asset stock, 119, 131, 214, 233, 247,
    248, 315, 326-328, 415, 494, 505
    city populations and densities, 139-
    141, 266, 273-280, 286, 327, 410, 411,
    479, 533
    corporate location, 157, 254, 391
    demographic transition, 233, 308,
    313, 319, 330, 411, 414, 415
    development assistance and FDI, 327,
    328, 331, 332, 348, 349, 361, 362, 388-
    392, 413, 513, 514
    investment transfers, 201, 330, 345,
    348, 371, 374, 392, 395, 504
    military expenditure and wars, 335-
    337, 350, 372, 375, 379, 408, 426
    path dependence, 25, 233, 234, 350
    planning standards, 240, 304, 315,
    326, 365, 414, 428, 463-467
    population growth, 309, 380
    share of GWP, 321, 323, 358, 393, 396,
    397, 505, 508, 511, 514
    workforce, 206, 423, 487, 502
Leverage, 8, 329, 374
Liberal, 112, 171, 175, 236, 373, 504
Life cycle,
    artefacts, 118, 129, 175, 531
    civilisation, 101, 273, 274, 307, 428,
    429, 536
    corporate, 101, 127, 156, 192, 231
    ecological, 92, 114, 129
    establishments, 76, 117, 118, 175
    human, 41, 92, 129, 175, 177, 207, 215,
    429, 531, 536
    infrastructure, 175, 200
    technological, 15, 89, 202, 218, 236,
    238

Life expectancy, 18, 28, 31, 33, 99, 106, 161, 212,
    309-316, 345, 346, 356, 357, 365, 370,
    373, 381, 388, 415, 419, 464, 512, 522,
    532, 533, 541
Life sciences & medicine, vi, 145, 214, 231, 303,
    377, 381, 411, 427, 512
Lifespan. See quinary lifespan division
Lifestage, 418, 427
    See also life cycle
Lifestyle, 34, 99, 106-111, 118, 175, 177, 193, 229,
    253, 357, 408, 418, 427, 508
Light railway. See under railway
Limits to growth, 9, 17, 18, 31, 89, 195, 434
Linkages, 24, 96, 101, 151, 182, 185-188, 222,
    233, 254, 311, 386, 393, 535
Literacy, 33, 315, 345, 346, 419
Livestock, 28, 299, 304, 325, 417, 489
Living standards, 110, 111, 163, 205, 213, 308,
    359, 380, 388
Location index, 149, 158, 159, 196, 458, 459,
    460
Location quotient, 149
Location theory, 115, 223, 434, 537
Lock-in, 231, 342, 532
London, 88, 113, 114, 138, 140-143, 147-151,
    158, 159, 164-167, 221, 247, 265, 266,
    269, 283, 284, 290-293, 297, 452-461,
    469, 484
Long-Range future. See future
Long-range scenario profiles. See scenarios
Lorenz, Edward, 14
Los Alamos Center for Nonlinear Studies, vi, 12,
    435
Los Angeles, 139, 272, 280, 283, 287, 290, 295,
    351, 474, 485
Losch, August, 15, 434
Lunar energy. See energy, tidal power

M

Macrolaws, 22, 26, 77, 98, 223-227, 366, 398, 521, 532

Macrostructure, 9, 13, 15, 22, 23, 26, 40, 77, 97, 117, 121-125, 144, 181, 191, 210, 342, 366, 401, 520, 523, 540

Macrosystem. *See* global macrosystem

Macrotrends. *See* global macrotrends

Madagascar, 470

Maintenance,
    infrastructure, 162, 173, 184, 224, 227, 229, 236-239, 266, 306, 506
    property, 126, 145, 155, 447
    technological systems, 237, 257, 542

Malaysia, 293, 337, 473, 485

Malnutrition, 129, 298, 304

Malthus, Thomas, 15, 531

Mammals, 57-59, 415

Manila, 272, 286, 287, 473, 485

Manufacturing industries, 87, **144-150**, 166, 179, 195, 205, 307, 332, 388, 391, 425, 447, 448, 452, 508, 540

Marshall, Alfred, 82

Mass transit. *See* railway, rapid transit

Massachusetts Institute of Technology (MIT), iv, 12, 14, 18

Material flows, 204, 306, 491, 497

McNamara, Robert, 12, 435

Media, 109, 118, 145, 146, 172, 174, 179, 193, 215, 253, 284, 312, 363, 387, 423-426, 450, 453, 511, 539

Megacity. *See* cities, megapolis

Megacity Region (MCR), 137

Megalopolis. *See under* cities

Megapolis. *See under* cities

Megatrends. *See* civil megatrends

Membership organisations, 150, 451, 453

Mercantile, 200, 201

Metapolis. *See under* cities

Metasystem, 39-41, 131, 227, 401, 417, 533, 534
    astrophysical, 19, 39, 41, 42-46, 401
    biological, 5, 39, 40, 56-60, 91, 406
    civil, 39, 40, 60-70, 408
    geophysical, 46-52, 128, 403
    physical, 52-56, 405

Meta-timescale, 20, 22, 68, 402, 444, 445

Metatrends. *See* planetary metatrends

Meteorite. *See* asteroid

Methane, 52

Methanol, 167, 244

Metro railway. *See* railway, rapid transit

Metropolis. *See under* cities

Metropolitan Statistical Area (MSA), 137, 139

Mexico, 47, 62, 288, 296, 305, 383, 415, 474, 485, 536

Mexico City, 140, 141, 266, 273, 286-288, 474, 485

Michigan Center for the Study of Complex Systems, vi

Micropolitan district centre, 138

Micropolitan local centre, 138

Microstructure, 9, 14, 22, 26, 40, 77, 97, 117, 121, 144, 181, 366, 401

Microsystems. *See* urban microsystems

Microtrends. *See* sectoral microtrends

Middle East,
    cities and urbanisation, 263, 268-271, 274, 470, 471, 476, 478, 483, 485
    civilisations, 61, 62, 441
    futures, 344, 353, 354, 476, 478, 487, 488, 497-499, 502
    GRP and investment, 492-495

Migration,
    cross-border, 55, 57, 99, 218
    rural-urban, 86, 206, 271, 274, 327, 368

Milankovitch cycle, 44, 45, 53, 81, 403, 434

Military deterrence, 130, 376-379, 426

Millennium, 19, 20, 22, 64, 315, 345, 372

    Second, 58

    Third, 9, 54, 55, 242, 276, 320, 383, 404, 405, 410, 421, 429, 431, 515

    Fourth, 45, 260, 402, 404, 515

Million years ago (MYA), ix, 45, 47-51, 56, 57, 60, 61

    million years forward (MYF), ix, 68, 70, 402

Mineral reserve, 307, 378, 534

Mining, 23, 122, 144, 146, 201, 264, 292, 320, 447, 452, 518, 538

Modelling, 3, 11-14, 17-24, 27, 31, 32, 40, 60, 65, 74-82, 85

    climate, 10, 29, 30

    economic, 13, 14, 78, 86-88

    emergent phenomena, 74, 79, 89, 95, 98, 226, 542

    See also International Futures

    See also parametric cost models

More developed countries (MDCs),

    asset stock, 119, 131, 224, 233, 326, 494, 495

    city populations and densities, 139, 266, 269, 273-275, 279-281, 412, 478-480, 533

    corporate location, 150-152, 240, 346, 363, 385, 391,

    ecologic transition, 161, 303, 313, 319, 410, 411, 414

    military expenditure and wars, 335-337,

    path dependence, 25, 234, 350

    planning standards, 154, 304, 314, 322, 414, 463-467

    population growth, 110, 204, 224

    redistribution of investment, 119, 131, 160, 227, 321, 327, 330, 348, 371, 373, 392, 414, 522, 528, 532, 539

    share of GWP, 30, 321, 323, 327, 358, 365, 371, 393, 395-397, 505, 508, 511, 514

    workforce, 184, 206, 211, 214, 252, 350, 380, 391, 422, 423, 505, 508

Morocco, 271, 287, 470, 484

Morphological, 21, 23, 121, 181, 376, 534

Mortality rates, 110, 123, 212, 274, 308-311, 314, 315, 346, 367, 368, 381, 504, 510, 513, 542

Moscow, 266, 270, 294-297, 469, 484

Motor trade, 150, 155, 229, 237, 246, 307, 447, 449

Mountains, 39, 47-51, 59, 128, 185, 267, 281, 287, 305, 403, 404

Mozambique, 470

Multimedia, 253, 423, 426

Multinational corporation. See transnational corporations

Multiplier effect, 151, 187, 223, 235

Multipolar, 373, 375, 378, 509

Mumbai, 141, 271, 286, 290, 471, 485

Municipal, 79, 137, 216, 235, 283, 320, 327, 339

Mutation, 4, 27, 60, 92, 93, 114, 127, 129, 230, 237, 512, 535

Myanmar, 272, 295, 471, 485

N

Nanotechnology, 231, 384, 420, 425, 512, 535

Nation state, 89, 130, 267, 312, 330, 335, 342, 343, 347, 363, 379, 413, 418

National Aeronautics and Space Administration (NASA), 258

Natural disasters, 23, 118, 184, 258, 287, 290, 298, 312, 316, 407, 409

    See also hazards

Natural resources, 24, 65, 100, 117, 131, 161, 179, 195, 210, 274, 313, 322, 354, 368, 383, 384, 390, 398, 408, 428, 509, 515, 535, 537, 541

Natural selection, 27, 57, 60, 84, 92, 94, 114, 117, 520, 535

Nepal, 292, 471

Net worth, 84, 119, 517, 518

Netherlands, 140, 208, 269, 282, 290, 468, 484, 527, 536

Network dynamics, *see under* dynamics

New England Complex Systems Institute (NECSI), vi

New Hampshire Complex Systems Research Center, vi

New York, 138, 139, 141, 256, 265, 272, 283, 284, 290, 291, 297, 412, 474, 485

Newspapers, 145, 172, 177, 185, 193, 423, 450

Niche, 4, 6, 61, 93, 105, 124, 125, 129, 180, 222, 229, 232, 342, 519, 536, 541, 542

Nigeria, 270, 278, 310, 336, 355, 470, 484

Non-Governmental Organisations (NGOs), 171, 172

Nonlinear science,
 See also dynamics, nonlinear

Normative scenarios. *See under* scenarios

North Africa,
 cities and urbanisation, 263, 268-271, 470, 476, 478, 483, 484
 civilisations, 62, 263, 441
 futures, 344, 353, 354, 380, 476, 478, 487, 488, 497-499, 502
 GRP and investment, 492-495

North America,
 cities and urbanisation, 248, 272, 274, 280, 474, 476, 478, 483, 485
 civilisations, 62, 443
 futures, 55, 246, 344, 380, 476, 478, 487, 488, 497-499, 502
 GRP and investment, 32, 33, 351, 358, 492-495

North America Free Trade Area (NAFTA), 350, 501

North, Douglass, C, 170

North Korea, 473

Northeastern University CIRCS, vi

Nuclear power, *see under* energy, nuclear

Nuclear weapons, 12, 257, 297, 337, 378, 405, 422

Nutrition, 314, 345, 346, 408, 419
 See also malnutrition

O

Oceania,
 cities and urbanisation, 268, 269, 272, 473, 476, 478, 483, 485
 civilisations, 443
 futures, 344, 476, 478, 487, 488, 497-499, 502
 GRP and investment, 492-495

Oceans, 5, 39, 46-48, 50, 52, 69, 80, 128, 243, 288, 308, 338, 344, 375, 383, 403, 410, 412, 421, 509

Offices, 122, 147, 148, 154, 158, 159, 201, 234, 297, 387, 454, 455-457

Official Development Assistance, 328, 362, 388

Oil, 54, 128, 175, 178, 229, 234, 241-247, 258, 292, 293, 306, 308, 350, 378, 384, 395, 405, 447, 506, 507, 515, 542, 545

Ontogenesis, 92, 429, 536

Open System, 3, 40, 90, 93-95, 131, 366, 536

Orbit,
 Low- earth, 255, 426
 orbital flight, 256, 257, 385, 386, 411, 412, 514

orbital missiles, 422

    planetary, 43-45, 67, 68, 80, 254, 259

Orbital route, 139, 143, 152, 166, 533

Organisation for Economic Cooperation

    and Development (OECD), iv, 28, 110,

    204, 233, 241, 378, 536

    OECD countries, 206, 241, 330, 536

    OECD economies, 33, 326, 374

Osaka-Kobe, 272, 280, 473, 485

Overshoot, 41, 217

Ozone layer, 9, 53, 55, 405, 543

P

Pacific. *See also* East Asia and Pacific

    Asian-Pacific region (including Japan

    and Australia), 344, 357, 358, 377

    coast, 272, 286, 335

    ring of fire, 287, 288

Pakistan, 62, 206, 271, 288, 292, 378, 471, 485

Pandemic, 20, 286, 298, 375, 381, 397, 406, 510,

    512

Pangea, 49, 57, 80

Paradigm, 23, 29, 32, 41, 78-88, 181, 191, 508, 513

Parametric,

    cost models, xi, 79, 428

    variables, 27, 28, 31, 366

Paris, 140, 247, 266, 269, 283, 468, 484

Passenger terminals, 147, 154, 155, 454, 455

Passenger travel.

    249, 291, 297, 465. *See also* aircraft

    248, 292. *See also* buses

    246, 292, 465. *See also* car

    248. *See also* light rail

    291, 297. *See also* sea transport

    247, 291, 297. *See also* rapid transit

    249, 292, 297, 465. *See also* railway

    256, 412. *See also* spaceline

Path dependence, 25, 233, 234, 350

Peru, 62, 288, 475, 485

Pesticides, 216, 292, 408, 528

Petroleum, 56, 244, 245, 292, 319, 353, 383,

    412

Phenotype, 92, 229, 516, 537

Philippines, 282, 288, 297, 473, 485

Photosynthesis, 52, 56, 69, 129, 375, 406

Physical metasystem. *See under* metasystem

Planetary,

    civilisation. *See under* civilisation

    governance, 365, 372-378, 413, 426

    metasystems. *See* metasystems

    metatrends, 401, 403, 405, 406, 408,

    416

Planetary terraforming. *See* terraforming

Planning standards.

    See under more developed countries

    See under less developed countries

Planning zones, 176, 178

Plants, vascular, 58

Plexus Institute, vi

Poland, 268, 293, 468, 477, 484, 536

Policy analysis, 8, 11, 12, 29, 537

Polis or regional city. *See under* cities

Pollution, 18, 28, 59, 128, 312, 345, 374, 419, 510

    abatement, 161, 319, 414

    air, 56, 163, 287, 320, 512

    control, 178, 316, 509

    toxic substances, 163

    water, 163, 306, 512

Popper, Sir Karl, 85

Population density, 139-141, 280-283, 352-357,

    368, 395-397, 430, 505, 508, 511, 537

Population growth. *See under* growth,

    demographic

Population policies, 376, 380

Portland State University Department of

    Systems Science, vi

Ports, 115, 182, 185, 233, 239, 241, 264-267, 290, 292, 378, 425

Postcodes, 107, 143, 452

Postindustrial, 144-148, 158-161, 185, 194, 196, 202, 205, 214, 220, 245, 251, 275, 281, 285, 420, 421

Potential,
> economic, 220, 227, 236, 410, 419
> ecostructure, 131, 227, 409, 532, 537
> evolutionary, 127-131, 210, 341, 349, 373, 374, 409, 504, 523
> hazards, 20, 100, 187, 211, 286, 288, 298, 375, 379, 380
> human development, 210, 118, 161, 342, 345, 350, 419, 505, 523
> territorial, 12, 367, 500, 537
> *See also* geopolitical potential

Potsdam Center for Dynamics of Complex Systems, vi

Poverty, 32, 33, 129, 162, 184, 214, 240, 298, 315-317, 341, 345, 356, 372, 373, 388, 389, 395, 419, 505

Power generation. *See under* energy

Power law, 278, 366, 538

Predator, 59, 126

Prediction, 11, 16, 27, 40, 74, 78-82, 88, 105, 140, 380, 527, 540

Prescience, 10

Prigogine, Ilya, 90, 91, 116, 435

Primary resources division, 124, 144-148, 194, 215, 320, 325, 345, 424, 516, 538

Principal city, 138

Prison, 155, 297, 298, 352, 357

Profit, **114-115**, 119, 127, 156, 157, 180-184, 192, 202, 219, 234, 264, 307, **330-332**, 387, 427, 517, 519

Prognostication, 11, 89, 276, 323, 437, 538

Property,

cycles, 114, 218, 219

investment, 143, 158-160, 175, 218, 219, 369, 458-460

profiles, 369, 454

sector, 144-147, 150-154, 159, 163, 166, 218-220, 254, 313, 425, 447, 452, 514, 540

> *See also under* development, property
> *See also under* maintenance, property

Provincial city, 138

Psychographics, 111, 427

Public health, 235, 347, 530

Public service facilities, 235, 530

Punctuated equilibrium, 120, 181, 225, 538

Purchasing Power Parity (PPP), 204, 313, 323, 325, 330, 357, 359, 463, 491, 538

Q

Quality of life, 33, 99, 163, 164, 176, 210, 212, 215, 313

Quarrying, 424

Quaternary information division, 82, 86, 125, 127, **144-153**, 160, 173, 181, 185, **194-196, 202-206**, 212, 215, 252, 332, 369, 422, 426, 462, 487, 518, 539

Quinary lifespan division, 145, 214, 215, 389, 427, 487

Quotas, 344, 524

R

Races, 295, 340

Radiation,
> cosmic, 42, 43, 68, 69, 543
> nuclear, 293, 405
> short-wave, 39, 55, 405
> solar, 47, 48, 53, 81, 404

Radioactive, 53, 243, 255, 258, 293, 421, 528

Radiological, 297, 337

Radio waves, 67

Railways,

    heavy rail, 64, 83, 125, 155, 167, 173, 174, 182, 185, 201, 233, 234, 238-241, 245, 248-250, 266, 271, 284, 291, 292, 297, 346, 357, 387, 449, 465

    high speed, 245, 249, 421

    light rail and trams, 247, 248, 250, 387

    rapid transit, 167, 179, 238, 245, 247, 284, 290-292, 297

Rain forest, 59, 415, 512

RAND Corporation, iii, iv, 11, 12, 17, 21, 34, 434

Rand Pardee Center, 19, 34

Random, 14, 23, 90, 92, 94, 112, 125, 521

Randstad, 138, 140, 269, 468, 484

Rapid transit. *See under* railway

Rapoport, Anatol, 3, 434

Rational, 81, 111, 112, 192, 193

Raw materials, 200, 202, 231, 320, 331

    See also material flows

Reciprocity, 7, 184, 348, 539

Recreational amenities, 119, 176, 205, 212, 216, 235, 389

Recycling, 127, 128, 205, 215, 222, 318, 322, 371, 389, 420, 448, 498, 506

Redistribution of investment. *See under* more developed countries

Reductionism, 76, 98, 539

Reforestation, 344, 410

Refugees, 298, 335, 338, 377-380, 383, 507

Regional Trade Blocs, 350, 508

Religion, 4-6, 76, 99, 177, 293, 294, 316, 340, 341, 379, 380, 389, 417, 418

Remote future. *See under* future

Renewable resource, 129, 161, 244, 539, 541

    See also energy, renewable

Rental values, 159

Replication, 114, 513

Reproduction, 99, 113, 114, 125, 129, 178, 313

Reptiles, 45, 57, 58

Resilience, 75, 125, 142, 180, 238, 285, 340, 366, 512, 519, 540

Resource conservation. *See* conservation

Restaurants, 113, 145, 150, 153, 177, 284, 296, 299, 369, 449, 454-457, 513

Retail, 108, **113-119, 125-126, 145-160**, 175-177, 186-188, 194, 199, 212, 220, 251, 284, 293, 294, 363, 369, 449, 452-459, 513

Rhine-Ruhr megapolitan region, 138, 140, 266, 270, 280, 468, 484

Ricardo, David, 363

RICI (Resource, Industrial, Commercial, and Informational) Sectoral Classification, 144, 146, 424, 447-451

Rio de Janeiro, 273, 280, 286, 296, 314, 474, 485

Riverine ports, 264-267, 271, 290, 468-475

Robotics, 70, 195, 384, 385, 421, 425

Rodinia, 47, 56

Romania, 268, 469, 477

Rural,

    areas, 142, 167, 172, 177, 182, 217, 243, 274, 327, 346, 387

    dwellings, 109, 324, 325, 370, 463

    population, 176, 274, 282, 316, 325, 326, 346, 366, 368

    settlements, 111, **137, 163-167,** 186, 274, 281, 282, 303, **324-326**, 368-370, 479-481

Russian Federation, 47, 55, 241, 268, 269, 296, 297, 340, 354. 378, 469, 484, 527

S

Santa Fe Institute, vi, 12, 25, 26, 73, 84, 435

Sao Paulo, 141, 266, 273, 280, 283, 475, 485

Satellite,

>communications, 83, 174, 201, 234, 239, 254-257, 284, 386, 426

>reconnaissance, 257, 375, 422

>technology, 173, 256

>television, 108, 212

Saudi Arabia, 206, 294, 470, 485

Savings, 106, 109, 111, 114, 119, 210, 327, 330, 427, 508, 517-519

Scenarios,

>exploratory, 28, 30, 31, 540

>long-range scenario profiles, 504-515

>normative, 27, 28, 30, 31, 375, 517, 540

Schools, 83, 107, 109, 166, 172, 240, 297, 314, 427

Schumpeter, Joseph, 25, 82, 83, 128, 199, 217, 251

Science of evolution and complexity. *See under* complexity science

Scotese, Christopher, 47

Sea levels, 35, 44-50, 54-57, 62, 211, 287-290, 344, 372, 403-405, 412, 515

Sea transport and shipping, 62, 64, 115, 146, 147, 200, 263, 291, 325, 383, 449

Secondary industrial division, 125, 144-149, 153, 194, 195, 215, 425, 462, 540

Sectoral microtrends, 401, 424

Security, 29, 31-35, 146, 155, 169, 216, 223, 254, 285-287, 317, **337-339**, 347, 348, 366, 372, 378-380, 388, 413, 450, 506, 509, 512, 515

Self-transformation, 540

Seoul, 266, 272, 279, 280, 286, 293, 473, 485

Service centre, 138, 143, 167, 276, 345, 356, 456

Shanghai, 141, 272, 280, 286, 472, 485

Silicon Valley, 232

Simon, Herbert, 3, 423, 434

Slums, 109, 287, 315, 316, 394, 395, 504-506

Smith, Adam, 83, 96, 363

Social equity. *See under* equity

Social hazards. *See under* hazards

Social security, 106, 170, 216, 330, 339, 388

Societal evolution. *See under* evolution

Societal transitions, 8, 23, 100, 209-216, 229, 369, 389, 542

Society of General Systems Research, 3

Software, 13, 99, 173, 181, 187, 252, 253, 363, 421, 425

Solar energy. *See under* energy

Solar system, 39, 43-44, 65-69, 128, 254-259, 445

South Africa, 271, 294-298, 356, 470, 484

South America,

>cities and urbanisation, 62, 264, 268, 273, 280, 474, 476-478, 483, 485

>civilisations, 62, 443

>futures, 352, 358, 476-478, 487, 488, 497-499, 502

>GRP and investment, 31, 352, 492-495

South Asia,

>cities and urbanisation, 268, 271-273, 282, 471, 476, 478, 483, 485

>civilisations, 340, 442

>futures, 246, 350, 356, 476, 478, 487, 497-499, 502

>GRP and investment, 492-495

South Korea, 282, 293, 296, 473, 485, 536

South-East Europe

>cities and urbanisation, 268, 270, 469, 484

>civilisations, 268, 441

>futures, 246, 309, 353, 378, 487, 488

>GRP and investment, 492-495

Space,

    exploration, 254-260, 514

    spacecraft, 67, 255, 257, 259, 402, 422

    spacelines, 255, 256, 280, 412

    spaceports, 255-257

Spatial integration, 366

Spatial structure, 3, 10, 18, 22, 24, 68, 73, 82, 85-92, 101, 115-119, 121, 185, 223, 233, 522-524, 530, 537

Species,

    species extinctions.
    See under extinctions, ecological

    species fitness, 114, 541

    human species, 41, 69, 260, 377, 402, 406, 430, 431, 512, 518

    natural species, 4, 25, 33, 39, 57-60, 347, 377, 402, 406, 407, 415, 416, 517, 518, 522-527, 536, 541

    sectoral species of establishments, 78, 84, 92-94, 105, **112-119**, 125-129, 144, 229- 232, 260

Spencer, Herbert, 125

Sports & leisure, 105-109, 113, 147, 149, 154, 155, 169, 175, 177, 283, 294-296, 324, 340, 369. 418, 426, 449, 455-457, 513

Sports Utility Vehicles (SUVs), 246

Sri Lanka, 471

Standard Industrial Classification (SIC), 144, 146, 147

Stanford Research Institute (SRI), iv

Stiglitz, Joseph, 86, 349

Structural adjustment programmes, 331, 353

Sub-Saharan Africa,

    cities and urbanisation, 268, 270, 278, 469, 470, 476, 478, 483, 484

    civilisations, 441

    futures, 310, 344, 355, 356, 380, 476, 478, 487, 488, 497-499, 502

GRP and investment, 492-495

Substitution,

    resources, 161, 307, 213, 245, 307, 424, 541

    technology, 15, 232, 238, 245

Suburban, 106, 111, 142, 166, 175, 178, 179, 216, 246, 247, 267, 296, 386, 387, 510, 545

Sudan, 271, 470, 484

Supersonic, 249, 256, 386

Superstore, 23, 126, 175, 194, 518

Surveillance technologies, 146, 254, 257, 337, 422, 426

Sustainable development. *See under* development

Swinburne University (Australia), iv

Synthetic ecostructures, 124, 152-156

Syria, 470

System. *See* open system

System growth parameter, 60, 94, 95, 365, 366, 525, 541

Systems science, 3, 8, 10-12, 23, 98, 394, 528

T

Taiwan, 473

Tanzania, 271, 297, 470, 484

Technological,

    development. *See under* development

    evolution. *See under* evolution

    hazards. *See under* hazards

    shift, 317-324, 371, 374, 393, 411, 498, 499

Tectonic plates, 39, 46-51, 69, 403, 404

Telecommunications, 141, 145-147, 151, 167, **170-175**, 181, **185-187**, 193, 201, 230, 233-235, 239-241, 245, **251-254**, 257, 312, 348, 363, 370-377, **384-387**, 398, 421-423, 426, 430, 431, 448, 450, 453, 535, 539

Teleology, 17, 98, 99, 542

Television, 65, 108, 109, 118, 141, 150, 172, 177, 188, 193, 212, 230, 234, 252, 387, 390, 419, 422-424, 431, 450, 467

Terraforming, 255, 260, 402, 431, 515

Terrestrial stocks, 39, 41, 114, 128, 403, 414, 526, 542

Territorial colonisation, 24, 55, 122-127, 520, 542

Terrorism, 20, 28, 243, 286, 293, 297, 299, 316, 335, 337, 366, 379, 408, 409, 426, 528

Tertiary commercial division, 144-148, 159, 194, 201, 215, 425, 542

Textiles, 150, 391, 447, 448

Thailand, 278, 295, 473, 485

The Cambridge Nonlinear Centre, vi

Thermodynamics, Laws of, 40, 89, 90, 226, 531

Tidal power. *See under* energy

Tobin, James, 329

Tobin tax on foreign exchange transactions, 329, 505

Tokyo, 138-141, 256, 265, 272, 279, 283, 286-290, 299, 383, 412

Tokyo-Yokohama, 272, 280, 473, 485

Tons (metric) or tonnes, 543

Tourism & travel, 14, 126, 147, 177, 312, 339, 426, 449, 511

Township, 138, 143, 167, 186, 264, 275-278, 368, 480

Toxic spills or waste, 163, 286, 292, 384, 395, 408, 506, 528

Toynbee, Arnold, 4, 64, 408

Traditional stage. *See* civil transitions

Transacting entities, 76, 77, 88, 98, 112, 120-122, 366, 427, 519, 525, 528, 543

Transactional complexity, 116, 120, 121, 128, 160, 180-185, 225, 227, 366, 369, 409, 518, 519, 523, 532

Transactional microstructure, 9, 14, 27, 36, 40, 97, 121, 135, 144, 401

Transnational Corporations (TNCs), 358-363

Transpart, 117, 229, 230, 516, 524, 543

Transport & distribution, 149, 179, 204, 420, 452

Travel, hospitality and consumer services, 145, 452, 542

Trends, 11, 13, 16, 21, 32, 35, 111, 151, 163, 184, 187, 192-195, 204-209, 212, 219, 229, 241, 247, 310, 321, 330, 332, 355, 359, 367, 376, 380, 384-386, 395, 396, 424, 437
  See also civil megatrends
  See also global macrotrends
  See also planetary metatrends
  See also sectoral microtrends

Trillion, 543

Troposphere, 39, 56, 405, 543

Turkey, 268, 270, 278, 287, 378, 469, 477, 484, 536

U

UCLA Human Complex Systems, vi

UCL Centre for Nonlinear Dynamics, vi

Uganda, 297, 470

Ukraine, 268, 270, 293, 469

UN Refugee Agency (UNHCR), 507

Underemployment, 211, 360

Unemployment, 106, 187, 211, 213, 221, 224, 295, 312, 317, 328-331, 353, 360, 380, 385, 391-393, 511

United Kingdom (UK), 110, 138, 200, 201, 206-208, 218, 246, 292, 295-299, 329, 336, 341, 343, 357-360, 363, 378, 469, 484, 527, 536

United Nations (UN), 16, 19, 27-35, 58, 110, 137, 141, 195, 212, 257, 268, 269, 275, 283, 284, 304, 310-314, 329, 335, 338, 363,

367-377, 388, 398, 413, 416, 422, 428, 435, 504, 507, 510, 513

United Nations Development Programme (UNDP), 314, 315, 504

United Nations Environment Programme (UNEP), 19, 32, 34, 58, 59, 306, 398, 435, 510

United Nations Human Settlements Programme (UN-HABITAT), 287, 315, 504

United States of America (USA), 9, 32, 87, 110, 137, 139, 142, 204, 206, 223, 242, 243, 246, 252, 265-269, 272, 275, 280, 282, 288-299, 304, 332, 343, 349-357, 363, 378, 380, 388. 391, 422, 443, 474, 485, 513, 527, 536

Universities, 147, 150, 154, 172, 427, 457

Uranium, 243, 293

Urban,

development. *See under* development

evolution. *See under* evolution

hierarchy, 96, 137-142, 147, 149, 160, 166, 186, 233, 275-279, 280, 285, 409, 430, 525

microsystems, 169, 170, 176, 191, 226, 417

system, 22, 24, 40, **75-78**, 87, 97, **116-121**, 125, 132, 138-143, 160, 175, 186, 191, 209, **224-227**, 233, **273-279**, 371, 374, 401, 409, 507, 517-533, 537, 543

transition, 159, 219, 323, 327, 365, 371, 398, 410

User charging, 167, 193, 239, 327, 329, 424

Utilities, 84, 108, 113, 137, 145, 155, 156, 170, 176, 182, 199, 204, 212, 234, 235, 275, 324, 368, 370, 383, 420, 525, 530

Uzbekistan, 269, 271, 471, 477

V

Vacancy rates, 218, 219

Venezuela, 296, 475, 485

Verhulst, Pierre-Francois, 11, 15, 93, 123, 436, 530, 542

Vietnam, 282, 336, 337, 473

Villages, 26, 62, 111, 117, 130, 137, 186, 274, 281, 282, 304, 368, 479-481, 488, 489, 543
See also rural, settlements

Viruses, 60, 298, 299, 381, 408, 510

Vivisystem, 13, 26, 77, 85, 92, 93, 112, 113, 121, 132, 181, 226, 253, 410, 423, 519, 524, 539, 543, 544

Volatile organic compounds (VOCs), 56, 163, 244

Voluntary sector, 145, 214-217, 361, 390

Voting rights, 372, 378, 413, 428

W

Warehouses, 115, 153, 154, 449, 454-457

Wars, 9, 11, 20, 23, 28, 64, 125, 169, 207, 257, 286, 297, 298, **335-337**, 350, 352, 372-381, 408, 422, 426, 506, 509, 528, 529

Washington Consensus, 171, 214, 349, 513

Waste,

combustible, 242, 319, 320, 421

nuclear, 243, 255, 258, 293, 383

solid, 162, 178, 213, 226, 315

toxic, 163, 307, 384, 425

water, 170, 177, 241, 306, 315, 381, 506, 515

Water scarcity, 305, 306, 415, 509

Wegener, Alfred, 49, 80, 87

Welfare, 117, 171, 212, 354, 381, 427

organisations, 145, 214, 451, 539

payments, 7, 107, 210, 225, 339

services, 28, 212, 216, 176, 389

Well-being, 6, 35, 113, 118, 176, 184, 313, 314

Wells, H.G., 9, 434

West Antarctic ice sheet, 54

Western Asia. *See* Middle East

Western Europe

 cities and urbanisation, 140, 267-270, 283, 354, 468, 476-478, 483, 484

 civilisations, 61, 64, 263, 440

 futures, 32, 344, 353, 354, 380, 476-478, 487, 488, 497-499, 502

 GRP and investment, 393, 492-495

Wholesale trade, 145, 150, 186, 449, 452, 542

Wildlife, 163, 281, 282, 304, 365, 368, 382, 395, 425, 488, 489, 506, 509, 511, 514

Wind power. *See under* energy

Wisdom, 5, 6, 79, 105, 161, 541, 544

Women, 106, 110, 206, 309, 311, 357, 377

Workforce, 6, 7, 25, 106, 111, 167, 204, **205-208, 213-215**, 330, 338, 345, 350, 358-362, 388, 392, 462, 487, 502, 503, 507, 516

Working-time, **205-209**

World Bank Group, iv, 12, 19, 28, 171, 202, 268, 319, 331, 366, 435, 504, 507, 528, 539

World cities. *See under* cities

World Economy, 25, 115, 151, 202, 205, 285, 311, 326, 377, 391-393, 514

 See also under economy, global

World Future Society (WFS), iv, 9, 424

World Futures Studies Federation, iv

World Health Organisation (WHO), 510

World Resources Institute (WRI), iv, 306

World Trade Organisation (WTO), 337, 341, 513

Worldwatch Institute, iv, 16

Y

Yemen, 471

Z

Zambia, 270, 470

Zimbabwe, 441, 470

Zipf's rank size rule, 87, 275, 278, 538

www.ingramcontent.com/pod-product-compliance
Lightning Source LLC
Chambersburg PA
CBHW081101170526
45165CB00008B/2287